AF689642

About Our Cover . . .

Our cover is a detail of an alchemical painting (oil on canvas, 43 x 59cm) by a Flemish artist, Matheus van Helmont, a versatile genre painter who worked in Antwerp in the middle of the 17th century.

Our chemist-collector really likes this painting because the alchemist's desk reminds him of his own — and, of course, the rich colors make it a fitting cover for Floyd Green's work on stains and dyes.

The Sigma-Aldrich Handbook of Stains, Dyes and Indicators

by
Floyd J. Green

Aldrich Chemical Company, Inc.
Milwaukee, Wisconsin

Printed in the United States of America

Library of Congress Catalog Card Number 90-060390
ISBN Number 0-941633-22-5

Table of Contents

Acknowledgements. . .

Sigma-Aldrich Corporation and Floyd Green acknowledge the efforts of the Aldrich Technical Services, Quality Control and Graphic Services Departments in the preparation of this book. Special thanks are extended to Jean Brockmeyer, Kevin Chesak and Joelle McDaniel.

We also thank Drs. Irwin Klundt and Thomas Wickersham for valuable comments, and Professor Howard W. Whitlock and Thomas Moertl for special computer programming .

Preface

Dr. Floyd Green has earned world-wide respect as a manufacturing chemist, and for his comprehensive knowledge in the area of stains and dyes. He has compiled this handbook on stains, dyes and indicators as a guide to the use of these compounds by both the chemist and non-chemist, in the laboratory and in industry. Most of the 525 compounds described here are listed in the Aldrich Catalog/Handbook or Sigma Chemical Company catalog.

The compounds are listed alphabetically. Each entry includes Dr. Green's description of dye type and use; technical data such as grade, CAS registry number, Colour Index number, molecular formula, molecular weight, melting point, boiling point, appearance, λ_{max} and solubilities in water, ethanol and methyl Cellosolve; history, preparation, synonyms, chemical names and references whenever applicable; and the spectra and structures when available. The technical data were determined by the Quality Control Laboratories of Aldrich Chemical Company.

The entries are cross-referenced using the following indices: alphabetical (including parenthetical names), CAS registry number, C.I. number and λ_{max} (shown with name and solubility). A table of the acronyms, abbreviations and definitions used throughout the book is included in the Introduction and a color chart describing the preparation and transition intervals of various indicators can be found at the back of the book.

Abbreviations, Acronyms and Definitions

A.C.S.	American Chemical Society
Aldrich Cat. No.	Number under which product is listed in the Aldrich Catalog/Handbook of Fine Chemicals
Aldrich FT-IR Library	Pouchert, C.J. *The Aldrich Library of FT-IR Spectra*; Aldrich Chemical Company, Inc.: Milwaukee, 1984.
Appearance	The color described is that of the actual compound, not its dyeing capability.
b.p.	Boiling point
BSC	Biological Stain Commission
Certified	Material has been certified by the Biological Stain Commission
C.I. No.	Number assigned to the dye in the *Colour Index,* 3rd ed.; Society of Dyers and Colourists: Bradford, West Yorkshire, and the American Association of Textile Chemists and Colorists: Research Triangle Park, 1971.
Colour Index	When referred to in text without citation: See above under C.I. No.
DNA	Deoxyribonucleic acid
EGME	2-Methoxyethanol; methyl Cellosolve; monomethyl ether of ethylene glycol
EtOH	Ethyl alcohol; ethanol
F.W.	Formula or molecular weight
g	Gram
GC	Gas chromatography
H_2O	Water
HPLC	High-performance liquid chromatography
IR Library	Pouchert, C.J. *The Aldrich Library of IR Spectra*; Aldrich Chemical Company, Inc.: Milwaukee, 1981.
λ_{max}	Wavelength, in nanometers, at which the maximum absorption of a stain or dye was observed. The secondary absorption appears in parentheses.
Merck Index	*The Merck Index,* 11th ed.; Budavari, S. *et al.*, Eds.; Merck: Rahway, 1989.
mg	Milligram
ml	Milliliter
Mol. Form.	Molecular formula
m.p.	Melting point
nm	Nanometer
RNA	Ribonucleic acid
Safety	*The Sigma-Aldrich Library of Chemical Safety Data,* 2nd ed.; Lenga, R.E., Ed.; Sigma-Aldrich Corp.: Milwaukee, 1987.
Sigma FT-IR Library	Keller, R.J. *The Sigma Library of FT-IR Spectra;* Sigma Chemical Co.: St. Louis, 1986.
Sigma Prod. No.	Number under which the product is listed in the catalog of the Sigma Chemical Company

Abbreviations, Acronyms and Definitions (cont.)

Solubilities	Solubility tests were run as follows: Add 1ml of appropriate solvent (water, ethanol or 2-methoxyethanol) to ~ 0.5g sample in a test tube. Cover the tube. Sonicate for 15 minutes; heat for 2 hours in a hot water bath (80-90°C); sonicate for another 15 minutes. Let mixtures stand overnight. Centrifuge. Withdraw 2.0ml of supernatant* and evaporate. Dry at 105°, then weigh the residue and calculate the quantity of product which was dissolved. * If no solids remained, the preceding steps were repeated with more sample. **Note:** Tests were run on products as available in the Aldrich inventory (*i.e.*, no purification, grinding or other sample pretreatment was done, nor was an attempt made to vary the times noted.) Sonication was the only mixing technique used. Some samples appeared to be "soluble" (suspended) until centrifugation.
USP	Indicates that material meets specifications as given in *The United States Pharmacopeia*, twentieth version; United States Pharmacopeial Convention; Mack: Easton, 1979.
~	Approximately

About the Author

Dr. Floyd Green is a member of and consultant to the Board of Trustees of the Biological Stain Commission. He joined Aldrich in the fall of 1977 expressly to establish and direct the company's Stain and Dye Division. The Division was created to fill the void which resulted from strangulation of many commercial dye sources by legislation and economic conditions.

Today Dr. Green enjoys world-wide respect for his record as a manufacturing chemist and for his comprehensive knowledge in the area of stains and dyes.

In college, Green had been leaning originally toward a career in medicine. He received his baccalaureate degree in biology from Maryville College, Maryville, Tennessee, in 1941. When World War II began, he joined the Navy which sent him to the Naval Medical School in Bethesda, Maryland. His training there centered on malaria and epidemiology. Subsequently, he was sent to the Pacific theater where he spent the next 26 months working with a team to obtain epidemiological profiles on both native and troop populations. It was there that he had his first exposure to stains. When the war ended, he went to work as an assistant to one of the U.S. pioneers in biological stains, Dr. Arthur Coleman, who was one of the early entrepreneurs in the field (Matheson, Coleman and Bell). It was during the five years that Green was working directly with Coleman that he became aware of the lack of reliable published information in this difficult area of stains and dyes.

All in all, Green invested 27 years with MCB, though he found time to attain an M.S. in biochemistry and chemistry in 1967 and a Ph.D. in 1969 from St. Thomas Institute of Advanced Studies. His Ph.D. thesis was based on work he had done with a University of Illinois pathologist. Their months of work together — alternate weekends in Illinois and Ohio — were directed toward cross-linking agents relative to stabilizing and preserving badly decayed teeth. Their initial work sought to use Procion dyes to stabilize collagen in teeth pulp, and it evolved into the development of colorless cross-linking agents to replace the Procions. Subsequently, papers were presented to the International Association for Dental Research and the North American Division of IADR, and thirteen patents were issued.

Although he has now "officially" retired, Dr. Green's first objective when he joined the company — to provide a reliable source of consistently pure stains and dyes for biological and scientific research — remains the prime objective of the Floyd Green Stain and Dye Division of Aldrich. His second intention was to accumulate and tabulate state-of-the-art data . This we now present in his book, *The Sigma/Aldrich Handbook of Stains, Dyes and Indicators.*

Reprint of article by Green, F.J. *Aldrichimica Acta* **1978**, *11*(1),3.

A Status Report on Biological Stains

INTRODUCTION:

The use of stains (dyes) as colorants for biological substrates has closely paralleled the use of dyes in the textile industry. In fact, microbiologists began using synthetic dyes shortly after Perkin's revolutionary discovery of a synthetic route to "mauve" in 1856. In less than twenty years some significant staining techniques had been developed, notably by Ehrlich.

While textile manufacturers were concerned with dyes strictly from an aesthetic standpoint, biologists were interested primarily in gaining information on the structure and function of tissues, cells and subcell components. This is not to say that differentially stained biological specimens cannot be strikingly beautiful.

With the advent of synthetic dyes, microbiologists were provided a new tool to assist them in their quest to release some of the secrets held captive in the cell. In fact, this was the first significant advance in the art since its inception in the 17th century when Leewenhoek opened up the startling new world of "animalcules" (tiny animals) with his invention and development of the microscope. (It should be noted that natural products such as saffron and carminic acid were used within their limits as biological stains prior to the development of synthetic dyes and some, including the two mentioned, are still used.) For the first time, tissue and cell components could be differentially colored effectively to provide contrasts heretofore unattained. (The basis of differential or selective staining of biological materials lies in the chemical and physical heterogeneity of tissue, cells and subcell components.) Thus, the geometry and chemical affinity of tissue, cells and subcell structures could be more readily observed and studied by utilizing skillfully tailored staining techniques.

FIXATIVES:

Unlike textiles, many biological substrates are water soluble. In addition, they are rendered useless for subsequent microscopic observation by temperatures commonly used in textile dyeing. They are also susceptible to degradation by bacteria and their own hydrolytic enzymes. These are a few of the problems that must be overcome before a biological substrate can be stained effectively and observed under a microscope. Therefore, the substrate must undergo a prestaining conditioning which frequently is quite elaborate. Denaturants known as fixatives are used to stabilize the substrate matrix. These vary with the substrate and the application. Organic solvents, reactive compounds (*e.g.*, aldehydes, osmium tetroxide and chlorotriazines) and heat are among the fixatives commonly used. Obviously, it is impossible to observe the substrate *in vivo* after fixation. The goal is to treat it as gently as possible in the stabilization process so as to minimize the resulting distortion.

A monocellular layer of substrate will afford the most information on cell detail to a microscopist. To approach this end, thin layers of cell suspensions are smeared on a slide, or tissue blocks are sliced thinly by a specialized slicing device known as a microtome. These tissue blocks must be made rigid prior to cutting, and this is accomplished by freezing or embedding with media such as warm paraffin "wax" which solidifies on cooling, or plastic monomers, thus providing the necessary support to the tissue to minimize mechanical distortion during the cutting operation.

DYE ATTACHMENT:

In order for a chromophore to be of

value as a colorant, it must become attached with some permanence to the substance being dyed. To a large degree, this holds true for biological stains. However, some staining techniques require stains that will respond to selective destaining.

The tenacity of the dye's attachment to the substrate and the concomitant color depth achieved varies, not only with the bonding mode in effect, but with chemical affinity, density and permeability of the substrate.

There are several bonding modes or forces that can come into play when a chromophore becomes attached to a biological substrate. These, of course, vary with the dye and substrate. The ionic bond appears to be one of the most prevalent means of attachment. However, other linkages occur in addition to ionic bonding. These include the covalent bond, hydrogen bond, van der Waals forces, hydrophobic bond and mordants. A combination of two or more of these forces, depending on the nature of the dye and the substrate, participate in the attachment.

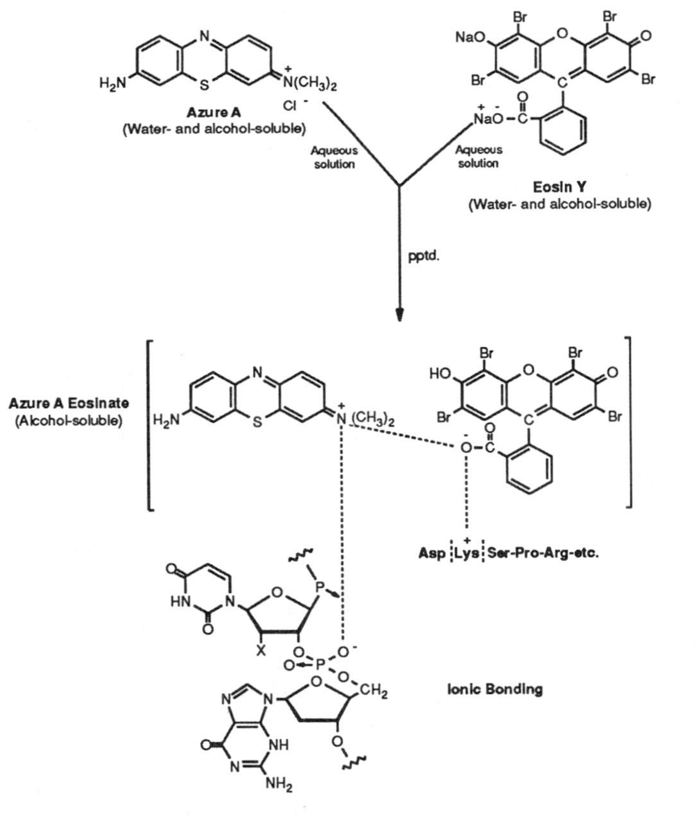

Fig. 1 ***Anionic dye*** *(Eosin Y bonded ionically with basic amino acid in hypothetical polypeptide chain)*
Cationic dye *(Azure A bonded through ionic bond with hypothetical section of nucleotide chain)*

IONIC BONDS:

The ionic bond formed between appropriately matched dyes and biological substrates is basically no different from the ionic bond demonstrated in other dye applications. The linkage occurs as the result of the Coulombic force initiated when a cation and an anion come into close proximity of each other. In other words, basic dyes (cations) are attracted to anionic sites in the substrate, while acid dyes (anions) are attracted to cationic sites.

The ionic bonding mode plays a large role in linkages demonstrated by the Romanowsky-type blood stains (Fig.1). Although the ionic bond should not be considered as the only type of bond that exists when these dye mixtures are utilized, it is certainly a major linking force in this application. Methylene Blue and its oxidation products (the azure dyes) serve as the basic dyes in these blends, while Eosin Y (yellowish) serves as the acid dye counterpart. These mixtures, containing both acid and basic dyes, stain thin blood films differentially from an "acid" bath. The optimum pH of the bath varies from 6.4-6.7 depending on the technique used. There are three different types of dry stain blends: (1) A mixture of dry powders of the acid dye Eosin Y and basic dyes (*e.g.*, Methylene Blue or Azure A). (2) Eosinates (Eosin Y ionically linked to a basic dye such as Methylene Blue or one of the azures) which are prepared by adding an aqueous solution of Eosin Y to an aqueous solution of the basic dye. The resulting precipitates are alcohol-soluble and water-insoluble. (3)

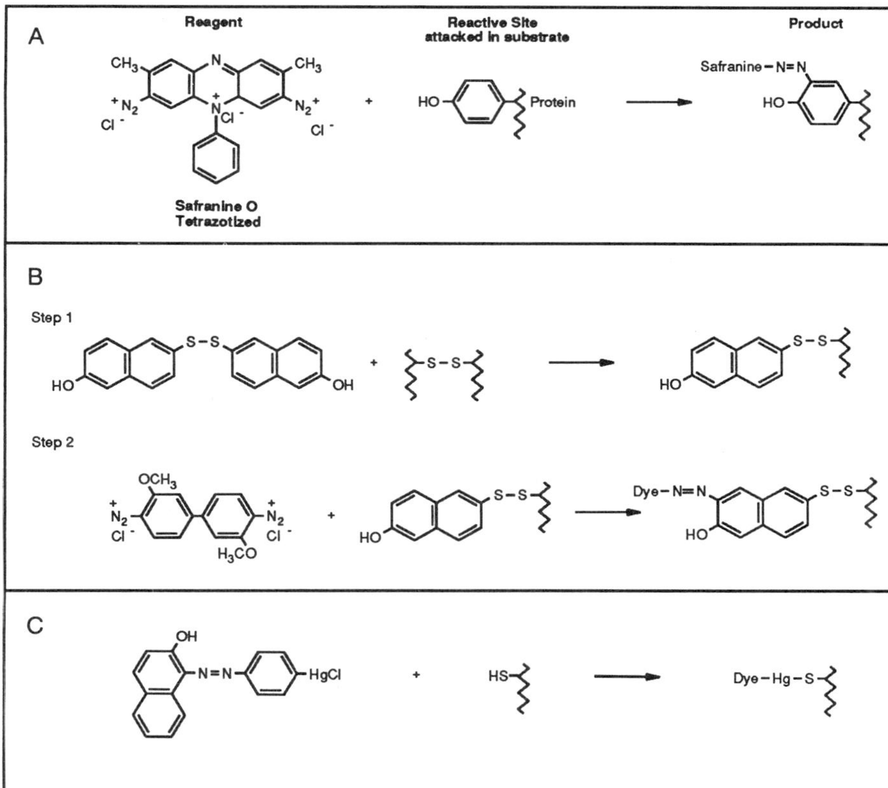

Fig. 2 Examples of covalent bonding between dye and substrate

A combination of 1 and 2. Methanolic solutions (0.3%) of any of the three dye blends will stain a fixed thin blood film satisfactorily, if not optimally, when diluted with an equal volume of buffer solution of pH 6.4. The resulting differentially stained blood film demonstrates that each dye component seeks out its ionic counterpart in the subcell structures of the blood cell milieu. Thus, regardless of whether the solution is prepared from dye blend 1, 2 or 3, the component dyes respond ionically in an aqueous system.

COVALENT BONDS:

Covalently bonded dyes, of course, offer the ultimate when permanence is desired since a dye linked in this manner becomes an integral part of the substrate with which it is linked. Naturally there are inherent disadvantages when these dyes are used in biological applications.

Dyes formed *in situ* from "colorless" components are important and useful representatives of the covalent bonding mode. These dyes are generally formed as a result of a coupling reaction between a diazonium salt and a phenolic or naphthoic protein moiety. Stable diazonium salts (zinc double salts) which are commercially available are frequently used in this application (Fig. 2A). Also, diazonium salts can be coupled to "colorless" naphthoic disulfides which have been linked *via* a disulfide bond to an amino acid site such as cystine in the substrate (Fig. 2B). This is another specific application.

Reactive dyes such as the Procion and Remazol dyes form covalent bonds with cellulosic and proteinaceous substrate entities and are quite effective in textile and other commercial dyeing. However, they have not proven to be very useful as biological stains due to their lack of selectivity, since the reactive chlorines on the cyanuric chloride moiety of the Procion dyes will react readily with, among other functions, primary and secondary amino, hydroxyl, carboxyl and sulfhydryl groups to form covalent bonds *via* condensation reactions.

The diazo chloromercury-type reactive dyes (Fig. 2C) are more suited to biological applications because they are more selective than the Procions and Remazols in their functional group attachment to proteins.

OTHER BONDING FORCES:

While there is less evidence to support hydrogen bonding, hydrophobic bonding and van der Waals forces as major linking modes, these forces certainly can act in supportive or reinforcing roles after a major linking mode is in progress. Indeed, it is conceivable that they initiate link-ups under proper conditions.

Hydrogen bonding is frequently suggested as a linking mode when the evidence does not suggest other forces. A plausible position can be established for an attraction between peptide hydrogens of tissue protein and amino-group hydrogens of the dye molecule.

Van der Waals attractions and hydrophobic bonding play somewhat competing roles in dye bonding and are difficult to quantitate. For this reason it is assumed that they generally play minor bonding roles. However, these bonding forces undoubtedly come into play in involved staining procedures where the solvent system is switched from water to an organic solvent medium such as alcohol, DMF or DMSO. The relative contribution of these two forces to the bonding in effect varies with the molecular size and polarity of the dye and the solvent system utilized.

The so-called instant Wright blood stain (staining time, 10 seconds) which utilizes one of the dry stain mixtures mentioned earlier demonstrates how the movement of the dye from bath to substrate can be accelerated by increased system entropy. This can be accomplished by adding approximately 5-10% of a polar solvent such as DMF or DMSO to the methanol normally used as the solvent. Advantage is also taken of the greater solubility of these polar dyes in the highly polar solvents mentioned. Thus the dye concentration may be increased two- to three-fold.

METACHROMASIA:

Certain pure dyes (free of related dyes or dyes formed by degradation), usually basic dyes, produce different colors with different tissue components. This phenomenon is termed metachromasia.

A substrate stained normally (without an absorbance shift) is said to be stained orthochromatically (orthochromasis), while a substrate causing an absorbance shift, usually hypsochromic, is said to be stained metachromatically. Current evidence indicates that this color shift is due to a dye-dye interaction which varies with the substrate site. In other words, the intensity of the bonding affinity between the dye molecule and the site will dictate the ability of the bonded dye molecule to interact with other dye molecules and conceivably form dimers or polymers. Basic dyes comprise the bulk of the metachromatic dyes known and, in all cases, these molecules contain either a primary or secondary amino group or groups. Dye molecules containing only tertiary amino functions are not metachromatic. However, it should be emphasized that not all basic dyes containing a primary or secondary amine are metachromatic. While the absorbance maximum shift of basic metachromatic dyes is normally hypsochromic, the few acid dyes (usually azo dyes such as Congo Rubin) that demonstrate metachromasia display a bathochromic shift.

Some unstable dye solutions stain a given substrate more than one color. However, this phenomenon, unlike metachromasia, is due to a new dye formed as the result of the degradation of the initial dye. For example, an aqueous solution of Nile Blue A (Fig. 3) can form a new dye molecule on standing. This new oxazone dye , Nile Red, stains neutral fat and alkylated fat sites red, while fatty acid-rich sites are stained blue by the unchanged dye portion. This phenomenon, involving two distinct dye molecules, is known as allochromasia.

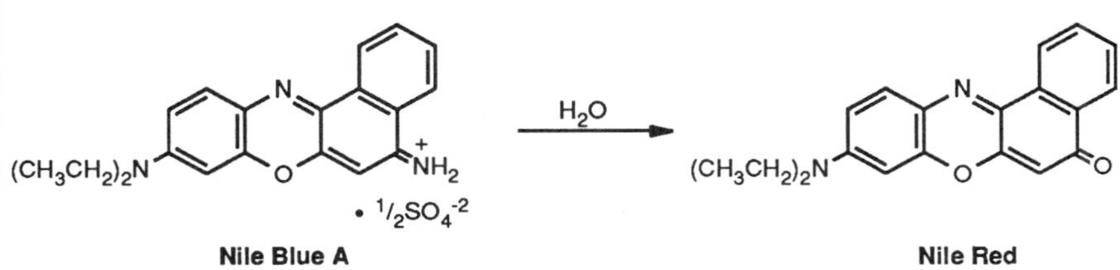

Fig. 3. Allochromatic

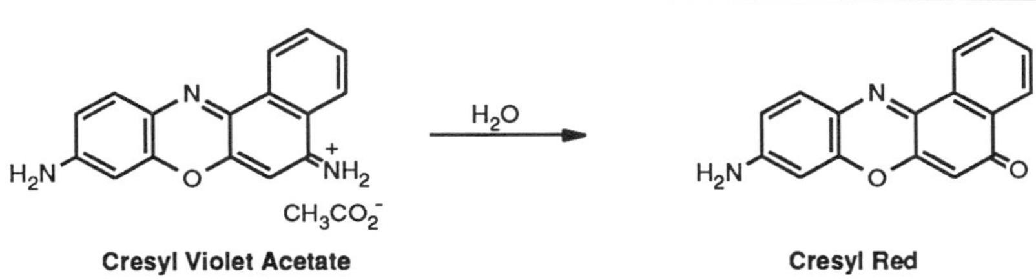

Fig. 4. Allochromatic, orthochromatic and metachromatic

Knaysi found that this phenomenon is not limited to the oxazine family of dyes, which Nile Blue A represents. Indeed, Neutral Red, an azine dye, and the thiazine dye, Methylene Blue, are specifically cited.

Cresyl Violet Acetate (Fig. 4), another oxazine dye, can act orthochromatically, metachromatically and allochromatically if enough of the oxazone impurity is present to function as the allochromic form. Normally, there is plenty of the oxazone present if the product is formed by the classic oxazine synthetic route. In fact, when used as a laser dye, Cresyl Violet has to be purified routinely by liquid-liquid extraction (H_2O and toluene) to eliminate the oxazone impurity.

CONCLUSION:

As noted earlier, biologists have been dependent upon commercial dye sources (dyes manufactured primarily for the paper, textile and leather trades) directly or indirectly since Perkin launched the industry.

Some of the early suppliers, such as the renowned Dr. Grübler and his successor, Dr. Hollborn, apparently selected dyes suitable for biological purposes from commercial dye lots while early American entrepreneurs in the biological-stain industry, such as Arthur Coleman, William Bell and Milton Hartman, got more involved in purifying suitable commercial dye products and synthesizing products not available from commercial sources. Their enterprises were started about the same time as The Biological Stain Commission which was founded in Geneva, NY, in 1922 by Dr. Harold J. Conn *et al.* The Commission was established to test and certify the suitability of dye lots for biological purposes. This organization has been and continues to be at the forefront of the struggle to provide quality stains for biological applications.

Today, the struggle to keep quality stains available intensifies as the biologists' reliance on commercial sources may have reached an impasse and their dependence on this industry is no longer assured. Technical problems have become secondary as the impact of recent legislation has emerged as the primary impediment to the availability of quality stains at prices confined to the general economy's inflationary spiral.

Indeed, within the last year, domestic commercial manufacturers curtailed or discontinued manufacture of four dyes of importance to biologists for three quarters of a century. Methylene Blue Chloride, Pararosaniline, Rosaniline and Safranine O have all but disappeared from commercial sources, putting an increased burden on the small dye manufacturers who, in the past, have produced special dyes solely for the biological market to supplement those available from textile sources.

Suggested Reading Material:

1) Baker, J.R. *The Principles of Biological Microtechnique*; Methuen and Co., Ltd: London, 1958.
2) Green, F.J. *Stain Technol.* **1966**, *41*, 115.
3) *Idem ibid.* **1966**, *41*, 223.
4) Harbin, R.W. J. *Soc. Dyers Colour.* **1975**, *91*, 4.
5) Thompson, J. *Selected Histochemical and Histopathological Methods*; Thomas: Springfield, 1966.
6) Pearse, A.G.E. *Histochemistry, Theoretical and Applied*, Vol. I and II; Little, Brown and Co.: Boston, 1953.
7) Ventakataraman, K. *The Chemistry of Synthetic Dyes*, Vols. I, II, III and IV; Academic Press: New York, 1952 and 1973.
8) Vickerstaff, T. *The Physical Chemistry of Dyeing*, 2nd ed.; Oliver and Boyd: London, 1954.
9) Conn, H.J. *Biological Stains*, 9th ed.; Lillie, R.D., Ed.; The Williams and Wilkins Co.: Baltimore, 1977.
10) Lillie, R.D.; Fullmer, H.M. *Histopathologic Technique and Practical Histochemistry*, 4th ed.; McGraw-Hill and Co.: New York, 1976.
11) Knaysi, J. *J. Bacteriol.* **1941**, *42*, 587.
12) *Idem Dairy Sci.* **1942**, *25*, 585.

ACID ALIZARIN VIOLET N

Mordant Violet 5

Aldrich Cat. No.: 21,100-1

CAS No.: [*2092-55-9*]
C.I. No.: 15670
Mol. Form.: $C_{16}H_{11}N_2NaO_5S$
F.W.: 366.33
Appearance: Brick-red powder
λ_{max}: 501nm in water
Solubility: H_2O 60mg/ml
EGME 10mg/ml
EtOH 1mg/ml

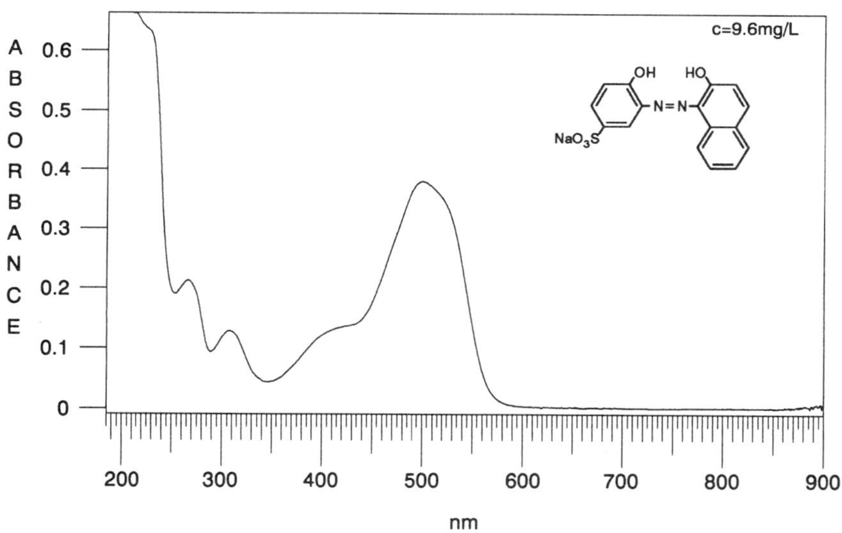

This anionic monoazo dye is made by coupling diazotized 2-amino-1-hydroxybenzene-4-sulfonic acid to 2-naphthol and converting the coupling product to the sodium salt.

Soluble in water, slightly soluble in acetone and very slightly soluble in ethanol, it is used to dye wool a bright violet from an acid bath, followed by afterchroming. A red can be produced if copper sulfate is substituted for the chromium.

Alizarin Violet N is one of several chelated azo dyes used by Hall and Hansen[1] as counterstains in a fluorescent-antibody technique.

Reference:

1) Hall, C.T.; Hansen, P.T. *Bacteriology* **1962**, 184, 548-554.

General references:

Aldrich FT-IR Library 1(2), 997C
Colour Index 3rd ed., 1971

ACID BLACK 24

Sulfone Cyanine Black BB

Aldrich Cat. No.:	21,045-5
CAS No.:	[*3071-73-6*]
C.I. No.:	26370
Mol. Form.:	$C_{36}H_{23}N_5Na_2O_6S_2$
F.W.:	731.72
Appearance:	Dark-brown powder
λ_{max}:	572nm in water
Solubility:	H_2O 20mg/ml
	EGME 20mg/ml
	EtOH 7mg/ml

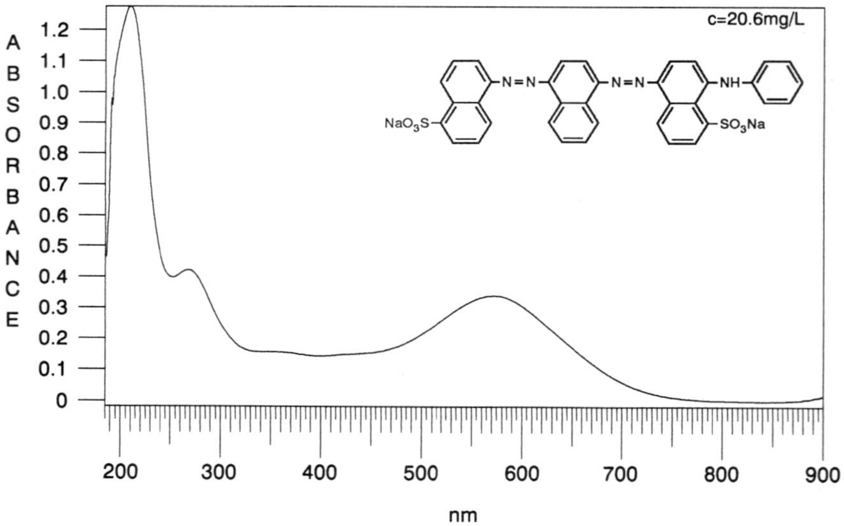

This anionic disazo dye is moderately soluble in water and methyl Cellosolve® and slightly soluble in ethanol.

Commercially, it is used to dye wool, silk and nylon from a neutral or acid bath, and fully chromed leather. Wool, silk, nylon and viscose are printed directly. Also, a pigment is prepared from its barium salt.

There is no evidence in the literature that it has been used as a biological stain.

®Registered trademark of Union Carbide Corporation.

General references:

Aldrich FT-IR Library	**1**(2), 992B
IR Library	**3**, 1465A
Colour Index	3rd ed., 1971

ACID BLACK 48

Alizarin Light Grey BBL • Coomassie® Grey 3G

Aldrich Cat. No.:	19,424-7
Sigma Prod. No.:	A4895
CAS No.:	[*1328-24-1*]
C.I. No.:	65005
Mol. Form.:	$C_{28}H_{18}N_3O_4 \cdot xSO_3Na$
m.p.:	275°C (dec.)
Appearance:	Black powder
λ_{max}:	663nm in water
Solubility:	H_2O 20mg/ml
	EGME 10mg/ml
	EtOH 0.8mg/ml

® Reg. trademark of Imperial Chemical Industries

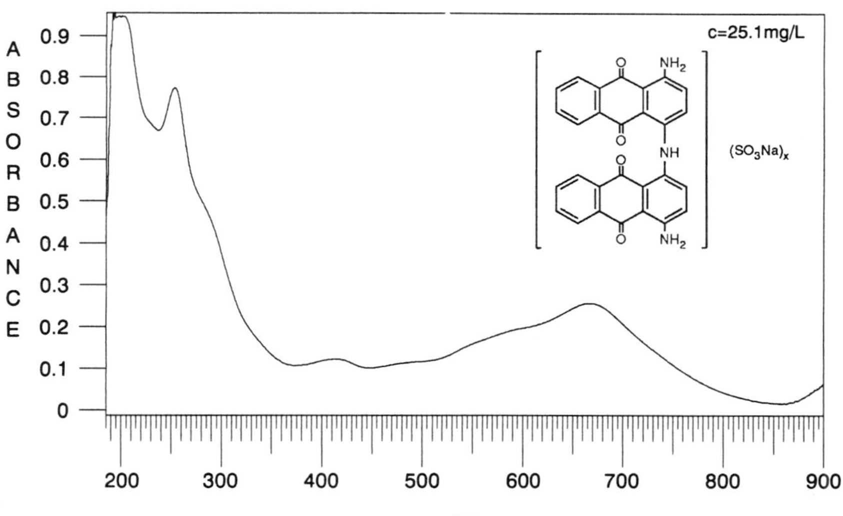

Acid Black 48 is an acid anthraquinone dye that is made by condensing 1,4-diaminoanthraquinone with 1-amino-4-bromoanthraquinone in the presence of copper and sodium acetate. The product of this reaction is sulfonated with oleum and converted to the sodium salt. It is soluble in water, slightly soluble in methyl Cellosolve and very slightly soluble in ethanol. Textile applications include the blue-grey dyeing of wool, nylon (acetic or formic acid bath), silk (neutral or acetic acid bath) and printing wool, silk and viscose directly. Although the penetration is low, leather is dyed a light bluish grey and fur is dyed a comparable color.

There is no evidence that it has been used as a biological stain.

General references:

Aldrich FT-IR Library	**1**(2), 1020D
Sigma FT-IR Library	**1**(2), 85A
IR Library	**3**, 1478D
Colour Index	3rd ed., 1971

ACID BLUE 25

Alizarin Saphirol A

Aldrich Cat. No.:	21,068-4
Sigma Prod. No.:	A5399
CAS No.:	[*6408-78-2*]
C.I. No.:	62055
Mol. Form.:	$C_{20}H_{13}N_2NaO_5S$
F.W.:	416.39
Appearance:	Deep-violet powder
λ_{max}:	600nm in water
Solubility:	H_2O 10mg/ml
	EGME 30mg/ml
	EtOH 4mg/ml

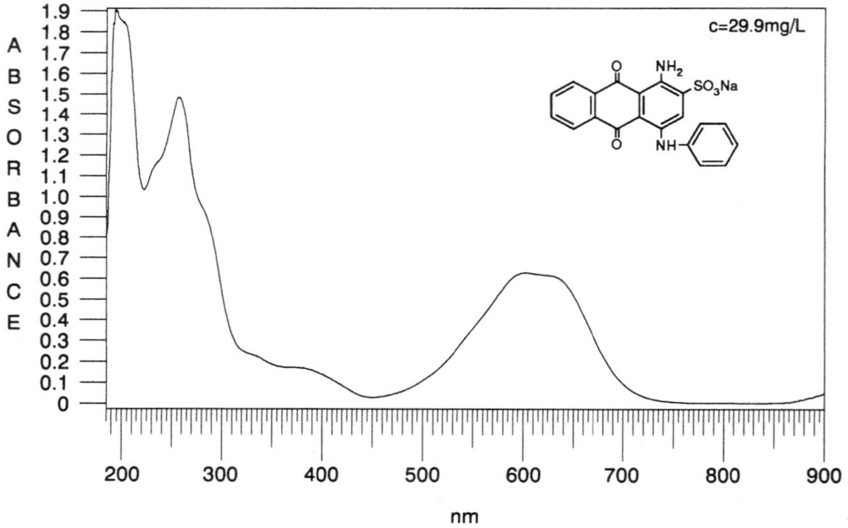

This is a blue, anionic anthraquinone dye that is made by condensing the sodium salt of 1-amino-4-bromo-2-anthraquinonesulfonic acid with aniline in the presence of a copper salt.

Slightly soluble in water and ethanol and soluble in acetone, it is used to dye wool, nylon (neutral or formic acid bath), silk (acetic, formic or sulfuric acid bath), and to print all of these substrates directly. Nontextile uses include coloring anodized aluminum, bath salts and soap and dyeing leather blue.

There is no evidence that it has been used as a biological stain.

General references:

Aldrich FT-IR Library	**1**(2), 1018B
Sigma FT-IR Library	**1**(2), 85B
IR Library	**3**, 1477C
Colour Index	3rd ed., 1971

ACID BLUE 29

Aldrich Cat. No.:	25,588-2
Sigma Prod. No.:	A8283

CAS No.:	*[5850-35-1]*
C.I. No.:	20460
Mol. Form.:	$C_{22}H_{14}N_6Na_2O_9S_2$
F.W.:	616.50
Appearance:	Bluish-black powder
λ_{max}:	602nm in water
Solubility:	H_2O 20mg/ml
	EGME 20mg/ml
	EtOH 2mg/ml

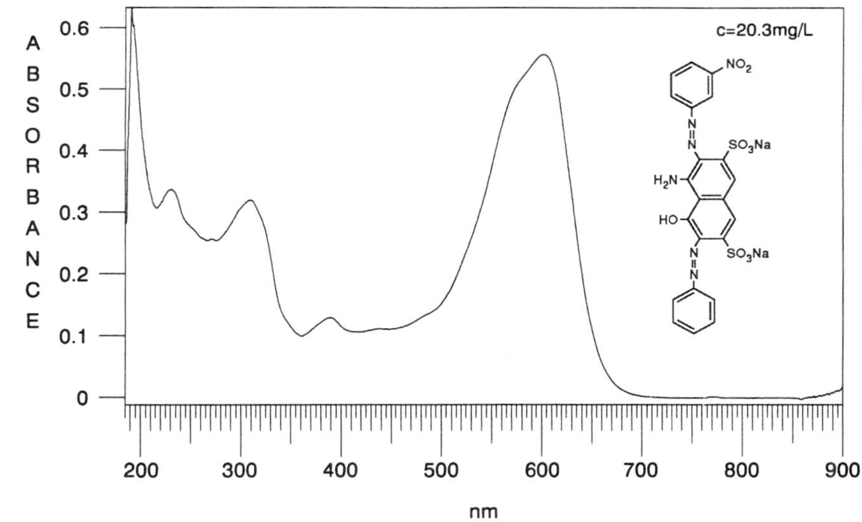

This anionic disazo dye is made by first coupling diazotized 3-nitroaniline to H acid (1-amino-8-hydroxy-3,6-naphthalenedisulfonic acid) in an acid medium, followed by an alkaline coupling of the reaction product to diazotized aniline.

Slightly soluble in methyl Cellosolve and water and very slightly soluble in ethanol, it is used to dye wool, nylon, silk, beater paper and chromed leather, to print wool and silk directly and to surface-color paper.

While requests have been made for this dye supposedly for its use as a biological stain, there are no literature references regarding specific uses in this area.

General references:

Aldrich FT-IR Library	**1**(2), 999B
Sigma FT-IR Library	**1**(2), 85C
Colour Index	3rd ed., 1971

ACID BLUE 40
Alizarin Direct Blue A2G

Aldrich Cat. No.: 21,054-4

CAS No.: *[6424-85-7]*
C.I. No.: 62125
Mol. Form.: $C_{22}H_{16}N_3NaO_6S$
F.W.: 473.44
Appearance: Blue-black powder
λ_{max}: 610(380)nm in water
Solubility: H_2O <0.1mg/ml
EGME 10mg/ml
EtOH 4mg/ml

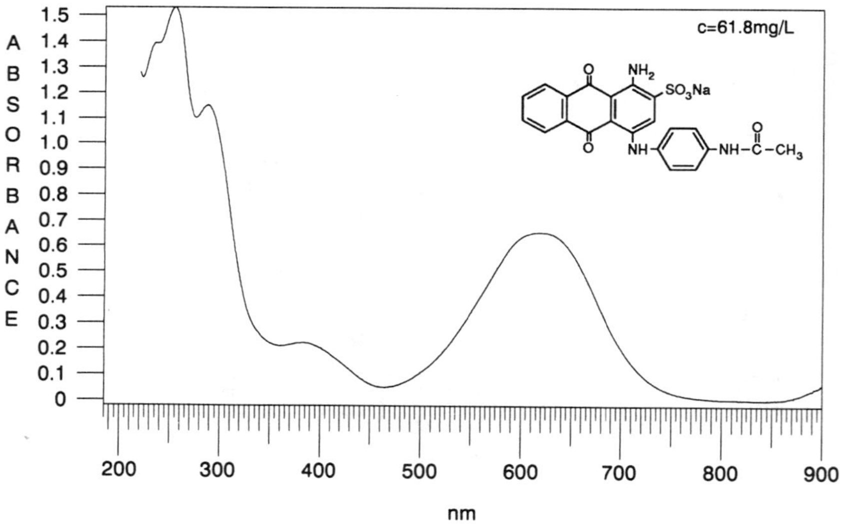

Acid Blue 40 is an acid anthraquinone dye that is soluble in acetone and slightly soluble in alcohol. It is made by condensing 1-amino-4-bromo-2-anthraquinonesulfonic acid with 4'-aminoacetanilide in the presence of a copper salt. The product is then converted to the sodium salt.

Commercially, it is used to dye wool a greenish blue. It dyes nylon from a formic acid bath and silk from acetic, formic or sulfuric acid. Silk, wool and nylon are printed greenish blue directly and fully chromed leather is dyed a greenish blue.

There is no evidence in the literature that it has been used as a biological stain.

General references:

Aldrich FT-IR Library 1(2), 1018D
IR Library **3**, 1464H
Colour Index 3rd ed., 1971

ACID BLUE 41

Alizarin Direct Blue AR • Anthralan Blue B

Aldrich Cat. No.: 21,055-2

CAS No.: [*2666-17-3*]
C.I. No.: 62130
Mol. Form.: $C_{23}H_{18}N_3NaO_6S$
F.W.: 487.47
Appearance: Purple powder
λ_{max}: 599nm in water
Solubility: H_2O 40mg/ml
EGME 8mg/ml
EtOH 0.9mg/ml

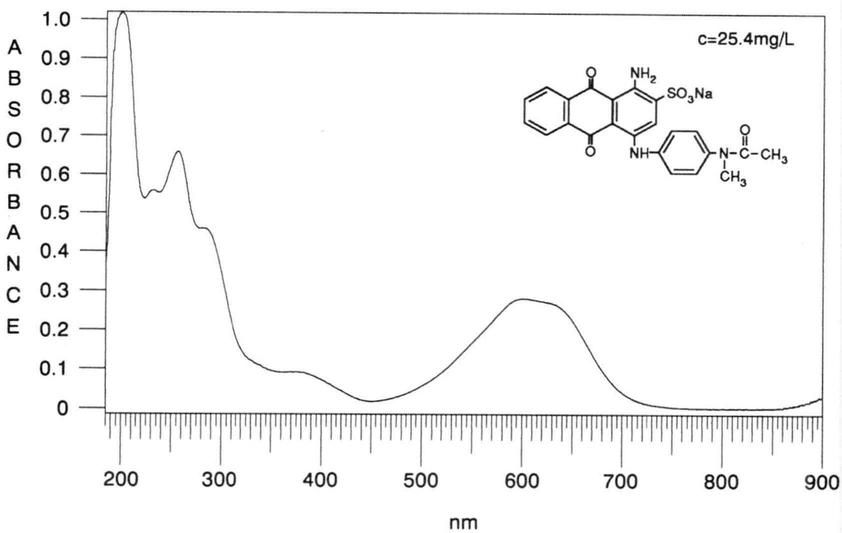

This anionic anthraquinone dye is made by first condensing molar equivalents of 1-amino-4-bromo-2-anthraquinonesulfonic acid and 4'-amino-*N*-methylacetanilide in the presence of a copper salt. The condensation product is then converted to the sodium salt.

Very slightly soluble in ethanol and soluble in acetone, it is used to dye wool, nylon (formic acid bath) and silk (acetic acid bath) and to print all three fabrics directly. Other industrial applications include coloring anodized aluminum and soap.

There is no evidence in the literature that it has been used as a biological stain.

General references:

Aldrich FT-IR Library **1**(2), 1018C
IR Library **3**, 1477D
Colour Index 3rd ed., 1971

ACID BLUE 45

Alizarin Saphirol B

Aldrich Cat. No.:	21,069-2
Sigma Prod. No.:	A5524

CAS No.:	*[2861-02-1]*
C.I. No.:	63010
Mol. Form.:	$C_{14}H_8N_2Na_2O_{10}S_2$
F.W.:	474.34
Appearance:	Medium-purple powder
λ_{max}**:**	594nm in water
Solubility:	H_2O 40mg/ml
	EGME 6mg/ml
	EtOH 2mg/ml

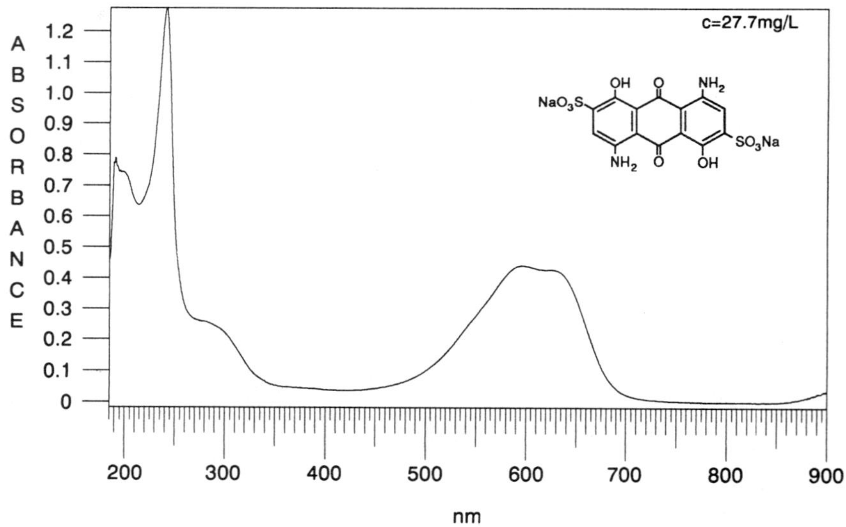

This blue amphoteric anthraquinone dye is soluble in dilute acid (pH 3.5) and slightly soluble in ethanol and methyl Cellosolve.

Industrial uses include dyeing wool, silk, beater paper and fully chromed leather, preparing pigments (heavy-metal salts) for printing inks, book cloths and wallpaper and coloring melamine, nitrocellulose plastics, soap and anodized aluminum.

Conn's[1] speculates that Alizarin Saphrol B can be substituted for the structurally similar but obsolete Alizarin Blue GR as a collagen-fiber stain (Volkmann and Strauss,[2] a variation of the Mallory, Heidenhain azan-collagen method).

References:

1) *Conn's Biological Stains,* 9th ed.; Lillie, R.D., Ed.; Williams and Wilkins: Baltimore, 1977; pp 446-447.

2) von Volkmann, R.; Strauss, F. *Z. Wiss. Mikrosk.* **1934,** *51,* 244-249.

General references:

Aldrich FT-IR Library	**1**(2), 1017D
Sigma FT-IR Library	**1**(2), 85D
IR Library	**3**, 1477A
Colour Index	3rd ed., 1971

ACID BLUE 80

Aldrich Cat. No.:	21,032-3
CAS No.:	*[4474-24-2]*
C.I. No.:	61585
Mol. Form.:	$C_{32}H_{28}N_2Na_2O_8S_2$
F.W.:	678.70
Appearance:	Dark-blue powder
λ_{max}:	625(580)nm in water
Solubility:	H_2O 50mg/ml
	EGME 40mg/ml
	EtOH 1mg/ml

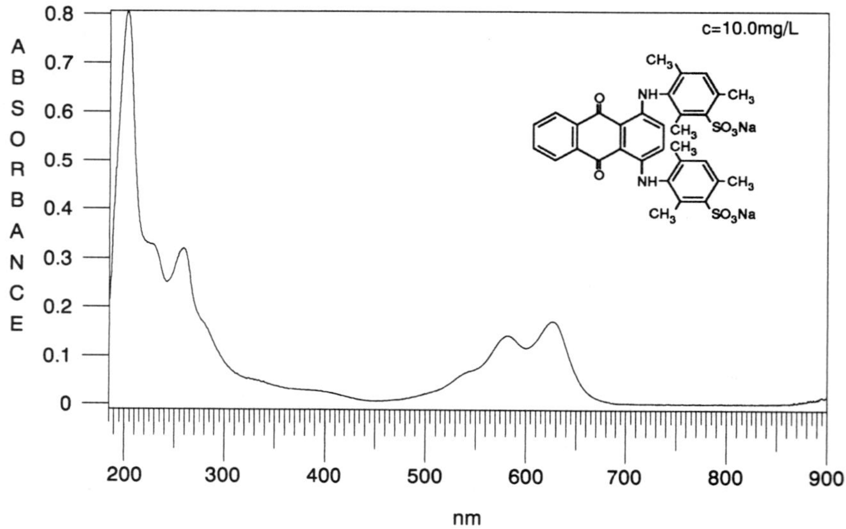

This blue, anionic anthraquinone dye is made by condensing one mole of 1,4-dichloroanthraquinone with two moles of mesidinesulfonic acid, sodium salt. The condensation product is converted to the disodium salt, which is soluble in water and methyl Cellosolve and very slightly soluble in ethanol .

Commercially, it is used to dye wool, silk (acetic acid bath) and nylon. Wool and silk can be printed directly.

There is no evidence in the literature that it has been used as a biological stain.

General references:

Aldrich FT-IR Library	**1**(2), 1020C
IR Library	**3**, 1478C
Colour Index	3rd ed., 1971

ACID BLUE 92

Anazolene sodium • Coomassie® Blue RL • Sulfone Acid Blue R

Aldrich Cat. No.: 21,042-0

CAS No.: *[3861-73-2]*
C.I. No.: 13390
Mol. Form.: $C_{26}H_{16}N_3Na_3O_{10}S_3$
F.W.: 695.59
Appearance: Dark-blue powder
λ_{max}: 571nm in water
Solubility: H_2O 40mg/ml
EGME 60mg/ml
EtOH 6mg/ml

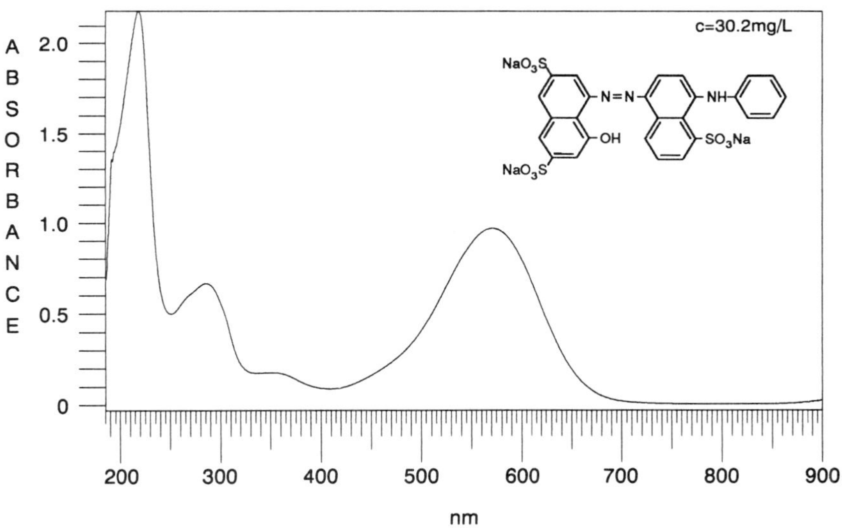

This anionic monoazo dye is made by coupling diazotized H acid (8-amino-1-naphthol-3,6-disulfonic acid) with *N*-phenyl-1-naphthylamine-8-sulfonic acid and converting the product to the trisodium salt.

Soluble in water and methyl Cellosolve and slightly soluble in ethanol and glycerol, it is used commercially to dye paper, soap and leather.

Merck[1] lists it, under the name of Anazolene sodium, as a diagnostic aid (plasma-volume indicator) and states "Coomassie Blue is cleared by the human organism faster than Evans Blue." British references are cited. Apparently it has no official status in the United States.

Coomassie Blue RL should not be confused with the triphenylmethane dyes, Coomassie Brilliant Blue 250 G (Acid Blue 90) and Coomassie Brilliant Blue 250 R (Acid Blue 83) which are used for coloring protein for electrophoresis separation on poly(acrylamide) gels. [See Brilliant Blue G (Aldrich Cat. No. 20,139-1) and Brilliant Blue R (Aldrich Cat. No. 20,140-5) monographs on pages 151 and 152, and 153 and 154 in this book.]

Reference:

1) *The Merck Index,* 11th ed.; Budavari, S. *et al.*, Eds.; Merck: Rahway, 1989; Monograph 662.

General references:

Aldrich FT-IR Library	**1**(2), 991B
IR Library	**3**, 1464E
Colour Index	3rd ed., 1971

ACID BLUE 113

Sulfone Cyanin 5R

Aldrich Cat. No.: 21,043-9

CAS No.: [*3351-05-1*]
C.I. No.: 26360
Mol. Form.: $C_{32}H_{21}N_5Na_2O_6S_2$
F.W.: 681.66
Appearance: Black powder
λ_{max}: 566nm in water
Solubility: H_2O 40mg/ml
EGME 30mg/ml
EtOH 4mg/ml

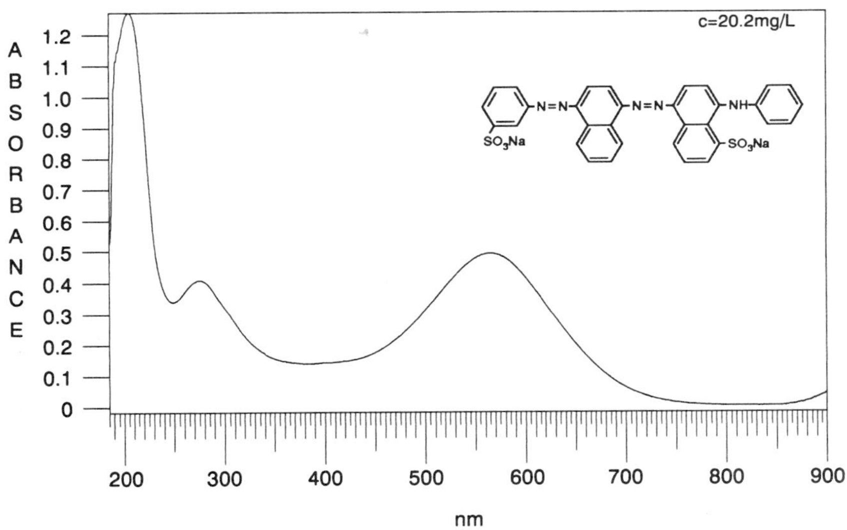

This anionic disazo dye is made by first coupling diazotized metanilic acid to 1-naphthylamine. The product of the coupling is diazotized and coupled to 8-amino-*N*-phenyl-1-naphthalenesulfonic acid, which is subsequently converted to the sodium salt.

Soluble in water and methyl Cellosolve and slightly soluble in ethanol and acetone, it has been used to dye both wool and silk out of neutral or acetic acid baths. Fully chromed leather is dyed navy blue.

Conn's[1] cites Puchtler's use of Acid Blue 113 along with the nuclear stain, Nuclear Fast Red, to demonstrate myofibrils and myoendothelial cells, particularly in the nervous system.

Reference:

1) *Conn's Biological Stains,* 9th ed.; Lillie, R.D., Ed.; Williams and Wilkins: Baltimore, 1977; pp 174-175.

General references:

Aldrich FT-IR Library	**1**(2), 991D
IR Library	**3**, 1464G
Colour Index	3rd ed., 1971

ACID BLUE 120
Sulfone Cyanine GR

Aldrich Cat. No.:	21,044-7
CAS No.:	[*3529-01-9*]
C.I. No.:	26400
Mol. Form.:	$C_{33}H_{23}N_5Na_2O_6S_2$
F.W.:	695.69
Appearance:	Dark-brown powder
λ_{max}:	524nm in water
Solubility:	H_2O 50mg/ml
	EGME 20mg/ml
	EtOH 10mg/ml

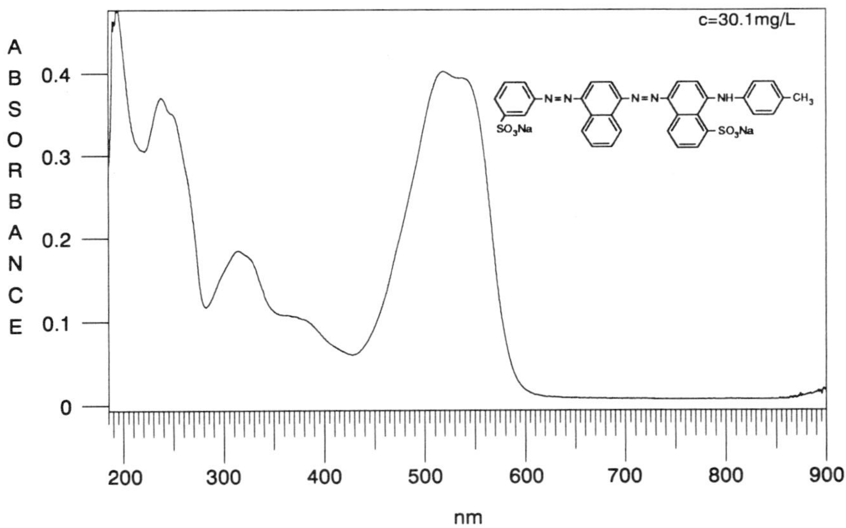

This anionic disazo dye is made by first coupling diazotized metanilic acid to 1-naphthylamine. The product of the coupling is diazotized and coupled to 8-amino-*N*-(*p*-tolyl)-1-naphthalenesulfonic acid, which is subsequently converted to the disodium salt.

Soluble in water and somewhat less soluble in ethanol, methyl Cellosolve and acetone, it has been used to dye wool, silk (acetic acid or Glauber's salt solution) and nylon (acetic acid or neutral bath). Fully chromed leather is dyed navy blue.

Puchtler[1] has used this compound, in combination with a 4-aminosalicylic acid reaction, as a blue stain for A and Z bands of striated muscle, for myofibrils in smooth muscle and for myoepithelial cells.

Reference:

1) Puchtler, H.; Waldrop, F.S.; Terry, M.S.; Conner, H.M. *J. Microsc.* **1969,** *89,* 329-338.

General references:

Aldrich FT-IR Library	**1**(2), 992A
IR Library	**3**, 1464H
Colour Index	3rd ed., 1971

ACID BLUE 129

Atanyl Blue PRL

Aldrich Cat. No.: 30,649-5
Sigma Prod. No.: A4545

CAS No.: [*6397-02-0*]
C.I. No.: 62058
Mol. Form.: $C_{23}H_{19}N_2NaO_5S$
F.W.: 458.47
m.p.: >300°C
Appearance: Dark-blue powder
λ_{max}: 629nm in water

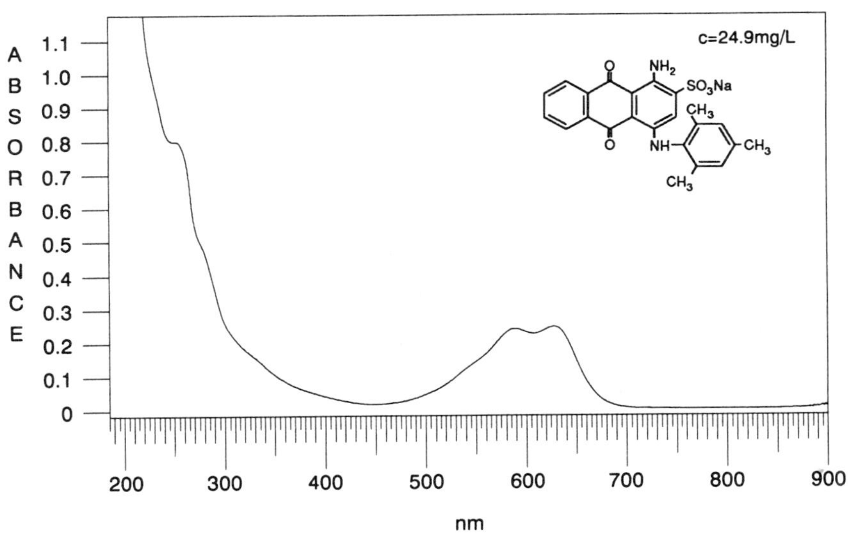

This bright-blue anthraquinone dye is made by condensing 1-amino-4-bromoanthraquinone-2-sulfonic acid with 2,4,6-trimethylaniline.

Soluble in water, it is used to dye wool, silk and nylon from an acetic acid bath and to print all three directly. Lightfastness is moderate (4-5).

There is no evidence in the literature that it has been used as a biological stain.

General references:

Safety	**2**(1), 52D
Colour Index	3rd ed., 1971

ACID BLUE 161

Palatine Fast Blue BN

Aldrich Cat. No.:	20,183-9
Sigma Prod. No.:	A4770
CAS No.:	[*12392-64-2*]
C.I. No.:	15706
Mol. Form.:	$C_{20}H_{13}N_2NaO_5S \cdot xCr$
F.W.:	416.39
Appearance:	Dark-blue powder
λ_{max}**:**	600(563)nm in water
Solubility:	H_2O 60mg/ml
	EGME 50mg/ml
	EtOH 50mg/ml

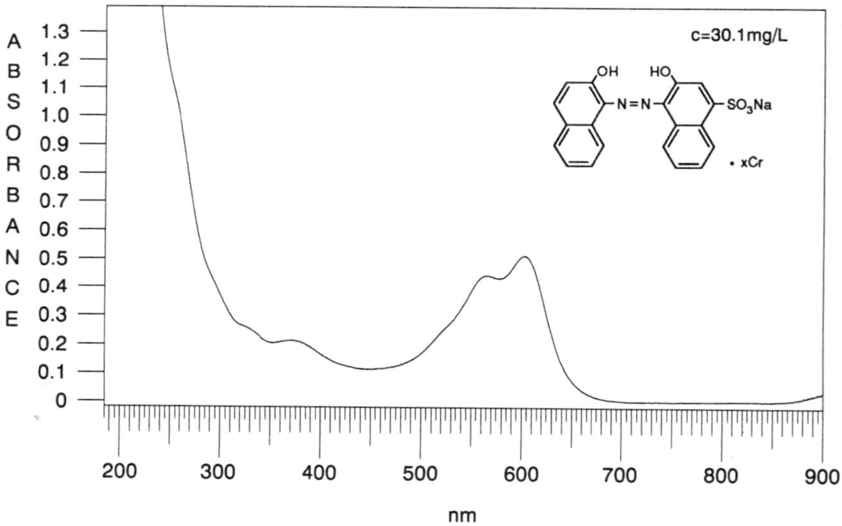

This anionic monoazo dye is made by complexing Palatine Chrome Black 6BN (Aldrich Cat. No. 21,673-9, C.I. 15705) with chromium. It is soluble in water and in ethanol. Commercial applications include dyeing wool directly and printing wool and silk directly.

There is no evidence in the literature that it has been used as a biological stain.

General references:

Aldrich FT-IR Library	**1**(2), 987B
Sigma FT-IR Library	**1**(2), 86A
IR Library	**3**, 1462G
Colour Index	3rd ed., 1971

ACID FUCHSIN, calcium salt

Aldrich Cat. No.: 85,740-8
Sigma Prod. No.: A3908
Grade: Certified

Mol. Form.: $C_{20}H_{17}CaN_3O_9S_3$
F.W.: 579.65
Appearance: Green crystalline powder
λ_{max}: 545nm in water +2.5ml 0.1*N* HCl
Solubility: H_2O 100mg/ml
EGME 10mg/ml
EtOH 1mg/ml

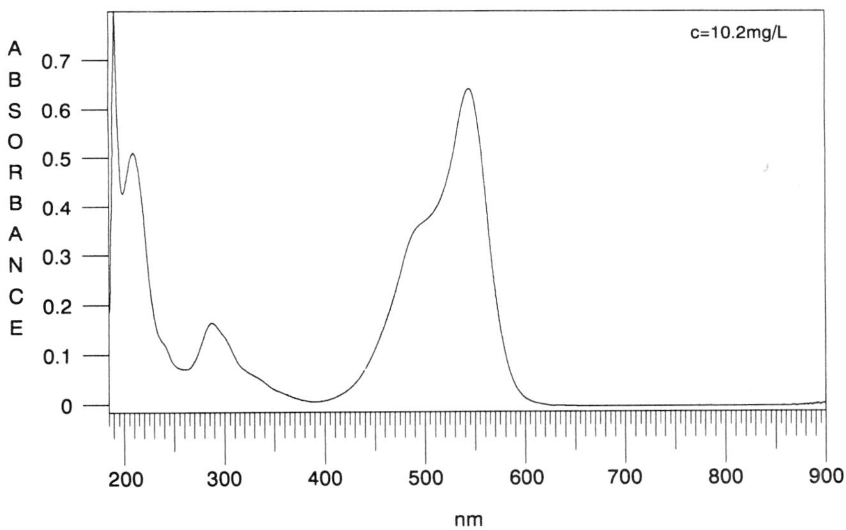

Acid Fuchsin is a bluish-red triphenylmethane dye that is made by the controlled sulfonation of Basic Fuchsin (preferably the Pararosaniline homolog). Sulfonation of the Basic Fuchsin is accomplished by adding the dye to the sulfonating reagent, oleum, with stirring, at such a rate that the slurry temperature does not exceed 150°C. After the exotherm has subsided, external heat is applied and the temperature is maintained at 150°C (no higher) for at least 30 minutes. Material made in this manner is essentially the trisulfonated product. It is free of decomposition products and contains only traces of the less desirable disulfonated compound.

There are three basic causes for variability in the manufacture of this product: 1) The raw material (Basic Fuchsin) may be composed either predominantly of Pararosaniline, Rosaniline, New Fuchsin, varying mixtures of two of these or varying mixtures of all three homologs; 2) incomplete sulfonation; 3) salt-form level and variability. The calcium, sodium and ammonium salts have been commercially available and all three perform satisfactorily with minor solubility variations. More important is the level of salt substitution. The disubstituted salt form is the ideal level and the trisubstituted the least desirable since it is colorless, although the color can be restored when the material is acidified and the chromophore recreated.

Acid Fuchsin was formerly used to dye wool red from an acid bath and silk a similar color from an acidified soap bath. Current commercial applications include the dyeing of crepe paper and soap and of photographic film, which is by far the largest application.

Acid Fuchsin is used widely in biological applications.[1] *Staining Procedures*[2] provides methods and bibliography for many uses involving plasma, tissue and collagen staining. These include Van Gieson's connective-tissue stain, in combination with picric acid and Hematoxylin; Ehrlich Biondi's triacid stain with Orange G and Methyl Green (this is a neutral blood stain) and Mallory's connective-tissue stain. Pianese[3] combined Acid Fuchsin with Martius Yellow and Malachite Green for use in plant histology, and Alexander[4] used a tristain composed of Orange G, Malachite Green and Acid Fuchsin to differentially stain aborted pollen grains green and unaborted grains red.

References:

1) Gray, P. *The Microtomist's Formulary and Guide;* Constable: London, 1954.

2) Clark, G. *Staining Procedures,* 4th ed.; Williams and Wilkins: Baltimore, 1981; pp 35, 38, 113, 114, 117, 119, 256, 262, 265, 336, 355, 358.

3) Pianese, G. *Beitr. Pathol. Anat.* **1896,** Suppl. I.

4) Alexander, M.P. *Stain Technol.* **1969,** *44*, 117-122.

General references:

Aldrich FT-IR Library	**1**(2), 964C
Sigma FT-IR Library	**1**(2), 86B
IR Library	**3**, 1448H
Safety	**2**(1), 53A
Colour Index	3rd ed., 1971

ACID FUCHSIN, sodium salt
Acid Violet 19

Aldrich Cat. No.: 33,270-4

CAS No.: [*3244-88-0*]
C.I. No.: 42685
Mol. Form.: $C_{20}H_{17}N_3Na_2O_9S_3$
F.W.: 585.55
Appearance: Dark-green crystalline powder
λ_{max}: 546nm in water + 2.5 ml 0.1*N* HCl

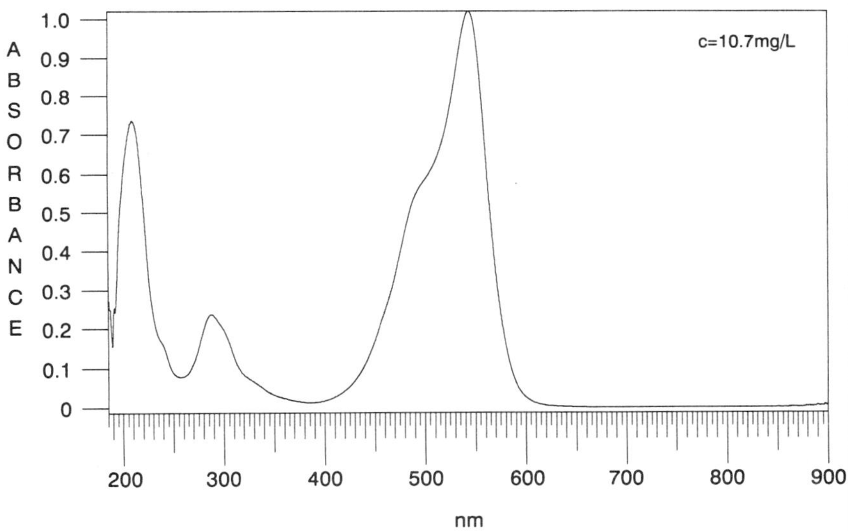

See description on pages 19 and 20 under Acid Fuchsin, calcium salt.

General reference:
Merck Index 11, 103

ACID GREEN 25

Alizarin Cyanin Green G

Aldrich Cat. No.:	21,456-6
CAS No.:	*[4403-90-1]*
C.I. No.:	61570
Mol. Form.:	$C_{28}H_{20}N_2Na_2O_8S_2$
F.W.:	622.59
m.p.:	235-238°C
Appearance:	Dark blue-green powder
λ_{max}:	642(608)nm in water
Solubility:	H_2O 60mg/ml
	EGME 30mg/ml
	EtOH 2mg/ml

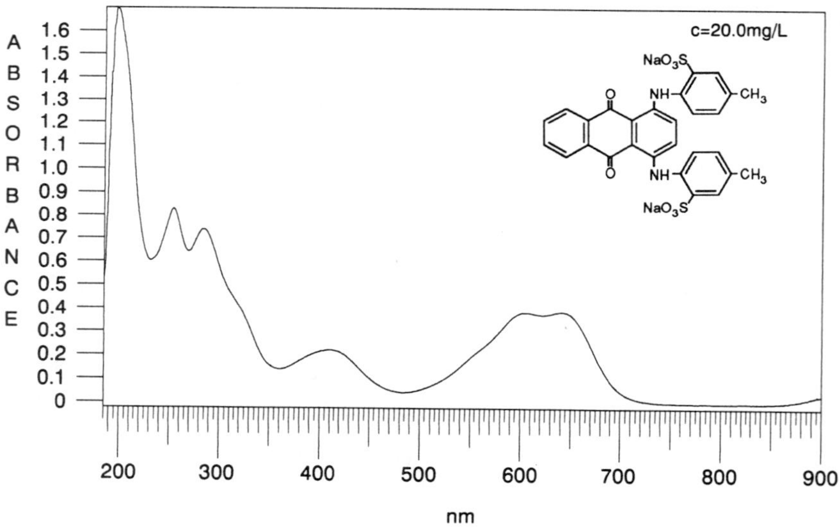

This anionic anthraquinone dye is made by disulfonating Solvent Green 3 (C.I. 61565) with oleum and converting the resulting product to the disodium salt.

Soluble in water and slightly soluble in ethanol, acetone and pyridine, it is used to dye paper, chromed leather, wool, silk and nylon (acid bath), to stain wood and to color anodized aluminum, urea-melamine resin and soap.

There is no evidence in the literature that it has been used as a biological stain.

General references:

Aldrich FT-IR Library	**1**(2), 1019D
IR Library	**3,** 1477H
Safety	**2**(1), 53B
Merck Index	**11**, 238
Colour Index	3rd ed., 1971

ACID GREEN 27

Aldrich Cat. No.:	21,033-1
CAS No.:	[*6408-57-7*]
C.I. No.:	61580
Mol. Form.:	$C_{34}H_{32}N_2Na_2O_8S_2$
F.W.:	706.75
m.p.:	258-260°C
Appearance:	Dark-green powder
λ_{max}:	605(642)nm in water
Solubility:	H_2O 30mg/ml
	EGME 20mg/ml
	EtOH 2mg/ml

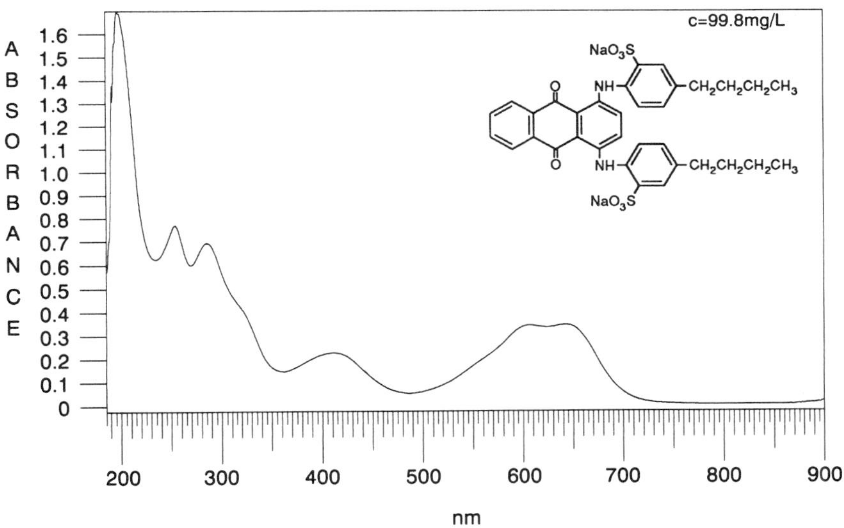

This anionic anthraquinone dye is made by treating leucoquinizarin with 4-butylaniline, followed by sulfonation and conversion to the sodium salt. Soluble in water, moderately soluble in methyl Cellosolve and very slightly soluble in ethanol, its textile uses include dyeing wool and silk from a neutral or acid bath and direct printing of wool, silk and nylon.

There is no evidence that it has been used as a biological stain.

General references:

Aldrich FT-IR Library	**1**(2), 1020A
IR Library	**3**, 1478A
Colour Index	3rd ed., 1971

ACID GREEN 41
Alizarin Cyanin Green 5G

Aldrich Cat. No.:	21,071-4
CAS No.:	[*4430-16-4*]
C.I. No.:	62560
Mol. Form.:	$C_{28}H_{20}N_2Na_2O_{10}S_2$
F.W.:	654.59
m.p.:	286-290°C
Appearance:	Dark-green powder
λ_{max}:	680(417)nm in water
Solubility:	H_2O 30mg/ml
	EGME 40mg/ml
	EtOH 5mg/ml

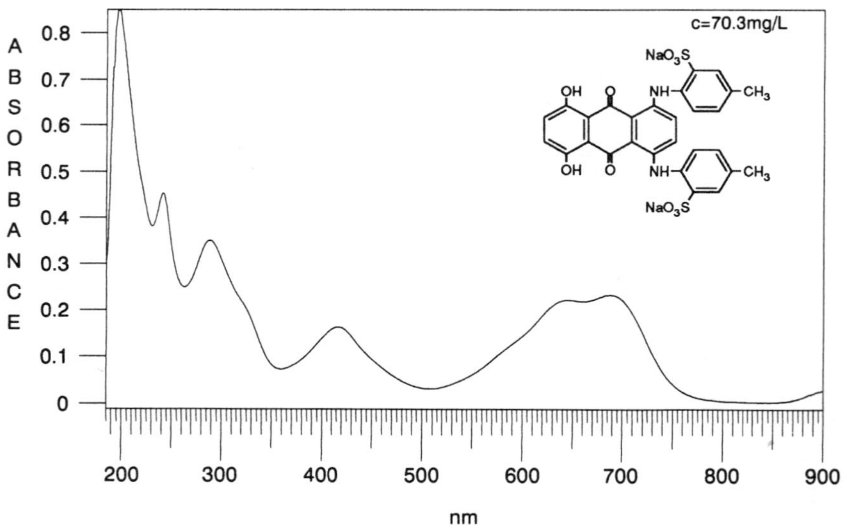

This anionic anthraquinone dye is made by first condensing one mole of 5,8-dichloroquinizarin with two moles of *p*-toluidine. Then the product of this reaction is disulfonated with oleum and converted to the disodium salt.

Soluble in water, acetone, methyl Cellosolve, 2-chlorophenol and pyridine and slightly soluble in ethanol, it is used to dye wool from an acid bath and silk from a broken degumming liquor, and to print wool and silk. Nontextile applications include dyeing paper and the production of pigments.

There is no evidence in the literature that it has been used as a biological stain.

General references:

Aldrich FT-IR Library	**1**(2), 1020B
IR Library	**3**, 1478B
Colour Index	3rd ed., 1971

ACID ORANGE 8
Orange ROF

Aldrich Cat. No.: 21,453-1
Sigma Prod. No.: A5649

CAS No.: *[5850-86-2]*
C.I. No.: 15575
Mol. Form.: $C_{17}H_{13}N_2NaO_4S_2$
F.W.: 364.36
Appearance: Dark-orange powder
λ_{max}: 490nm in water
Solubility: H_2O 30mg/ml
EGME 40mg/ml
EtOH 2mg/ml

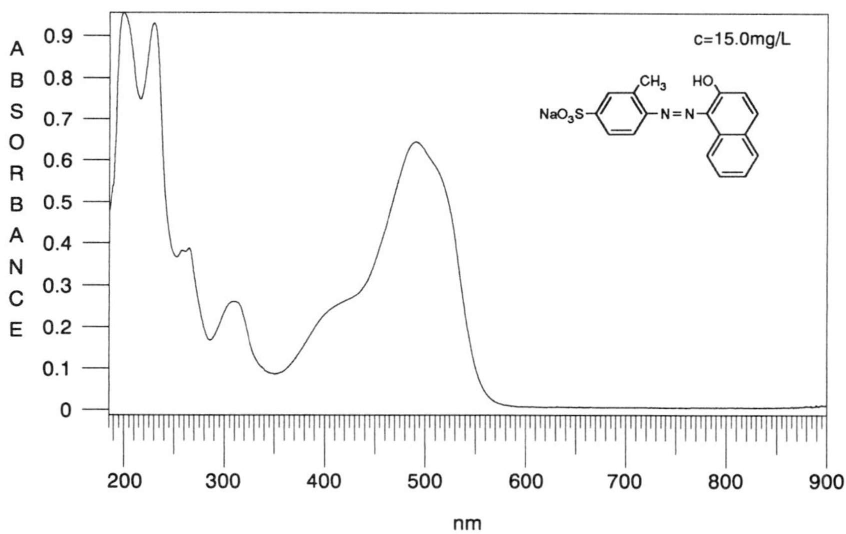

This anionic monoazo dye is soluble in water and methyl Cellosolve and slightly soluble in alcohol and acetone. It is made by coupling diazotized 4-amino-*m*-toluenesulfonic acid to 2-naphthol and converting the coupling product to the sodium salt.

Industrial uses include dyeing wool (acid bath), silk (broken degumming liquor), nylon (acetic or formic acid bath), jute (acid or neutral bath), beater paper and chromed leather. Wool and silk are printed directly, while heavy-metal salts are used as pigments.

There is no evidence in the literature that it has been used as a biological stain.

General references:

Aldrich FT-IR Library	**1**(2), 981D
Sigma FT-IR Library	**1**(2), 86C
IR Library	**3**, 1459E
Colour Index	3rd ed., 1971

ACID ORANGE 51

Supranol Brown 4R

Aldrich Cat. No.:	21,459-0
Sigma Prod. No.:	A5774

CAS No.:	[*8003-88-1*]
C.I. No.:	26550
Mol. Form.:	$C_{36}H_{26}N_6Na_2O_{11}S_3$
F.W.:	860.81
Appearance:	Dark-brown powder
λ_{max}:	446nm in water
Solubility:	H_2O 30mg/ml
	EGME 50mg/ml
	EtOH 2mg/ml

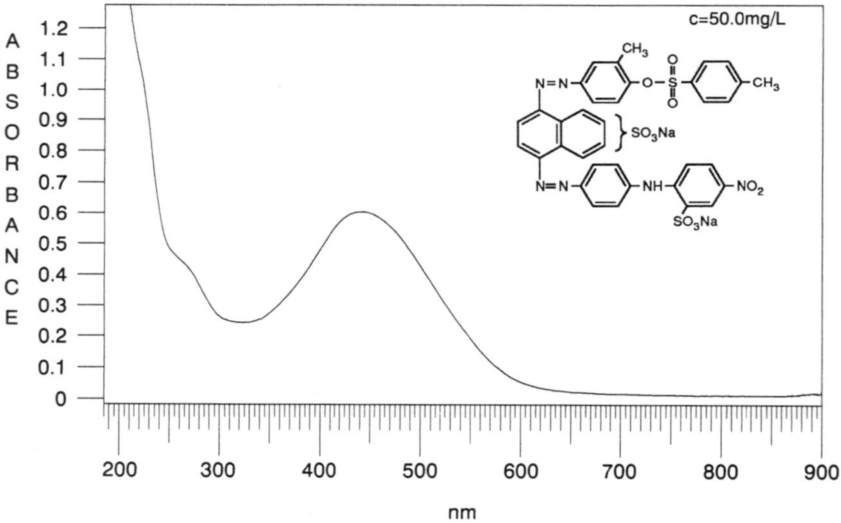

This anionic disazo dye is soluble in water and methyl Cellosolve and very slightly soluble in ethanol.

Textile applications include dyeing wool (neutral or acid bath), nylon (neutral, acid, or formic acid bath), silk (neutral or acetic acid bath), and printing wool, silk, nylon and viscose directly. Other industrial applications include dyeing paper pulp and chromed leather (brown) and coloring phenol-formaldehyde casting resins and soap.

There is no evidence that it has been used as a biological stain.

General references:

Aldrich FT-IR Library	**1**(2), 992D
Sigma FT-IR Library	**1**(2), 86D
IR Library	**3**, 1465D
Colour Index	3rd ed., 1971

ACID ORANGE 63

Aldrich Cat. No.:	20,197-9
Sigma Prod. No.:	A6145
CAS No.:	[*15792-50-4*]
C.I. No.:	22870
Mol. Form.:	$C_{35}H_{26}N_6Na_2O_{10}S_3$
F.W.:	832.80
Appearance:	Red powder
λ_{max}:	424nm in water
Solubility:	H_2O 50mg/ml
	EGME 30mg/ml
	EtOH 6mg/ml

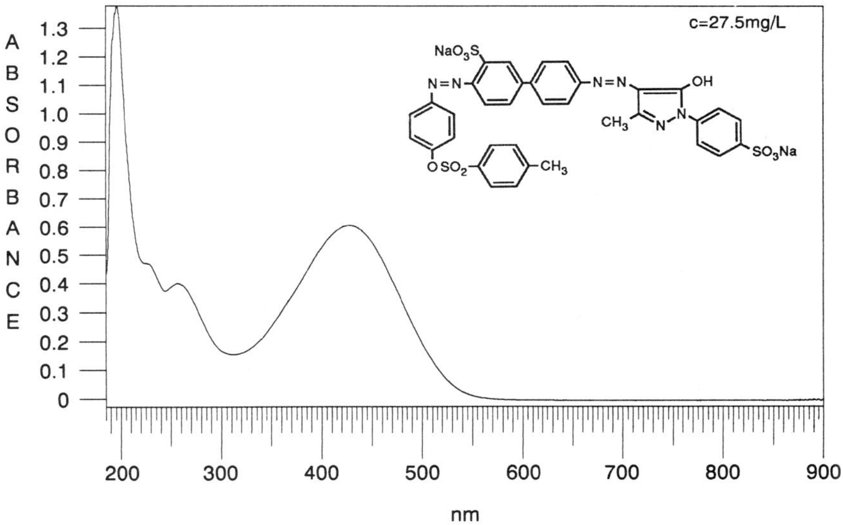

This orange, anionic disazo dye is soluble in water and methyl Cellosolve and slightly soluble in ethanol. It is prepared by coupling one mole of phenol and one mole of 3-methyl-1-(4-sulfophenyl)-5-pyrazolone with tetraazotized 3-sulfobenzidine. The reaction product is then treated with 4-toluenesulfonyl chloride. Industrial applications include dyeing wool, silk and nylon directly and the direct printing of wool, silk, nylon and viscose.

There is no evidence that it has been used as a biological stain.

General references:

Aldrich FT-IR Library	**1**(2), 1003C
Sigma FT-IR Library	**1**(2), 87A
IR Library	**3**, 1470A
Colour Index	3rd ed., 1971

ACID ORANGE 74

GEN • Palatine Fast Orange GN • Zapon Fast Orange G

Aldrich Cat. No.:	20,181-2
Sigma Prod. No.:	A5270
CAS No.:	[*10127-27-2*]
C.I. No.:	18745
Mol. Form.:	$C_{16}H_{12}N_5NaO_7S \cdot xCr$
F.W.:	493.38
Appearance:	Red-brown powder
λ_{max}**:**	455nm in water
Solubility:	H_2O 20mg/ml
	EGME 9mg/ml
	EtOH 10mg/ml

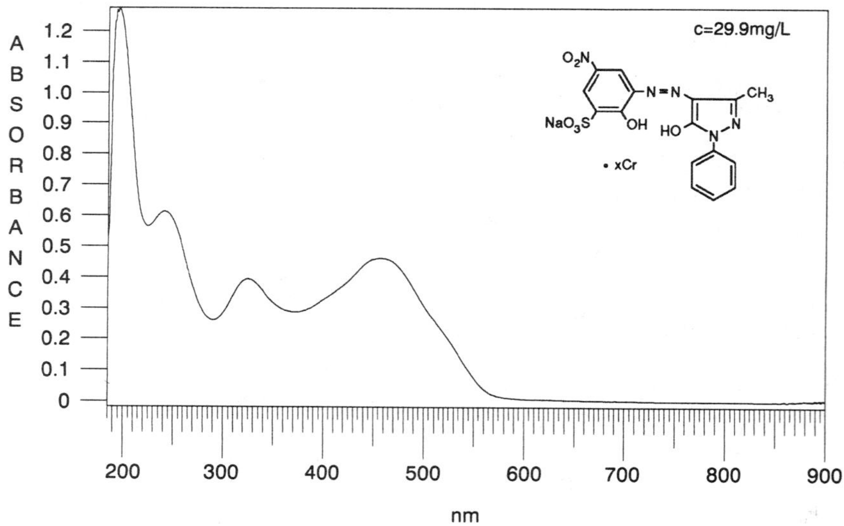

This anionic, monoazo chromium-complexed dye is the sodium salt of a reaction product prepared by treating a chromium salt with Mordant Orange 29 (C.I. 18744). Moderately soluble in water and less soluble in ethanol, it is used to dye wool (acid bath), nylon (formic acid bath) and silk (formic or sulfuric acid bath). Silk and wool are printed directly. Other industrial applications include coloring anodized aluminum and dyeing fully chromed leather orange.

There is no evidence in the literature that it has been used as a biological stain.

General references:

Aldrich FT-IR Library	1(2), 1000D
Sigma FT-IR Library	1(2), 87B
IR Library	**3**, 1468F
Colour Index	3rd ed., 1971

ACID RED 1

Amido Naphthol Red G • Azophloxine

Aldrich Cat. No.: 21,063-3
Sigma Prod. No.: A2775

CAS No.: [*3734-67-6*]
C.I. No.: 18050
Mol. Form.: $C_{18}H_{13}N_3Na_2O_8S_2$
F.W.: 509.43
Appearance: Red powder
λ_{max}: 532(506)nm in water
Solubility: H_2O 60mg/ml
EGME 10mg/ml
EtOH 2mg/ml

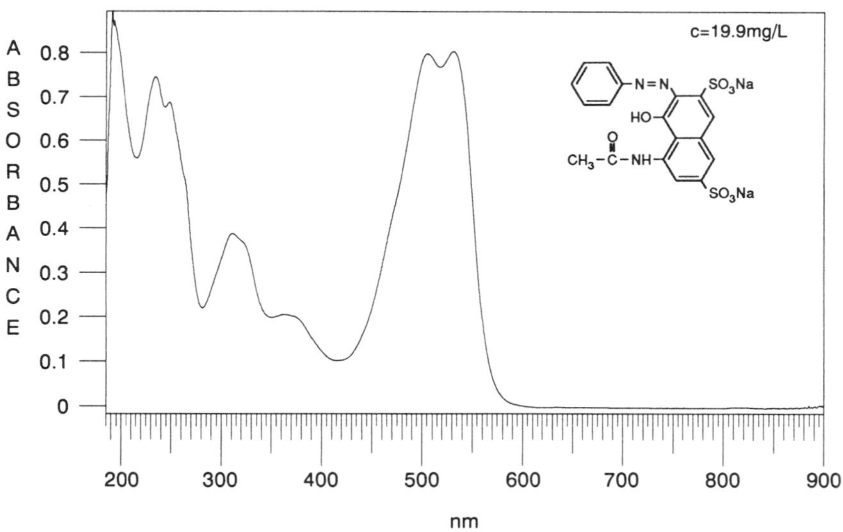

This anionic monoazo dye is made by coupling diazotized 4-aminoaniline to *N*-acetyl H acid (*N*-acetyl-8-amino-1-naphthol-3,6-disulfonic acid) and converting the coupling product to the disodium salt.

Soluble in water and slightly soluble in ethanol, acetone and methyl Cellosolve, it dyes wool from a sulfuric acid bath, nylon from a formic acid bath and silk from either an acetic acid or sulfuric acid bath. Nylon, silk and wool are printed directly. Other industrial applications include coloring paper, plastics, inks and soap, staining wood, preparing pigments (barium salts) and dyeing fully chromed leather bluish red.

Conn's[1] describes its use as a counterstain, as a component in the Masson trichrome stain and, in combination with Naphthazarin or Gewacin Blue, as a substitute for the Hematoxylin and Eosin (H & E) stain.

Reference:

1) *Conn's Biological Stains,* 9th ed.; Lillie, R.D., Ed.; Williams and Wilkins: Baltimore, 1977; p 130.

General references:

Aldrich FT-IR Library	**1**(2), 984B
Sigma FT-IR Library	**1**(2), 87C
IR Library	**3**, 1460F
Colour Index	3rd ed., 1971

ACID RED 4

Aldrich Cat. No.:	21,074-9
Sigma Prod. No.:	A5899
CAS No.:	[*5858-39-9*]
C.I. No.:	14710
Mol. Form.:	$C_{17}H_{13}N_2NaO_5S$
F.W.:	380.36
Appearance:	Orange powder
λ_{max}:	508(376)nm in water
Solubility:	H_2O 40mg/ml
	EGME 10mg/ml
	EtOH 0.9mg/ml

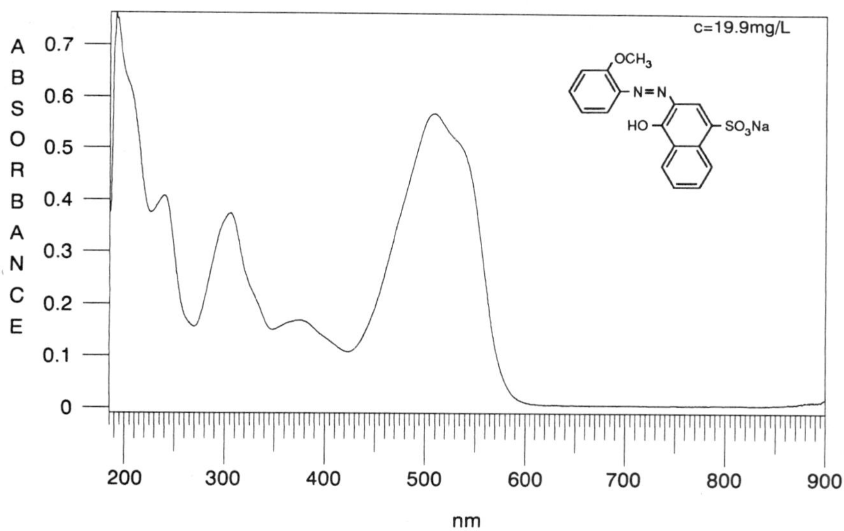

This red, anionic dye with a monoazo chromophore is made by coupling diazotized *o*-anisidine with Nevile and Winther's acid (1-naphthol-4-sulfonic acid) and converting the coupling product to the sodium salt.

Soluble in water and acetone and very slightly soluble in ethanol and toluene, it is used to dye wool from a sulfuric acid bath and to print nylon and wool directly. Nontextile, industrial applications include dyeing paper and chromed leather, coloring sealing wax and, converted to the barium salt, use as a red pigment.

There is no evidence in the literature that it has been used as a biological stain.

General references:

Aldrich FT-IR Library	**1**(2), 981C
Sigma FT-IR Library	**1**(2), 87D
IR Library	**3**, 1459C
Colour Index	3rd ed., 1971

ACID RED 8

Cochenillenscarlach PS

Aldrich Cat. No.: 21,452-3

CAS No.: [*4787-93-3*]
C.I. No.: 14900
Mol. Form.: $C_{18}H_{14}N_2Na_2O_7S_2$
F.W.: 480.43
Appearance: Dark-red powder
λ_{max}: 508nm in water
Solubility: H_2O 30mg/ml
EGME 30mg/ml
EtOH 3mg/ml

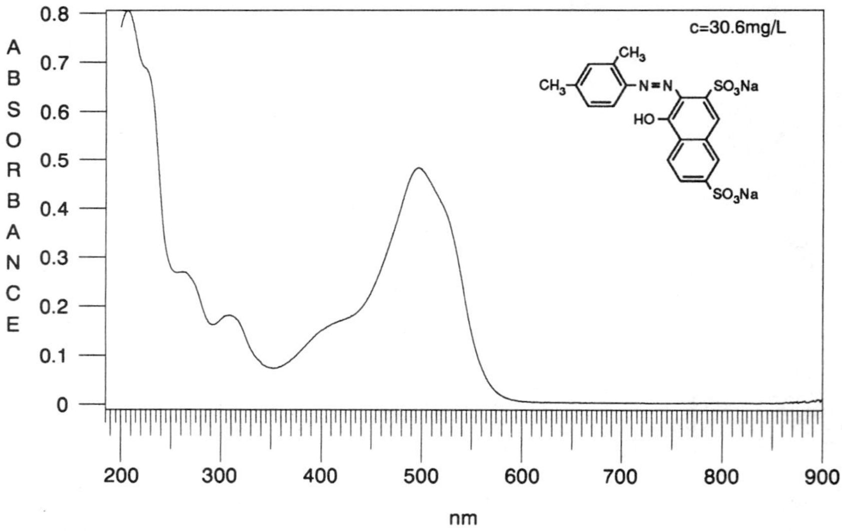

This bright yellowish-red, anionic monoazo dye is made by coupling diazotized 2,4-xylidine to 1-naphthol-3,6-disulfonic acid and converting the coupling product to the sodium salt.

Soluble in water and slightly soluble in ethanol and acetone, its utility has been confined largely to textile dyeing. Wool is dyed from a sulfuric acid bath, while silk is processed from a degumming liquor, or formic or sulfuric acid baths. Silk and wool are printed directly.

There is no evidence in the literature that it has been used as a biological stain.

General references:

IR Library **3**, 1459F
Colour Index 3rd ed., 1971

ACID RED 37

Anthranal Red G • Fast Light Red B

Aldrich Cat. No.: 21,064-1

CAS No.: [*6360-07-2*]
C.I. No.: 17045
Mol. Form.: $C_{18}H_{14}N_4Na_2O_8S_2$
F.W.: 524.44
Appearance: Dark-red powder
λ_{max}: 513nm in water
Solubility: H_2O 30mg/ml
EGME 20mg/ml
EtOH 0.4mg/ml

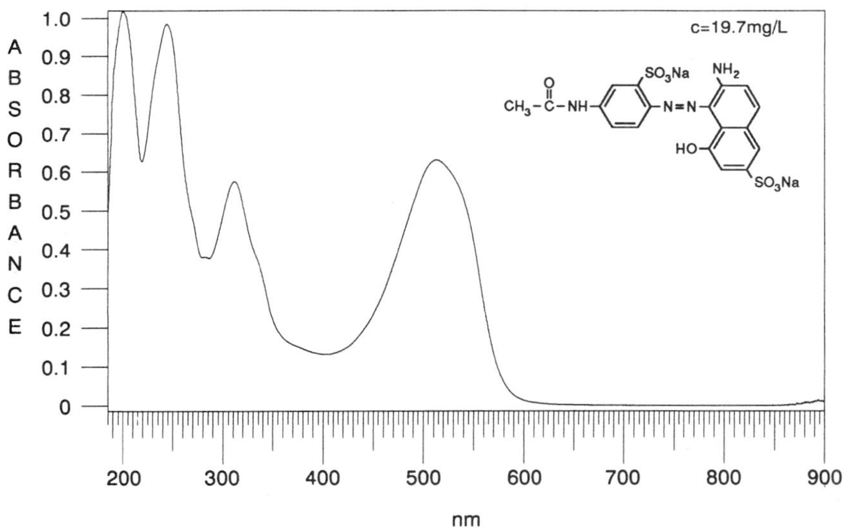

This bluish-red, anionic monoazo dye is made by the acid coupling of diazotized 5-acetamido-2-aminobenzenesulfonic acid to *gamma* acid (7-amino-1-naphthol-3-sulfonic acid) and converting the coupling product to the disodium salt.

Soluble in water and very slightly soluble in ethanol, it dyes wool from a sulfuric acid bath, nylon from an acetic acid bath and silk from either a formic acid or sulfuric acid bath. Wool and silk are printed directly. Other industrial applications include dyeing paper and fully chromed leather and coloring soap and anodized aluminum.

There is no evidence in the literature that it has been used as a biological stain.

General references:

Aldrich FT-IR Library 1(2), 985B
IR Library 3, 1461B
Colour Index 3rd ed., 1971

ACID RED 40
Calcoid Ceresine

Aldrich Cat. No.:	21,455-8
CAS No.:	[*12167-45-2*]
C.I. No.:	18070
Mol. Form.:	$C_{20}H_{17}N_3Na_2O_8S_2$
F.W.:	537.48
Appearance:	Maroon powder
λ_{max}**:**	516(538)nm in water
Solubility:	H_2O 80mg/ml
	EGME 20mg/ml
	EtOH 2mg/ml

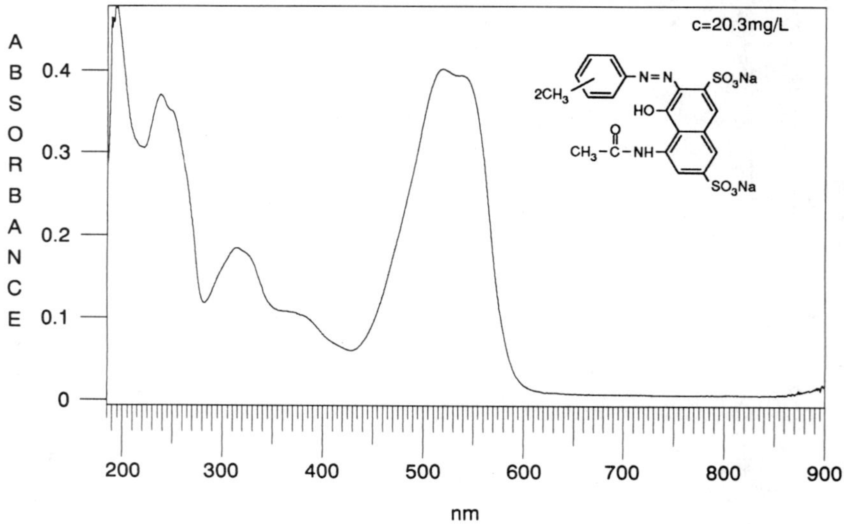

This bright bluish-red, anionic monoazo dye is made by coupling diazotized mixed xylidines with *N*-acetyl H acid (8-acetylamino-1-naphthol-3,6-disulfonic acid) and converting the coupling compound to the disodium salt.

Soluble in water and slightly soluble in ethanol, it dyes wool from a neutral, weak-acid or strong-acid bath while nylon is processed from a bath containing either formic or acetic acid with aluminum chloride.

There is no evidence that it has been used as a biological stain.

General references:

Aldrich FT-IR Library	**1**(2), 984C
IR Library	**3**, 1460G
Colour Index	3rd ed., 1971

ACID RED 88

Aldrich Cat. No.:	19,522-7
Sigma Prod. No.:	A5145
CAS No.:	[*1658-56-6*]
C.I. No.:	15620
Mol. Form.:	$C_{20}H_{13}N_2NaO_4S$
F.W.:	400.39
m.p.:	280°C (dec.)
Appearance:	Maroon powder
λ_{max}:	505nm in water
Solubility:	H_2O 30mg/ml
	EGME 20mg/ml
	EtOH 1mg/ml

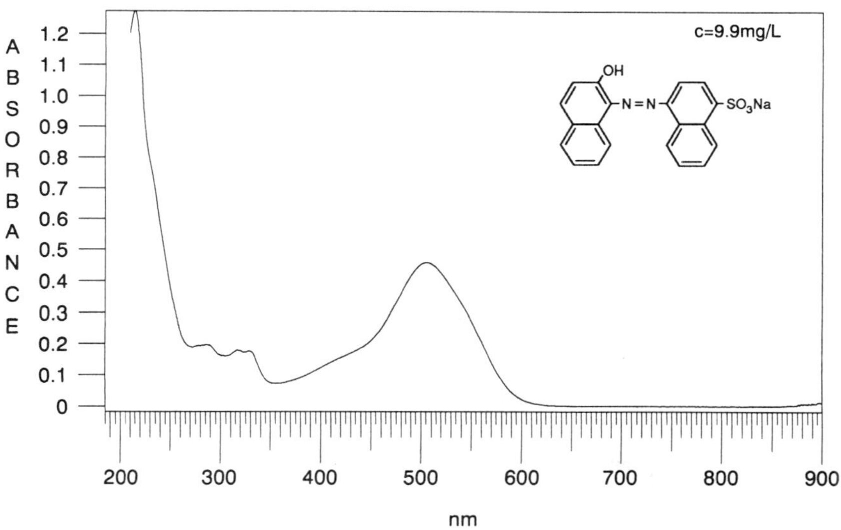

Acid Red 88 is a monoazo dye made by coupling diazotized naphthionic acid to 2-naphthol. It is soluble in water and methyl Cellosolve and very slightly soluble in ethanol. Wool is dyed from either a weak- or strong-acid bath, nylon from a formic acid bath and silk from either a formic acid or acetic acid bath as are jute and other ligneous fibers. Silk and wool are printed directly. Other industrial applications include dyeing chromed leather and paper, coloring soap and hydrated creams and staining wood.

There is no evidence that it has been used as a biological stain.

General references:

Aldrich FT-IR Library	**1**(2), 986D
Sigma FT-IR Library	**1**(2), 88A
IR Library	**3**, 1462E
Colour Index	3rd ed., 1971

ACID RED 97
Acid Anthracene Red G

Aldrich Cat. No.: 21,039-0

CAS No.: [*10169-02-5*]
C.I. No.: 22890
Mol. Form.: $C_{32}H_{20}N_4Na_2O_8S_2$
F.W.: 698.65
Appearance: Maroon powder
λ_{max}: 498nm in water
Solubility: H_2O 40mg/ml
EGME 20mg/ml
EtOH 2mg/ml

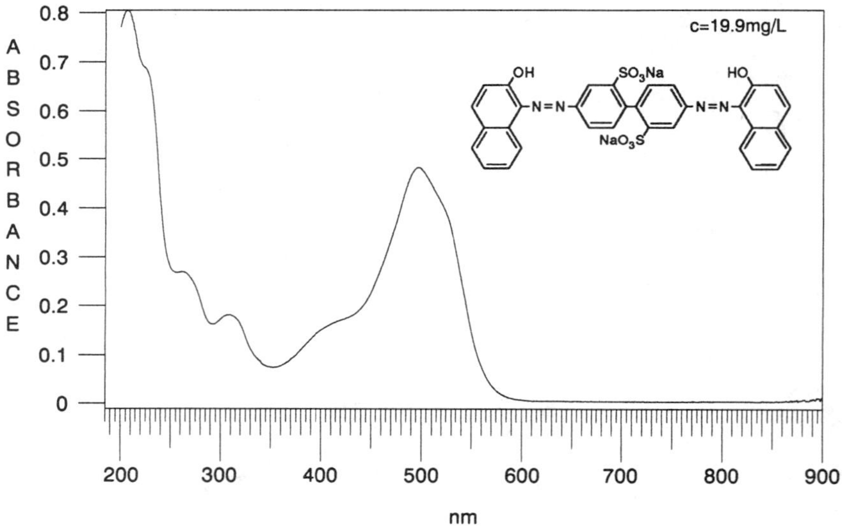

This red, anionic disazo dye is made by coupling two moles of 2-naphthol to one mole of tetraazotized 2,2'-disulfobenzidine.

Soluble in water and methyl Cellosolve and slightly soluble in ethanol, it dyes wool from a neutral or weakly acid bath, nylon from an acetic acid bath and silk from either an acetic acid bath or a Glauber's salt solution. Wool and silk are printed directly. Other industrial applications include dyeing chromed leather, paper and jute; heavy-metal salts are used as pigments.

There is no evidence in the literature that it has been used as a biological stain.

General references:

Aldrich FT-IR Library	**1**(2), 993D
IR Library	**3**, 1465G
Colour Index	3rd ed., 1971

ACID RED 106

Aldrich Cat. No.:	21,076-5
Sigma Prod. No.:	A6024
CAS No.:	[*6844-74-2*]
C.I. No.:	18110
Mol. Form.:	$C_{23}H_{17}N_3Na_2O_9S_3$
F.W.:	621.58
Appearance:	Deep-red powder
λ_{max}:	532(505)nm in water
Solubility:	H_2O 20mg/ml
	EGME 30mg/ml
	EtOH 3mg/ml

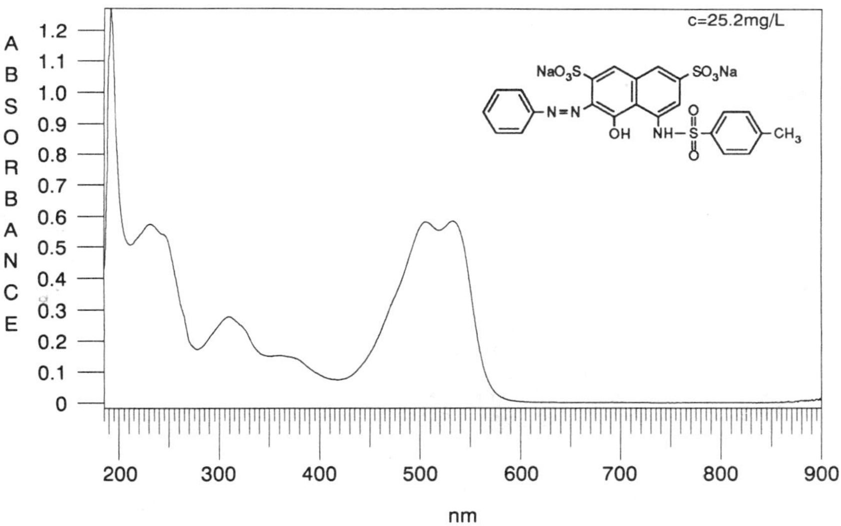

This bright-red, anionic monoazo dye is prepared by coupling diazotized aniline to *N*-(*p*-tolylsulfonyl) H acid [8-(*p*-tolylsulfonylamino)-1-naphthol-3,6-disulfonic acid] and converting the coupling product to the disodium salt.

Soluble in water and slightly soluble in ethanol and acetone, it dyes wool from either weakly or strongly acid baths, nylon from either acetic acid or formic acid baths, silk from an acetic acid bath or a Glauber's salt solution, prints silk and wool directly, colors paper and anodized aluminum, and forms heavy-metal salts that are used as pigments.

There is no evidence in the literature that it has been used as a biological stain.

General references:

Aldrich FT-IR Library	**1**(2), 990D
Sigma FT-IR Library	**1**(2), 88B
IR Library	**3**, 1464C
Colour Index	3rd ed., 1971

ACID RED 114

Aldrich Cat. No.:	21,031-5
Sigma Prod. No.:	A6149
CAS No.:	[*6459-94-5*]
C.I. No.:	23635
Mol. Form.:	$C_{37}H_{28}N_4Na_2O_{10}S_3$
F.W.:	830.83
Appearance:	Deep-maroon powder
λ_{max}:	514(365)nm in 50% ethanol

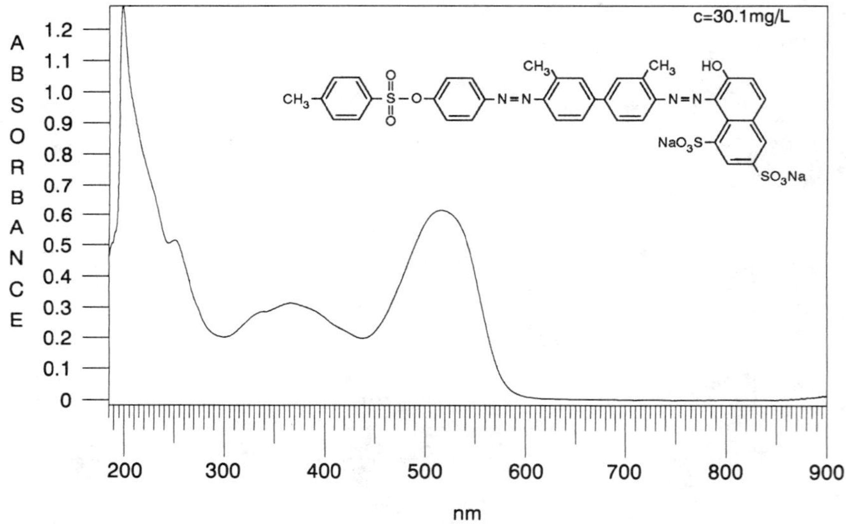

This bright-red, anionic disazo dye is prepared by coupling one mole each of phenol and G acid (2-naphthol-6,8-disulfonic acid) to one mole of tetraazotized *o*-tolidine and finally esterifying the phenol-hydroxy function with *p*-tolylsulfonyl chloride.

Soluble in water and very slightly soluble in ethanol, Acid Red 114 dyes wool (from a weak acid bath), silk (from either a neutral or acetic acid bath), jute and chromed leather. Wool and silk are printed directly.

There is no evidence in the literature that this dye has been used as a biological stain.

General references:

Aldrich FT-IR Library	**1**(2), 993B
Sigma FT-IR Library	**1**(2), 88C
IR Library	**3**, 1465E
Colour Index	3rd ed., 1971

ACID RED 151

Aldrich Cat. No.:	21,078-1
Sigma Prod. No.:	A2650
CAS No.:	*[6406-56-0]*
C.I. No.:	26900
Mol. Form.:	$C_{22}H_{15}N_4NaO_4S$
F.W.:	454.44
Appearance:	Maroon powder
λ_{max}:	512(356)nm in water

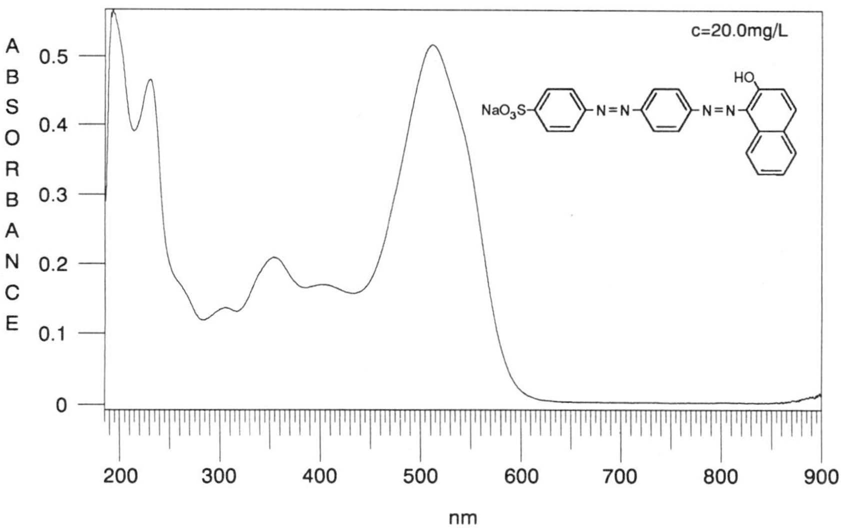

This anionic disazo dye is prepared by coupling diazotized sulfanilic acid to aniline. Then the product of this reaction is diazotized, coupled to 2-naphthol and finally converted to the sodium salt.

Soluble in methyl Cellosolve, it is used to dye wool, nylon and silk from either neutral or acetic acid baths, while jute and cotton require an alum mordant. Silk and wool are printed directly. Nontextile applications include coloring casein plastics and dyeing paper.

There is no evidence in the literature that it has been used as a biological stain.

General references:

Aldrich FT-IR Library	**1**(2), 989C
Sigma FT-IR Library	**1**(2), 89A
IR Library	**3**, 1463F
Colour Index	3rd ed., 1971

ACID RED 183
Palatine Fast Red GREN

Aldrich Cat. No.:	20,182-0
Sigma Prod. No.:	A5020
CAS No.:	*[6408-31-7]*
C.I. No.:	18800
Mol. Form.:	$C_{16}H_{11}ClN_4Na_2O_8S_2 \cdot xCr$
F.W.:	584.87
Appearance:	Dark-rust powder
λ_{max}:	494nm in water
Solubility:	H_2O 70mg/ml
	EGME 20mg/ml
	EtOH 20mg/ml

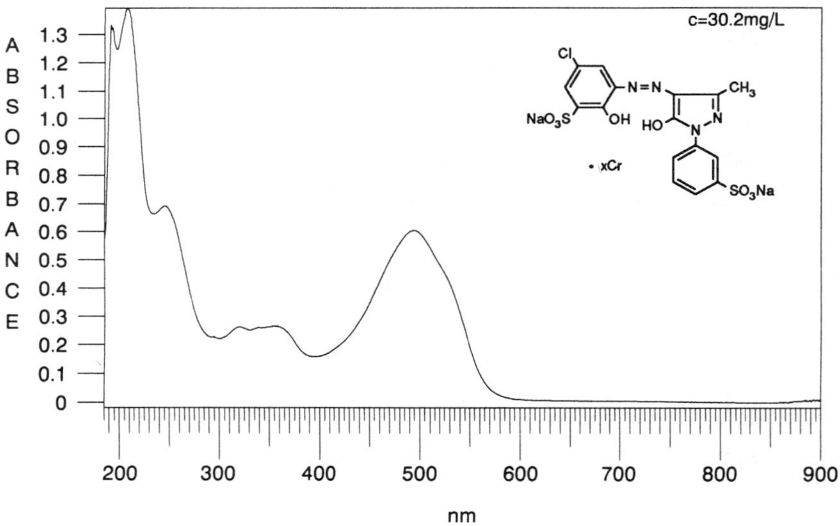

This yellowish-red, anionic monoazo dye is made by coupling diazotized 6-amino-4-chloro-1-phenol-2-sulfonic acid to 3-methyl-1-(*m*-sulfophenyl)-5-pyrazolone. Finally, an aqueous solution of the reaction product is treated with chromium formate to form a chromium complex which is soluble in water.

Wool is dyed from a sulfuric acid bath, nylon from a formic acid bath and silk from a degumming liquor. Silk and wool are printed directly. Chromed leather is dyed a bright orange.

There is no evidence in the literature that it has been used as a biological stain.

General references:

Aldrich FT-IR Library	**1**(2), 1001B
Sigma FT-IR Library	**1**(2), 89B
IR Library	**3**, 1468H
Colour Index	3rd ed., 1971

ACID VIOLET 5

Lanafuchsin 6B

Aldrich Cat. No.:	21,080-3
Sigma Prod. No.:	A6274
CAS No.:	[*10130-48-0*]
C.I. No.:	18125
Mol. Form.:	$C_{25}H_{20}N_4Na_2O_{10}S_3$
F.W.:	678.63
Appearance:	Black powder
λ_{max}:	527nm in water
Solubility:	H_2O 40mg/ml
	EGME 30mg/ml
	EtOH 10mg/ml

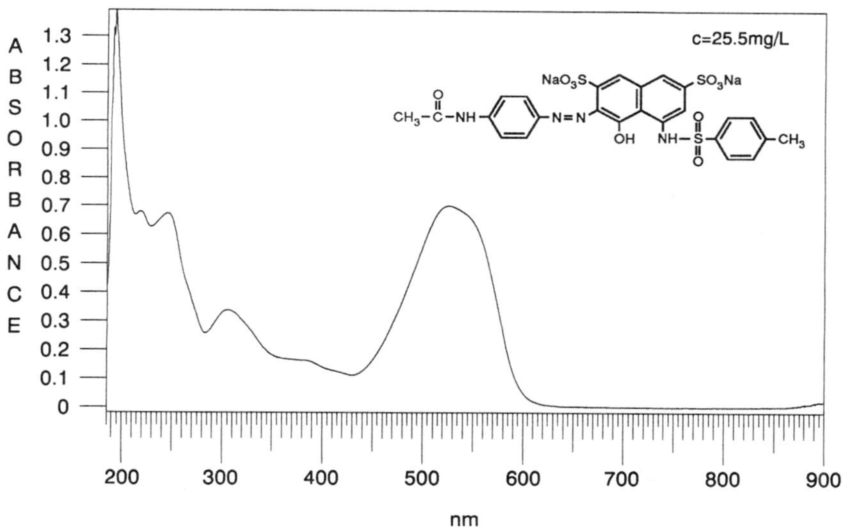

This reddish-violet, anionic monoazo dye is made by coupling diazotized 4'-aminoacetanilide to *N*-(*p*-tolylsulfonyl) H acid and converting the coupling product to the disodium salt.

Soluble in water and somewhat soluble in ethanol, it dyes wool from a sulfuric acid bath and prints it directly. Nontextile applications include coloring inks and soap, and staining satin shoes.

There is no evidence in the literature that it has been used as a biological stain.

General references:

Aldrich FT-IR Library	**1**(2), 991A
Sigma FT-IR Library	**1**(2), 89C
IR Library	**3**, 1464D
Colour Index	3rd ed., 1971

ACID VIOLET 7

Amido Naphthol Red 6B

Aldrich Cat. No.:	21,066-8
Sigma Prod. No.:	A6399
CAS No.:	[*4321-69-1*]
C.I. No.:	18055
Mol. Form.:	$C_{20}H_{16}N_4Na_2O_9S_2$
F.W.:	566.48
Appearance:	Very dark red powder
λ_{max}**:**	520nm in water
Solubility:	H_2O 30mg/ml
	EGME 20mg/ml
	EtOH 2mg/ml

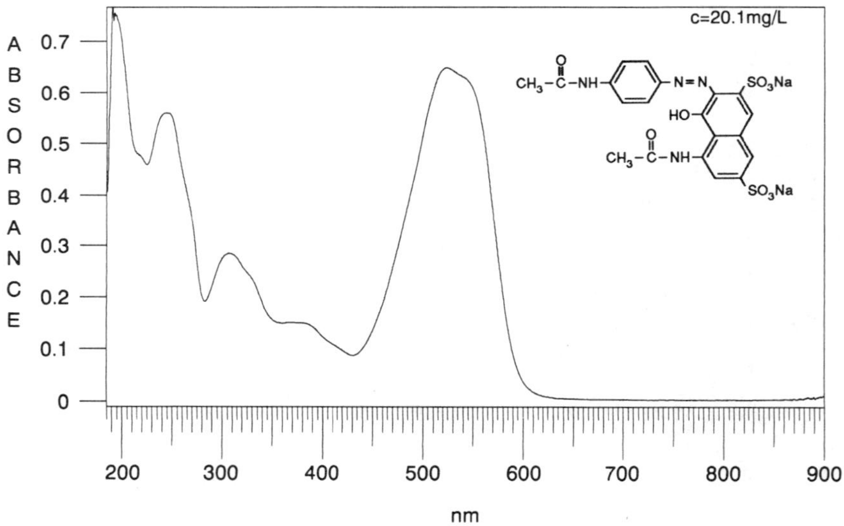

This reddish-violet, anionic monoazo dye is prepared by coupling diazotized 4'-aminoacetanilide to *N*-acetyl H acid (*N*-acetyl-8-amino-1-naphthol-3,6-disulfonic acid) and converting the coupling product to the disodium salt.

Soluble in water, somewhat less soluble in methyl Cellosolve, slightly soluble in acetone and very slightly soluble in ethanol, it dyes wool from a sulfuric acid bath, nylon from a formic acid bath and silk either from a formic acid or sulfuric acid bath. Industrially, it is used to dye chromed leather a reddish violet, color soap and paper (surface) and stain wood.

Conn's[1] suggested it as a replacement for Biebrich Scarlet (BS) or Ponceau de Xylidine (PDX) in Masson's trichrome stains. Gurr[2] indicated that Pontacyl Carmine 6B, a dye closely related to Acid Violet 7, might be used as a substitute for the same two dyes, BS and PDX, as a counterstain.

References:

1) *Conn's Biological Stains,* 9th ed.; Lillie, R.D., Ed.; Williams and Wilkins: Baltimore, 1977; p 131.

2) Gurr, E. *Encyclopedia of Microscopic Stains* ; Hill: London, 1960; p 333.

General references:

Aldrich FT-IR Library **1**(2), 984D
Sigma FT-IR Library **1**(2), 89D
IR Library **3**, 1460H
Colour Index 3rd ed., 1971

ACID VIOLET 17
Formyl Violet S4B

Aldrich Cat. No.: 21,057-9
Sigma Prod. No.: A6524

CAS No.: [*4129-84-4*]
C.I. No.: 42650
Mol. Form.: $C_{41}H_{44}N_3NaO_6S_2$
F.W.: 761.94
Appearance: Dark-purple powder
λ_{max}: 545nm in water
Solubility: H_2O 30mg/ml
EGME 30mg/ml
EtOH 6mg/ml

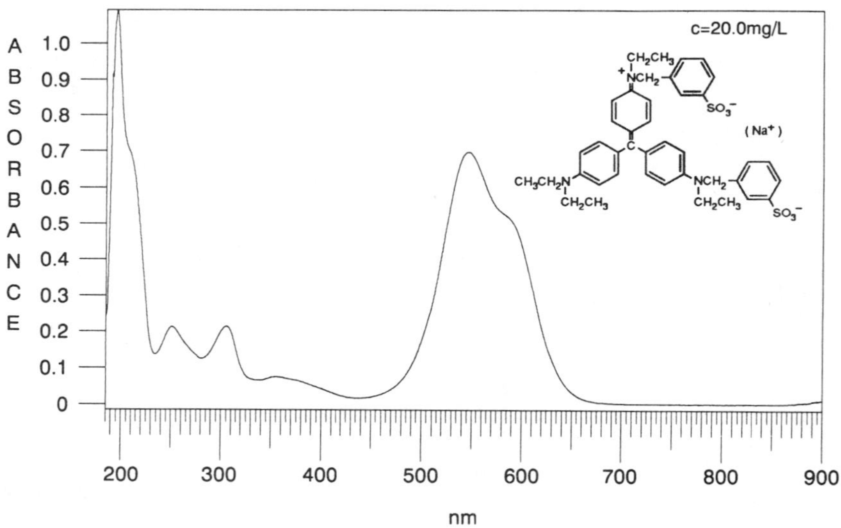

This bright, bluish-violet, anionic triphenylmethane dye is made by condensing two moles of α-(*N*-ethylanilino)-*m*-toluenesulfonic acid with one mole of formaldehyde. After the product of this reaction is oxidized to the hydrol, it is condensed with *N,N*-diethylaniline, the second product is oxidized and then converted to the sodium salt.

Soluble in water (both hot and cold) and slightly soluble in ethanol, it dyes wool from either an acetic acid or sulfuric acid bath, and silk from an acetic acid bath. Wool, silk, nylon and viscose are printed directly. Other industrial applications include coloring paper, straw, soaps, ink and anodized aluminum, dyeing chromed leather a reddish blue and preparing metal salts for use as pigments.

Apparently, Formyl Violet S4B has not been used widely as a biological stain. Lillie[1] reported it to be too red a violet, when substituted in the Bensely Safranin-Acid Violet neutral stain, to compare with the Maximow and Bloom[2] colored illustration.

References:

1) Lillie, R.D. *Histopathological Technic*; Blakiston: Philadelphia, 1948; p 98.

2) Maximow, A.A.; Bloom, W.A.A. *Text Book of Histology;* Sanders: Philadelphia, 1930; p 692.

General references:

Aldrich FT-IR Library	**1**(2), 1030D
Sigma FT-IR Library	**1**(2), 90A
IR Library	**3,** 1484B
Safety	**2**(1), 53C
Colour Index	3rd ed., 1971

ACID YELLOW 17

Aldrich Cat. No.:	20,198-7
Sigma Prod. No.:	A4645

CAS No.:	[*6359-98-4*]
C.I. No.:	18965
Mol. Form.:	$C_{16}H_{10}Cl_2N_4Na_2O_7S_2$
F.W.:	551.30
Appearance:	Yellow powder
λ_{max}:	400nm in water
Solubility:	H_2O 40mg/ml
	EGME 7mg/ml
	EtOH 2mg/ml

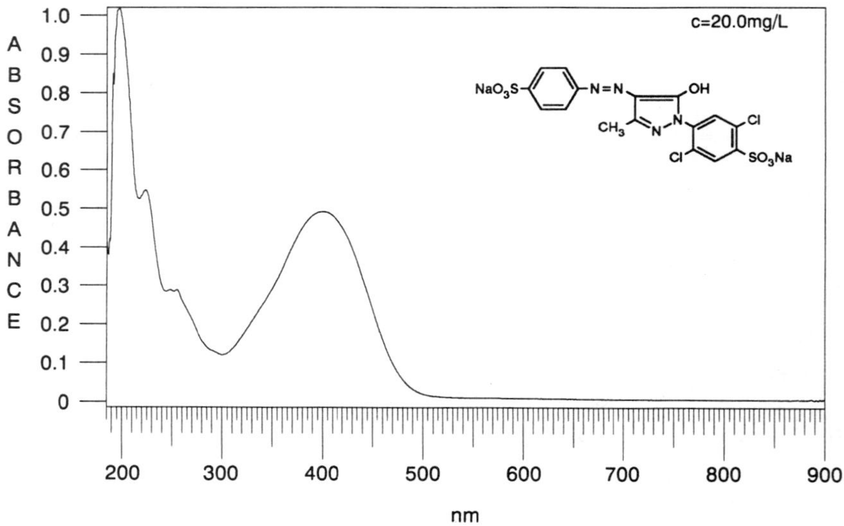

This bright-yellow, acid monoazo dye is made by coupling diazotized sulfanilic acid with 1-(2,5-dichloro-4-sulfophenyl)-3-methyl-5-pyrazolone. It is soluble in water, slightly soluble in methyl Cellosolve and very slightly soluble in ethanol and acetone. Wool is dyed from a sulfuric acid bath, silk from either formic acid or sulfuric acid bath and nylon from a formic acid bath. Wool and silk are printed directly. Other industrial applications include the coloring of paper surface, anodized aluminum, crepe paper, and formaldehyde and casein resins. The dye's barium salt is used as a pigment for paper coatings. Chromed leather is dyed a greenish yellow.

A highly purified form was formerly used as a food color, Food Yellow 5. However, there is no evidence that it has been used as a biological stain.

General references:

Aldrich FT-IR Library	**1**(2), 1001C
Sigma FT-IR Library	**1**(2), 90B
IR Library	**3**, 1469A
Colour Index	3rd ed., 1971

ACID YELLOW 25

Supramine Yellow R

Aldrich Cat. No.:	20,196-0
Sigma Prod. No.:	A4520
CAS No.:	[*6359-85-9*]
C.I. No.:	18835
Mol. Form.:	$C_{23}H_{20}N_5NaO_6S_2$
F.W.:	549.56
Appearance:	Mustard-brown powder
λ_{max}:	392nm in water
Solubility:	H_2O 60mg/ml
	EGME 40mg/ml
	EtOH 3mg/ml

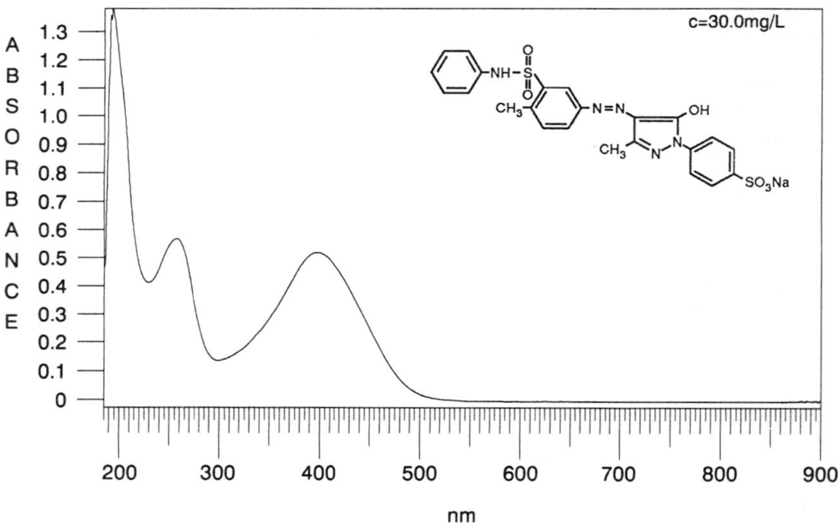

This yellow, anionic monoazo dye is prepared by coupling diazotized 5-amino-*o*-toluenesulfonanilide to 3-methyl-1-(4-sulfophenyl)-5-pyrazolone. It is soluble in water and slightly soluble in ethanol and acetone.

Industrial uses include dyeing wool from a sulfuric acid bath, nylon from a formic acid bath and silk from a broken degumming solution; printing wool, silk and nylon and coloring anodized aluminum and, occasionally, paper coatings.

There is no evidence in the literature that it has been used as a biological stain.

General references:

Aldrich FT-IR Library	**1**(2), 1002A
Sigma FT-IR Library	**1**(2), 90C
IR Library	**3**, 1469C
Colour Index	3rd ed., 1971

ACID YELLOW 29

Supramine Yellow 3GL

Aldrich Cat. No.: 21,036-6
Sigma Prod. No.: A6649

CAS No.: [*6359-91-7*]
C.I. No.: 18900
Mol. Form.: $C_{22}H_{17}ClN_5NaO_6S_2$
F.W.: 569.98
m.p.: 238°C (dec.)
Appearance: Yellow-gold powder
λ_{max}: 407nm in water
Solubility: H_2O 40mg/ml
EGME 40mg/ml
EtOH 20mg/ml

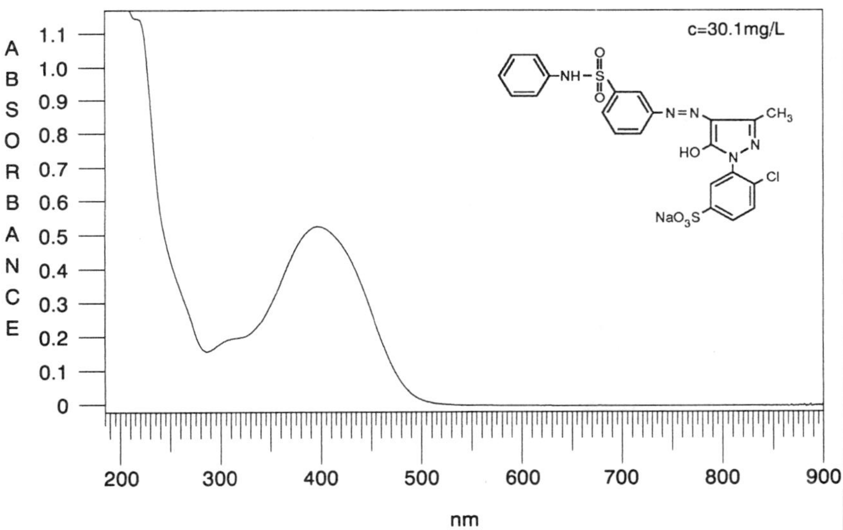

This yellow, anionic monoazo dye is prepared by coupling diazotized metanilanilide to 1-(2-chloro-5-sulfophenyl)-3-methyl-5-pyrazolone.

Soluble in water, ethanol and acetone, it is used to dye wool either from an acetic or sulfuric acid bath, silk from an acid or neutral bath and nylon from an acid bath. Wool and silk are printed directly and chromed leather is dyed a greenish yellow.

There is no evidence in the literature that it has been used as a biological stain.

General references:

Aldrich FT-IR Library **1**(2), 1002D
Sigma FT-IR Library **1**(2), 90D
IR Library **3**, 1469F
Colour Index 3rd ed., 1971

ACID YELLOW 34

Fast Light Yellow D3GA

Aldrich Cat. No.:	20,200-2
Sigma Prod. No.:	A4395
CAS No.:	[*6359-90-6*]
C.I. No.:	18890
Mol. Form.:	$C_{16}H_{12}ClN_4NaO_4S$
F.W.:	414.81
Appearance:	Yellow powder
λ_{max}**:**	408nm in water
Solubility:	H_2O 40mg/ml
	EGME 90mg/ml
	EtOH 10mg/ml

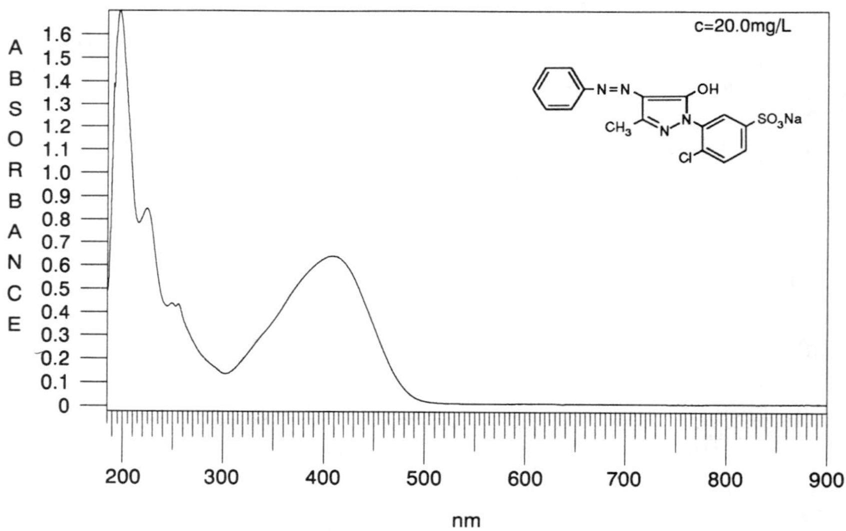

This yellow, anionic, monoazo pyrazolone dye is prepared by the coupling of diazotized aniline to 1-(2-chloro-5-sulfophenyl)-3-methyl-5-pyrazolone. It is soluble in water and methyl Cellosolve and slightly soluble in ethanol.

Wool, silk and nylon are dyed from a sulfuric acid bath and wool and silk are printed directly. Occasionally, it is used to dye beater paper and a barium salt pigment is used to coat casein plastics.

There is no evidence in the literature that it has been used as a biological stain.

General references:

Aldrich FT-IR Library	**1**(2), 1000B
Sigma FT-IR Library	**1**(2), 91A
IR Library	**3,** 1468D
Colour Index	3rd ed., 1971

ACID YELLOW 38

Aldrich Cat. No.:	20,194-4
Sigma Prod. No.:	A4270
CAS No.:	[*13390-47-1*]
C.I. No.:	25135
Mol. Form.:	$C_{28}H_{24}N_4Na_2O_8S_3$
F.W.:	686.70
Appearance:	Gold powder
λ_{max}:	368nm in water

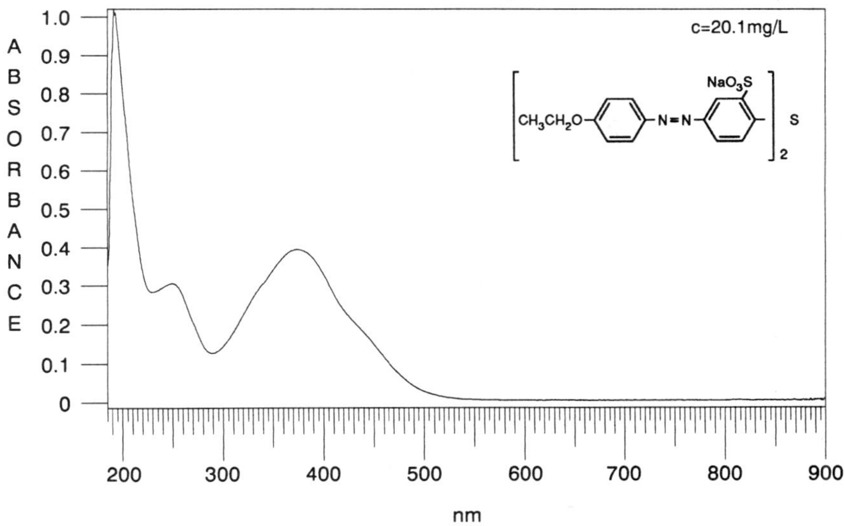

This yellow, anionic disazo dye is soluble in water and slightly soluble in ethanol and acetone. The preparation involves coupling one mole of tetra-azotized 6,6'-thiodimetanilic acid with two moles of phenol and ethylating the reaction product's hydroxy functions with ethyl chloride under pressure.

Industrial uses include dyeing silk, wool, nylon, beater paper, cellulose acetate (in injection molding) and chromed leather. Both wool and silk are printed directly.

There is no evidence that it has been used as a biological stain.

General references:

Aldrich FT-IR Library	**1**(2), 980D
Sigma FT-IR Library	**1**(2), 91B
IR Library	**3**, 1458G
Colour Index	3rd ed., 1971

ACID YELLOW 40

Aldrich Cat. No.:	20,193-6
Sigma Prod. No.:	A4145
CAS No.:	[*6372-96-9*]
C.I. No.:	18950
Mol. Form.:	$C_{23}H_{18}ClN_4NaO_7S_2$
F.W.:	584.99
m.p.:	147°C (dec.)
Appearance:	Rust powder
λ_{max}**:**	412nm in methanol
Solubility:	H_2O 30mg/ml
	EGME 50mg/ml
	EtOH 40mg/ml

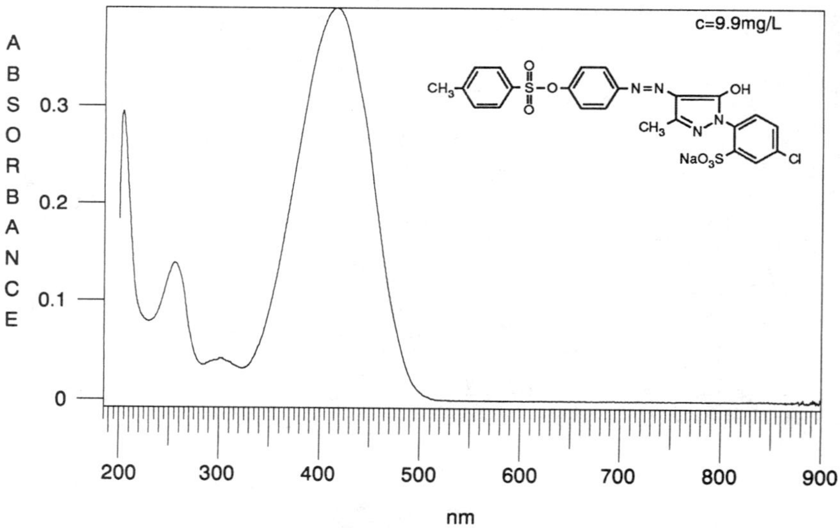

This yellow, anionic monoazo pyrazolone dye is prepared by coupling diazotized 4-aminophenol to 1-(4-chloro-2-sulfophenyl)-3-methyl-5-pyrazolone. The product is treated with *p*-toluenesulfonyl chloride to esterify the phenolic hydroxy group and then converted to the sodium salt. It is soluble in water, ethanol and acetone.

Wool is dyed from an acetic acid bath, silk from either a neutral or acetic acid bath and nylon from an ammonium sulfate solution. Lightfastness on nylon is excellent (7).

There is no evidence that it has been used as a biological stain.

General references:

Aldrich FT-IR Library	**1**(2), 1002C
Sigma FT-IR Library	**1**(2), 91C
IR Library	**3**, 1469E
Colour Index	3rd ed., 1971

ACID YELLOW 42

Sulfone Yellow R

Aldrich Cat. No.: 21,037-4

CAS No.: [*6375-55-9*]
C.I. No.: 22910
Mol. Form.: $C_{32}H_{24}N_8Na_2O_8S_2$
F.W.: 758.71
m.p.: 263°C (dec.)
Appearance: Orange powder
λ_{max}: 410nm in water
Solubility: H_2O 30mg/ml
EGME 10mg/ml
EtOH <0.1mg/ml

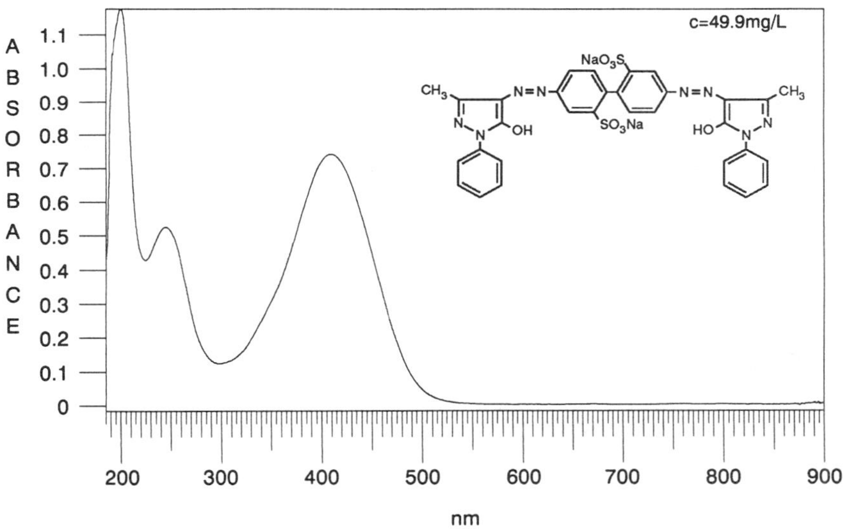

This yellow, anionic disazo pyrazolone dye is made by coupling one mole of tetraazotized benzidine-2,2'-disulfonic acid with two moles of 3-methyl-1-phenyl-5-pyrazolone and then converting the product to the disodium salt.

Soluble in water and slightly soluble in acetone, it dyes wool from a neutral or acetic acid bath and prints wool and silk directly. In other industrial applications, chromed leather is dyed a greenish yellow; straw, casein and soap are colored yellow and metal salts are used to prepare printing inks and paper coatings.

There is no evidence that it has been used as a biological stain.

General references:

Aldrich FT-IR Library	**1**(2), 1003B
IR Library	**3**, 1469H
Colour Index	3rd ed., 1971

ACID YELLOW 65

Aldrich Cat. No.: 20,201-0
Sigma Prod. No.: A4020

CAS No.: [*6408-90-8*]
C.I. No.: 14170
Mol. Form.: $C_{25}H_{19}N_4NaO_8S_2$
F.W.: 590.57
Appearance: Brown powder
λ_{max}: 414nm in water
Solubility: H_2O 40mg/ml
EGME 80mg/ml
EtOH 6mg/ml

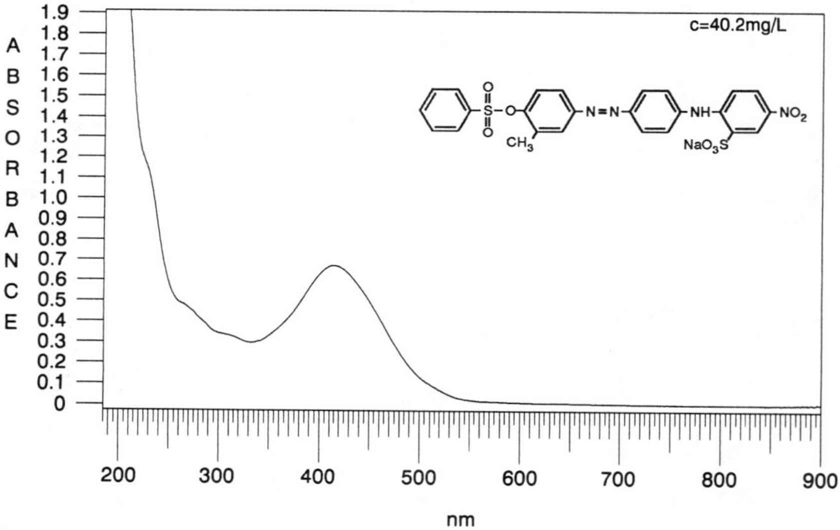

This is a reddish-yellow, anionic monoazo dye that is prepared by coupling diazotized 2-(4-aminoanilino)-5-nitrobenzenesulfonic acid to *o*-cresol. The cresol hydroxy function is then esterified with benzenesulfonyl chloride and the product is converted to the sodium salt. The resulting dye is soluble in water and slightly soluble in ethanol, acetone and toluene.

Industrial uses include dyeing wool and silk from either a neutral or acetic acid bath, nylon from an acid bath, printing silk directly and coloring soaps.

There is no evidence that it has been used as a biological stain.

General references:

Aldrich FT-IR Library **1**(2), 980B
Sigma FT-IR Library **1**(2), 91D
IR Library **3**, 1458E
Colour Index 3rd ed., 1971

ACID YELLOW 76

Aldrich Cat. No.: 21,458-2

CAS No.: [*6359-88-2*]
C.I. No.: 18850
Mol. Form.: $C_{23}H_{19}N_4NaO_7S_2$
F.W.: 550.55
Appearance: Dark-gold powder
λ_{max}: 393nm in 50% ethanol
Solubility: H_2O 40mg/ml
EGME 40mg/ml
EtOH 9mg/ml

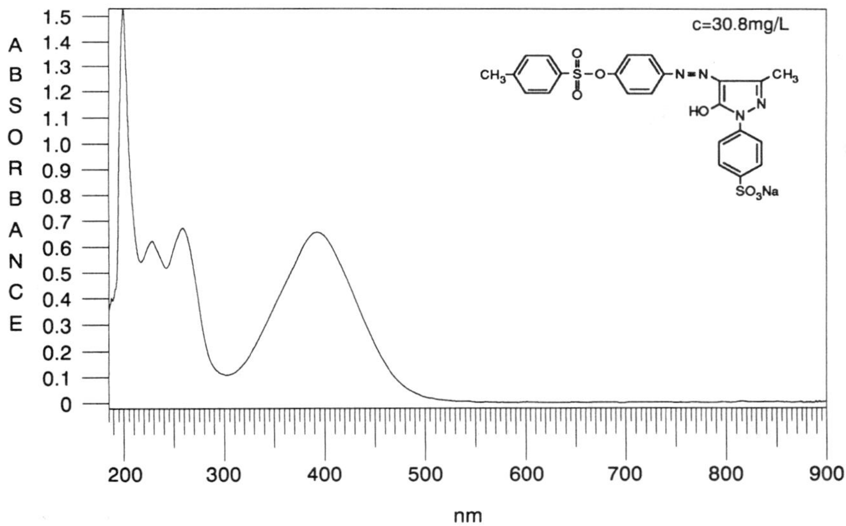

This yellow, anionic monoazo pyrazolone dye is prepared by first coupling diazotized 4-aminophenol to 3-methyl-1-(*p*-sulfophenyl)-5-pyrazolone. The phenolic hydroxy group of the resulting product is then esterified with *p*-toluenesulfonyl chloride and the final product is converted to the sodium salt.

Soluble in water and slightly soluble in ethanol, it is used to dye wool moderately fast from an acetic acid bath, while silk treated comparably is somewhat less fast.

There is no evidence that it has been used as a biological stain.

General references:

Aldrich FT-IR Library **1**(2), 1002B
IR Library **3**, 1469D
Colour Index 3rd ed., 1971

ACID YELLOW 99

Palatine Fast Yellow GRN • Zapon Fast Yellow GR

Aldrich Cat. No.: 20,180-4
Sigma Prod. No.: A3895

CAS No.: [*10343-58-5*]
C.I. No.: 13900
Mol. Form.: $C_{16}H_{13}N_4NaO_8S \cdot Cr$
F.W.: 496.35
Appearance: Dark-yellow solid
λ_{max}: 445nm in water
Solubility: H_2O 60mg/ml
EGME 20mg/ml
EtOH 10mg/ml

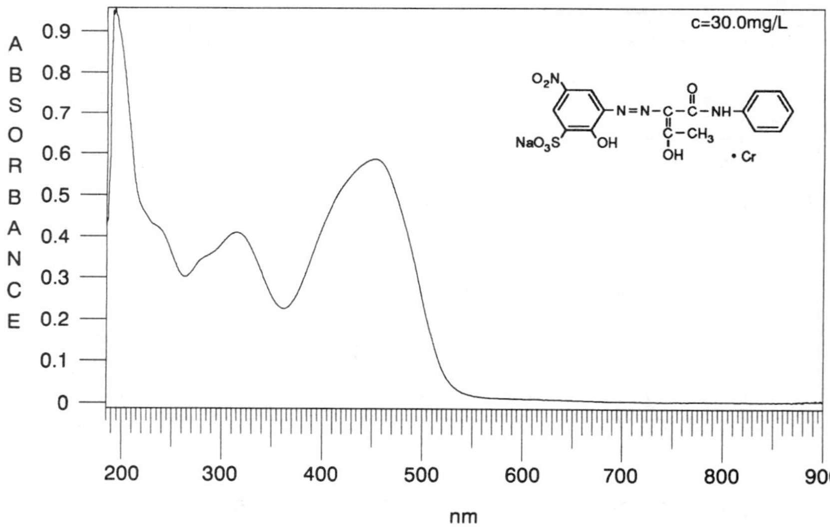

This dye is a complex composed of one mole atom of chromium and one mole of the anionic monoazo dye prepared in the following way: Diazotized 6-amino-4-nitro-1-phenol-2-sulfonic acid is coupled to acetoacetanilide and converted to the sodium salt. The complex is then formed by heating combined aqueous solutions of the dye and chromium fluoride.

Wool is dyed from a sulfuric acid bath, silk from either a sulfuric acid or formic acid bath and nylon from a formic acid bath. Wool and silk are printed directly. Chromed leather, especially for gloves and clothing, is dyed a bright yellow. The dye is also used to color anodized aluminum. The free acid form is Solvent Yellow 19 (C.I. 13900:1).

There is no evidence in the literature that it has been used as a biological stain.

General references:

Aldrich FT-IR Library	**1**(2), 978B
Sigma FT-IR Library	**1**(2), 92A
IR Library	**3**, 1457F
Colour Index	3rd ed., 1971

ACRIDINE

Aldrich Cat. No.: A2,360-9
Sigma Prod. No.: A9640
Purity: 99%

CAS No.: [*260-94-6*]
Mol. Form.: $C_{13}H_9N$
F.W.: 179.22
m.p.: 107-110°C
b.p.: 346°C/750mm Hg
Appearance: Yellow powder
λ_{max}: 358nm in hexane
Solubility: H_2O <0.1mg/ml
EGME 15mg/ml
EtOH 0.8mg/ml

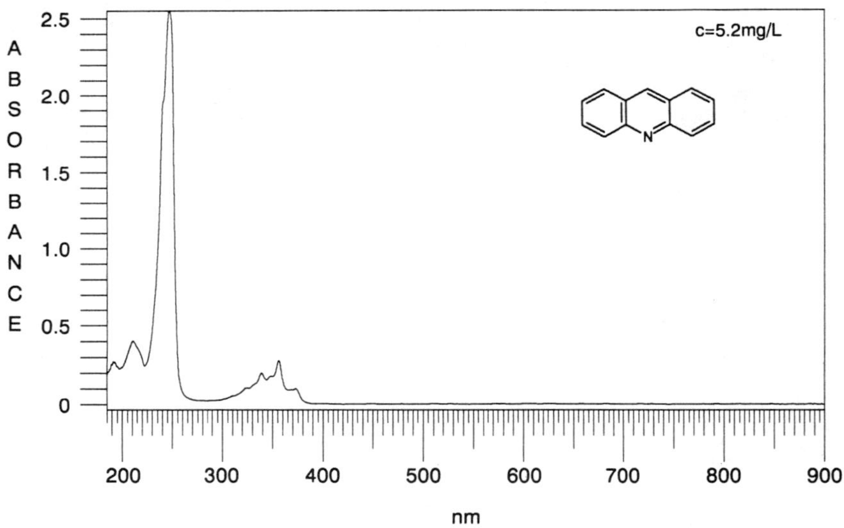

This heterocyclic aromatic is derived from coal tar, which is the source of much of the commercial supply. However, it can be synthesized readily, if not cost competitively. When pure, it forms small, colorless needles from alcohol, although the acridine of commerce is supplied, usually, as a yellow to yellowish-orange, amorphous powder which is soluble in ether, carbon disulfide and the common hydrocarbons and very slightly soluble in alcohol. It is sparingly soluble in *hot* water.

The acridine molecule serves as the nucleus for dyes of the acridine series and, reportedly, acridine is used as an intermediate in the production of certain dyes and the antimalarial drug atabrine. However, the *Colour Index* does not list it as a dye intermediate. In fact, Acridine Orange, Acridine Yellow, Proflavine, Acriflavine and Atabrine are usually made by direct synthesis and acridine is not utilized as the intermediate.

Acridine is a weak base that forms salts with mineral acids. *Merck*[1] indicates that the hydrochloride has been used as an analytical reagent for cobalt, iron and zinc.

Reference:

1) *The Merck Index,* 11th ed.; Budavari, S. *et al.*, Eds.; Merck: Rahway, 1989; Monograph 117.

General references:

Aldrich FT-IR Library	**1**(2), 874A
Sigma FT-IR Library	**1**(2), 246A
IR Library	**3**, 1399E
NMR Library	**2**(2), 749C
Safety	**2**(1), 55A
Colour Index	3rd ed., 1971

ACRIDINE ORANGE

Basic Orange 14

Aldrich Cat. No.:	15,855-0
Sigma Prod. No.:	A6014
Grade:	Biological stain
CAS No.:	[*10127-02-3*]
C.I. No.:	46005
Mol. Form.:	$C_{17}H_{19}N_3 \cdot {}^1/_2\ ZnCl_2 \cdot HCl$
F.W.:	369.94
Appearance:	Red-brown powder
λ_{max}:	489nm in water
Solubility:	H_2O 6mg/ml
	EGME 4mg/ml
	EtOH 2mg/ml

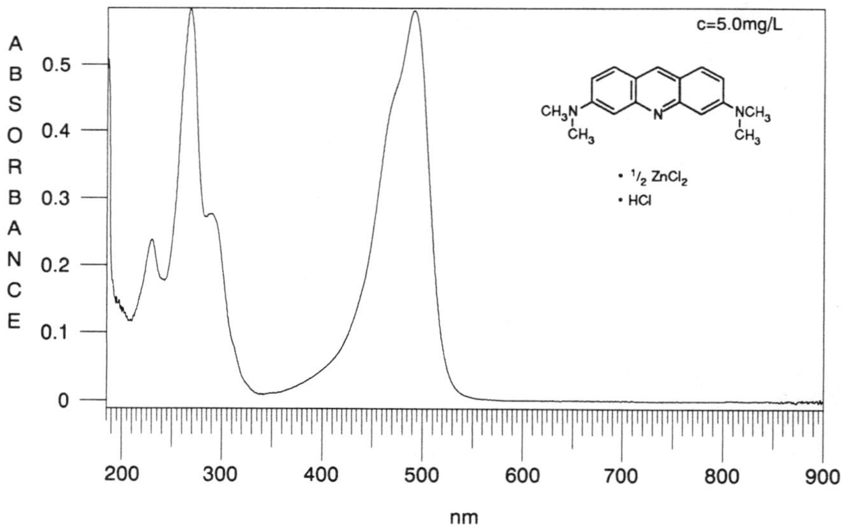

Acridine Orange (AO) is a metachromatic, fluorescent, cationic acridine dye that is available commercially as the zinc chloride double salt (Aldrich Cat. No. 15,855-0) with a dye content of approximately 90%. Slightly soluble in water and even less soluble in the lower alcohols, this salt form is used commercially in lithographic applications, for dyeing leather, in the manufacture of printing inks and in the preparation of lakes for pigments.

Lewis and Goland[1] found that AO retarded cancer-cell growth in a study that compared its antineoplastic efficacy with that of a number of other dyes.

More recently, due to its capacity to differentially stain single- and double-stranded nucleic acids, AO has found wide acceptance as a stain in flow-cytometry applications. Tryer *et al.*[2] used Acridine Orange for differentiating proliferating and latent cells and for sorting sputum and cancer cells by quantitating the cellular RNA.

McCarthy and Senne[3] used AO for the rapid (same-day) detection of bacteria in blood, that previously had not demonstrated growth in blood cultures. Sensitivity equivalent to a slower sub-culture method was claimed.

Staining Procedures[4] provides a bibliography and procedures for the following applications: demonstrating RNA in virus infected plant cells, AO method for nucleic acids, a fluorescence screening procedure for determining the amount of cytoplasmic contamination in nuclear fractions, fluorescent AO technique for acidic mucins and fungi, Saunder's (1964) AO method for acid mucopolysaccharides, AO stain for lyposomes in living cells, differential fluorescence of live and dead yeast cells with AO, Love's (1979) supravital fluorochroming method with AO for isolated Mast cells, Smith-Sonneborn's (1974) rapid supravital fluorochroming method for cytological structures in *Paramecia* using AO, procedure for staining membranes for the counting of cells by epifluorescent microscopy. All life-science and other precise applications should be performed with Acridine Orange hydrochloride hydrate, 98% by HPLC, (Aldrich Cat. No. 31,833-7) or its equivalent.

Acridine Orange base (Aldrich Cat. No. 23,547-4) is used in the preparation of printing inks and it appears to have some utility as a stain for fats and lipids in biological applications. The dye content of this product is approximately 75%.

References:

1) Lewis, M.R.; Goland, P.G. *Am. J. Med. Sci.* **1948,** *215,* 282-289.
2) Tryer, H.; Golden, J.F.; Vansickel, M.H.; Echols, C.K.; Frost, J.K.; West, S.S.; Pressman, N.J.; Albright, C.D.; Adams, L.A.; Gill, G.W. *J. Histochem. Cytochem.* **1979,** *27,* 552.
3) McCarthy, L.R.; Senne, J.E. *J. Clin. Microbiol. II* **1980**, *3*, 281-285.
4) Clark, G. *Staining Procedures,* 4th ed.; Williams and Wilkins: Baltimore, 1981; pp 48, 57, 61, 71, 72, 86, 87, 89, 90, 429.

General references:

Aldrich FT-IR Library	**1**(2), 876C
Sigma FT-IR Library	**1**(2), 92B
IR Library	**3**, 1400F
Safety	**2**(1), 55D
Colour Index	3rd ed., 1971

ACRIDINE ORANGE base

Solvent Orange 15

Aldrich Cat. No.: 23,547-4
Sigma Prod. No.: A6529

CAS No.: [*494-38-2*]
C.I. No.: 46005:1
Mol. Form.: $C_{17}H_{19}N_3$
F.W.: 265.36
Appearance: Brown powder
λ_{max}: 488nm in methanol
Solubility: H_2O 0.7mg/ml
EGME 20mg/ml
EtOH 7mg/ml

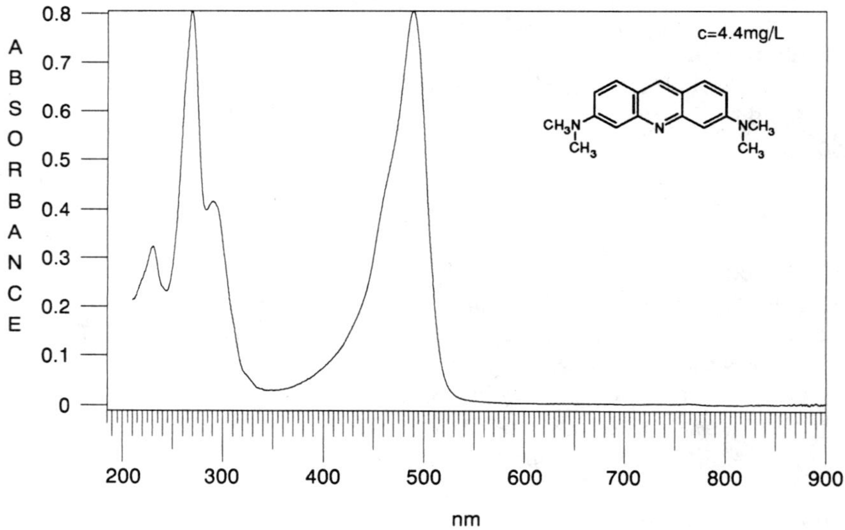

See description on pages 60 and 61 under Acridine Orange.

General references:

Aldrich FT-IR Library	**2**(1), 55D
Sigma FT-IR Library	**1**(2), 92C
NMR Library	**2**(2), 751B
Safety	**2**(1), 56A
Colour Index	3rd ed., 1971

ACRIDINE ORANGE hydrochloride hydrate

Aldrich Cat. No.: 31,833-7
Sigma Prod. No.: A4921

Mol. Form.: $C_{17}H_{19}N_3 \cdot HCl \cdot xH_2O$
F.W.: 301.82
Appearance: Burnt-orange powder
λ_{max}: 492nm in water
Solubility: H_2O 40mg/ml
EGME 40mg/ml
EtOH 30mg/ml

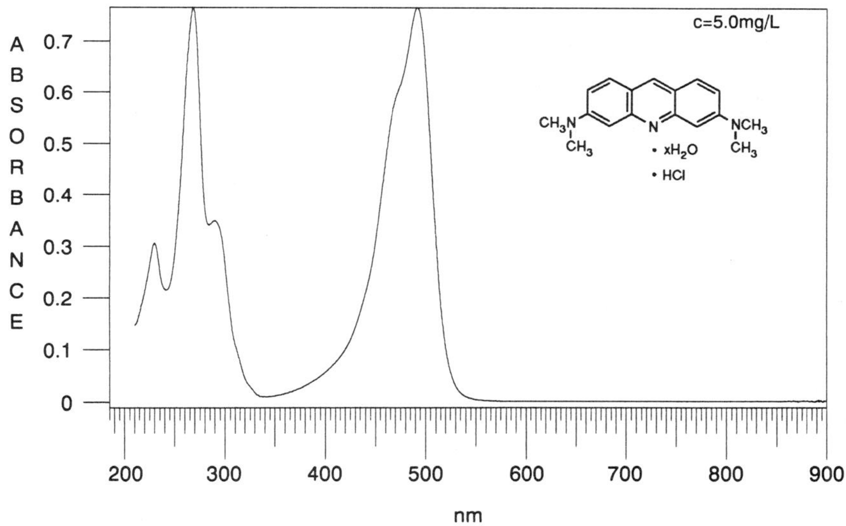

See description on pages 60 and 61 under Acridine Orange.

ACRIDINE YELLOW G

Aldrich Cat. No.:	19,950-8
Sigma Prod. No.:	A2520
Grade:	Biological stain
CAS No.:	[*135-49-9*]
C.I. No.:	46025
Mol. Form.:	$C_{15}H_{15}N_3 \cdot HCl$
F.W.:	273.77
Appearance:	Copper crystalline powder
λ_{max}:	442nm in water
Solubility:	H_2O 4mg/ml
	EGME 3mg/ml
	EtOH 1mg/ml

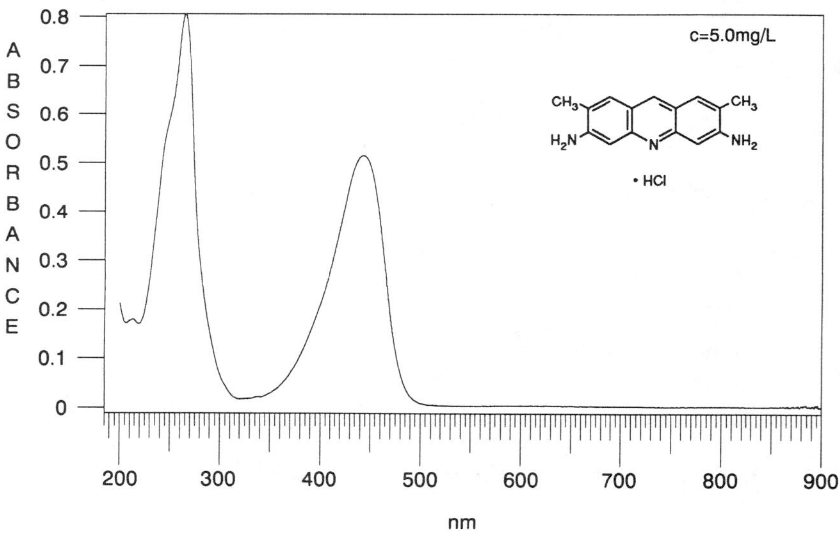

Acridine Yellow G is a cationic acridine dye that is slightly soluble in water and the lower aliphatic alcohols producing yellow solutions with a green fluorescence. Although the *Colour Index* indicates no known producer, Acridine Yellow is commercially available as the chloride salt with a dye concentration of approximately 90%. The *Colour Index* states that it was formerly used in colored discharges or to give white discharges with chlorate.

Conn's[1] indicates that it is rarely used in microtechniques, citing uses such as a fluorochrome in insect histology, Schiff reagent after periodic acid oxidation for demonstration of periodate-positive fungi in tissue by fluorescence microscopy and for the fluorochroming of mast cells at pH 1 using ultraviolet to blue light.

Reference:

1) *Conn's Biological Stains,* 9th ed.; Lillie, R.D., Ed.; Williams and Wilkins: Baltimore, 1977; p 358.

General References:

Aldrich FT-IR Library	**1**(2), 1045B
Sigma FT-IR Library	**1**(2), 92D
IR Library	**3**, 1490A
Safety	**2**(1), 56B
Colour Index	3rd ed., 1971

ACRIFLAVINE

Acriflavine neutral

Aldrich Cat. No.: 22,844-3
Sigma Prod. No.: A8126

CAS No.: *[8048-52-0]*
Appearance: Rust-orange powder
λ_{max}: 462(262)nm in methanol
Solubility: H_2O 20mg/ml
EGME 7mg/ml
EtOH 4mg/ml

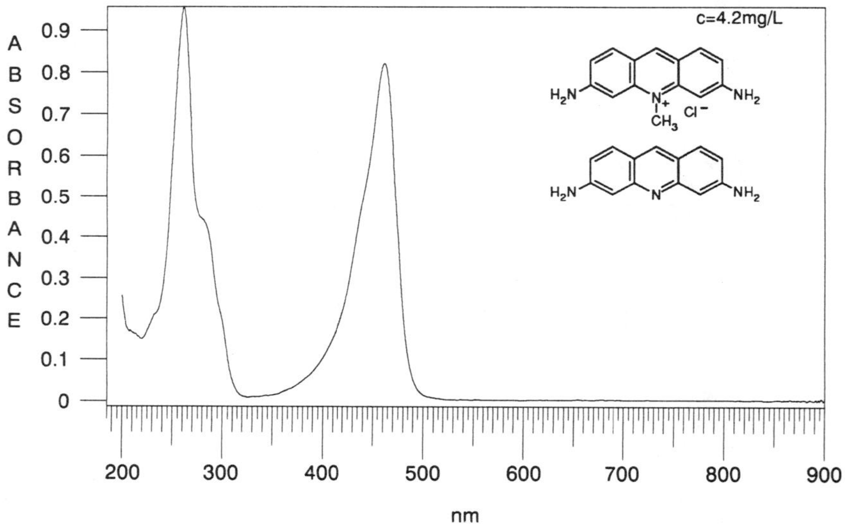

Acriflavine is a yellow, fluorescent, cationic acridine dye. The "neutral" form (Aldrich Cat. No. 22,844-3) is typically composed of an approximate 8 to 1 mixture of 3,6-diamino-10-methylacridinium chloride and its precursor 3,6-diaminoacridine. The ratio of the dye components is dependent on the efficiency of the methylation reaction.

Acriflavine hydrochloride (Aldrich Cat. No. 12,126-6) is composed of 3,6-diamino-10-methylacridinium dihydrochloride and 3,6-diaminoacridinium dihydrochloride in the same approximate ratio.

Both the "neutral" and the hydrochloride forms are acidic. *Merck*[1] indicates that a 1% aqueous solution of the "neutral" product exhibits a pH of 3.5 and a comparable solution of the hydrochloride displays a pH of 1.5. Aqueous solutions are yellow with a green fluorescence, while the alcoholic solutions are orange in transmitted light and yellowish green in reflected light.

Conn's[1] reports that Acriflavine has limited utility as an ordinary biological stain with some use as a fluorochrome and fluorochrome Schiff reagent.

References:

1) *The Merck Index*, 11th ed.; Budavari, S. *et al.*, Eds.; Merck: Rahway, 1989; Monograph 118.

2) *Conn's Biological Stains*, 9th ed.; Lillie, R.D., Ed.; Williams and Wilkins: Baltimore, 1977; p 355.

General references:

Aldrich FT-IR Library	**1**(2), 876A
Sigma FT-IR Library	**1**(2), 93B
Safety	**2**(1), 57C

ACRIFLAVINE hydrochloride

Aldrich Cat. No.:	12,126-6
Sigma Prod. No.:	A8251
CAS No.:	[*8063-24-9*]
m.p.:	260°C (dec.)
Appearance:	Red-orange powder
λ_{max}:	465(262)nm in methanol
Solubility:	H_2O 20mg/ml
	EGME 6mg/ml
	EtOH 7mg/ml

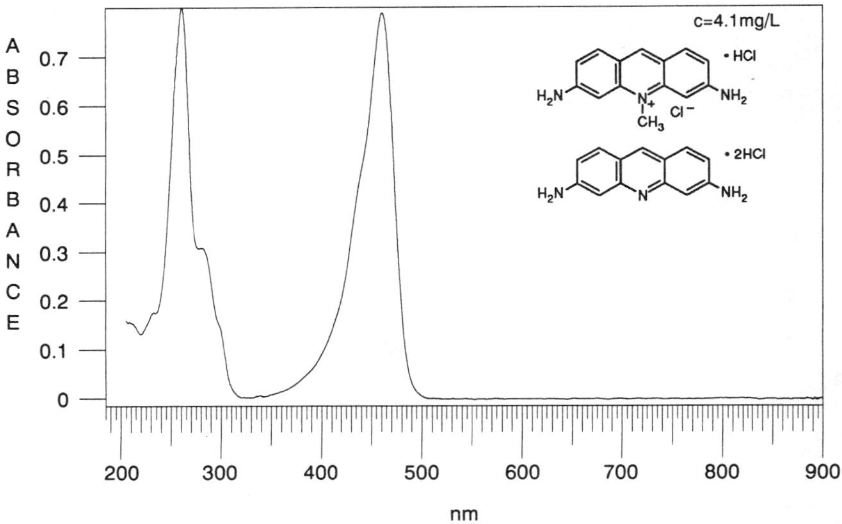

See description on pages 66 and 67 under Acriflavine.

General references:

Aldrich FT-IR Library	**1**(2), 875D
Sigma FT-IR Library	**1**(2), 93A
IR Library	**3**, 1400D
Safety	**2**(1), 57B

ALCEC BLUE™

Aldrich Cat. No.: 30,903-6

Appearance: Violet powder
λ_{max}: 611nm, pH 5.7 acetate buffer
Solubility: H_2O 3mg/ml
EGME 6mg/ml
EtOH 1mg/ml

™ Trademark of Professor J.E. Scott and Aldrich Chemical Co., Inc.

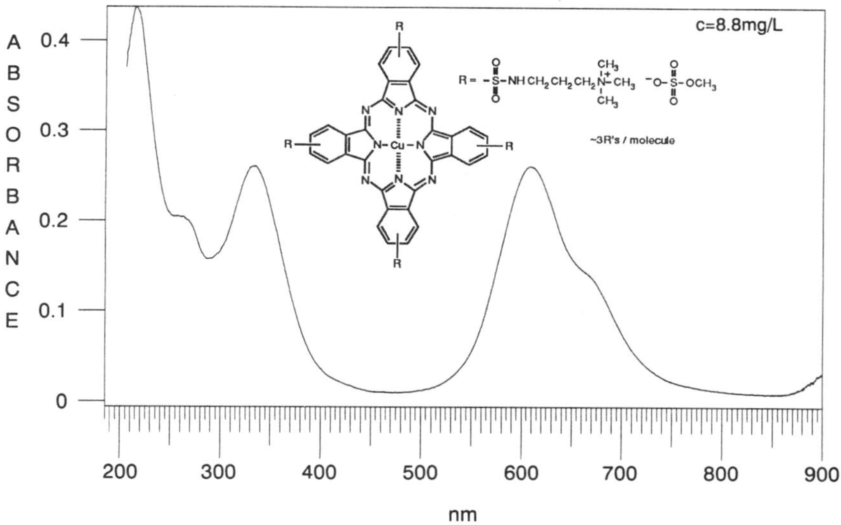

Alcian Blue has been widely used to stain and quantify biological polyanions of all kinds,[1] particularly mucins and connective tissue glycosaminoglycans. When manufacture of the dye ceased in 1973,[2] stocks began to run out quickly. Moreover, the dye is of limited stability since it is an ingrain dye, designed to break down under mild conditions.

Alcec Blue (*Al* from Alcian Blue, *cec* from critical electrolyte concentration)[3,4] was developed by Scott[3] to replace Alcian Blue in most of its biochemical applications. The structure of Alcec Blue is based on molecular-biological principles; it is quaternized Astrablau™ and is stable over a wider pH range than is Alcian Blue.[5]

The dye can replace Alcian Blue in acetic acid as a mucin stain, in hydrochloric acid as a connective-tissue stain, in ethanol-water solutions to stain cellulose acetate, electrophoretograms and chromatograms and in electrolyte solutions in CEC methods.[3,4] Cellulose acetate strips can be dissolved in dimethyl sulfoxide to quantify the stained bands, exactly as with Alcian Blue.[6]

™ Trademark of Hoechst

Alcec Blue has the same CEC pattern as Alcian Blue with polycarboxylates, DNA and RNA at < 0.2*M* $MgCl_2$, and with sulfated proteoglycans (*e.g.*, in cartilage) at 0.5-0.8*M* $MgCl_2$.[7-9]

References and notes:

1) Mowry, R.W. *Ann. N.Y. Acad. Sci.* **1963,** *105*, 402.

2) Scott, J.E. *Histochemie* **1973,** *37*, 379.

3) *Idem J. Microsc.* **1980,** *119*, 373.

4) Scott, J.E.; Dorling, J. *Histochemie* **1965,** *5*, 221.

5) The tetramethylisothiouronium group of Alcian Blue is unstable in alkali (pH >8), especially on warming; the quaternary ammonium group of Alcec Blue is stable to at least pH 12, with no tendency to decompose on standing as the solid or in solution.

6) Newton, D.J.; Scott, J.E.; Whiteman, P. *Anal. Biochem.* **1974,** *62*, 268.

7) Alcec Blue chloride is slightly less soluble than Alcian Blue; therefore there may be precipitation of the dye chloride at high $MgCl_2$ (or other chloride) concentrations. This can be avoided by staining in 20% DMSO (instead of straight aqueous) solutions, with the added advantage that differentiation between polycarboxylate-polysulfate or sulfated polyanions is improved.

8) Excellent CEC patterns are obtained using $Mg(OAc)_2$ with improved polycarboxylate-polysulfate differentiation, without dye precipitation (Scott, J.E., unpublished results).

9) Scott, J.E.; Hughes, E.W. *J. Microsc.* **1983,** *123*, 209.

ALCIAN BLUE 8GX

Ingrain Blue 1

Aldrich Cat. No.:	23,655-1	19,978-8
Sigma Prod. No.:	A3157	A2899
Grade:	Certified	—
Dye content:	~75%	~25%
CAS No.:	[*33864-99-2*]	
C.I. No.:	74240	
Mol. Form.:	$C_{56}H_{68}Cl_4CuN_{16}S_4$	
F.W.:	1298.88	
Appearance:	Dark blue-violet powder	Dark-blue powder
λ_{max}**:**	615nm in water	
Solubility:	H_2O 8mg/ml	
	EGME 8mg/ml	
	EtOH 1mg/ml	

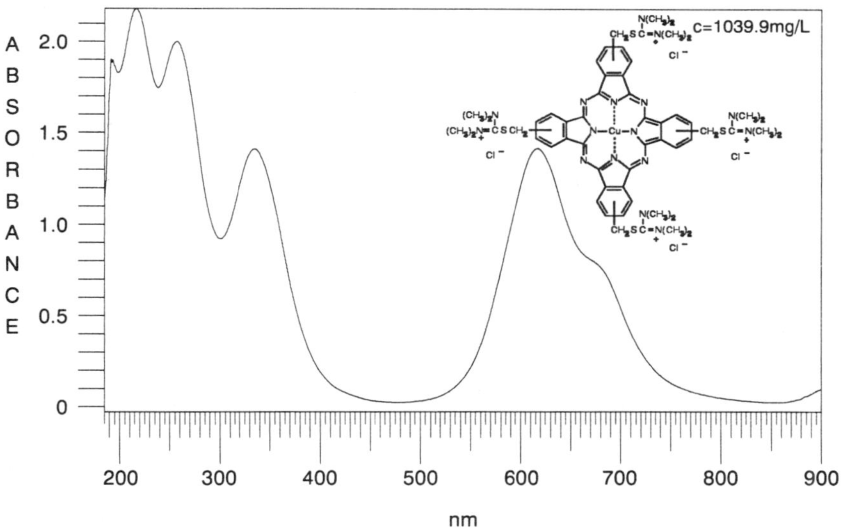

The chemical constitution of Alcian Blue, which was developed in 1944 by Haddock and Wood, was a closely guarded trade secret for a number of years. When its identity became essential to the understanding and interpretation of its histologic behavior, Scott[1] undertook to establish its structure utilizing, among other things, nuclear magnetic resonance and mass spectrometry. After performing an elegant piece of detective work, he concluded the structure to be as shown above and the authenticity of this structure was later substantiated by the manufacturer. At the same time, the manufacturer revealed to Scott[2] that sizeable percentages of boric acid or a combination of boric acid and sodium borate were added to the commercial product to act as a stabilizer in storage. The manufacturer also revealed the structure of Alcian Yellow with the foregoing disclosures.

Alcian Blue is no longer produced on a commercial scale. Apparently, this is due to both the toxic nature of the chloromethylating agent required for its manufacture and the lack of demand for the finished product in industrial applications. Ventkataraman[3] states that it was used commercially to dye and print cellulosic substrates and as a gelling agent for lubricating oils.

Steedman[4] was the first biological investigator to use Alcian Blue. He reported it to be a differential stain for acidic polysaccharides. When the staining pH is variably controlled, sulfomucins, sialomucins and uronic acid mucins can be distinguished. Further work in this area was done by Lison[5] and Mowry.[6]

Alcian Blue is one of the most widely utilized cationic dyes for the demonstration of glycosaminoglycans in tissue sections. The staining mechanism appears to involve the formation of a reversible electrostatic bond between the cationic dye molecules and the anionic sites on the glycosaminoglycan. These electrostatic bonds can be readily broken, revealing what appears to be a variation in bonding among different types of glycosaminoglycans. The concentration of added electrolyte required to break the bond is called the "critical electrolyte concentration" (CEC) and Scott[7] reported that it is a function of the type and density of the charge. Knowing that ionization depolarizes glycosaminoglycans, Morgan *et al.*[8] carried out experiments which showed that γ-irradiation reduced the CEC when used with Alcian Blue.

References:

1) Scott, J.E. *J. Histochem. Cytochem.* **1972,** *20*, 387.

2) *Idem* *Histochemie* **1973,** *37*, 379-380.

3) Ventkataraman, K. *The Chemistry of Synthetic Dyes;* Academic: New York and London, 1952.

4) Steedman, H.F. *Microsc. Sci.* **1950,** *91*, 477-479.

5) Lison, L. *Stain Technol.* **1954,** *29*, 131-138.

6) Mowry, R. *J. Histochem. Cytochem.* **1956,** *4*, 407.

7) Scott, J.E.; Dorling, J. *Histochemie* **1965,** *5*, 281.

8) Morgan, R.E.; Moore, J.S.; Phillips, G.O. *J. Histochem. Cytochem.* **1972,** *20*, 831.

General references:

Aldrich FT-IR Library	**1**(2), 1042A
Sigma FT-IR Library	**1**(2), 94B
Merck Index	**11**, 211
Colour Index	3rd ed., 1971

ALCIAN YELLOW

Ingrain Yellow 1

Aldrich Cat. No.:	12,128-2
Sigma Prod. No.:	A7394
CAS No.:	*[61968-76-1]*
C.I. No.:	12840
Mol. Form.:	$C_{40}H_{46}Cl_2N_8S_4$
F.W.:	838.03
m.p.:	169°C (dec.)
Appearance:	Rust-colored powder
λ_{max}**:**	388nm in methanol
Solubility:	H_2O 20mg/ml
	EGME 30mg/ml
	EtOH 10mg/ml

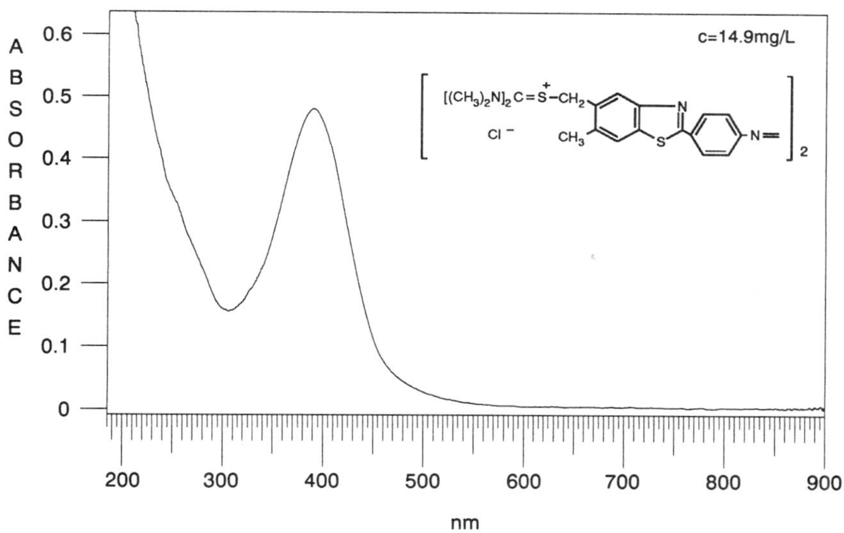

Alcian Yellow, unlike the similarly named phthalocyanine dye Alcian Blue 8GX, contains a monoazo chromophore.[1] However, the two closely resemble each other in functionality and they are aptly named ingrain dyes. Indeed, in the dyeing process, their identical, water-solubilizing cationic auxochromes (*S*-methylenetetramethylisothiouronium chloride) are hydrolytically cleaved from their dissimilar chromogens resulting, in each instance, in the deposition of a pigment within the matrix of the substrate. Alcian Yellow contains two of the foregoing cationic functions while Alcian Blue contains four.

TLC and HPLC analyses indicate that Alcian Green (C.I. Ingrain Green 2) is a mixture of Alcian Yellow and Alcian Blue.

Maxwell[2] differentiated between gastric-surface epithelium mucin and metaplastic intestinal mucin by first staining the stain-receptive metaplastic mucin with either Alcian Yellow or Alcian Blue. Intense staining of the surface tissue, previously unreceptive to staining, was facilitated by oxidizing the section with periodic acid and finally treating with the other dye.

References:

1) Scott, J.E. *Histochemie* **1973,** *37*, 379-380.

2) Maxwell, A. *Stain Technol.* **1963,** *38*, 286-287.

General references:

Aldrich FT-IR Library **1**(2), 1004A

Sigma FT-IR Library **1**(2), 94C

IR Library **3**, 1470D

ALIZARIN

Mordant Red 11

Aldrich Cat. No.:	12,277-7
Sigma Prod. No.:	A7378
Purity:	97%
CAS No.:	[*72-48-0*]
C.I. No.:	58000
Mol. Form.:	$C_{14}H_8O_4$
F.W.:	240.21
m.p.:	287-289°C
Appearance:	Burnt-orange powder
λ_{max}:	567(609)nm in 0.1*N* sodium hydroxide
Solubility:	H_2O 0.4mg/ml
	EGME 7mg/ml
	EtOH 2mg/ml

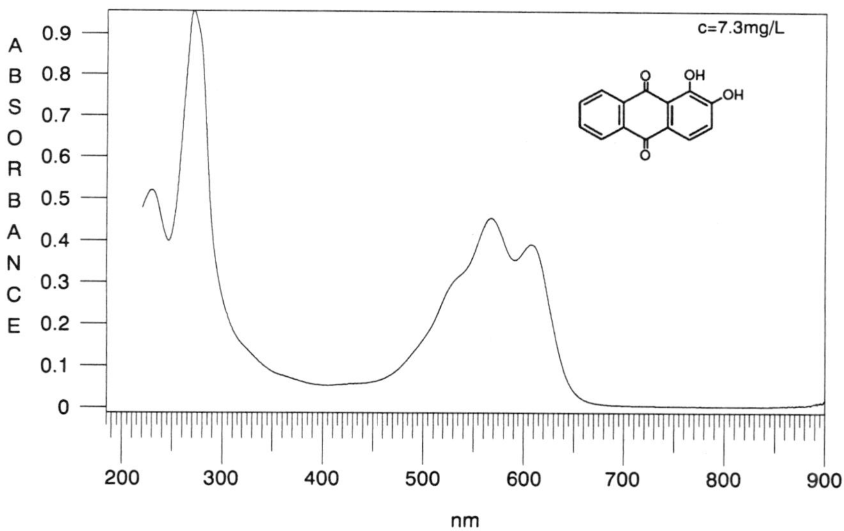

Alizarin and Purpurin, two closely related anthraquinone dyes, were originally prepared from the roots of *Rubia tinctorium*, madder. The use of madder as a dye source has been traced to the ancient Egyptians who used it to dye goat skins for clothing, among other uses. Alizarin, as such, does not exist in the madder roots. It is present in the form of ruberythrinic acid, a glucoside which, on boiling with hydrochloric acid, yields the dye and glucose. Synthetically, it is prepared by treating anthraquinone with sulfuric acid at 120°C to obtain anthraquinone-2-sulfonic acid. Then this product is fused with an alkali to form 2-hydroxyanthraquinone. Since 2-hydroxyanthraquinone is susceptible to nucleophilic hydroxylation in the α-position, exposure to air oxidation or to an oxidant such as sodium chlorate, readily yields Alizarin. Alizarin is virtually insoluble in water, soluble in alkalis and acetone and slightly soluble in ethanol, methyl Cellosolve, benzene and carbon tetrachloride. It is used with various metal mordants to dye wool, silk and cotton. Cotton is printed using aluminum, chromium or iron as a mordant. Pigment Red 83 (C.I. 58000:1) is a metal complex of Alizarin.

In the laboratory, Alizarin is used as a pH indicator and spot-test reagent. Kolthoff[1] reported visual-transition intervals of pH 5.5 (yellow) to pH 6.8 (violet) and pH 10.1 (violet) to pH 12.1 (purple). Feigl[2] describes spot tests using Alizarin for the detection of aluminum, indium, mercury, zinc and zirconium.

Conn's[3] reported that it stains tissue a feeble yellowish red if used directly. He indicated that more intense colors are formed in the presence of various metal salts: aluminum (red), iron (bluish violet) and chromium (brownish violet). Additional applications, including the vital staining of protozoa and nervous tissue, were also reported.

References:

1) Kolthoff, I.M.; Rosenblum, C. *Acid-Base Indicators*; Macmillan: New York, 1937; p 159.

2) Feigl, F. *Spot Tests in Inorganic Analysis*; Elsevier: New York, 1958.

3) *Conn's Biological Stains,* 9th ed.; Lillie, R.D., Ed.; Williams and Wilkins: Baltimore, 1977; p 431.

General references:

Aldrich FT-IR Library	**1**(2), 86D
Sigma FT-IR Library	**1**(2), 94D
IR Library	**3**, 900B
NMR Library	**2**(2), 90D
Safety	**2**(1), 93C
Merck Index	**11**, 237
Colour Index	3rd ed., 1971

ALIZARIN BLUE BLACK B

Mordant Black 13

Aldrich Cat. No.:	23,403-6
Sigma Prod. No.:	A3175
CAS No.:	[*1324-21-6*]
C.I. No.:	63615
Mol. Form.:	$C_{26}H_{16}N_2Na_2O_9S_2$
F.W.:	610.53
Appearance:	Black powder
λ_{max}:	548nm in methanol
Solubility:	H_2O 0.8mg/ml
	EGME 8mg/ml
	EtOH 0.8mg/ml

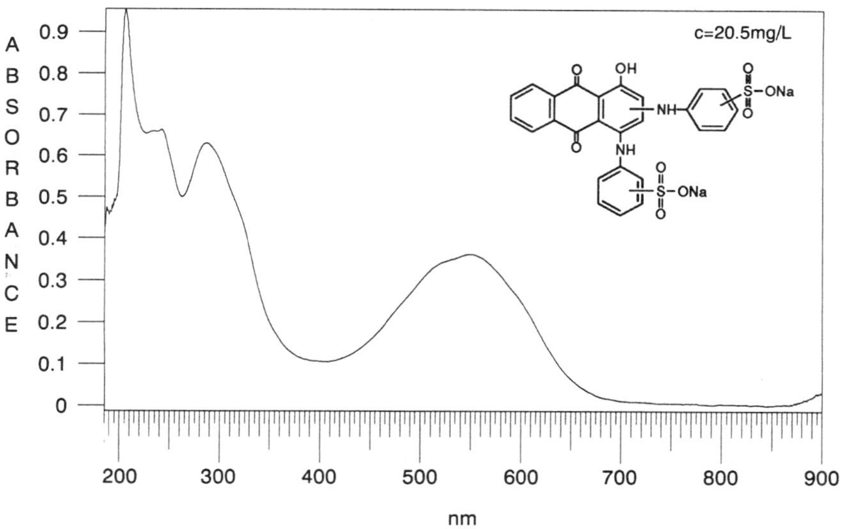

This bluish-grey, anionic anthraquinone dye is made by first condensing one mole of Purpurin with two moles of aniline in the presence of boric acid. The product of this reaction is disulfonated and converted to the disodium salt.

Slightly soluble in methyl Cellosolve, methanol, acetone, pyridine and 2-chlorophenol and very slightly soluble in water and ethanol, it is used to dye wool, silk, nylon and leather (using chromium as a mordant) and to print wool, silk and cotton.

There is no evidence that it has been used as a biological stain.

General references:

Aldrich FT-IR Library	**1**(2), 511C
Sigma FT-IR Library	**1**(2), 95A
Colour Index	3rd ed., 1971

ALIZARIN COMPLEXONE dihydrate

Aldrich Cat. No.:	12,765-5
Sigma Prod. No.:	A3882
Grade:	Indicator

CAS No.:	*[3952-78-1]*
Mol. Form.:	$C_{19}H_{15}NO_8 \cdot 2H_2O$
F.W.:	421.36
m.p.:	190°C (dec.)
Appearance:	Orange powder
λ_{max}:	427nm in methanol
Solubility:	H_2O 0.6mg/ml
	EGME 5mg/ml
	EtOH 2mg/ml

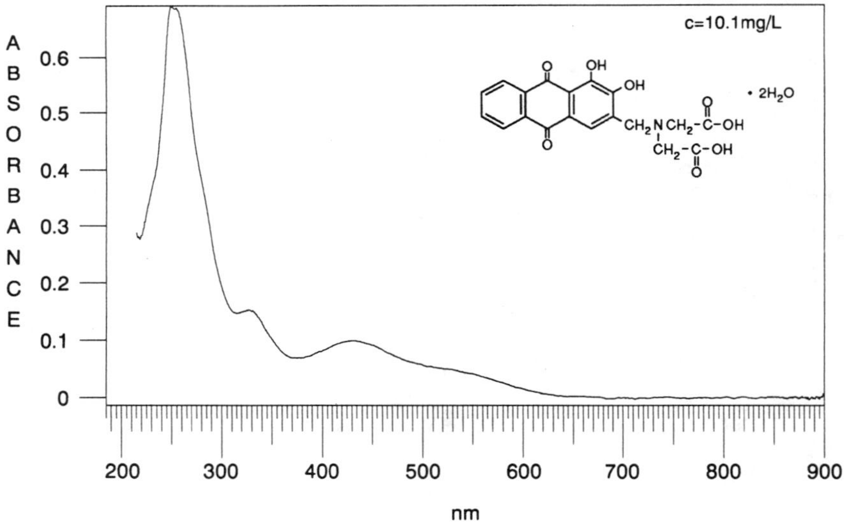

Alizarin Complexone is an acid anthraquinone dye that is used almost exclusively as a colorimetric reagent for the determination of rare-earth metals and fluoride ion. It is very slightly soluble in water (displaying a yellow solution up to pH 4.5, red in the range of pH 6-10 and blue-violet above pH 13), somewhat more soluble in ethanol and insoluble in ether. Complexes formed with lanthanum, cesium and zirconium are red at pH 4.5, red-violet in the range of pH 7-10 and violet above pH 11.5.[1-8]

There is no evidence in the literature that it has been used as a biological stain.

References:

1) Belcher, R.; Leonard, M.A.; West, T.S. *J. Chem. Soc.* **1958,** 2390-2393.

2) *Idem ibid.* **1959,** 3577.

3) *Idem ibid.* **1960,** 4477-4485.

4) *Idem Talanta* **1959,** *2*, 92-97.

5) Johnson, C.A.; Leonard, M.A. *Analyst* **1961,** *86*, 101-104.

6) Wade, M.A.; Yamamura, S.S. *Anal. Chem.* **1965,** *37*, 1276-1278.

7) Hashitani, H.; Muto, H. *Japan Analyst* **1965,** *14*, 1114-1119.

8) Taskaesu, Y. *Jpn. J. Oral Hyg.* **1966,** *16* (1), 1-24.

General references:

Aldrich FT-IR Library	**1**(2), 247A
Sigma FT-IR Library	**1**(2), 95B
IR Library	**3,** 991B
Safety	**2**(1), 93D

ALIZARIN RED S monohydrate
Mordant Red 3

Aldrich Cat. No.:	11,996-2
Sigma Prod. No.:	A5533
Grade:	Certified
CAS No.:	[*130-22-3*]
C.I. No.:	58005
Mol. Form.:	$C_{14}H_7NaO_7S \cdot H_2O$
F.W.:	360.28
Appearance:	Orange-brown powder
λ_{max}:	556(596)nm in 0.1*N* sodium hydroxide
Solubility:	H_2O 20mg/ml
	EGME 10mg/ml
	EtOH 1mg/ml

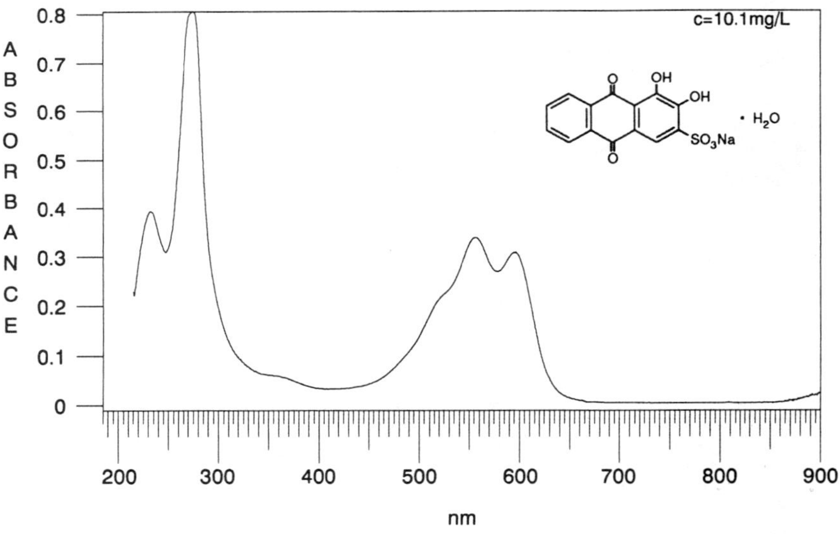

Alizarin Red S is a red, anionic anthraquinone dye. Gurr[1] reported that its industrial uses included the preparation of lakes for pigments. The *Colour Index* states that it dyes wool mordanted with aluminum a scarlet red, wool mordanted with chromium a bordeaux or that mordanted with iron, dark violet. It has been used as an indicator for the estimation of alkali carbonates, ammonia, total alkali in soaps and the alkalinity of water. As an acid-base indicator, it displays a visual transition from pH 3.7 (yellow) to pH 5.2 (purple). Welcher[2] reported it as a reagent for the determination of aluminum, boron and fluoride.

Clark[3] provides references and methods for the following biological applications: staining developing bones in small vertebrates, minute bones (fetal ossification), in combination with Toluidine Blue O for distinguishing between bone and cartilage (relative amount of ossification) and calcium deposits in bone sections.

Murgatroyd and Horobin[4] replaced Carmine with Alizarin Red S in a variation of the Best glycogen stain.

References:

1) Gurr, E. *Synthetic Dyes*; Academic: London, 1971; pp 234-235.

2) Welcher, F.J. *Organic Analytical Reagents*; Van Nostrand: New York, 1949; Vol. 4.

3) Clark, G. *Staining Procedures,* 4th ed.; Williams and Wilkins: Baltimore, 1981; pp 183-186.

4) Murgatroyd, L.B.; Horobin, R.W. *Stain Technol.* **1960,** *44,* 59-62.

General references:

Aldrich FT-IR Library	1(2), 510D
Sigma FT-IR Library	1(2), 95D
IR Library	**3**, 1165G
Merck Index	11, 8516
Colour Index	3rd ed., 1971

ALIZARIN VIOLET 3R

Acid Violet 34

Aldrich Cat. No.: 22,783-8
Sigma Prod. No.: A4278
Dye content: ~50%

CAS No.: [*6408-63-5*]
C.I. No.: 61710
Mol. Form.: $C_{28}H_{20}N_2Na_2O_8S_2$
F.W.: 622.59
Appearance: Dark-blue powder
λ_{max}: 540nm in methanol

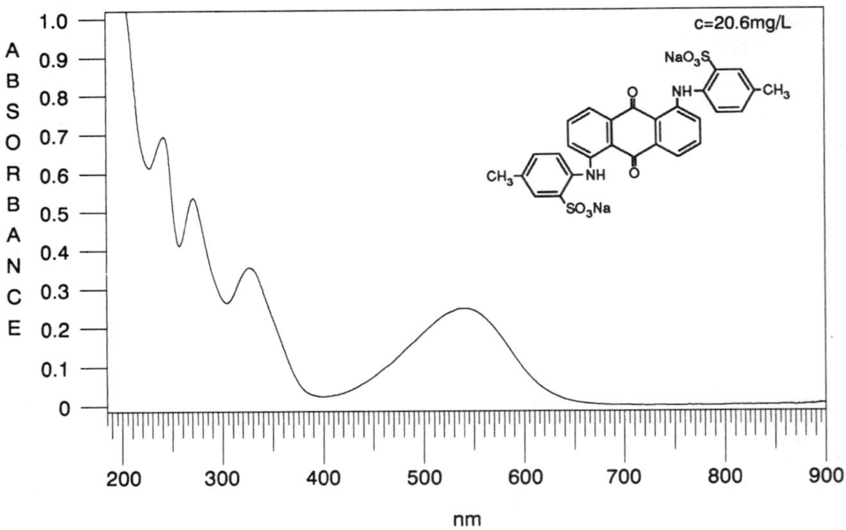

This violet, anionic anthraquinone dye is prepared by condensing one mole of either 1,5-dinitro- or 1,5-dichloroanthraquinone with two moles of *p*-toluidine-2-sulfonic acid and converting the condensation product to the disodium salt.

Slightly soluble in water, methyl Cellosolve, chloroform, 2-chlorophenol, pyridine, toluene, acetone, benzene and carbon tetrachloride and very slightly soluble in ethanol, it is used to dye wool from either an acetic acid or sulfuric acid bath, silk from acetic acid, formic acid or sulfuric acid baths, and nylon from an acetic acid bath. Nontextile industrial applications include dyeing chromed leather a dull purple and coloring anodized aluminum and soap.

While there are no specific applications cited in the literature for its use as a biological stain, *Conn's*[1] believes Alizarin Violet 3R should work satisfactorily as a plasma stain.

Reference:

1) *Conn's Biological Stains,* 9th ed.; Lillie, R.D., Ed.; Williams and Wilkins: Baltimore, 1977; p 444.

General references:

Sigma FT-IR Library 1(2), 96A
Colour Index 3rd ed., 1971

ALIZARIN YELLOW GG

Mordant Yellow 1 • 5-(3-Nitrophenylazo)salicylic acid, sodium salt

Aldrich Cat. No.: 20,670-9

CAS No.: [*584-42-9*]
C.I. No.: 14025
Mol. Form.: $C_{13}H_8N_3NaO_5$
F.W.: 309.22
Appearance: Orange-brown powder
λ_{max}: 362nm in methanol

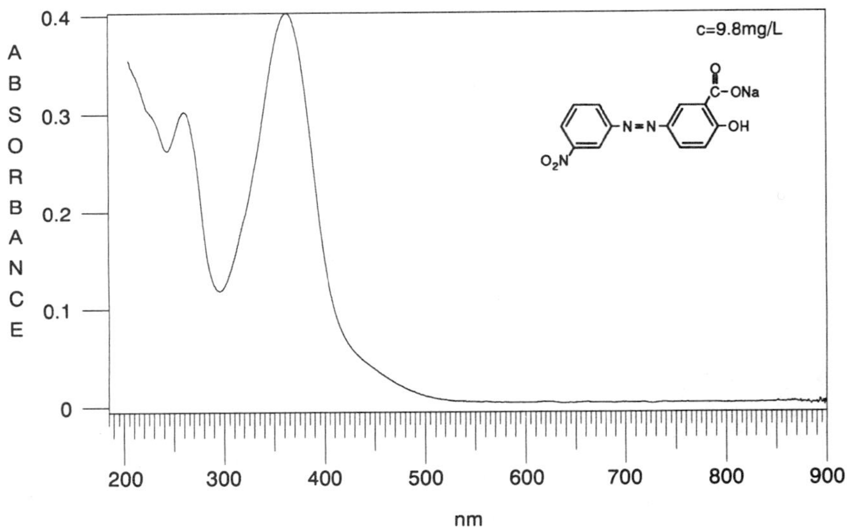

This dull-yellow, anionic monoazo dye is prepared by coupling diazotized 3-nitroaniline to salicylic acid. It is soluble in water and methyl Cellosolve and slightly soluble in ethanol and acetone.

Using chromium as a mordant, it is used to dye wool, silk and nylon, while wool, silk and cotton are printed using the same mordant. Other industrial applications include dyeing chromed leather a greenish yellow and coloring anodized aluminum yellow.

Alizarin Yellow GG has useful indicator properties with a visual-transition interval at pH 10.0 (yellow) to pH 12.0 (orange).

There is no evidence in the literature that it has been used as a biological stain.

General references:

Aldrich FT-IR Library **1**(2), 972A
IR Library **3**, 1454G
Colour Index 3rd ed., 1971

ALKALI BLUE 6B

Acid Blue 110

Aldrich Cat. No.: 30,219-8

CAS No.: [*30586-13-1*]
C.I. No.: 42750
Mol. Form.: $C_{32}H_{28}N_3NaO_4S$
F.W.: 573.65
Appearance: Dark-blue powder
λ_{max}: 603nm in methanol + 1ml 1*N* hydrochloric acid
Solubility: H_2O 90mg/ml
EGME 20mg/ml
EtOH 9mg/ml

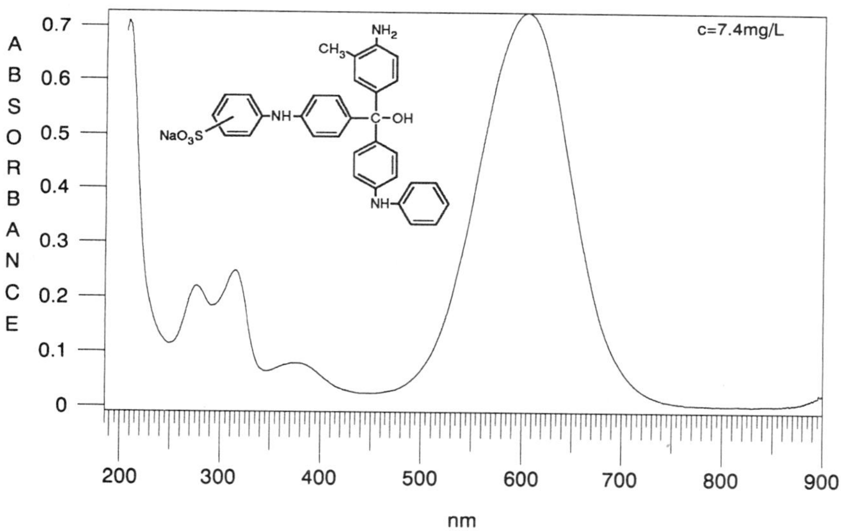

This blue, acid triphenylmethane dye is made by heating (180°C) one mole of Rosaniline (C.I. 42510) with two moles of aniline, in the presence of benzoic acid, to form C.I. Solvent Blue 3 (C.I. 42775). This in turn is monosulfonated with concentrated sulfuric acid to obtain the final product. While it is used to dye wool and silk and to print wool, silk and viscose directly, it is light-fugitive (1-2). Probably it is more useful commercially for dyeing paper and as a pigment for paints. The free acid form (C.I. 42750:1, Pigment Blue 19) is bright blue with a high tinctorial strength. It is used as a pigment in printing inks, coloring paper, carbon paper and typewriter ribbons. Alkali Blue 6B has indicator properties with a visual-transition interval of pH 9.4 (blue) to pH 14.0 (red).

Conn's[1] cites Cook and Lamb who, in 1969, reported the use of Alkali Blue (C.I. 42750) as an elastin stain. This procedure subjects the elastin substrate to an aqueous solution of the stain for a short time, after which it is afterchromed with iron alum.

Reference:

1) *Conn's Biological Stains,* 9th ed.; Lillie, R.D., Ed.; Williams and Wilkins: Baltimore, 1977; pp 288-289.

General references:

Safety	**2**(1), 94A
Colour Index	3rd ed., 1971

ALPHAZURINE A

C.I. Acid Blue 7 • Patent Blue A

Aldrich Cat. No.: 19,814-5
Sigma Prod. No.: A2770

CAS No.: [*3486-30-4*]
C.I. No.: 42080
Mol. Form.: $C_{37}H_{35}N_2NaO_6S_2$
F.W.: 690.82
m.p.: 290°C (dec.)
Appearance: Maroon powder
λ_{max}: 637(409)nm in water
Solubility: H_2O 20mg/ml
EGME 20mg/ml
EtOH 20mg/ml

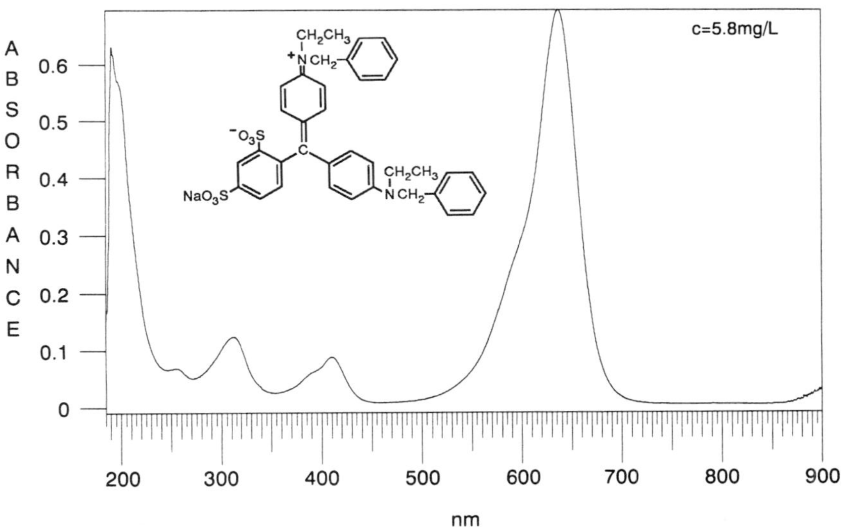

Alphazurine A is a bright, greenish-blue, anionic triphenylmethane dye with a high affinity for proteinaceous substrates and a weak attraction to cellulosic materials. It dyes wool from a weakly acid (pH 3.5-5.5) or a sulfuric acid bath (pH 3.5), Nylon from an acetic or formic acid bath and silk from an acetic, formic or sulfuric acid bath. Wool and silk are printed directly. However, the most important industrial application appears to be the production of its barium salt which is used as a pigment for rubber, paper and surface coloring.

The *Colour Index*[1] indicates C.I. 42080 has been used as a biological stain. No specific area is cited. There is no evidence in the recent literature indicating that it has been used in this application. It is noted that Gurr[2] refers to a similar compound, which he called Patent Blue A with a structure identical to C.I. 42052, "for neurological tissue".

References:

1) *Colour Index*, 3rd ed.; The Society of Dyers and Colourists: Bradford, Yorkshire, 1971; p1298

2) Gurr, E. *Synthetic Dyes*; Academic Press: London and New York, 1971; p 407.

General references:

Aldrich FT-IR Library	**1**(2), 1031A
Sigma FT-IR Library	**1**(2), 96D
IR Library	**3,** 1484D
Safety	**2**(1), 1031A
Colour Index	3rd ed., 1971

ALUM CARMINE

Aldrich Cat. No.: 23,404-4

Appearance: Dark-red powder

λ_{max}: 531(334)nm in 0.1*N* sodium hydroxide

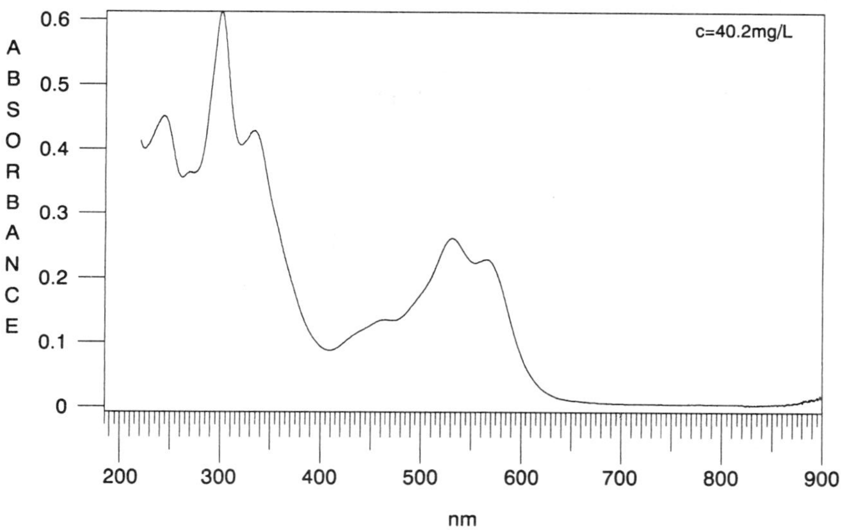

Alum Carmine is a 50/50 mixture (w/w) of ammonium alum (aluminum ammonium sulfate) and Carmine (aluminum-calcium lake of carminic acid).

In order to prepare a working solution equivalent to the Grenacher[1] recipe, 2 grams of Alum Carmine powder is dissolved, with stirring, in hot, deionized water and filtered. Lee[2] described it as a nuclear stain suitable for sections and for bulk staining. Emmel and Cowdry[3] reported that it does not penetrate very well and it is not suitable for large bulk items, although, in their view, it is useful and does not overstain.

References:

1) Grenacher, H. *Arch. Mikrosk. Anat.* **1879,** *16*, 463-471.
2) Lee, A.B. *The Microtomists Vade-Mecum*, 10th ed.; Blackiston's: Philadelphia, 1937; p 140.
3) Emmel, V.; Cowdry, E.V. *Laboratory Technique in Biology and Medicine*, 4th ed.; Williams and Wilkins: Baltimore, 1964; p 10.

ALUMINON

Aurintricarboxylic acid, triammonium salt

Aldrich Cat. No.:	A3,688-3
Sigma Prod. No.:	A0885
Grade:	A.C.S. reagent

CAS No.:	[*569-58-4*]
Mol. Form.:	$C_{22}H_{23}N_3O_9$
F.W.:	473.44
m.p.:	220-225°C (dec.)
Appearance:	Dark-red crystalline powder
λ_{max}**:**	542nm in 0.1*N* sodium hydroxide
Solubility:	H_2O 5mg/ml
	EGME 2mg/ml
	EtOH 1mg/ml

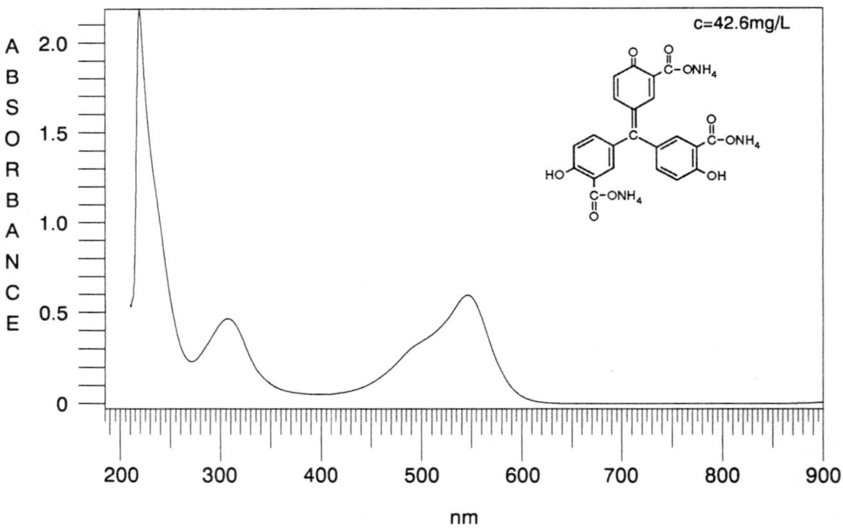

Aluminon is the triammonium salt of aurintricarboxylic acid. It is slightly soluble in water and very slightly soluble in alcohol. *Reagent Chemicals*[1] lists it as a reagent for the detection of aluminum.

Aurintricarboxylic acid, which is soluble in alcohol and slightly soluble in water, is used extensively as a powerful inhibitor of cellular processes that are dependent on the formation of protein-nucleic acid complexes.[2,3] The solubility properties of aurintricarboxylic acid, trisodium salt (C.I. Mordant Violet 39) are very similar to those of Aluminon, although it is more soluble in water. Aluminon previously was used, with chromium as a mordant, for printing calico and dyeing cotton, wool and silk. Currently, there is very little demand for this dye in industrial applications.

Conn's[4] cites a method for the demonstration of aluminum in tissue and a reference that utilizes anionic and cationic lakes from aluminum-chromic acid mixtures to stain plasma (anionic) and certain nuclei (cationic).

References:

1) *Reagent Chemicals,* 7th ed.; American Chemical Society: Washington, DC, 1986; p 133.

2) Sarma, M.H.; Feman, E.R.; Baglioni, C. *Biochim. Biophys. Acta.* **1976,** *418*, 29.

3) González, R.G.; Haxo, R.S.; Schleich, T. *Biochemistry* **1980,** *19*, 4299.

4) *Conn's Biological Stains,* 9th ed.; Lillie, R.D., Ed.; Williams and Wilkins: Baltimore, 1977; p 304.

General references:

Aldrich FT-IR Library	**1**(2), 275B
Sigma FT-IR Library	**1**(2), 101A
IR Library	**3**, 1015E
Merck Index	**11**, 320

AMARANTH

Azorubin S • C.I. Acid Red 27

Aldrich Cat. No.: 12,056-1
Sigma Prod. No.: A1016

CAS No.: *[915-67-3]*
C.I. No.: 16185
Mol. Form.: $C_{20}H_{11}N_2Na_3O_{10}S_3$
F.W.: 604.48
Appearance: Dark reddish-brown powder
λ_{max}: 521nm in water
Solubility: H_2O 60mg/ml
EGME 10mg/ml
EtOH 0.7mg/ml

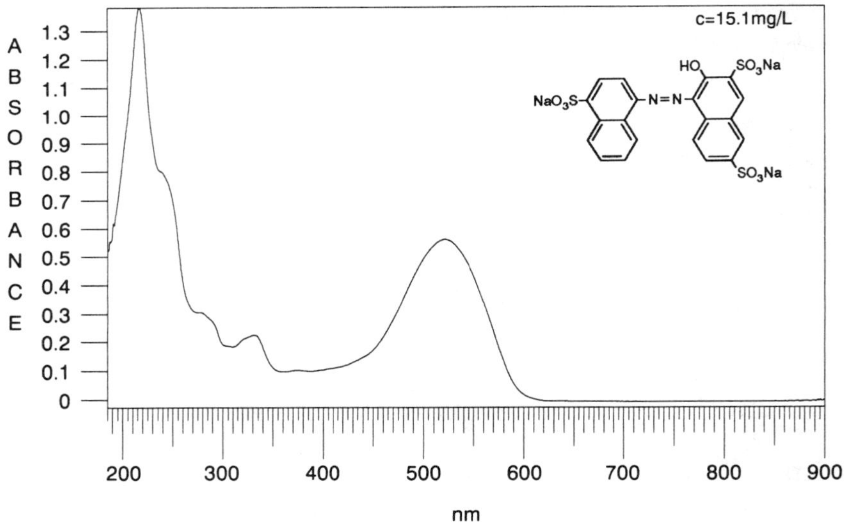

Amaranth is a red, water-soluble, anionic monoazo dye that displays a strong affinity for proteinaceous substrates and a weak attraction to cellulosic materials. Wool and silk are dyed a bluish red from an acid Amaranth bath, while both are readily printed. These same materials discharge white when treated with hydrosulfite. Other industrial applications include dyeing leather, staining wood, coloring paper coatings and coloring photographs.

Prior to 1977, a highly purified grade was used widely in the United States and various other countries to color foods, drugs and cosmetics. Marmion[1] reported that it was delisted for these purposes in the United States in 1977.

Welcher[2] describes the use of Azorubin S as a reagent for the detection of nitrites and nitrates.

Conn's[3] observed that Amaranth finds limited use as a biological stain.

References:

1) Marmion, D.M. *Handbook of U.S. Colorants for Foods, Drugs and Cosmetics;* Wiley: New York, 1979; pp 50-51.

2) Welcher, F.J. *Organic Analytical Reagents*; Van Nostrand: New York, 1949; Vol. 4.

3) *Conn's Biological Stains,* 9th ed.; Lillie, R.D., Ed.; Williams and Wilkins: Baltimore, 1977; p 121.

General references:

Aldrich FT-IR Library	**1**(2), 988B
Sigma FT-IR Library	**1**(2), 97A
Safety	**2**(1), 118A
Merck Index	**11**, 383
Colour Index	3rd ed., 1971

9-AMINOACRIDINE hydrochloride monohydrate

Acramine Yellow • Monacrin

Aldrich Cat. No.: A3,840-1
Sigma Prod. No.: A1135
Purity: 98%

CAS No.: [*52417-22-8*]
Mol. Form.: $C_{13}H_{10}N_2$ • HCl • H_2O
F.W.: 248.71
Appearance: Yellow powder
λ_{max}**:** 400(420)nm in water
Solubility: H_2O 8mg/ml
EGME 8mg/ml
EtOH 7mg/ml

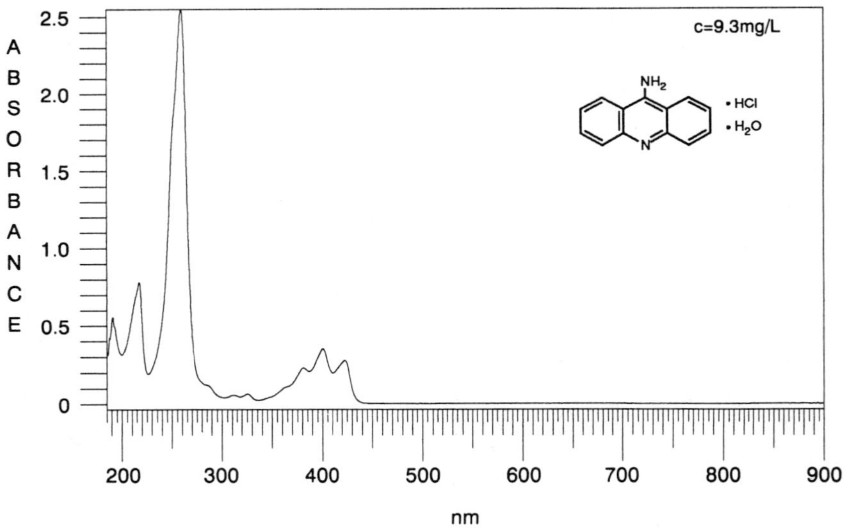

This highly fluorescent, slightly water-soluble acridine dye is made by heating *N*-phenylanthranilic acid with phosphorus pentachloride to form 9-chloroacridine. This in turn is reacted with ammonium carbonate to obtain the dye.[1]

Merck[2] states that it is used as an anti-infective and as an antiseptic in veterinary medicine. No industrial or biological-stain applications are reported.

Seeman *et al.*[3] describe it as a mutagen; however, further investigations have failed to confirm this.

References:

1) Albert, A.; Richie, B. *Org. Synth.* **1942,** *22*, 5.

2) *The Merck Index,* 11th ed.; Budavari, S. *et al.*, Eds.; Merck: Rahway, 1989; Monograph 418.

3) Seeman, N.C.; Day, R.O.; Rich, A. *Nature* **1975,** *253*, 324.

General references:

Aldrich FT-IR Library	**1**(2), 875A
Sigma FT-IR Library	**1**(2), 254C
IR Library	**3**, 1400B
NMR Library	**2**(2), 750D
Safety	**2**(1), 133C

ANILINE BLUE, water soluble

Acid Blue 93 • Methyl Blue

Aldrich Cat. No.: 86,102-2
Sigma Prod. No.: M5528
Grade: Certified

C.I. No.: 42780
Appearance: Brownish-purple powder
λ_{max}: 600nm in water
Solubility: H_2O 70mg/ml
EGME 20mg/ml
EtOH 0.4mg/ml

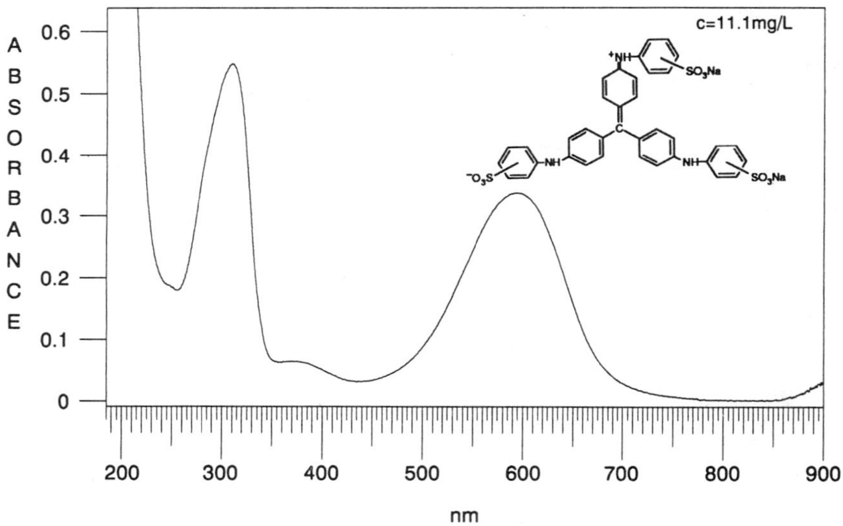

The trivial name Aniline Blue, water soluble (WS), traditionally has been associated with Acid Blue 22 (C.I. 42755). However, before the advent and common use of the more precise analytical tools, such as HPLC, Acid Blue 93, Acid Blue 22, mixtures of the two and various mixtures of 22 and 93 with other closely related dyes were sold and used interchangeably with Aniline Blue WS. This could easily have been done, unknowingly, by persons not privy to the manufacturing pedigrees of the dyes they were handling. *Conn's*[1] reported that they (22 and 93) seem to give equally good results in the Mallory connective-tissue stain; and, understandably, the BSC visible spectrophotometric specification for Aniline Blue WS is liberal enough [λ_{max} 594(610)nm] to embrace both products plus other closely related variations. Pararosaniline, the starting material for Solvent Blue 23 (precursor for Acid Blue 93), can be made in a pure form at a relatively low cost, while Rosaniline, the starting material for Acid Blue 22, requires expensive, chromatographic separations to reach the same purity level. Since Acid Blue 93 can be produced consistently in a reproducible and purer state than its Acid Blue 22 rival which currently has few if any producers, it appears to be largely replacing, if not superceding, the latter as the Aniline Blue WS used in biological applications.

Aniline Blue WS is widely used as a contrast stain in histology and cytology, for both animal and plant tissue. *Staining Procedures*[2] provides numerous methods and a bibliography for both disciplines.

References:

1) *Conn's Biological Stains,* 9th ed.; Lillie, R.D., Ed.; Williams and Wilkins: Baltimore, 1977; p 289.

2) Clark, G. *Staining Procedures,* 4th ed.; Williams and Wilkins: Baltimore, 1981.

General references:

Aldrich FT-IR Library	**1**(2), 1033A
Sigma FT-IR Library	**1**(2), 99A
IR Library	**3**, 1485B
Safety	**2**(1), 271C

ARSENAZO I

2-(1,8-Dihydroxy-3,6-disulfo-2-naphthylazo)benzenearsonic acid, trisodium salt

Aldrich Cat. No.:	10,798-0
Sigma Prod. No.:	A9050
Grade:	Indicator
CAS No.:	*[520-10-5]*
Mol. Form.:	$C_{16}H_{10}AsN_2Na_3O_{11}S_2$
F.W.:	614.28
Appearance:	Red-brown powder
λ_{max}:	499nm in water
Solubility:	H_2O 70mg/ml
	EGME 10mg/ml
	EtOH 0.4mg/ml

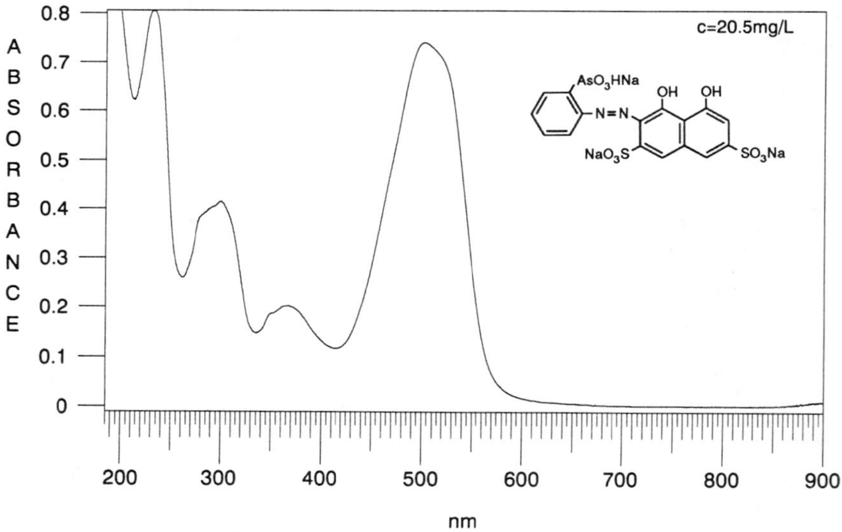

The metallochromic indicator Arsenazo I was first prepared by Kuznetsov and Vasunina.[1] This compound, containing both a salt-forming and a chelate-forming functional group *ortho* to an azo linkage, provides a favored environment for metal complexing. With its discovery, interest was stimulated in searching for the optimum application for Arsenazo I. Subsequently, Arsenazo III with two such chelating groups was described by Savvin[2] and isomers of it followed.

Arsenazo I is readily soluble in water and very slightly soluble in ethanol.

References:

1) Kuznetsov, V.I.; Vasyunina, N.A. *J. Gen. Chem. (U.S.S.R.)* **1940,** *10,* 1203-1209

2) Savvin, S.B. *Dokl. Akad. Nauk. S.S.S.R.* **1959,** *127,* 1231-1234.

General reference:

Safety **2**(1), 308A

ARSENAZO III

2,2'-(1,8-Dihydroxy-3,6-disulfonaphthylene-2,7-bisazo)bisbenzenearsonic acid

Aldrich Cat. No.:	A9,277-5
Sigma Prod. No.:	A4657

CAS No.:	[*1668-00-4*]
Mol. Form.:	$C_{22}H_{18}As_2N_4O_{14}S_2$
F.W.:	776.37
m.p.:	>320°C
Appearance:	Dark-green powder
λ_{max}:	560nm in methanol
Solubility:	H_2O 20mg/ml
	EGME 8mg/ml
	EtOH 3mg/ml

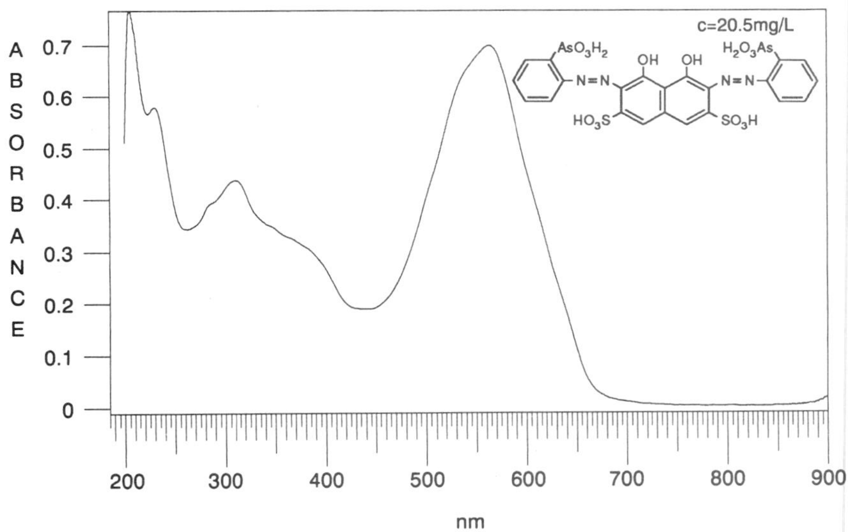

Arsenazo III is an acid disazo dye that is made by the (3,6) coupling of two moles of diazotized *o*-arsanilic acid to one mole of chromotropic acid (4,5-dihydroxynaphthalene-2,7-disulfonic acid).

The pure free-acid form (free of all sodium) is soluble in alcohol and slightly soluble in water, while the disodium salt is readily soluble in water and slightly soluble in alcohol. Both forms have been used widely to determine thorium, uranium, zirconium, cadmium, zinc and calcium.[1-3]

Bauer [4] describes Arsenazo III as a metallochromic indicator that is a useful tool for assaying micromolar calcium concentrations in biological systems for the reasons that follow: "(a) A high and specific calcium affinity in the neutral pH range; (b) a maximal spectral sensitivity for calcium around 652nm where most biological pigments do not absorb; (c) a large difference between the extinction coefficients of the calcium-bound and the calcium-free forms of Arsenazo III (differential extinction coefficient of about 2.8×10^4 L $mole^{-1}$ cm^{-1} at λ_{max} 652nm and pH 7.5); (d) a high chemical stability."

While Arsenazo III has no apparent industrial application, it is a useful *in vitro* diagnostic reagent that is used in both manual and automated systems to determine calcium levels in body fluids.

References:

1) Mikhailova, V.; Ilkova, P. *Anal. Chim. Acta* **1971,** *53*, 194.

2) Mikhailova, V.; Yurukova, L. *ibid.* **1974,** *68*, 73.

3) Kiriyama, T.; Kuroda, R. *ibid.* **1974,** *71*, 375.

4) Bauer, P.J. *Anal. Biochem.* **1981,** *110,* 61-72.

ARSENAZO III, sodium salt hydrate

Aldrich Cat. No.: 28,516-1
Sigma Prod. No.: A8891
Purity: 98%

Mol. Form.: $C_{22}H_{16}As_2N_4Na_2O_{14}S_2 \cdot xH_2O$
Appearance: Brownish-black crystalline powder
λ_{max}: 533nm in water

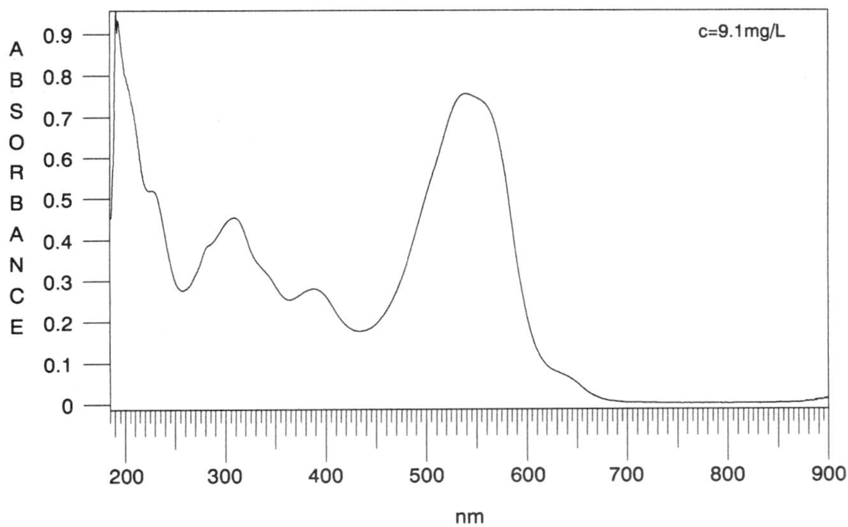

See description on pages 99 and 100 under Arsenazo III.

General references:

Sigma FT-IR Library	**1**(2), 100B
Safety	**2**(1), 308C

ASTRAZON ORANGE G

Basic Orange 21

Aldrich Cat. No.:	27,421-6
CAS No.:	[*3056-93-7*]
C.I. No.:	48035
Mol. Form.:	$C_{22}H_{23}ClN_2$
F.W.:	350.89
Appearance:	Brick-red powder
λ_{max}**:**	490nm in methanol
Solubility:	H_2O 30mg/ml
	EGME 50mg/ml
	EtOH 20mg/ml

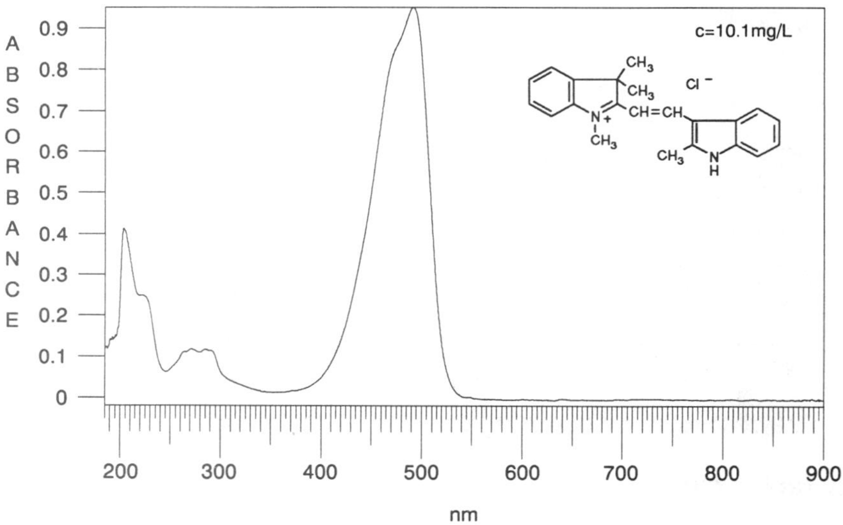

This orange, cationic polymethine dye is prepared by condensing 1,3,3-trimethyl-Δ2,α-indoleacetaldehyde with 2-methylindole in hydrochloric acid at 90°C. The melt is quenched in water and converted to the chloride.

Soluble in water, ethanol, methyl Cellosolve and pyridine and slightly soluble in acetone and ethylene glycol, it is used to dye acrylics and as a photosensitizing dye in photography.

Kass[1] describes it as a quaternary, metachromatic dye that, in the absence of any other dye, can be used for the differential determination of leucocytes. He claims a simple yet accurate means of distinguishing between B-cells and T-cells. The latter reportedly display red granules in the cytoplasm when stained with Astrazon Orange G.

Reference:

1) Kass, L. U.S. Patent 4 400 370, 1983.

General references:

Safety	**2**(1), 315D
Colour Index	3rd ed., 1971

AURAMINE O

C.I. Basic Yellow 2 • 4,4'-(Imidocarbonyl)bis(*N,N*-dimethylaniline) monohydrochloride

Aldrich Cat. No.:	86,103-0	85,653-3
Sigma Prod. No.:	A9655	A1396
Grade:	Certified	Biological stain
Dye content:	~85%	~80%

CAS No.:	[*2465-27-2*]	
C.I. No.:	41000	
Mol. Form.:	$C_{17}H_{21}N_3 \cdot HCl$	
F.W.:	303.84	
m.p.:	>250°C (dec.)	
Appearance:	Yellow powder	
λ_{max}:	432(370) in water	434(370)nm in water
Solubility:	H_2O 10mg/ml	
	EGME 60mg/ml	
	EtOH 20mg/ml	

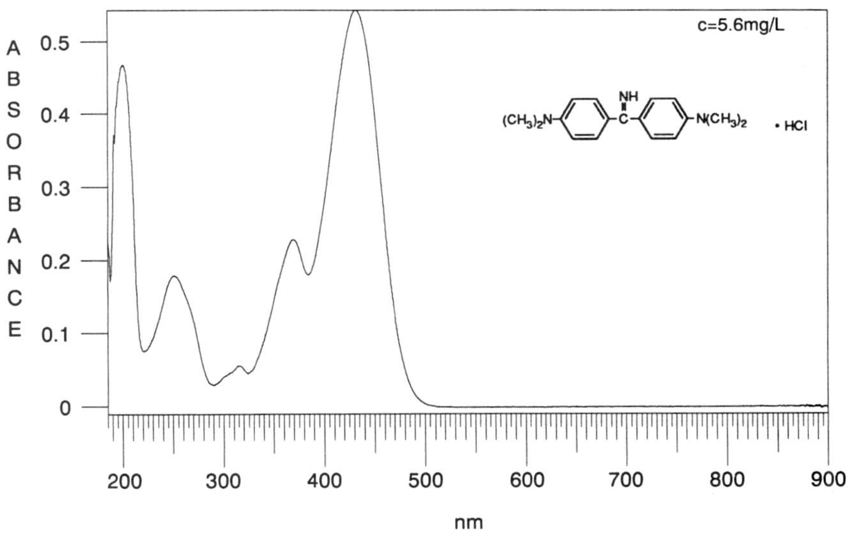

This yellow, cationic diphenylmethane dye is made by heating Michler's ketone [4,4'-bis(dimethylamino)benzophenone] with ammonium chloride. It readily reverts to the precursor, with the evolution of ammonia, when it is subjected to either the oxidative effect of a hot acid or hydrolysis in hot water. Lubs[1] claims that the hydrolysis occurs at 60°C; fortunately, dissolution can be accomplished easily at lower temperatures.

Soluble in alcohol and slightly soluble in water, commercial Auramine O usually contains dextrin, or other diluents, that are added to standardize the dye. The BSC will approve samples with a dye content of 80% or greater, if the sample passes all other qualifying tests.

Auramine O has an affinity for a wide spectrum of fibers. In fact, it has been used to dye materials as diverse as paper, leather, cotton, wool, silk, jute and coco fiber. Substrates exposed only to Auramine are dyed a golden yellow. However, in combination with other dyes, it creates or enhances blues and reds. Other industrial applications include the manufacture of lakes and photographic filters.

Auramine O is an excellent fluorochrome, and this quality has been utilized to advantage in *in vitro* biological applications. Kasten[2] found it to be one of the best of the group of fluorochrome Schiff reagents that he investigated. *Staining Procedures*[3] describes the following fluorochromatic-staining methods: acid-fast bacteria in smears, bacterial spore stain and acid-fast bacteria in tissue. Conrad *et al.*[4] and VanderMeulen and Govindjee[5] observed that Auramine O forms highly fluorescent complexes with horse-liver alcohol dehydrogenase.

Kisser,[6] in one of the few nonfluorescent staining applications described in the literature, used Auramine O in combination with Cresyl Violet as a tissue stain.

References:

1) Lubs, H.A. *The Chemistry of Synthetic Dyes & Pigments;* Reinhold: New York, 1955.

2) Kasten, F.H. *Histochemie* **1959,** *1*, 466-509.

3) Clark, G. *Staining Procedures,* 4th ed.; Williams and Wilkins: Baltimore, 1981; 382, 426.

4) Conrad, R.H.; Heitz, J.R.; Brand, L. *Biochemistry* **1970,** *9*, 1540.

5) VanderMeulen, D.L.; Govindjee *Biochim. Biophys. Acta,* **1974,** *368*, 61.

6) Kisser, J. *Mikr. Naturfr.* **1931,** *9*, 95.

General references:

Aldrich FT-IR Library	**1**(2), 1045D	——
Sigma FT-IR Library	**1**(2), 100C	——
IR Library	**3**, 1490C	**3**, 1104A
NMR Library	——	**2**(2), 386A
Safety	**2**(1), 320C	
Colour Index	3rd ed., 1971	

AURINTRICARBOXYLIC ACID

Aldrich Cat. No.:	12,326-9
Sigma Prod. No.:	A1895

CAS No.:	[*4431-00-9*]
Mol. Form.:	$C_{22}H_{14}O_9$
F.W.:	422.35
m.p.:	300°C
Appearance:	Dark-red powder
λ_{max}**:**	545nm in 0.1*N* sodium hydroxide
Solubility:	H_2O 7mg/ml
	EGME 100mg/ml
	EtOH 60mg/ml

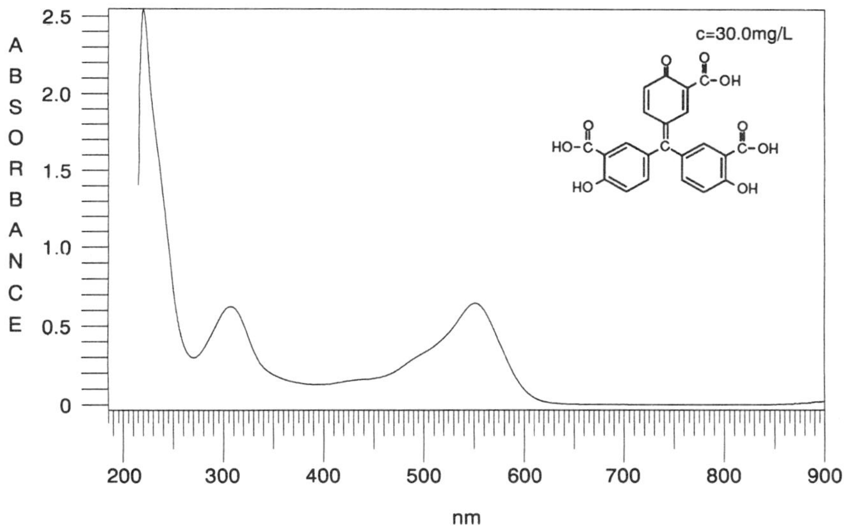

See description on pages 90 and 91 under Aluminon.

General references:

Aldrich FT-IR Library	**1**(2), 237A
Sigma FT-IR Library	**1**(2), 100D

AURINTRICARBOXYLIC ACID, TRISODIUM SALT

C.I. Mordant Violet 39

Aldrich Cat. No.: 20,150-2

CAS No.: [*13186-45-3*]
C.I. No.: 43810
Mol. Form.: $C_{22}H_{11}Na_3O_9$
F.W.: 488.30
Appearance: Black powder
λ_{max}: 545nm in 0.1*N* sodium hydroxide
Solubility: H_2O 30mg/ml
EGME 7mg/ml
EtOH 0.2mg/ml

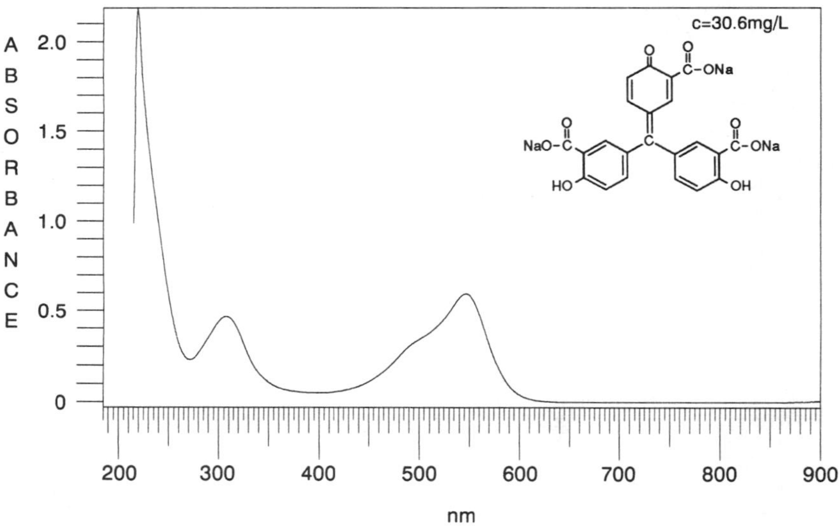

See description on pages 90 and 91 under Aluminon.

General references:

Aldrich FT-IR Library	**1**(2), 1031B
IR Library	**3**, 1484E

AZOCARMINE B

Acid Red 103

Aldrich Cat. No.: 28,456-4
Sigma Prod. No.: A1529

CAS No.: [*25360-72-9*]
C.I. No.: 50090
Mol. Form.: $C_{28}H_{17}N_3Na_2O_9S_3$
F.W.: 681.63
Appearance: Brown powder
λ_{max}: 516nm in water
Solubility: H_2O 20mg/ml
EGME 50mg/ml
EtOH 0.8mg/ml

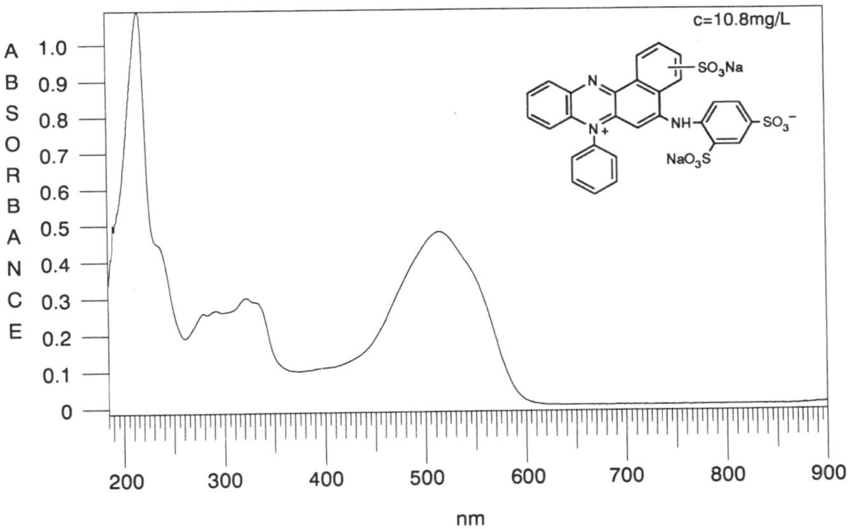

This red, anionic rosinduline (diazine) dye is prepared by heating 4-phenylazo-1-naphthylamine with aniline and aniline hydrochloride to form phenyl-rosinduline (Rhodindine). This product is trisulfonated by heating with 24° oleum at 70-80°C, isolated and converted to the disodium salt.

Soluble in water, it is used to dye and print both wool and silk with moderate lightfastness (3-4).

Azocarmine B is used in Heidenhain's "Azan" stain, which is a variation of Mallory's Aniline Blue connective-tissue stain. *Staining Procedures*[1] provides methods for Heidenhain's "Azan" modification and a modification of

Gomori's stain for crustaceans especially suited for staining marine and freshwater shrimp tissue.

Reference:

1) Clark, G. *Staining Procedures,* 4th ed.; Williams and Wilkins: Baltimore, 1981; pp 117, 269.

General references:

Sigma FT-IR Library	1(2), 101C
Colour Index	3rd ed., 1971

AZURE A

Aldrich Cat. No.:	86,104-9
Sigma Prod. No.:	A6270
Grade:	Certified
CAS No.:	[*531-53-3*]
C.I. No.:	52005
Mol. Form.:	$C_{14}H_{14}ClN_3S$
F.W.:	291.80
m.p.:	290°C (dec.)
Appearance:	Dark-green powder
λ_{max}:	633nm in water
Solubility:	H_2O 40mg/ml
	EGME 20mg/ml
	EtOH 10mg/ml

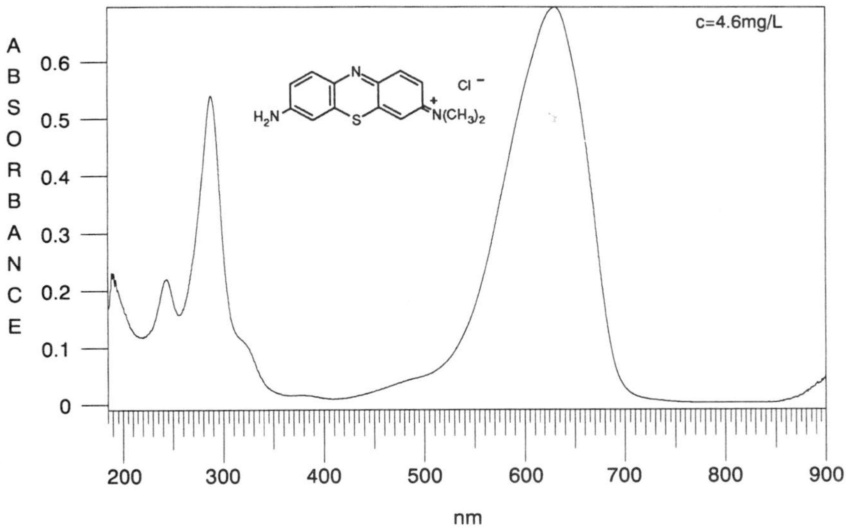

Azure A (asymmetrical dimethylthionin), a cationic thiazine dye, has a symmetrical isomeric counterpart, symmetrical dimethylthionin. The latter has not been named as a member of the Azure series. Both isomers are soluble in water, methanol and glycerol and slightly soluble in ethanol; they display comparable staining qualities. Each is present in Polychromed Methylene Blue which is made by the oxidative demethylation of Methylene Blue, although Azure A is usually the larger fraction by twofold.

MacNeal[1] considered Azure A the most important constituent present in Polychromed Methylene Blue and included it in his three-component tetrachrome-stain formulation for differentially staining blood cells and blood parasites. Wittekind[2] found Azure A displayed the Romanowsky-Giemsa effect, which he considered necessary to formulate a good blood stain. However, unlike Azure B, it required the presence of small amounts of Methylene Blue to avoid an inferior cytoplasmic blue and to provide optimum staining quality.

Apparently neither Azure A nor symmetrical dimethylthionin has been used in industrial applications.

References:

1) MacNeal, W.J. *J. Infect. Dis.* **1925,** *36*, 538-546.

2) Wittekind, D. *Clin. Lab. Haematol.* **1979,** *1*, 247-262.

General references:

IR Library **3**, 1486G
Merck Index **11**, 940

AZURE A eosinate

Aldrich Cat. No.: 20,021-2
Sigma Prod. No.: A5770

CAS No.: [*62298-43-5*]
Appearance: Green crystalline powder
λ_{max}: 628(522)nm in methanol
Solubility: H_2O 2mg/ml
EGME 6mg/ml
EtOH 1mg/ml

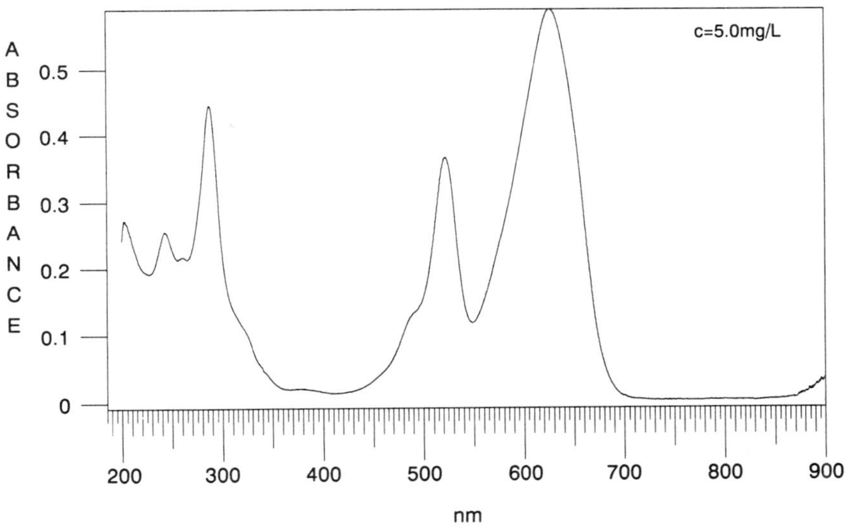

Azure A eosinate is a "neutral" dye made by reacting the cationic dye Azure A with the anionic dye Eosin Y (homologous to Azure B eosinate). This results in a coulombic attachment between the two dyes. As with similar eosinates, Azure A eosinate is practically insoluble in water. This dye is also only slightly soluble in methanol, ethanol and glycerol.

Apparently, Azure A eosinate is used almost exclusively as a component of Romanowsky-type blood stains. Lillie[1] included it as a minor component in his modified-Giemsa-stain recipe.

Reference:

1) Lillie, R.D. *Am. J. Public Health* **1948**, *33*, 948-951.

General references:

Aldrich FT-IR Library **1**(2), 1044B
Sigma FT-IR Library **1**(2), 101A
IR Library **3**, 1489E

AZURE B

Azure 1

Aldrich Cat. No.:	86,105-7	22,793-5
Sigma Prod. No.:	——	A1029
Grade:	Certified	Prepared by direct synthesis

CAS No.:	*[531-55-5]*		
C.I. No.:	52010		
Mol. Form.:	$C_{15}H_{16}ClN_3S$		
F.W.:	305.83		
m.p.:	205°C (dec.)		
Appearance:	Dark-green powder		
λ_{max}:	648nm in ethanol		
Solubility:	H_2O	50mg/ml	50mg/ml
	EGME	20mg/ml	20mg/ml
	EtOH	10mg/ml	8mg/ml

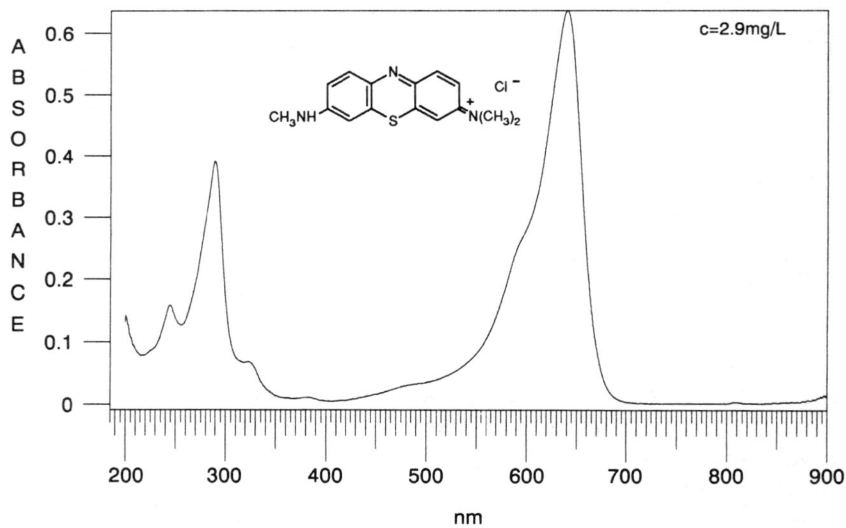

Azure B is a cationic, metachromatic, thiazine dye that is usually available as either the chloride or the bromide salt.

Until recently, Azure B was made almost exclusively by the oxidative demethylation of Methylene Blue, utilizing acid dichromate as the oxidizing agent. Unfortunately, material made in this manner typically contains no more than 35% Azure B (determined by HPLC). Methylene Blue comprises the bulk of the remaining dye with lesser amounts of the other azures, Thionin and Methylene Violet Bernthsen are usually present. Inorganic salts, incidental to the manufacturing process, may comprise as much as 15% of some lots.

Direct-synthesis Azure B chloride with an Azure B content of approximately 90% has been available commercially since 1980 (Aldrich Cat. No. 22,793-5). The direct-synthesis tetrafluoroborate salt is also available and this form will be elaborated on later.

Giemsa[1] and Lillie[2] recognized the importance of Azure B as a component of Romanowsky-type stains for blood and blood protozoa. However, their observations were not substantially reinforced until the mid-1970s. Lohr *et al.*,[3] utilizing column chromatography, separated pure Azure B along with the other azures from Polychrome Methylene Blue, while Marshall and Lewis purified Azure B and Methylene Blue by solvent extraction and recrystallization techniques. Using pure dyes, Marshall *et al.*[4] found that a well-balanced Romanowsky-type stain could be prepared by combining pure Azure B, Methylene Blue and Eosin Y. Wittekind *et al.*[5] concluded that only Azure B and Eosin Y were necessary to provide a balanced Romanowsky-type stain.

With these developments, it became apparent that a standardized Romanowsky-type-stain solution could be realized.[4,6]

The International Committee for Standardization in Haematology (ICSH)[7] provided definitions, component specifications and a recipe for preparing a standard stain with pure Azure B and Eosin Y. The Committee specified Azure B with a minimum dye content of 95% established by the molar extinction coefficient [9.4×10^4 L $mole^{-1}$ cm^{-1} at λ_{max} 641 nm (in ethanol)] and chromatography. (Excellent separation of Azure B from related dyes has been achieved with HPLC in our laboratory.) The Committee stated that several salt forms are available commercially which were sufficiently pure to meet ICSH requirements. The salt forms cited were the thiocyanate, the perchlorate and the tetrafluoroborate. These salts, unlike the halide salts, can be isolated free of water of crystallization without difficulty. They are very slightly soluble in water, a property that facilitates their purification. Unfortunately, they are also less soluble in the lower alcohols. This quality precludes their use in Romanowsky-stain solutions when either neat methanol or neat ethanol is used. In order to increase the solubility of these salts, a second, more polar solvent must be incorporated into the recipe. The ICSH recommends a solvent system composed of 40% methyl sulfoxide (DMSO) and 60% methanol (v/v).

In our laboratory, we have observed that Azure B made by the classical oxidative method, on aging, tends to oxidize further when incorporated in a Romanowsky-stain solution. This phenomenon has not been observed in comparable solutions prepared with direct-synthesis Azure B.

References:

1) Giemsa, G. *Zentralbl. Bakterial* **1902,** *31,* 429-430.

2) Lillie, R.D. *Public Health Rep.* **1944,** *178,* Suppl., 17-21.

3) Lohr, W.; Sohmer, J.; Wittekind, D. *Stain Technol.* **1974,** *49*, 359.

4) Marshall, P.N.; Bentley, S.A.; Lewis, S.M. *J. Clin. Pathol.* **1975,** *28*, 920-923.

5) Wittekind, D.; Kretschmer, V.; Lohr, W. *Blut* **1976,** *2,* 71-78.

6) Lapen, D. *Cytometry* **1982,** *2*, 309-315.

7) International Committee for Standardization in Haematology *Br. J. Haematol.* **1984,** *57*, 707-710.

General references:

Aldrich FT-IR Library	**1**(1), 1038A
Sigma FT-IR Library	**1**(2), 102A
Safety	**2**(1), 329C
Merck Index	**11**, 941

AZURE B eosinate

Aldrich Cat. No.: 20,023-9
Sigma Prod. No.: A5645

CAS No.: [*62298-42-4*]
Appearance: Dark-green crystalline powder
λ_{max}: 636(523)nm in methanol
Solubility: H_2O 1mg/ml
EGME 2mg/ml
EtOH 1mg/ml

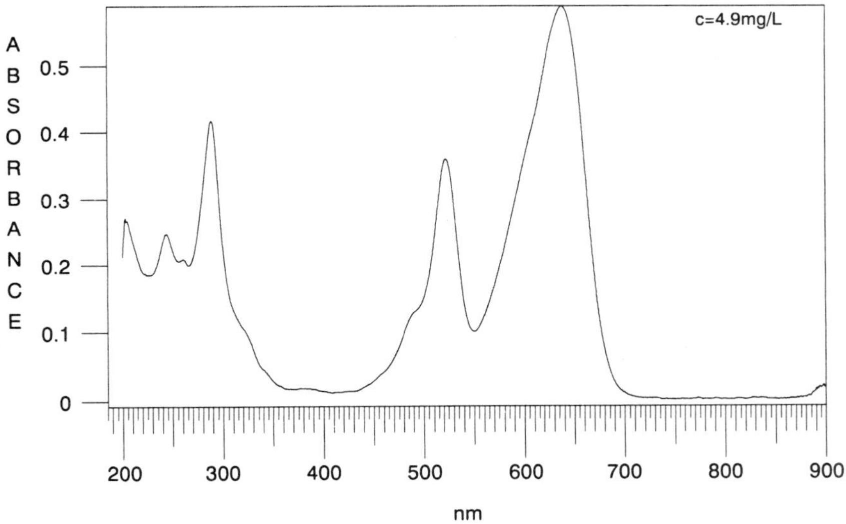

Azure B eosinate is a neutral dye made by the controlled addition of an aqueous Eosin Y solution to an aqueous solution of Azure B. It is practically insoluble in water and ethanol and slightly soluble in methanol and glycerol.

Azure B eosinate is used as a component in Romanowsky-type blood stains. In fact, a satisfactory blood-stain solution can be prepared using Azure B eosinate as the only dye component.

With the inception of World War II and the loss of Giemsa stain of German origin, the U.S. Armed Forces prevailed on the U.S. Public Health Service to develop a domestic product comparable in quality to the revered German product. Lillie,[1] a physician turned researcher at the Center, was assigned the task. He responded with a recipe that utilized Azure B eosinate as a major component. The Lillie-modified Giemsa was used widely during the hostilities by U.S. and allied military physicians as a blood stain for identifying malaria parasites. Today, the military continues to use the Lillie variation and it has found worldwide civilian acceptance as well.

Methanolic blood-stain solutions utilizing Azure B eosinate, prepared from Azure B made by direct synthesis, initially will stain better than solutions formulated from Azure B eosinate using Azure B manufactured by the classical demethylation of Methylene Blue. However, a stock solution of the former normally is only stable for a few hours, while a comparable solution prepared with the latter may last several months.

Apparently, Azure B eosinate is not being used in any industrial application at the present time.

Reference:

1) Lillie, R.D. *Am. J. Public Health* **1943,** *33*, 948-951.

General references:

Aldrich FT-IR Library **1**(2), 1044A
Sigma FT-IR Library **1**(2), 102B
IR Library **3,** 1489D

AZURE B tetrafluoroborate

Aldrich Cat. No.:	30,261-9
Sigma Prod. No.:	A4792
CAS No.:	[*79288-94-1*]
Mol. Form.:	$C_{15}H_{16}BF_4N_3S$
F.W.:	357.18
m.p.:	244°C (dec.)
Appearance:	Brownish-green powder
λ_{max}:	639nm in methanol
Solubility:	H_2O 3mg/ml
	EGME 2mg/ml
	EtOH 0.9mg/ml

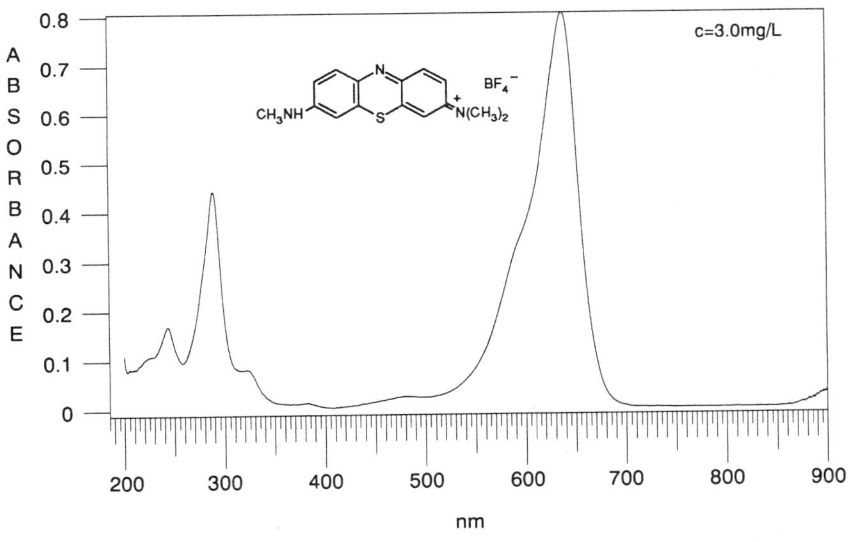

See description on pages 112-114 under Azure B.

General reference:

Safety **2**(1), 329D

AZURE C

Aldrich Cat. No.:	24,218-7
Grade:	Biological stain
CAS No.:	*[531-57-7]*
Mol. Form.:	$C_{13}H_{12}ClN_3S$
F.W.:	277.78
Appearance:	Black powder
λ_{max}**:**	616nm in water
Solubility:	H_2O 20mg/ml
	EGME 30mg/ml
	EtOH 7mg/ml

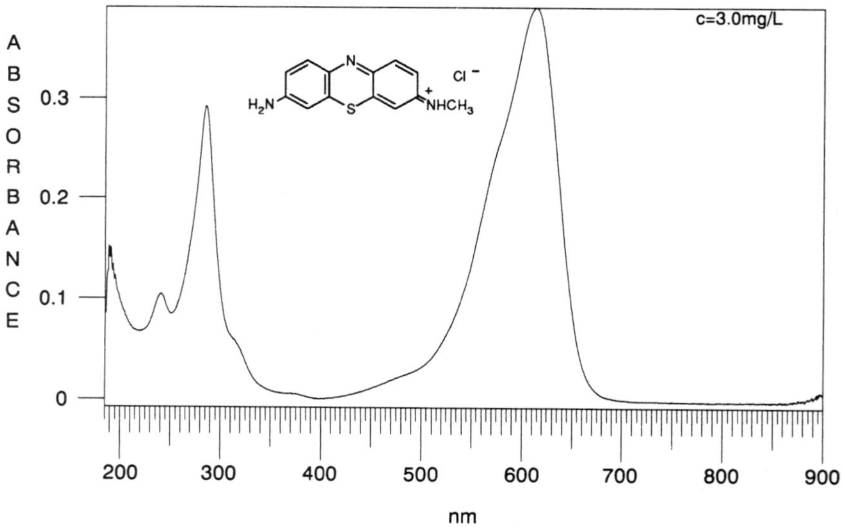

Azure C is a water-soluble, cationic thiazine dye which is slightly soluble in alcohol. It finds limited use as a biological stain. It is commercially available in a crude form made by the oxidative demethylation of Methylene Blue by acid dichromate. This method results in a product that is difficult to distinguish from Azure A, a homolog which contains an additional methyl group. (Separation of the two products from the same reaction medium is not ideal, since it is controlled by visible absorption spectra with λ_{max} ranges within 8nm of each other.) A direct synthesis probably could be developed if the demand and usefulness of Azure C merited the necessary funding and development time.

In a study using pure dyes, Wittekind and Lohr[1] found that Azure C does not display the Romanowsky-Giemsa effect (RGE) ideally embodied in Azure B and, to a much lesser degree, in Azure A. This study utilized an Azure C fraction separated from Polychromed Methylene Blue by column chromatography.

There is no mention in the literature of any industrial applications.

Reference:

1) Wittekind, D.; Lohr, W. In *Quality Control in Haematology;* Lewis, S.M.; Coster, F.J., Eds.; Purification, standardization and quality control of Romanowsky dyes; Symposium of the International Committee for Standardization in Haematology; Academic: New York, 1975; Chapter 12.

AZURE II

Aldrich Cat. No.:	86,106-5
Sigma Prod. No.:	A2507
CAS No.:	*[37247-10-2]*
Appearance:	Dark-green powder
λ_{max}:	657nm in water
Solubility:	H_2O 40mg/ml
	EGME 30mg/ml
	EtOH 20mg/ml

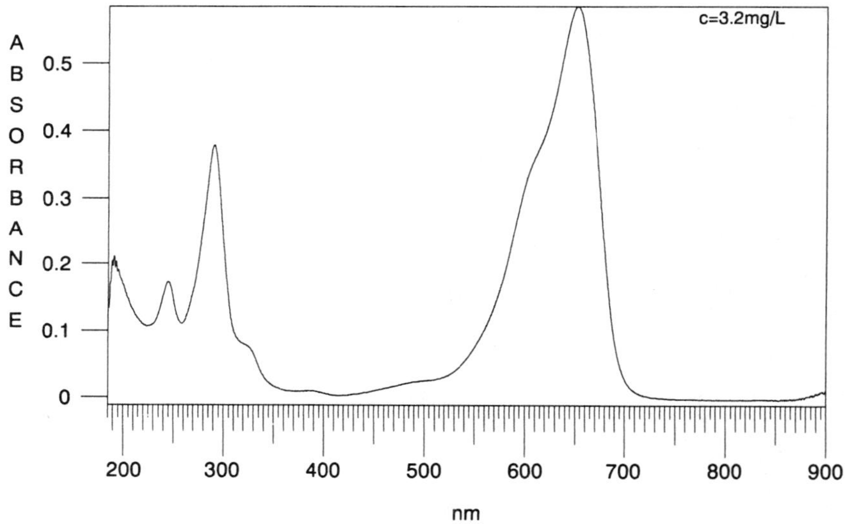

Azure II is a deliberate mixture of equal parts of Azure I (Azure B)[1] and Methylene Blue. Giemsa listed it as a component (21%) of his original blood stain recipe used for differentiating leucocytes and identifying blood protozoa. It is currently used in the preparation of Giemsa-Stain solutions and, to a limited degree, in other blood-stain formulations.

Azure II has no industrial application.

Reference:

1) *Conn's Biological Stains,* 9th ed.; Lillie, R.D., Ed.; Williams and Wilkins: Baltimore, 1977; pp 496-497.

General references:

Sigma FT-IR Library	**1**(2), 1020
Safety	**2**(1), 330A

AZURE II eosinate

Aldrich Cat. No.: 20,022-0
Sigma Prod. No.: A4396

CAS No.: [*53092-85-6*]
Appearance: Dark-green powder
λ_{max}**:** 647(524)nm in methanol
Solubility: H_2O 6mg/ml
EGME 10mg/ml
EtOH 3mg/ml

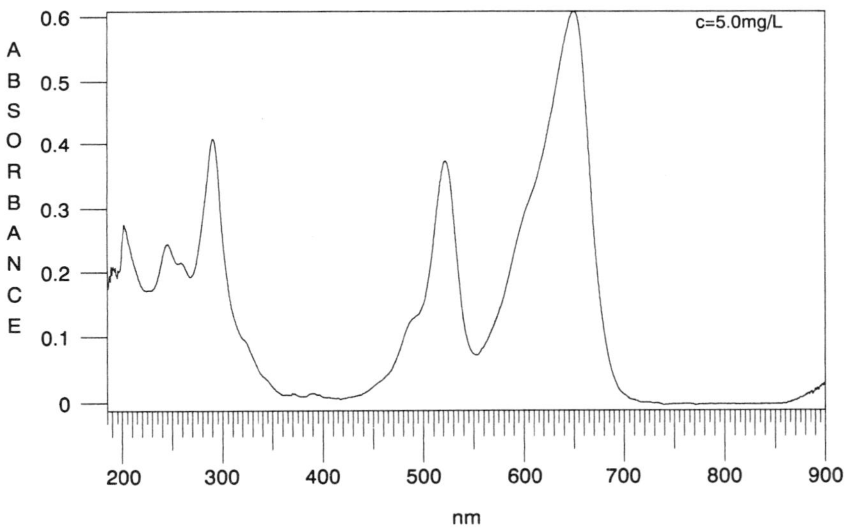

Azure II eosinate is a neutral salt formed by combining aqueous solutions of equimolar equivalents of the cationic dyes Azure I (Azure B) and Methylene Blue and the anionic dye Eosin Y. Unlike the individual free-dye components, the precipitate formed from this combination is slightly soluble in water and less soluble in methanol, ethanol and glycerol than each individual dye. However, the dried precipitate is soluble enough (up to 0.3% w/v at 25°C) in each of the three organic solvents mentioned to function satisfactorily as a Romanowsky-type stain. In an Azure II eosinate solution prepared using an aprotic organic solvent, the individual dyes used to form the solid Azure II eosinate revert to ionic species and as such are prone to make new coulombic attachments to the substrate. Thus, a compatible functional group (with a charge opposite to the charge on the dye function) present on an appropriately structured substrate exposed to the solution, provides a possible site of attachment for the dye. The ultimate choice is dependent upon the thermodynamic forces that prevail in the environment at the time of the exposure.

Azure II eosinate is used to prepare Romanowsky-type stains. In fact, it is a balanced Romanowsky-type stain in its own right, and the chief component (79%) of classical Giemsa stain. *Conn's*[1] describes the preparation of Giemsa stain solution as recommended by Giemsa.

Azure II eosinate has no known industrial use.

Reference:

1) *Conn's Biological Stains,* 9th ed.; Lillie, R.D., Ed.; Williams and Wilkins: Baltimore, 1977; p 497.

General references:

Aldrich FT-IR Library	**1**(2), 1044C
Sigma FT-IR Library	**1**(2), 102D
IR Library	**3**, 1489F

BASIC BLUE 3

Rhodulinreinblau 3G • Zaponechtblau FIAT 764

Aldrich Cat. No.: 21,191-5
Sigma Prod. No.: B3885

CAS No.: [*33203-82-6*]
C.I. No.: 51004
Mol. Form.: $C_{20}H_{26}ClN_3O$
F.W.: 359.90
Appearance: Dark-green powder
λ_{max}: 654nm in water
Solubility: H_2O 30mg/ml
EGME 8mg/ml
EtOH 5mg/ml

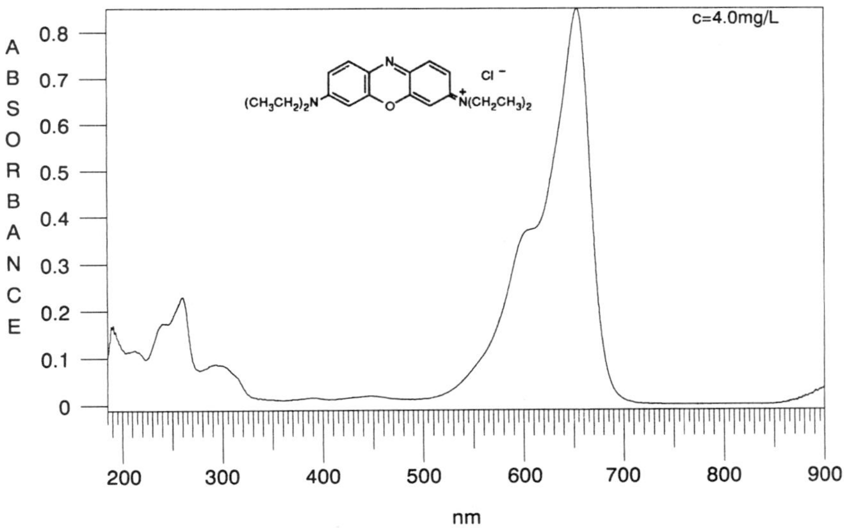

This blue-green, cationic oxazine dye is moderately soluble in water and slightly soluble in methanol and ethanol.

Industrial applications include dyeing acrylic fibers a greenish blue and cellulosic substrates the same color when tannin-mordanted, dyeing chromed leather and coloring organic solvents. Stability is good under pH variations (even under alkaline conditions) and lightfastness is moderately good (5-6).

There are no reported life-science applications. However, the structure prompts speculation that it may have possibilites as a nuclear stain and/or be susceptible to oxidative deethylation with the concomitant formation of polychromatic end products.

General references:

Aldrich FT-IR Library **1**(2), 1027B
Sigma FT-IR Library **1**(2), 103A
Colour Index 3rd ed., 1971

BASIC BLUE 47

Aldrich Cat. No.: 30,648-7
Sigma Prod. No.: B1146

CAS No.: [*12217-43-5*]
C.I. No.: 61111
Mol. Form.: $C_{23}H_{21}N_3O_2$
F.W.: 371.44
Appearance: Dark-blue powder
λ_{max}: 610nm in methanol

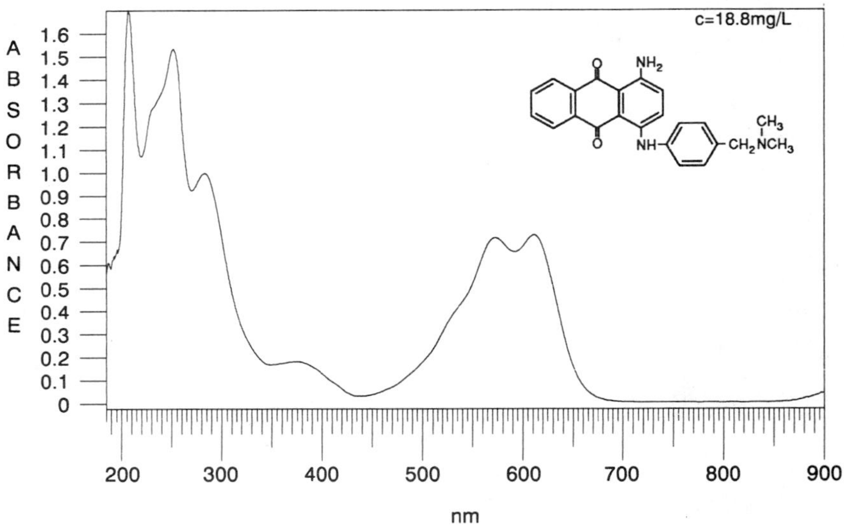

This reddish-blue, cationic anthraquinone dye is slightly soluble in ethanol and water.

Commercially, it is used to dye acrylic with very good fastness to light (7-8) and to print acetate (5-6) and silk (4-5).

There is no evidence that it has been used as a biological stain.

General references:

Safety	**2**(1), 338B
Colour Index	3rd ed., 1971

BASIC BLUE 66

Aldrich Cat. No.: 30,644-4
Sigma Prod. No.: B1148

CAS No.: *[94233-04-2]*
C.I. No.: 11075
Mol. Form.: $C_{22}H_{28}ClN_5O_2S \cdot {}^1/_2\ ZnCl_2$
F.W.: 530.16
m.p.: 196°C (dec.)
Appearance: Dark-blue powder
λ_{max}: 615nm in water
Solubility: H_2O 30mg/ml
EGME 10mg/ml
EtOH 3mg/ml

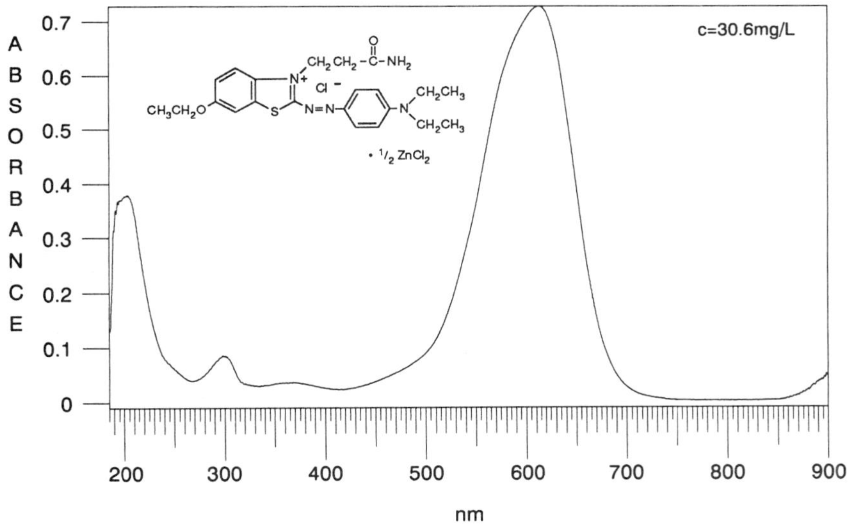

This bright-blue, cationic monoazo dye is soluble in water and slightly soluble in ethanol. Industrially, it is used to dye acrylics with good lightfastness.

There is no evidence in the literature that it has been used as a biological stain.

General reference:

Colour Index 3rd ed., 1971

BASIC FUCHSIN

Basic Red 9 • Pararosaniline chloride

Aldrich Cat. No.:	85,734-3	21,559-7
Sigma Prod. No.:	P1528	——
Grade:	Certified	——
Dye Content:	95%	90%

CAS No.:	*[569-61-9]*		
C.I. No.:	42500		
Mol. Form.:	$C_{19}H_{18}ClN_3$		
F.W.:	323.83		
m.p.:	268-270°C (dec.)		
Appearance:	Dark-green crystalline powder		
λ_{max}:	544nm in 50% ethanol		
Solubility:	H_2O	3mg/ml	2mg/ml
	EGME	70mg/ml	50mg/ml
	EtOH	25mg/ml	2mg/ml

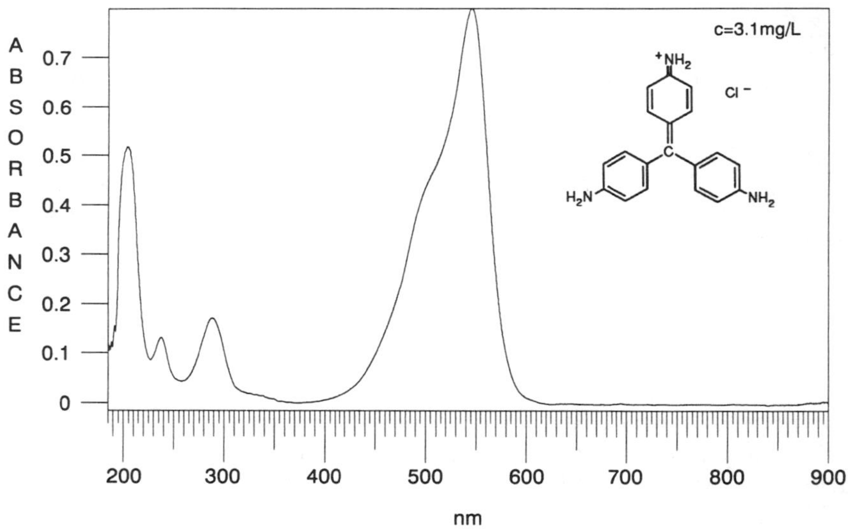

Basic Fuchsin, a cationic triphenylmethane dye, is commonly, though not in all cases, a homologous mixture of dyes and in a given lot any homolog may be dominant. However, the BSC has determined that, for a lot to perform satisfactorily in all of the usual applications embraced by their protocol, Pararosaniline chloride (C.I. 42500) must comprise not less than 50% of the total dye present. Other components that may be found are Rosaniline (C.I. 42510), Magenta II and Magenta III (C.I. 42520). Both Pararosaniline and Magenta III can be produced directly as pure products. However, to obtain pure Rosaniline and Magenta II, free of the other homologs, one has to resort to chromatographic separation.

The visible absorption spectrum is pH dependent. When the pH of a Basic Fuchsin solution approaches either extreme of the pH scale, the solution becomes practically colorless with a predictable concomitant drop in absorption. Heinemann,[1] working with acidified Basic Fuchsin at high hydrogen-ion concentration, found that there was binding of an acid group (chloride ion or sulfite ion) to the central methine carbon and the addition of hydrogen to each amino group. This was demonstrated by ultraviolet- and infrared-absorption spectra.

Basic Fuchsin is an excellent nuclear stain. It is also a valuable stain for mucin, elastic tissue, fuchsinophil granules and in the Ziehl-Nielsen method for demonstrating acid-fast bacteria such as the tubercule bacillus. The cell walls of certain bacteria and fungi are stained a deep red with this dye.

Aoki[2] differentially stained cell components of spermatozoa from rat and hamster testes and epididymides with a Basic Fuchsin-Toluidine Blue stain. In the germinal epithelium of hamster testes, this dye combination gives unusually clear cytoplasmic detail with the sperm cells easily differentiated from the surrounding tissue.

A rapid-staining method for epoxy-embedded tissue using a Basic Fuchsin-Methylene Blue stain has been reported by Huber *et al.*[3] Vivid contrasts are perceived between the cellular- and connective-tissue elements. In human skin, the collagen or the dermis stains a brilliant blue, readily distinguished from the pink hues of the cytoplasm in epidermal, vascular and connective tissues. Cell nuclei are stained purple. Other examples of the staining of human nerve, rabbit skin, kidney, blood vessel, cartilage and muscle are given in this reference.

A stain made from a complex of Basic Fuchsin and paraldehyde called "Aldehyde-Fuchsin", was introduced by Gomori[4] who first recommended Aldehyde-Fuchsin for staining pancreatic islets. Later, Scott[5] noted that the staining was more rapid and selective following oxidation by an acidified solution of potassium permanganate. Evidence has been presented to show that the degree of staining is related to the amount of insulin in the pancreas. A number of other substances are stainable by Aldehyde-Fuchsin, including elastic fibers, mast cells and mucin. Rosenblum *et al.*,[6] in a more recent study, found that, though Aldehyde-Fuchsin is specific for insulin, this stain is not sufficiently sensitive to detect normal serum levels of the hormone.

References:

1) Heinemann, K.L. *Stain Technol.* **1970,** *45*, 165.

2) Aoki, A.; Gutierrez, L.S. *ibid.* **1968,** *43*, 307.

3) Huber, J.D.; Parker, F.; Oland, G.E. *ibid.* **1969,** *44,* 87.

4) Gomori, G. *Am. J. Clin. Pathol.* **1950,** *20*, 265.

5) Scott, H.R. *Stain Technol.* **1952,** *27*, 267.

6) Rosenblum, A.L.; Rennert, O.M. *ibid.* **1970,** *45*, 25.

General references:

Aldrich FT-IR Library	**1**(2), 963B	**1**(2), 1030B
Sigma FT-IR Library	**1**(2), 191B	——
IR Library	**3**, 1448F	**3**, 1483G
Safety	**2**(1), 338C	

BASIC FUCHSIN

Basic Violet 14 • Rosaniline

Aldrich Cat. No.:	86,202-9
Sigma Prod. No.:	R5126
Grade:	Biological stain
CAS No.:	[*632-99-5*]
C.I. No.:	42510
Mol. Form.:	$C_{20}H_{20}ClN_3$
F.W.:	337.86
m.p.:	250°C (dec.)
Appearance:	Dark-green crystalline powder
λ_{max}:	549nm in 50% ethanol
Solubility:	H_2O 4mg/ml
	EGME 30mg/ml
	EtOH 30mg/ml

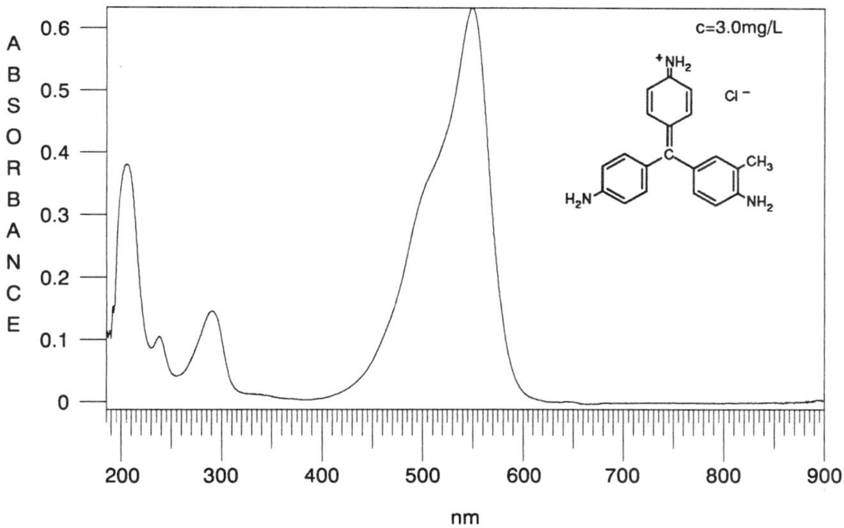

See description on pages 126-128 under Basic Fuchsin (Pararosaniline chloride).

General references:

IR Library	**3**, 1488H
Safety	**2**(1), 339A
Merck Index	**11**, 5528

BASIC FUCHSIN

Aldrich Cat. No.: 86,108-1
Grade: Certified
Special for flagella

CAS No.: *[58969-01-0]*
Appearance: Green crystalline powder
λ_{max}: 544nm in 50% ethanol
Solubility: H_2O 20mg/ml
EGME 60mg/ml
EtOH 50mg/ml

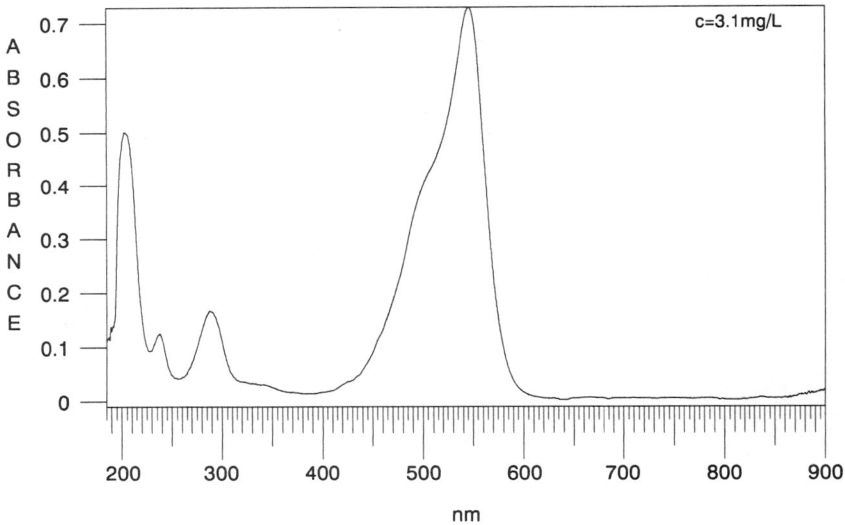

This is a mixture of three parts Pararosaniline acetate and one part Pararosaniline chloride (Basic Fuchsin). Leifson[1] (1951) used Basic Fuchsin as a bacterial flagella stain.

For a further description, see pages 126-128 under Basic Fuchsin.

Reference:

1) Leifson, E. *J. Bacteriol.* **1951**, *62*, 387-389.

General references:

IR Library **3**, 1488H
Safety **2**(1), 338D

BASIC RED 29
Basacryl Red GL

Aldrich Cat. No.: 30,652-5
Sigma Prod. No.: B1271

CAS No.: [*42373-04-6*]
C.I. No.: 11460
Mol. Form.: $C_{19}H_{17}ClN_4S$
F.W.: 368.89
Appearance: Brown powder
λ_{max}: 511(285)nm in water

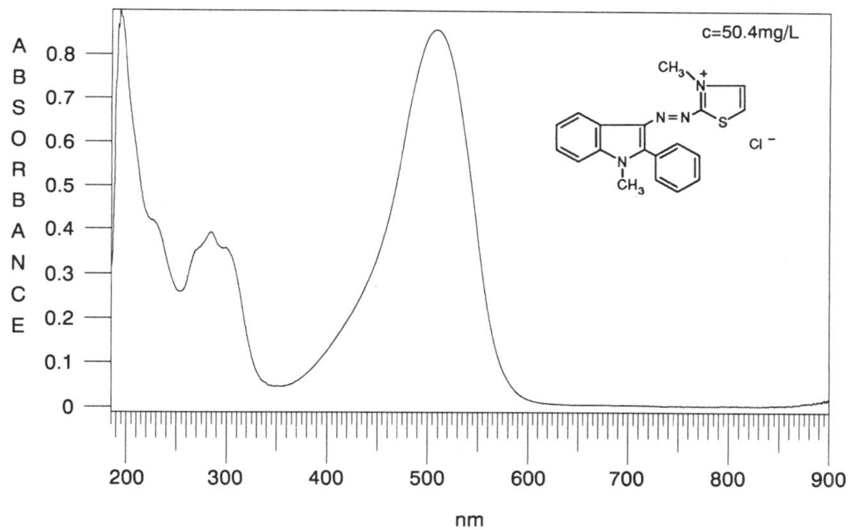

Basic Red 29 is a red, cationic monoazo dye. Commercially, it is used to dye acrylics. Fastness to light is excellent (7-8).

There is no evidence that it has been used as a biological stain.

General references:

Safety **2**(1), 339B
Colour Index 3rd ed., 1971

BASIC YELLOW 11

Astrazon Yellow 4G

Aldrich Cat. No.:	20,186-3
Sigma Prod. No.:	B7133

CAS No.:	[*4208-80-4*]
C.I. No.:	48055
Mol. Form.:	$C_{21}H_{25}ClN_2O_2$
F.W.:	372.90
m.p.:	180°C (dec.)
Appearance:	Burnt-orange powder
λ_{max}:	395nm in methanol

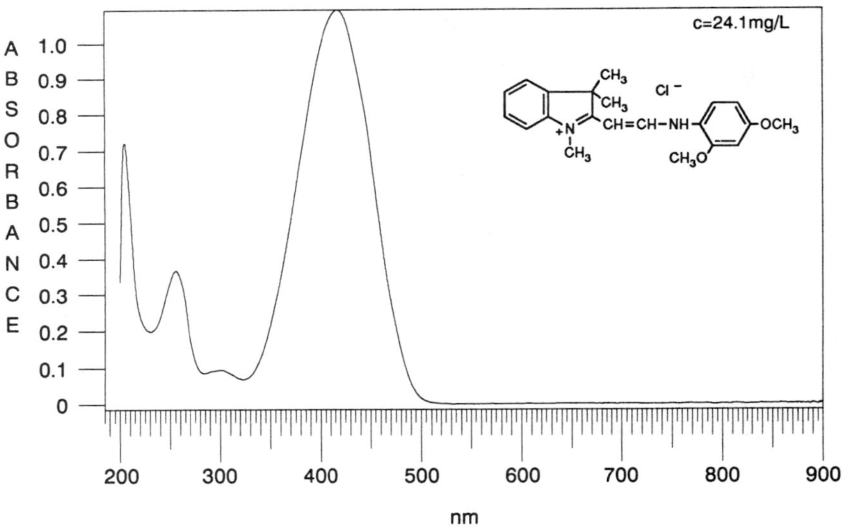

This cationic, indoline methine dye is soluble in water and the lower aliphatic alcohols. The commercial product has a dye content of approximately 60% and is used to dye acrylic fibers a bright-yellow color. Resistance to light-initiated fading is moderate (4-5).

There is no evidence that it has been used as a biological stain.

General references:

Aldrich FT-IR Library	**1**(2), 1041C
Sigma FT-IR Library	**1**(2), 103C
IR Library	**3**, 1488C
Colour Index	3rd ed., 1971

BATHOPHENANTHROLINEDISULFONIC ACID, disodium salt hydrate

Aldrich Cat. No.: 14,661-7
Sigma Prod. No.: B1375

CAS No.: [*52746-49-3*]
Mol. Form.: $C_{24}H_{14}N_2Na_2O_6S_2 \cdot xH_2O$
F.W.: 536.50
Appearance: Light-orange powder
λ_{max}: 276nm in water
Solubility: H_2O 6mg/ml
EGME 20mg/ml
EtOH 6mg/ml

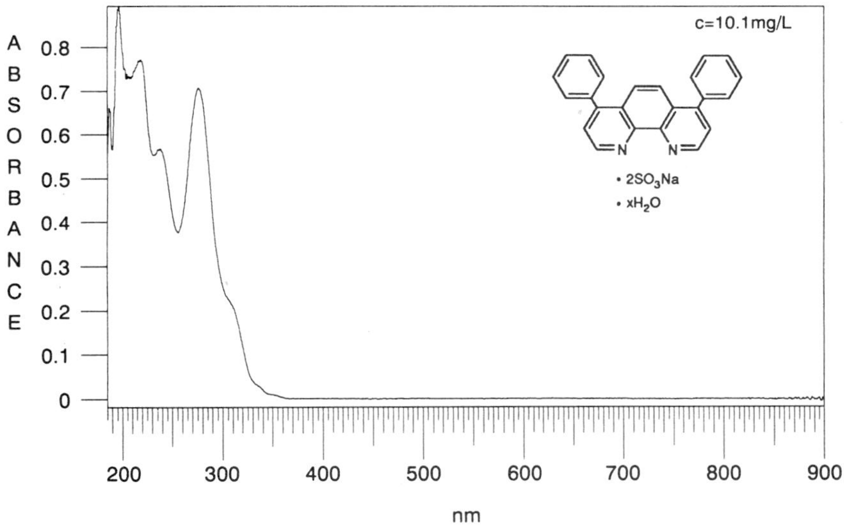

This slightly water-soluble 1,10-phenanthroline derivative is not a dye but a colorimetric reagent for the selective determination of iron(II). It reacts with iron(II) to form a red, water-soluble complex with a maximum absorption at 535nm and a molar absorptivity of 22,140. Beer's Law is obeyed in the range of 0.1-3.6 micrograms/ml of iron(II). This product is especially useful for the determination of iron(II) in serum, urine and other biological materials.[1-5]

References:

1) Trinder, P. *J. Clin. Pathol.* **1958,** *9*, 170.

2) Blair, D.; Diehl, H. *Talanta* **1961,** *7*, 163.

3) Zak, B. *Am. J. Clin. Pathol.* **1958,** *29*, 590.

4) Vours, I.M.; Voronen, E.R.; Hartiala, K. *J. Clin. Lab. Invest.* **1963,** *15*, 323.

5) Matsubara, T. *Acta Haematol. Jpn.* **1961,** *24*, 18-36.

General references:

Aldrich FT-IR Library	**1**(2), 892B
Sigma FT-IR Library	**1**(2), 300A
IR Library	**3,** 1412A
NMR Library	**2**(2), 766A

BENZOPURPURIN 4B

Direct Red 2

Aldrich Cat. No.: 22,882-6
Sigma Prod. No.: B0761

CAS No.: [*992-59-6*]
C.I. No.: 23500
Mol. Form.: $C_{34}H_{26}N_6Na_2O_6S_2$
F.W.: 724.73
Appearance: Dark-red powder
λ_{max}: 500nm in water
Solubility: H_2O 20mg/ml
EGME 70mg/ml
EtOH 3mg/ml

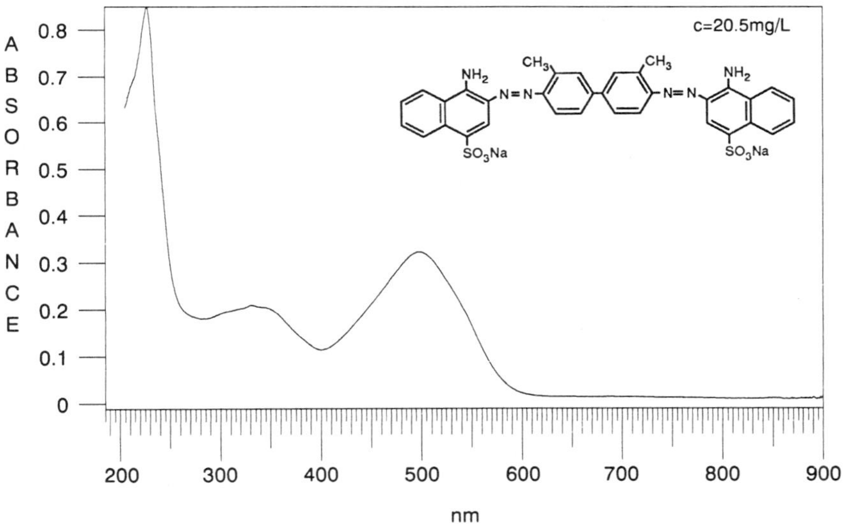

Benzopurpurin 4B is a bright-red, anionic disazo dye that is made by coupling one mole of tetraazotized *o*-tolidine with two moles of naphthionic acid (a pH of 8-10 is maintained during coupling). Soluble in water, acetone and methyl Cellosolve and slightly soluble in ethanol, it is used to dye chromed leather, cotton, wool, silk, nylon, linen, beater paper and jute and to print cellulose, wool and nylon. However, its lightfastness is poor (1-2).

Kolthoff[1] reported a visual-transition interval of pH 1.3 (blue-violet) to pH 4.0 (red). However, he stated that it has high salt and protein errors, making it unsuitable as an indicator.

Gurr[2] reported the use of Benzopurpurin 4B in a number of biological-stain applications by various investigators. Generally, it was considered inferior to certain other dyes available for the same applications.

References:

1) Kolthoff, I.M.; Rosenblum, C. *Acid-Base Indicators*; MacMillan: New York, 1937; p 149.

2) Gurr, E. *Encyclopedia of Microscopic Stains* ; Hill: London, 1960; p 75.

General references:

Aldrich FT-IR Library	**1**(2), 994A
Sigma FT-IR Library	**1**(2), 104B
Safety	**2**(1), 366A
Merck Index	**11**, 1111
Colour Index	3rd ed., 1971

BIEBRICH SCARLET, water soluble

Acid Red 66 • Ponceau B

Aldrich Cat. No.:	85,833-1
Sigma Prod. No.:	B6008
CAS No.:	[*4196-99-0*]
C.I. No.:	26905
Mol. Form.:	$C_{22}H_{14}N_4Na_2O_7S_2$
F.W.:	556.49
Appearance:	Deep-red powder
λ_{max}:	505nm in methanol
Solubility:	H_2O 30mg/ml
	EGME 7mg/ml
	EtOH 1mg/ml

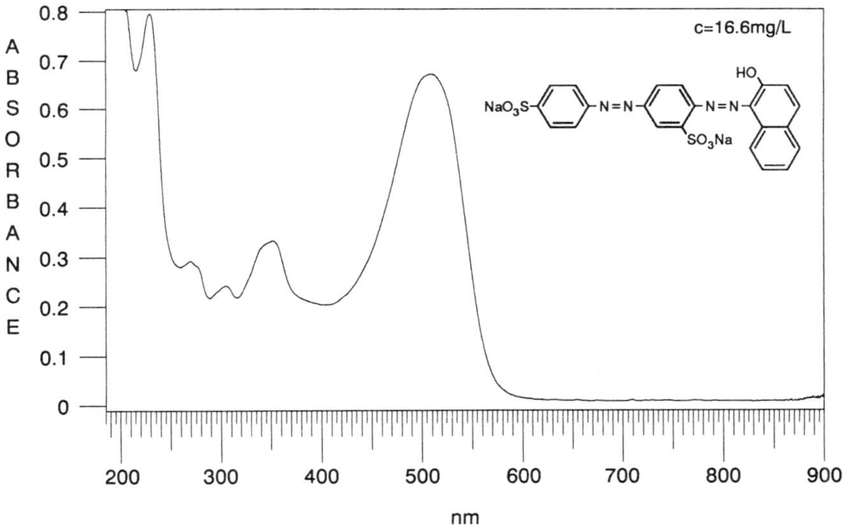

This red, anionic disazo dye is made by coupling diazotized 4-amino-1,1'-azobenzene-3,4'-disulfonic acid (Acid Yellow 9, C.I. 13015) to 2-naphthol.

Soluble in water, slightly soluble in methyl Cellosolve and very slightly soluble in ethanol, it is used to dye wool, silk, nylon and paper; its aluminum salt is used as a pigment.

Water-soluble Biebrich Scarlet must not be confused with Biebrich Scarlet Red (C.I. Solvent Red 24, C.I. 26105), which is a fat stain and has been used medicinally to stimulate wound healing. The latter is now listed as a cancer suspect agent.

Water-soluble Biebrich Scarlet is widely used as a counterstain. Gurr[1] stated "it is second only to Eosin as a counterstain after Hematoxylin." Lillie[2] regarded it as a useful plasma stain. *Staining Procedures*[3] provides methods for the use of Lillie's Biebrich Scarlet/Picric Aniline Blue to show the staining of collagen, reticulum, muscle and plasma, the Lillie modification of Masson's trichrome stain, the differential staining of various tissues with a change in pH and the use of a Hematoxylin/Biebrich Scarlet/Fast Green trichrome stain designed to show mantle tissues and glandular, muscle, connective and nervous tissues.

References:

1) Gurr, E. *Encyclopedia of Microscopic Stains*; Hill: London, 1960; p 80.

2) Lillie, R.D. *Histopathologic Technic;* Blakiston: Philadelphia and Toronto, 1948.

3) Clark, G. *Staining Procedures,* 4th ed.; Williams and Wilkins: Baltimore, 1981; pp 114, 120, 207, 301.

General references:

Aldrich FT-IR Library	**1**(2), 986B
Sigma FT-IR Library	**1**(2), 105B
IR Library	**3**, 1462B
Safety	**2**(1), 413A
Colour Index	3rd ed., 1971

3,3'-(4,4'-BIPHENYLENE)BIS(2,5-DIPHENYL-2*H*-TETRAZOLIUM CHLORIDE)

Neotetrazolium chloride

Aldrich Cat. No.:	B3,510-5
Sigma Prod. No.:	N2251
CAS No.:	*[298-95-3]*
Mol. Form.:	$C_{38}H_{28}Cl_2N_8$
F.W.:	667.60
m.p.:	230°C (dec.)
Appearance:	Tan powder
λ_{max}:	250nm in methanol
Solubility:	H_2O 20mg/ml
	EGME 30mg/ml
	EtOH 40mg/ml

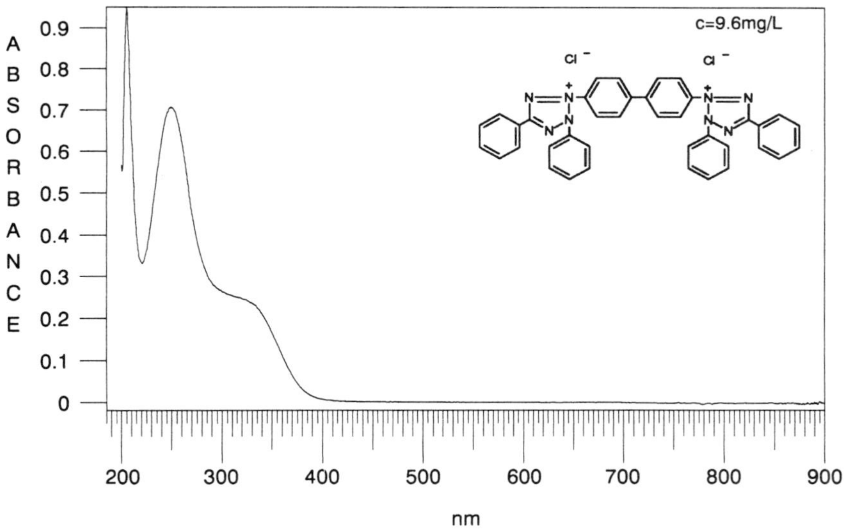

Neotetrazolium chloride is soluble in water and ethanol and insoluble in ether.

A colorless dye precursor, it forms a red-violet diformazan dye on reduction. Thus it has been used to advantage as a histochemical and cytochemical stain for the demonstration of dehydrogenase enzyme systems.[1,2]

References:

1) Oda, T. *J. Histochem. Cytochem.* **1961,** *9*, 605-606.
2) Ogama, A.; Suzuki, Y.; Okui, S. *J. Biochem. (Tokyo)* **1966,** *60, 197-203.*

General references:

Aldrich FT-IR Library	**1**(2), 632D
Sigma FT-IR Library	**1**(2), 183A
IR Library	**3**, 1227E
NMR Library	**2**(2), 501A
Safety	**2**(1), 417D
Merck Index	**11**, 6381

BIS-*N*-METHYLACRIDINIUM NITRATE
Lucigenin

AldrichCat. No.:	B4,920-3
Sigma Prod. No.:	M8010
Purity:	97%

CAS No.:	*[2315-97-1]*
Mol. Form.:	$C_{28}H_{22}N_4O_6$
F.W.:	510.51
m.p.:	245-250°C (dec.)
Appearance:	Green-gold powder
λ_{max}:	430(452)nm in methanol
Solubility:	H_2O 8mg/ml
	EGME 4mg/ml
	EtOH 10mg/ml

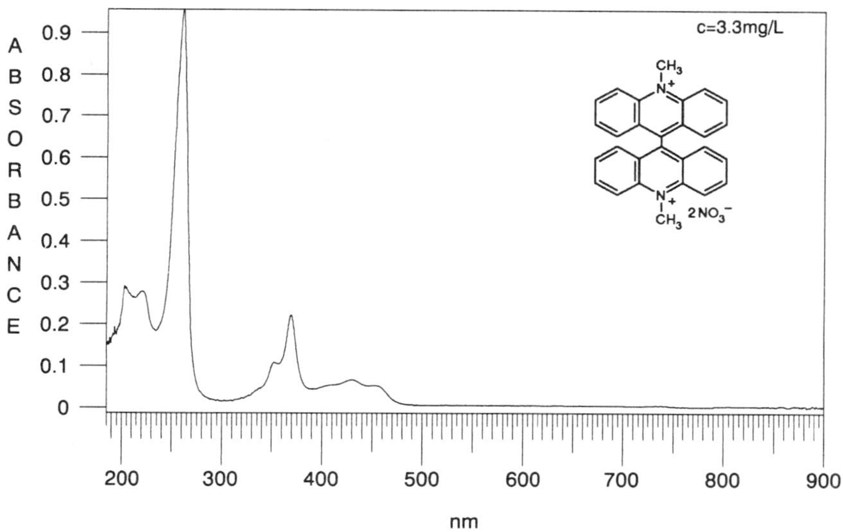

This chemiluminescence reagent has been used as an indicator for analytical pathway research.[1] It is slightly soluble in water and ethanol.

Reference:

1) Erdey, L.; Buzas, I. *Anal. Chim. Acta* **1956,** *15*, 322.

General references:

Aldrich FT-IR Library	**1**(2), 874B
Sigma FT-IR Library	**1**(2), 719B
IR Library	**3**, 1399G
NMR Library	**2**(2), 750A

BISMARCK BROWN R
Basic Brown 4

Aldrich Cat. No.: 21,222-9
Sigma Prod. No.: B2520

CAS No.: [*5421-66-9*]
C.I. No.: 21010
Mol. Form.: $C_{21}H_{24}N_8$
F.W.: 461.40
m.p.: 222°C (dec.)
Appearance: Brownish-black powder
λ_{max}: 468nm in 50% ethanol
Solubility: H_2O 70mg/ml
EGME 10mg/ml
EtOH 7mg/ml

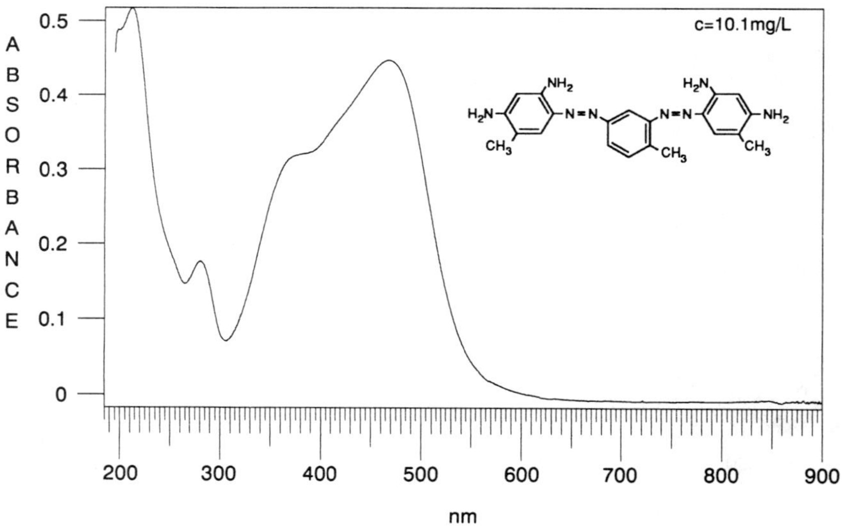

This reddish-brown, cationic disazo dye is made by treating three moles of 2,4-diaminotoluene hydrochloride with two moles of nitrous acid.

Soluble in water and slightly soluble in ethanol, methyl Cellosolve and acetone, it is used to dye silk, cotton, paper, acrylic fibers and leather and to stain wood. The free base (C.I. Solvent Brown 12) is used as a solvent stain, while the phosphomolybdic acid salts are used as pigments (*e.g.*, C.I. Pigment Brown 3). Lightfastness is poor (1-2).

Normally, Bismarck Brown R is not used in biological-stain applications; however, when it has been substituted for Bismarck Brown Y, it appears to have performed equally well albeit slightly redder in hue.

General references:

Aldrich FT-IR Library	**1**(2), 979D
Sigma FT-IR Library	**1**(2), 105D
IR Library	**3**, 1458C
Merck Index	**11**, 1264
Colour Index	3rd ed., 1971

BISMARCK BROWN Y

Basic Brown 1 • Vesuvin

Aldrich Cat. No.:	86,111-1
Sigma Prod. No.:	B5263
Grade:	Certified

CAS No.:	[*10114-58-6*]
C.I. No.:	21000
Mol. Form.:	$C_{18}H_{18}N_8 \cdot 2HCl$
F.W.:	419.32
Appearance:	Red-brown powder
λ_{max}:	457nm in 50% ethanol + 5ml 1*N* hydrochloric acid
Solubility:	H_2O 50mg/ml EGME 30mg/ml EtOH 5mg/ml

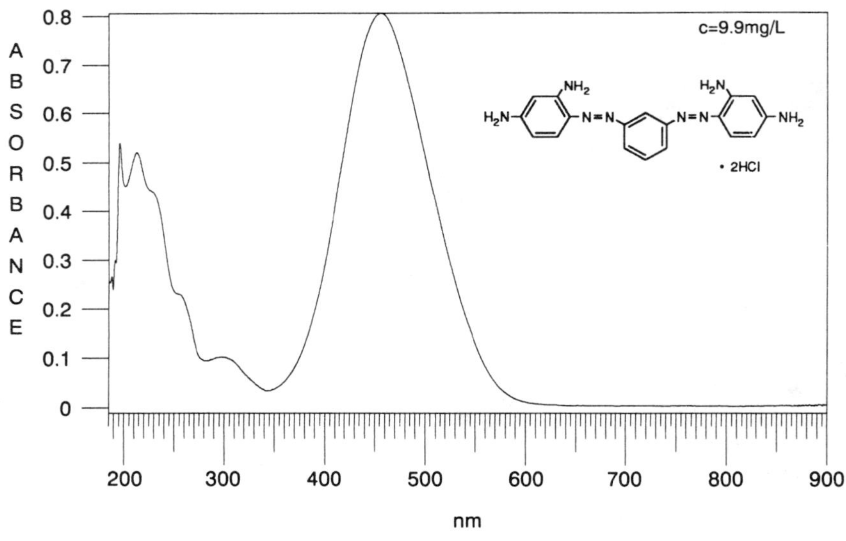

Bismarck Brown Y (BBY) is a brown, cationic disazo dye made by treating three moles of 1,3-phenylenediamine with two moles of nitrous acid.

Soluble in water and methyl Cellosolve, slightly soluble in ethanol and insoluble in acetone, benzene and carbon tetrachloride, it is used to dye cotton (with a tannin mordant), silk, wool, paper and leather, to stain wood and to prepare pigments. The free base is a solvent dye (C.I. Solvent Brown 41, C.I. 21000:1) that is soluble in ethanol, acetone and benzene. It is used to color oils, fats and waxes, including shoe polishes.

As a biological stain, BBY is used widely as a contrast stain for both animal and plant tissues. It is a component of Papanicolaou's stain for vaginal smears. *Staining Procedures*[1] provides methods for the following: use as a component of a displacement tissue stain; use as a displacement tissue stain for acidophilic structures; as a component of Papanicolaou stain. With Methyl Green, it is used on embryonic tissue, tracheae and intestines to demonstrate mucin, cartilage and goblet cells and to demonstrate plant cellulose walls and mucin.

Reference:

1) Clark, G. *Staining Procedures,* 4th ed.; Williams and Wilkins: Baltimore, 1981; pp 121, 208, 209, 211, 322.

General references:

Aldrich FT-IR Library	**1**(2), 979C
Sigma FT-IR Library	**1**(2), 106B
Merck	**11**, 1265
Colour Index	3rd ed., 1971

BLUE TETRAZOLIUM

Ditetrazolium Chloride • Tetrazolium Blue

Aldrich Cat. No.:	B5,480-0
Sigma Prod. No.:	T4375

CAS No.:	[*1871-22-3*]
Mol. Form:	$C_{40}H_{32}Cl_2N_8O_2$
F.W.:	727.66
m.p.:	255°C (dec.)
Appearance:	Yellow powder
λ_{max}:	253nm in methanol
Solubility:	H_2O 9mg/ml
	EGME 4mg/ml
	EtOH 7mg/ml

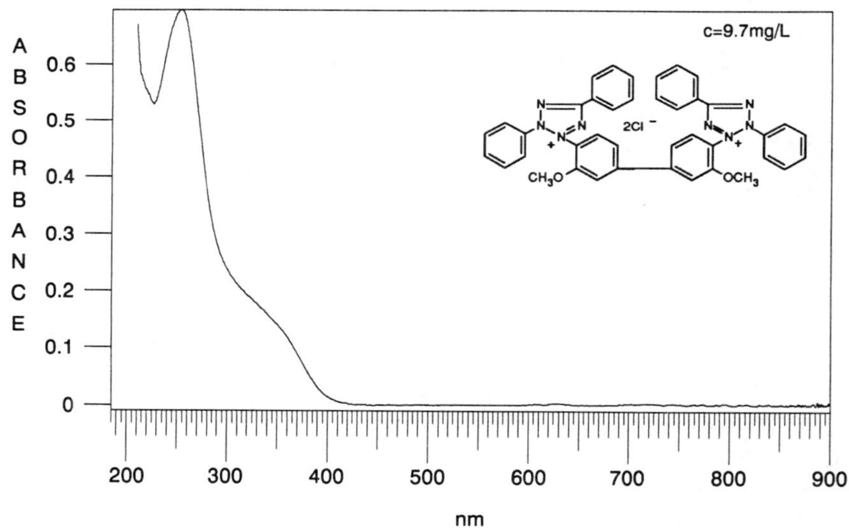

This colorless tetrazolium compound forms a blue diformazan pigment, with a violet-red fat-soluble monoformazan contaminant, on reduction. It is slightly soluble in water, methanol, ethanol and chloroform and insoluble in ethyl acetate and ether.

Reduced by dehydrogenases, it is used in histochemistry and cytochemistry for the detection and determination of dehydrogenase system.[1,2] However, a sluggish response limits its usefulness in this area. Berttok[3] used it for the detection of corticosteroids in urine. *Merck*[4] states that it is used for research in seed germination and as a stain for bacteria and mold.

References:

1) Rutenberg, A.M.; Gofstein, R.; Seligman, A.M. *Cancer Res.* **1950,** *10*, 113.

2) Mader, W.J.; Buck, R.R. *Anal. Chem.* **1952,** *24*, 666.

3) Berttok, E. Z. *ibid.* **1964,** *204*, 118-121.

4) *The Merck Index,* 11th ed.; Budavari, S. *et al.*, Eds.; Merck: Rahway, 1989; Monograph 9173.

General references:

Aldrich FT-IR Library	**1**(2), 633A
Sigma FT-IR Library	**1**(2), 220A
IR Library	**3**, 1227F
NMR Library	**2**(2), 501B

BORDEAUX R
Acid Red 17

Aldrich Cat. No.:	20,962-7
Sigma Prod. No.:	B0759
Grade:	FIAT

CAS No.:	*[5858-33-3]*
C.I. No.:	16180
Mol. Form.:	$C_{20}H_{12}N_2Na_2O_7S_2$
F.W.:	502.44
Appearance:	Dark-red powder
λ_{max}:	518nm in water
Solubility:	H_2O 10mg/ml
	EGME 10mg/ml
	EtOH 1mg/ml

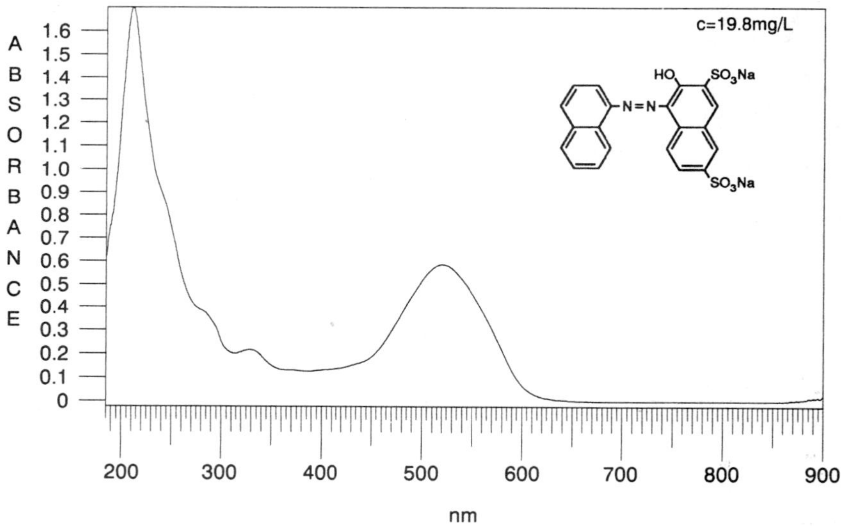

Bordeaux R is a red anionic dye with a monoazo chromophore. Somewhat soluble in water and very slightly soluble in the lower alcohols, it has a strong affinity for proteinaceous substrates but is ineffective in staining cellulosic materials, although it has been used to color the surface of paper. Other industrial applications include dyeing leather and the formation of lakes (barium and heavy-metal salts) for use as pigments to color paints, lacquers, alkyd resins, plastics and enamels.

Conn's[1] considered Bordeaux R an effective cytoplasmic stain when used in conjunction with nuclear stains such as Hematoxylin, Thionin and Methyl Green. The same source gave it a lesser rating as a collagen stain when compared to Acid Fuchsin, Methyl Blue and Amido Black 10B.

Reference:

1) *Conn's Biological Stains,* 9th ed.; Lillie, R.D., Ed.; Williams and Wilkins: Baltimore, 1977; p 120.

General references:

Aldrich FT-IR Library	1(2), 999D
Sigma FT-IR Library	1(2), 106C
Colour Index	3rd ed., 1971

BRILLIANT BLACK BN

Black BN

Aldrich Cat. No.: 21,184-2
Sigma Prod. No.: B8384

CAS No.: *[2519-30-4]*
C.I. No.: 28440
Mol. Form.: $C_{28}H_{17}N_5Na_4O_{14}S_4$
F.W.: 867.69
Appearance: Black powder
λ_{max}: 570(407)nm in water
Solubility: H_2O 30mg/ml
EGME 20mg/ml
EtOH 0.7mg/ml

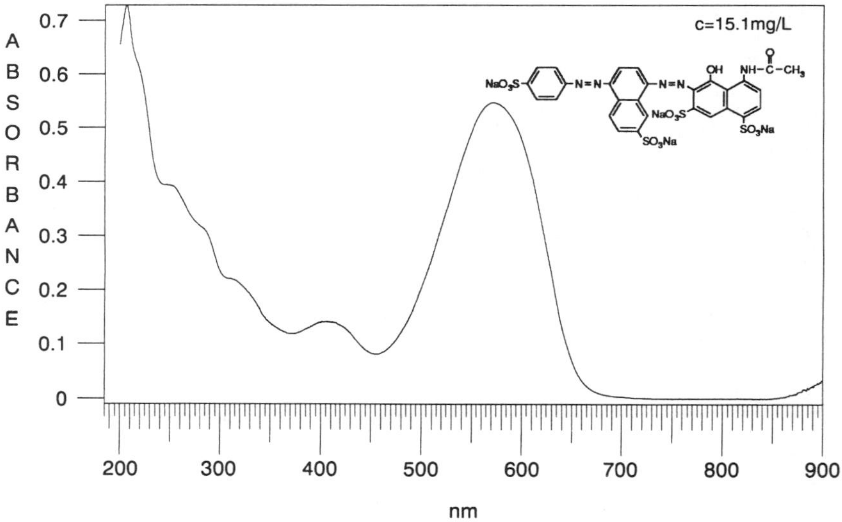

This bluish-black, anionic disazo dye is soluble in water, very slightly soluble in ethanol and insoluble in vegetable oil. It is made by first coupling diazotized sulfanilic acid to 1,7-Cleve's acid (1-naphthylamine-7-sulfonic acid). The product of this reaction is isolated, diazotized, coupled to *N*-acetyl K acid (8-acetylamino-1-naphthol-3,5-disulfonic acid) and then converted to the tetrasodium salt.

The *Colour Index* does not list any industrial applications for this dye. However, it is listed as Food Black 1 and was formerly on the approved list in Great Britain. Where, it was used largely in food-dye mixtures.

There is no evidence that it has been used as a biological stain.

General references:

Aldrich FT-IR Library **1**(2), 1998A
Sigma FT-IR Library **1**(2), 106D

BRILLIANT BLUE G

Acid Blue 90 • Coomassie® Brilliant Blue G 250

Aldrich Cat. No.:	20,139-1	23,464-8
Sigma Prod. No.:	B1131	B5133
Grade:	—	Chromatographically purified and recrystallized
Dye content:	60%	90%

CAS No.:	*[6104-58-1]*	
C.I. No.:	42655	
Mol. Form.:	$C_{47}H_{48}N_3NaO_7S_2$	
F.W.:	854.04	
Appearance:	Dark-purple powder	Violet-brown powder
λ_{max}:	610nm in 50% ethanol	
Solubility:	H_2O 50mg/ml EGME 30mg/ml EtOH 40mg/ml	

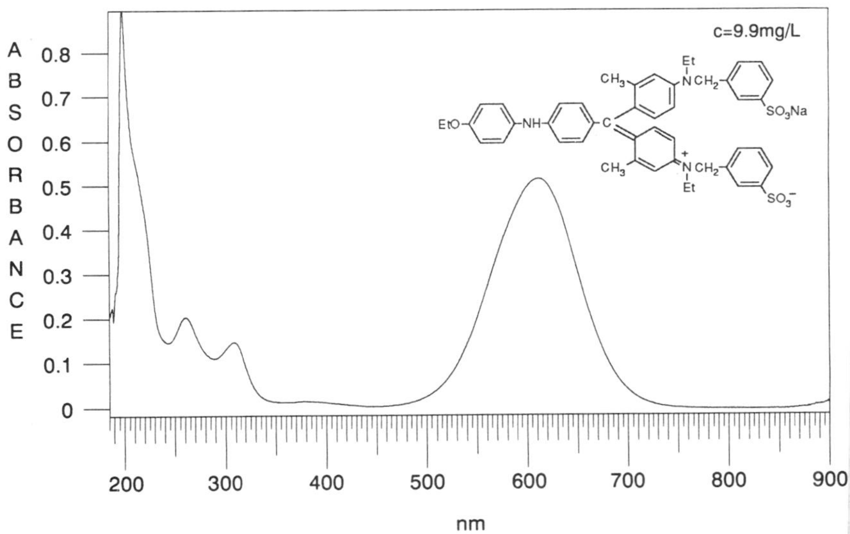

This blue triphenylmethane dye is slightly soluble in cold water, moderately soluble in 10% trichloroacetic acid and soluble in hot water and ethanol. Although not very lightfast (2-3), it is used to dye wool, silk, nylon and leather, to print wool, silk and viscose directly and to coat and surface-color paper.

Using Coomassie Brilliant Blue, Vesterberg[1] reported a rapid and reproducible method for the detection of protein zones after gel-electrophoresis focusing. Bradford[2] described "*A Rapid and Sensitive Method for the Quantitation of Microgram Quantities of Protein Utilizing the Principle of Protein-Dye Binding.*"

References:

1) Vesterberg, O. *Biochim. Biophys. Acta* **1971,** *243*, 345-348.

2) Bradford, M.M. *Anal. Biochem.* **1976,** *72*, 248-254.

General references:

Aldrich FT-IR Library	——	**1**(2), 1040A
Sigma FT-IR Library	**1**(2), 107B	**1**(2), 107A
IR Library	**3**, 1487F	——
Safety	**2**(1), 489A	
Colour Index	3rd ed., 1971	

BRILLIANT BLUE R

Acid Blue 83 • Coomassie® Brilliant Blue R

Aldrich Cat. No.: 20,140-5
Sigma Prod. No.: B0630

CAS No.: [*6104-59-2*]
C.I. No.: 42660
Mol. Form.: $C_{45}H_{44}N_3NaO_7S_2$
F.W.: 825.99
Appearance: Dark purplish-brown powder
λ_{max}: 585nm in ethanol
Solubility: H_2O 70mg/ml
EGME 20mg/ml
EtOH 10mg/ml

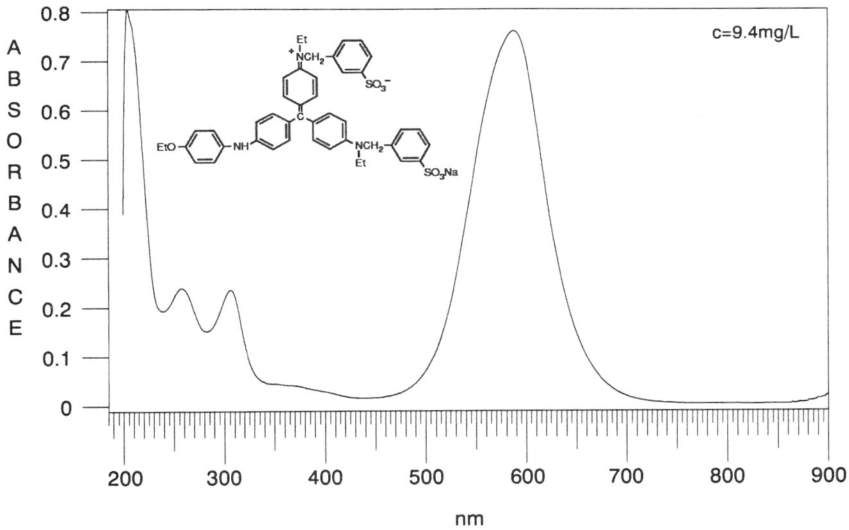

This bright-blue triphenylmethane dye is soluble in hot water and slightly soluble in ethanol. Despite marginal light-fastness (2-3), it has been used to dye wool, silk, nylon, feathers and chromed leather (bright blue), to print wool, silk and viscose directly and to color paper surfaces and coatings.

Diezel *et al.*[1] used it for the determination of protein by gel electrophoresis.

Reference:

1) Diezel, W.; Kopperschlager, G.; Hoffmann, E. *Anal. Biochem.* **1972,** *48*, 617-620.

General references:

Aldrich FT-IR Library	**1**(2), 1040B
Sigma FT-IR Library	**1**(2), 107C
IR Library	**3**,1487G
Safety	**2**(1), 489B
Colour Index	3rd ed., 1971

BRILLIANT CRESYL BLUE ALD

Aldrich Cat. No.:	86,086-7	85,837-4
Sigma Prod. No.:	B5388	——
Grade:	Certified	Indicator
Dye content:	~65%	~40%

CAS No.:	*[81029-05-2]*
C.I. No.:	See below
Mol. Form.:	$C_{17}H_{20}ClN_3O \cdot {}^1/_2\ ZnCl_2$
F.W.:	385.96
m.p.:	233-236°C
Appearance:	Green crystalline powder
λ_{max}:	622nm in 50% ethanol

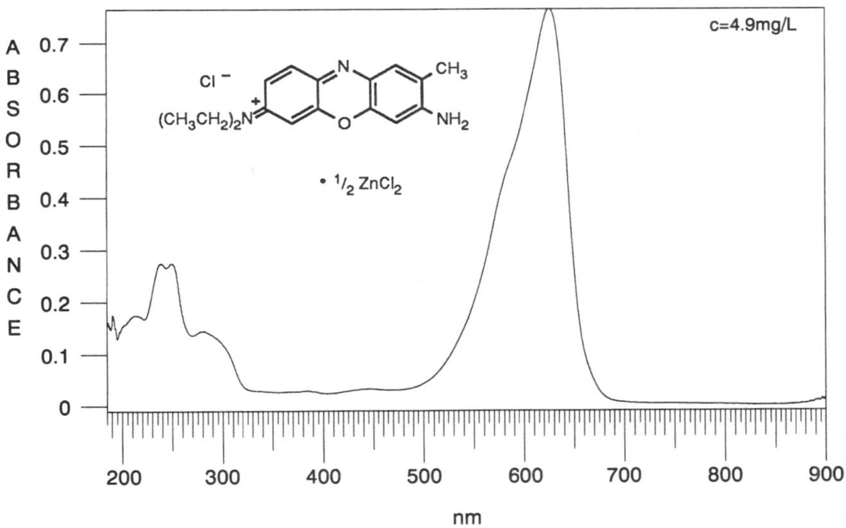

Brilliant Cresyl Blue (BCB) is a cationic oxazine dye that was formerly used commercially to print calico and to dye cotton blue using tannin as a mordant (*Colour Index*). Prior to World War II, practically all of the commercial BCB was manufactured in Europe, primarily in Germany.

The first edition of the *Colour Index* [1] indicates BCB was made, in general, by reacting a 5-dialkylamino-2-nitroso-*p*-cresol hydrochloride with either *m*- or *p*-phenylenediamine or their N,N-dialkyl derivatives. The Structure I (C.I. 877) shown in both this publication and in Gurr,[2] except for minor discrepancies (salt-form bonding and choice of the predominant resonance form), appears to be correct using the intermediates indicated, 5-diethylamino-2-nitroso-*p*-cresol hydrochloride and *m*-phenylenediamine. The second (1956) and third (1971) editions of *Colour Index*, listing the same intermediates, show an improbable structure, different from that in the first edition. This Structure II (C.I. 51010) has an extra primary amino group in the one position.

As far as domestic producers were concerned, 5-diethylamino-2-nitroso-*p*-cresol hydrochloride, a necessary intermediate in the production of BCB, became a casualty of World War II. With the loss of this intermediate in addition to the end product BCB, domestic manufacturers resorted to a similar intermediate, 5-diethylamino-2-nitrosophenol hydrochloride for the manufacture of BCB. The latter intermediate differs from the original in that it lacks a methyl group. To compensate for this deficiency, domestic producers substituted 2,4-diaminotoluene for *m*-phenylenediamine as the other BCB intermediate. Thus, the resulting product retained the same empirical formula as the European product, the net difference being that the methyl group was shifted from position seven to position two in the molecule. Since the change was subtle and proprietary, it was closely guarded, and the name BCB was retained even though the new product was not a cresyl derivative. As a result, the domestic product erroneously became associated with the C.I. 51010 structure.

Conn's[3] indicates that other structures eventually were suggested.

To summarize the foregoing, an examination of spectral data indicates that practically all of the BCB certified by the BSC in the last 30-40 years has been of the domestic variety (see Structure III). It is doubtful if any material represented by Structure II (C.I. 51010) has ever been available commercially. The material manufactured prior to World War II and, on a limited scale, up until the present decade in Europe, is represented by Structure I.

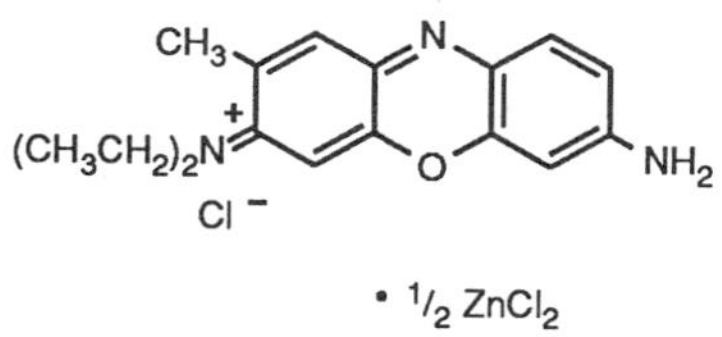

Structure I

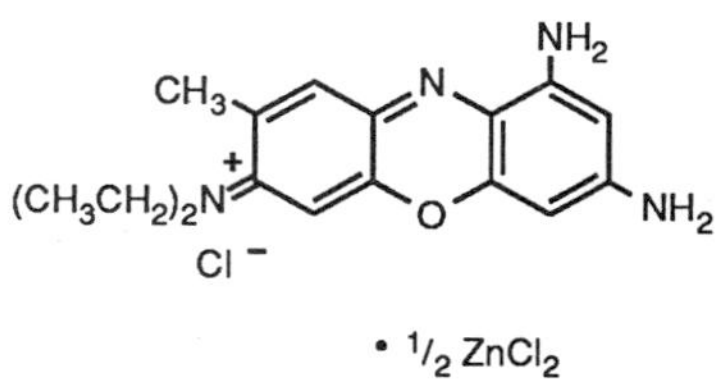

Structure II

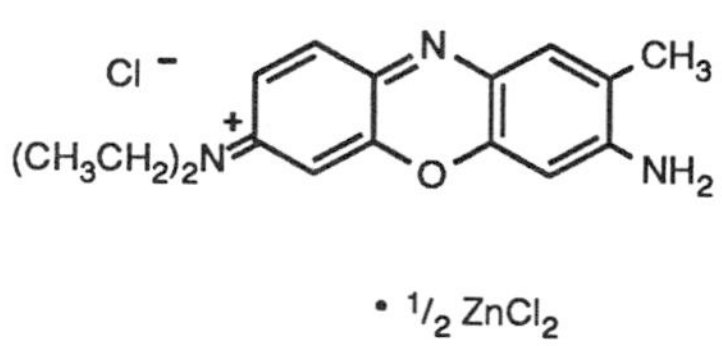

Structure III

Presently, BCB is used almost exclusively as a biological stain with wide application in hematology for demonstrating and counting reticulocytes and platelets. *Staining Procedures*[4] provides a method for this application, and also a method, according to Bang and Bang,[5] for use as a vital stain for demonstrating the mucous secretions of the marine coelomates.

References:

1) *Colour Index,* 1st ed.; Society of Dyers and Colourists: Bradford, Yorkshire, 1924.
2) Gurr, E. *Encyclopedia of Microscopic Stains*; Williams and Wilkins: Baltimore, 1960; pp 86-87.
3) *Conn's Biological Stains,* 9th ed.; Lillie, R.D., Ed.; Williams and Wilkins: Baltimore, 1977; pp 393-395.
4) Clark, G. *Staining Procedures,* 4th ed.; Williams and Wilkins: Baltimore, 1981; pp 176-275.
5) Bang, B.G.; Bang, F.B. *Am. J. Pathol.* **1972,** *68*, 407-417.

General references:

Aldrich FT-IR Library	——	**1**(2), 899B
Sigma FT-IR Library	**1**(2), 107D	——
IR Library	**2**, 1236A	**3,** 1417D

BRILLIANT CROCEIN MOO

Acid Red 73 • Woodstain Scarlet

Aldrich Cat. No.:	21,075-7
CAS No.:	[*5413-75-2*]
C.I. No.:	27290
Mol. Form.:	$C_{22}H_{14}N_4Na_2O_7S_2$
F.W.:	556.49
Appearance:	Maroon powder
λ_{max}:	510nm in water
Solubility:	H_2O 50mg/ml
	EGME 50mg/ml
	EtOH 5mg/ml

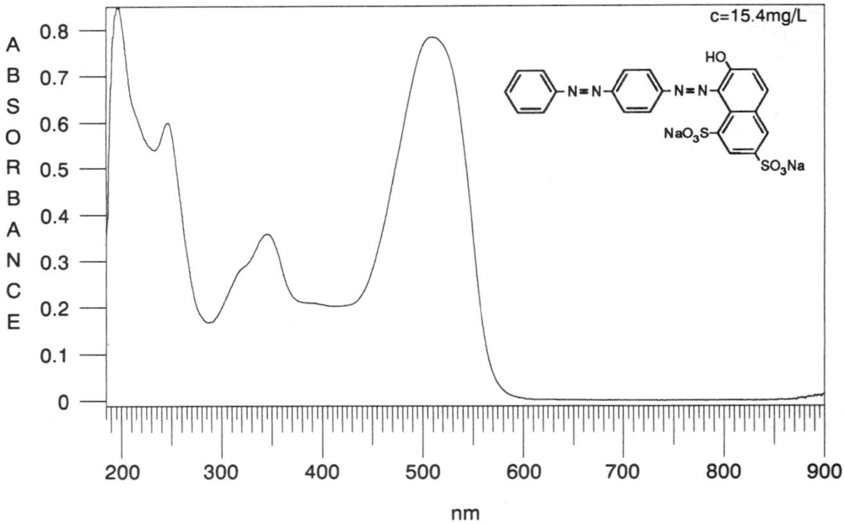

Brilliant Crocein MOO is a red, water-soluble, anionic disazo dye that stains proteinaceous substrates from an acid bath, but requires a mordant such as alum or Glauber salt to dye cellulosic materials.

Industrial applications include dyeing cotton, silk, wool and chromed leather, staining wood and producing lakes for use as pigments.

Conn's[1] describes Acid Red 73, under various trivial names, as a substitute for other unrelated dyes normally used in certain plasma and collagen stain solutions.

Reference:

1) *Conn's Biological Stains,* 9th ed.; Lillie, R.D., Ed.; Williams and Wilkins: Baltimore, 1977; pp 184-185.

General references:

Aldrich FT-IR Library	**1**(2), 989D
IR Library	**3**, 1463G
Colour Index	3rd ed., 1971

BRILLIANT GREEN

Basic Green 1

Aldrich Cat. No.:	86,087-5
Sigma Prod. No.:	B4014
Grade:	Certified

CAS No.:	[*633-03-4*]
C.I. No.:	42040
Mol. Form.:	$C_{27}H_{34}N_2O_4S$
F.W.:	482.65
m.p.:	210°C (dec.)
Appearance:	Green crystalline powder
λ_{max}:	625(428)nm in 50% ethanol
Solubility:	H_2O 40mg/ml
	EGME 60mg/ml
	EtOH 50mg/ml

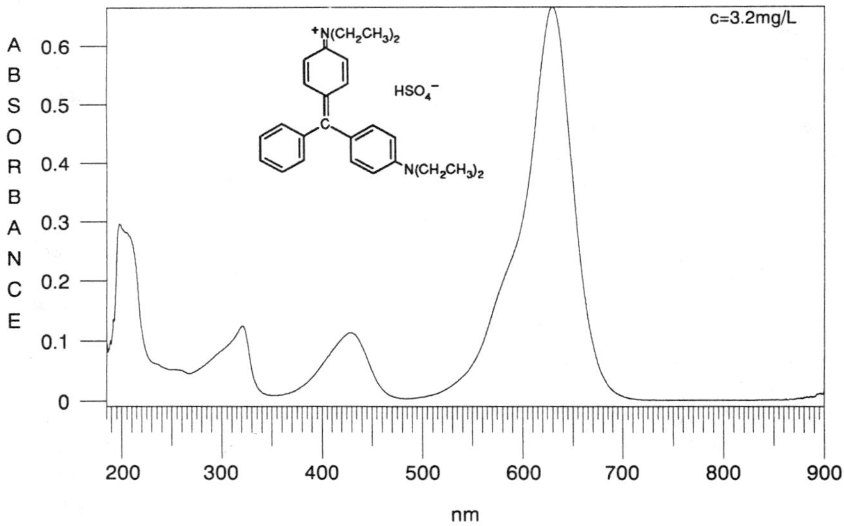

Brilliant Green is a cationic triphenylmethane dye which, like Malachite Green, displays two absorption peaks in the visible region of the spectrum. The minor peak occurs at approximately 428nm while the major peak, with an absorbance approximately four times as great as the minor peak, appears at approximately 625nm. As with its homolog, Malachite Green, the minor peak in Brilliant Green is due to the asymmetric geometry of the molecule which results in two axes of polarizability that exist at right angles to each other.

Brilliant Green is made by condensing one mole of benzaldehyde with two moles of diethylaniline, in the presence of either hydrochloric or sulfuric acid, to form the leuco dye. The leuco dye is oxidized and converted to the acid sulfate. This salt form is soluble in both hot and cold water and in ethanol.

Industrial applications include dyeing wool, acrylic, cotton, bast fibers, acetate, nylon, PVC fibers and paper (poor lightfastness) and preparing spirit inks, wood stains and pigments (phosphotungstomolybdic acid salts). Brilliant Green has indicator properties with a visual-transition interval of pH 0.0 (yellow) to pH 2.6 (green).

Brilliant Green is used along with Gentian Violet and Proflavine hemisulfate to prepare a triple-dye, broad-spectrum, antiseptic solution for neonatal nursery use.[1-3] It finds limited uses as a stain for bacteria;[4] its most important use in bacteriology is as a component of culture media. Since Brilliant Green does not alter the growth of *Salmonella typhosa* but selectively inhibits the growth of the more plentiful *Escherichia coli*, it is used in the preparation of culture media and subsequently inoculated with stool samples to facilitate the separation and enrichment of *S. typhosa.*

References:

1) Remington, J.P. *Pharmaceutical Sciences*, 13th ed.; Mack: Easton, 1965; p 1257.

2) Rogers, T.H. *J. Soc. Chem. Ind.* **1940,** *59*, 34-39.

3) *Tests and Standards. New and Unofficial Remedies*, American Medical Association; Lippincott: Philadelphia, 1953; p 4.

4) Krajian, A.A. *Arch. Pathol.* **1941,** *32*, 825.

General references:

Aldrich FT-IR Library	**1**(2), 1029B
Sigma FT-IR Library	**1**(2), 108A
IR Library	**3**, 1483C
Safety	**2**(1), 489C
Merck Index	**11**, 1367
Colour Index	3rd ed., 1971

BRILLIANT SULPHAFLAVINE

Aldrich Cat. No.:	12,131-2
Sigma Prod. No.:	B4141
Mol. Form.:	$C_{20}H_{15}N_2NaO_5S$
F.W.:	418.41
Appearance:	Yellow-orange powder
λ_{max}:	422nm in water
Solubility:	H_2O 20mg/ml
	EGME 8mg/ml
	EtOH 0.6mg/ml

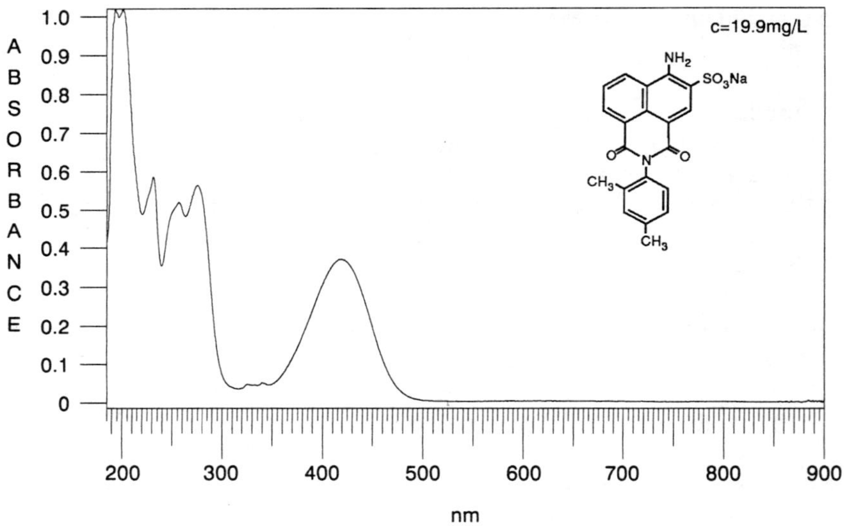

Brilliant Sulphaflavine is a water-soluble, yellow, fluorescent, acid dye with an aminonaphthalimide chromophore. It is very slightly soluble in ethanol. Although it is not listed in the *Colour Index*, to a large degree it has replaced Acid Yellow 7 (Brilliant Yellow, C.I. 56205), a homolog, in industry. The commercial applications include dyeing wool, direct printing of wool and silk, illuminating formaldehyde sulfoxylate discharges, the surface coloring of paper, beater dyeing and coating of paper, production of daylight-fluorescent pigments and inks and coloring of melamine and urea resins.

It should be emphasized that Brilliant Sulphaflavine cannot be substituted for Acid Yellow 7 as an intermediate for preparing the reactive dyes Lucifer Yellow CH and Lucifer Yellow VS.[1]

Reference:

1) Stewart, W.W. *J. Am. Chem. Soc.* **1981,** *103* , 7615-7620.

General reference:

Aldrich FT-IR Library **1**(2), 955C

BRILLIANT YELLOW

Direct Yellow 4 • Paper Yellow

Aldrich Cat. No.: 20,137-5
Sigma Prod. No.: B7507

CAS No.: [*3051-11-4*]
C.I. No.: 24890
Mol. Form.: $C_{26}H_{18}N_4Na_2O_8S_2$
F.W.: 624.56
Appearance: Dark rust-orange powder
λ_{max}: 497nm in 0.1*N* sodium hydroxide
Solubility: H_2O <0.1mg/ml
EGME 30mg/ml
EtOH 6mg/ml

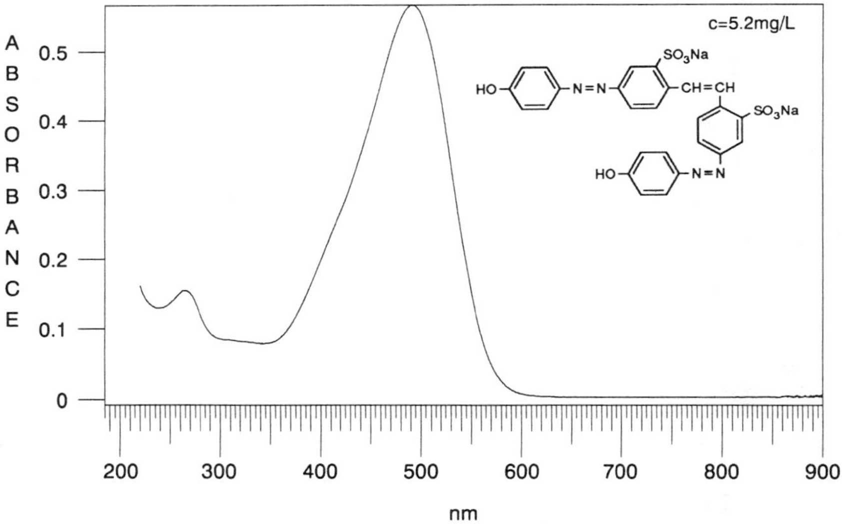

Brilliant Yellow is an anionic disazo dye that is slightly soluble in ethanol, methoxyethanol and acetone. It is very fast to light and heat but not to alkalis. The dye concentration of the commercial product is standardized at 70%, but a nonstandardized product with a dye content of approximately 85% is available.

Commercial applications include beater dyeing and surface coloring of paper, dyeing of nylon in the presence of acetic acid and dyeing leather. Brilliant Yellow is rarely used as a textile dye. However, it is a useful indicator with a visual-transition interval from pH 6.4 (yellow) to pH 8.0 (red-orange) and is usually used in the form of an indicator paper.

Except on rare occasions, Brilliant Yellow has not been used as a biological stain. Kutlik,[1] on such an occasion, used it as a cytoplasmic counterstain in a study involving bilirubin and some of its oxidation products.

Reference:

1) Kutlik, I.E. *Acta Histochem.* **1957,** *4*, 141-157.

General references:

Aldrich FT-IR Library	**1**(2), 980A
Sigma FT-IR Library	**1**(2), 108B
IR Library	**3**, 1458D
Colour Index	3rd ed., 1971

BROMOCHLOROPHENOL BLUE, water soluble

3',3"-Dibromo-5',5"-dichlorophenolsulfonephthalein, sodium salt

Aldrich Cat. No.:	21,298-9
Sigma Prod. No.:	B4010
CAS No.:	[*102185-52-4*]
Mol. Form.:	$C_{19}H_9Br_2Cl_2NaO_5S$
F.W.:	580.07
Appearance:	Black powder
λ_{max}:	590nm in water
Solubility:	H_2O 20mg/ml
	EGME 30mg/ml
	EtOH 10mg/ml

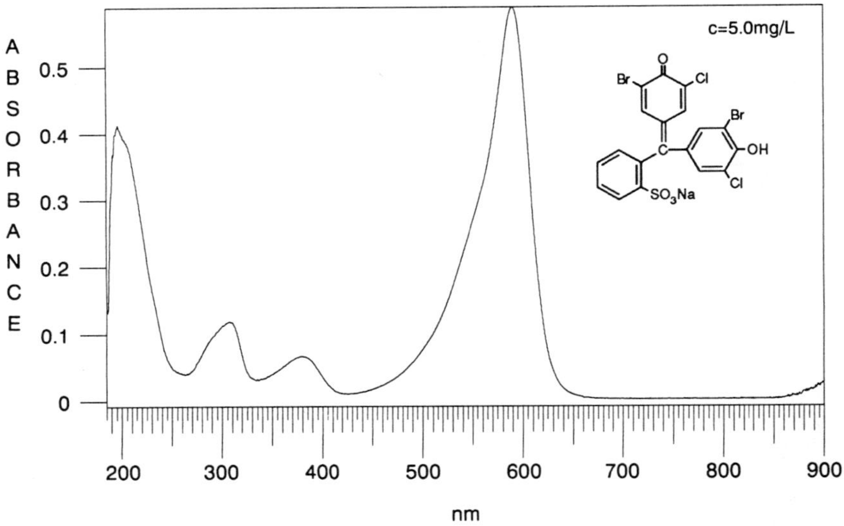

Bromochlorophenol Blue, water soluble, belongs to a group of pH-sensitive, colored products known as sulfonephthalein indicators. They are made by a multi-step synthesis starting with saccharin. While they fulfill the necessary requisites of a dye structurally, their sensitivity and/or instability to pH, oxidation and light render them of little practical value as conventional dyes. They are not listed in the *Colour Index;* however, there continue to be specialized industrial applications for which their acid-base-indicator properties make them uniquely qualified. Contrary to predictions of their ultimate demise when pH meters were developed, sulfonephthalein uses are significantly greater now than they were before the invention.

Bromochlorophenol Blue is made by first condensing two moles of 2-chlorophenol with one of 2-sulfobenzoic acid anhydride. The resulting product (5',5"-dichlorophenolsulfonephthalein) is then dibrominated, with the bromine atoms bonding at positions 3' and 3".

The Bromochlorophenol Blue visual-transition interval (pH 3.2-4.8, yellow to purple) is only slightly different from the range displayed by Bromophenol Blue (Tetrabromophenol Blue) which occurs at pH 3.0-4.6, with a similar color change.[1]

Since Bromophenol Blue can be produced at about 15% the cost of Bromochlorophenol Blue, the demand for the latter is confined largely to research studies.

Reference:

1) Kolthoff, I.M. *Acid-Base Indicators*; Macmillan: New York, 1937; p 131.

General references:

Aldrich FT-IR Library **1**(2), 1023B
Sigma FT-IR Library **1**(2), 108D
IR Library **3**, 1480A

BROMOCRESOL GREEN

3',3",5',5"-Tetrabromo-*m*-cresolsulfonephthalein

Aldrich Cat. No.:	11,435-9
Sigma Prod. No.:	B7396
Grade:	A.C.S. reagent

CAS No.:	*[76-60-8]*
Mol. Form.:	$C_{21}H_{14}Br_4O_5S$
F.W.:	698.04
m.p.:	225°C (dec.)
Appearance:	Beige powder
λ_{max}:	423nm in methanol
Solubility:	H_2O 6mg/ml
	EGME 200mg/ml
	EtOH 40mg/ml

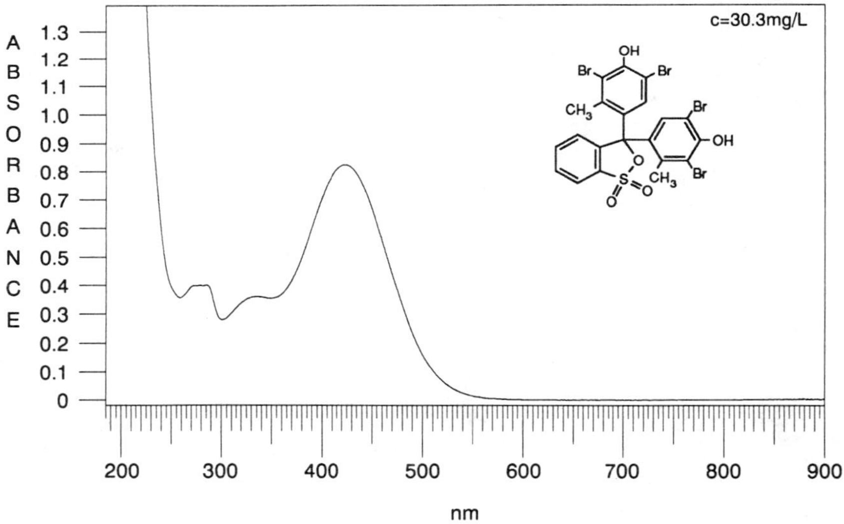

Bromocresol Green is a sulfonephthalein indicator (see description on pages 166 and 167 under Bromochlorophenol Blue). Specifications are described in *Reagent Chemicals*[1] for both the sultone (Aldrich Cat. No. 11,435-9) and sodium salt (Aldrich Cat. No. 11,436-7) forms. The former, devoid of a chromophore with the sultone ring closed, is practically colorless, while the latter is a greenish-black powder. The color results from the monosodium salt formation and the concomitant sultone ring opening, causing a structure rearrangement that displays a paraquinoidal chromophore.

The sultone is readily soluble in alcohol, ether and ethyl acetate and slightly soluble in benzene and acetic acid. It is only slightly soluble in water.

Solvents and the required concentrations for the use of both forms are provided on the pH Indicator Chart in the back of this book. Both forms display the same visual-transition interval, pH 3.8 (yellow) to pH 5.4 (blue-green).

Industrially, Bromocresol Green, sodium salt is used in highly specialized lithographic applications.

Reference:

1) *Reagent Chemicals*, 7th ed.; American Chemical Society: Washington, DC, 1981; p 160.

General references:

Aldrich FT-IR Library	**1**(2), 1024C
Sigma FT-IR Library	**1**(2), 109A
IR Library	**3**, 1480F
Merck Index	**11**, 1375

BROMOCRESOL GREEN, water soluble

3',3'',5',5''-Tetrabromo-*m*-cresolsulfonephthalein, sodium salt

Aldrich Cat. No.: 11,436-7
Sigma Prod. No.: B6771
Grade: A.C.S. reagent

CAS No.: [*62625-32-5*]
Mol. Form.: $C_{21}H_{13}Br_4NaO_5S$
F.W.: 720.02
m.p.: 230°C (dec.)
Appearance: Greenish-black crystalline powder
λ_{max}: 617(400)nm in water
Solubility: H_2O 40mg/ml
EGME 50mg/ml
EtOH 60mg/ml

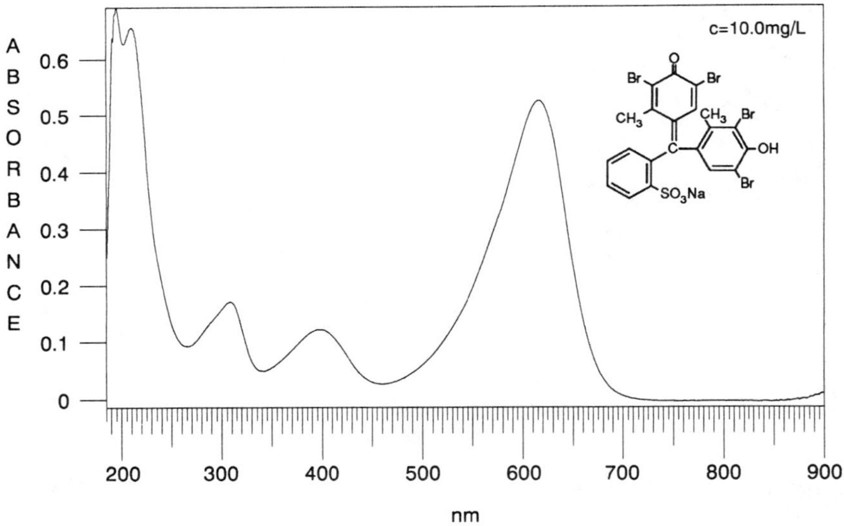

See description on pages 168 and 169 under Bromocresol Green.

General references:

Aldrich FT-IR Library	**1**(2), 1024D
Sigma FT-IR Library	**1**(2), 109B
IR Library	**3**, 1480G
Safety	**2**(1), 518C

BROMOCRESOL PURPLE

5',5"-Dibromo-*o*-cresolsulfonephthalein

Aldrich Cat. No.:	11,437-5
Sigma Prod. No.:	B5880
Grade:	Indicator

CAS No.:	[*115-40-2*]
Mol. Form.:	$C_{21}H_{16}Br_2O_5S$
F.W.:	540.24
m.p.:	240°C (dec.)
Appearance:	Light-purple powder
λ_{max}:	419nm in methanol
Solubility:	H_2O 20mg/ml
	EGME 300mg/ml
	EtOH 80mg/ml

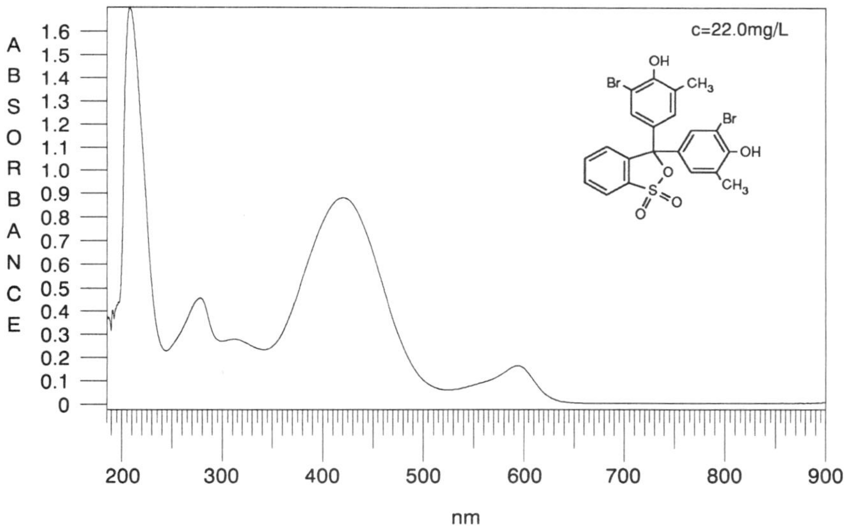

Bromocresol Purple is a sulfonephthalein indicator (see the description in this book on pages 166 and 167 under Bromochlorophenol Blue). It is made by dibrominating Cresol Red. The indicator grade, sultone form, is soluble in ethanol, ether and ethyl acetate and slightly soluble in water, benzene and acetic acid.

Bromocresol Purple, water soluble, is made from the colorless sultone form by introducing one mole of sodium hydroxide. Information on solvent systems and indicator concentration for both forms, for use as acid-base indicators, is included in the pH Indicator Chart in back of this book. The visual-transition interval is pH 5.2-6.8 (yellow to purple), with a pK of 6.3.[1]

The water-soluble product is used as an indicator in dairy laboratories.

Reference:

1) Orndorff, W.R.; Purdy, A.C. *J. Am. Chem. Soc.* **1926**, *48*, 2216.

General references:

Aldrich FT-IR Library	**1**(2), 1025D
Sigma FT-IR Library	**1**(2), 109C
IR Library	**3**, 1481B
Merck Index	**11**, 1376

BROMOCRESOL PURPLE, water soluble

5',5"-Dibromo-*o*-cresolsulfonephthalein, sodium salt

Aldrich Cat. No.: 86,089-1
Sigma Prod. No.: B4263
Grade: Indicator

CAS No.: [*62625-30-3*]
Mol. Form.: $C_{21}H_{15}Br_2NaO_5S$
F.W.: 562.22
m.p.: 255°C (dec.)
Appearance: Brownish-black crystalline powder
λ_{max}: 585(379)nm in 0.1*N* sodium hydroxide
Solubility: H_2O 80mg/ml
EGME 90mg/ml
EtOH 20mg/ml

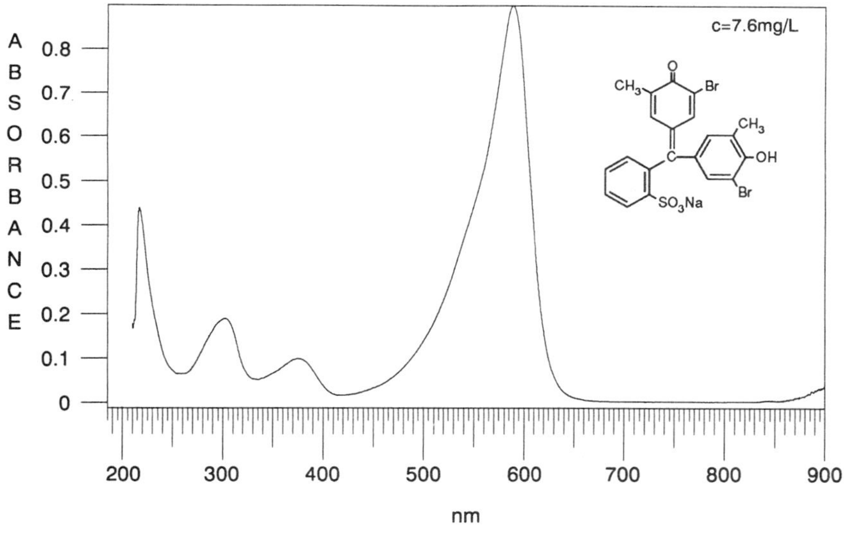

See description on pages 171 and 172 under Bromocresol Purple.

General references:

Aldrich FT-IR Library	**1**(2), 1022A
Sigma FT-IR Library	**1**(2), 109D
IR Library	**3**, 1479C
Safety	**2**(1), 518D

BROMOPHENOL BLUE

3',3",5',5"-Tetrabromophenolsulfonephthalein

Aldrich Cat. No.: 11,439-1
Sigma Prod. No.: B6896
Grade: A.C.S. reagent

CAS No.: [*115-39-9*]
Mol. Form.: $C_{19}H_{10}Br_4O_5S$
F.W.: 669.98
m.p.: 273°C
Appearance: Peach powder
λ_{max}: 598nm in 0.005*N* sodium hydroxide in methanol.
Solubility: H_2O 3mg/ml
EGME 20mg/ml
EtOH 9mg/ml

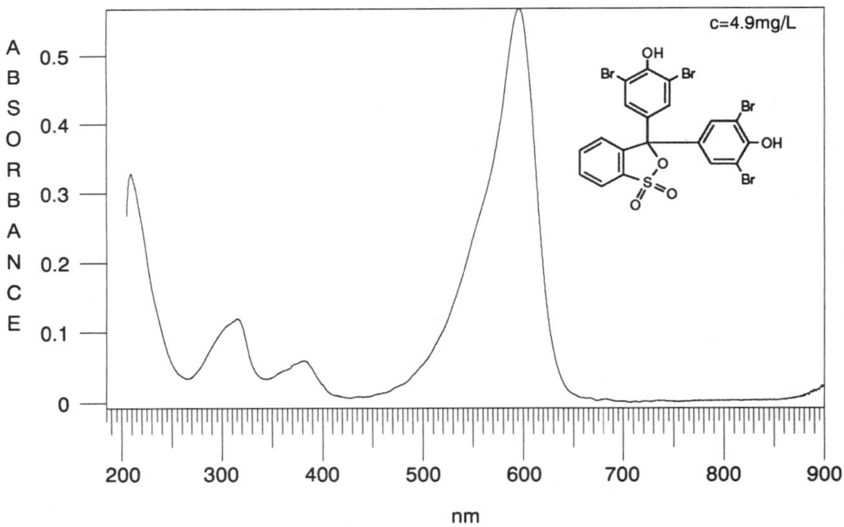

Bromophenol Blue is a member of the sulfonephthalein series of indicators (see Bromochlorophenol Blue on pages 166 and 167). Typical of this group, the sultone form (Aldrich Cat. No. 11,439-1) is practically colorless and the sodium salt (water-soluble product, Aldrich Cat. No. 11,440-5), is colored; the Bromophenol Blue, sodium salt powder, is typically blue-black, indicating the presence of a chromophore in the molecule.

Reagent Chemicals[1] lists specifications for both products which include the visual-transition interval, pH 3.0 (yellow) to pH 4.6 (blue). The pK is 4.0. Information on solubilities and the preparation of solutions for the use of these products as acid-base indicators is included in the pH Indicator Chart in the back of this book.

Probably more Bromophenol Blue is used than all of the other sulfonephthalein indicators combined. Uses range from indicator in plastics and adhesives applications to sensitizer in the photocopying industry. The photocopying application has been the largest consumer of this product in the past; however, this use is in decline as more recent high-technology methods rapidly replace it.

Reference:

1) *Reagent Chemicals,* 7th ed.; American Chemical Society: Washington, DC, 1981; p 161.

General references:

Aldrich FT-IR Library	**1**(2), 1023D
Sigma FT-IR Library	**1**(2), 111B
IR Library	**3**, 1480D
Safety	**2**(1), 565A
Merck Index	**11**, 1434

BROMOPHENOL BLUE, water soluble

3',3",5',5"-Tetrabromophenolsulfonephthalein, sodium salt

Aldrich Cat. No.: 11,440-5
Sigma Prod. No.: B7021
Grade: A.C.S. reagent

CAS No.: [*62625-28-9*]
Mol. Form.: $C_{19}H_9Br_4NaO_5S$
F.W.: 691.97
Appearance: Blue-black crystalline powder
λ_{max}: 589(383)nm in water
Solubility: H_2O 30mg/ml
EGME 30mg/ml
EtOH 7mg/ml

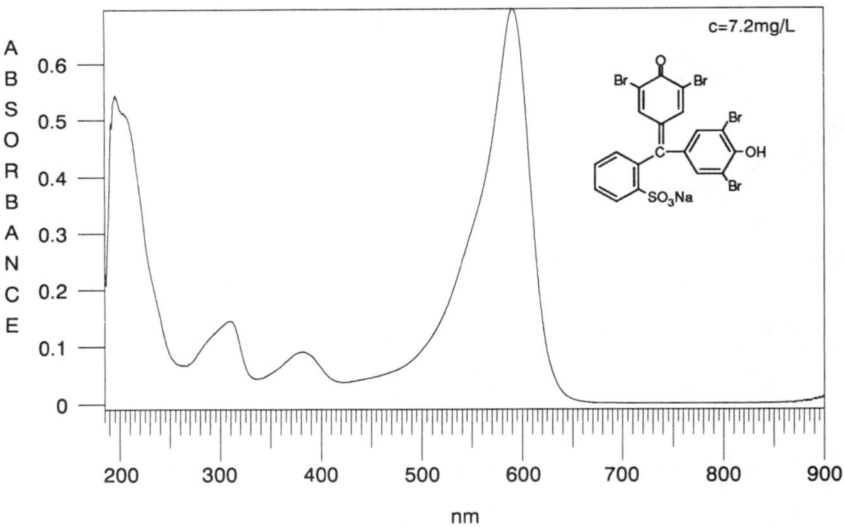

See description on pages 174 and 175 under Bromophenol Blue.

General references:

Aldrich FT-IR Library **1**(2), 1024A
Sigma FT-IR Library **1**(2), 111C
IR Library **3**, 1480E
Safety **2**(1), 565B

2-(5-BROMO-2-PYRIDYLAZO)-5-(DIETHYLAMINO)PHENOL

Bromo-PADAP

Aldrich Cat. No.:	18,001-7
Sigma Prod. No.:	B7383
Purity:	97%

CAS No.:	[*14337-53-2*]
Mol. Form.:	$C_{15}H_{17}BrN_4O$
F.W.:	349.24
m.p.:	157-158°C
Appearance:	Red powder
λ_{max}:	443nm in methanol
Solubility:	H_2O 1mg/ml
	EGME 20mg/ml
	EtOH 20mg/ml

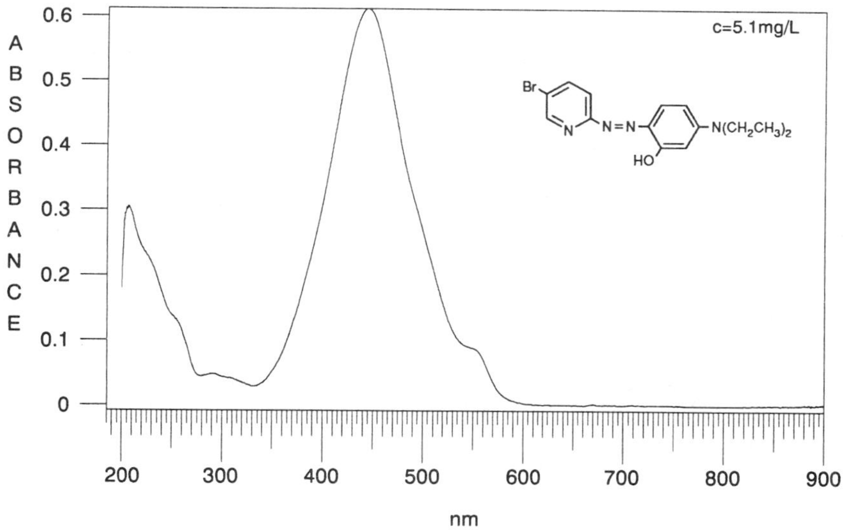

This yellow monoazo dye is soluble in methanol, ethanol and hydrochloric acid and very slightly soluble in water. It is made by coupling diazotized 2-amino-5-bromopyridine with 3-diethylaminophenol.

Bromo-PADAP is used as a reagent for the determination of uranium(VI).[1]

Reference:

1) Johnson, D.A.; Florence, T.M. *Anal. Chim. Acta* **1971,** *53*, 73.

General references:

Aldrich FT-IR Library	1(2), 973C
Sigma FT-IR Library	1(2), 110B
IR Library	**3**, 1455E

BROMOPYROGALLOL RED

Aldrich Cat. No.: 12,195-9
Sigma Prod. No.: B1256
Grade: Indicator

CAS No.: [*16574-43-9*]
Mol. Form.: $C_{19}H_{12}Br_2O_9S$
F.W.: 576.18
m.p.: 300°C
Appearance: Brown powder
λ_{max}: 552nm in water
Solubility: H_2O 5mg/ml
EGME 10mg/ml
EtOH 7mg/ml

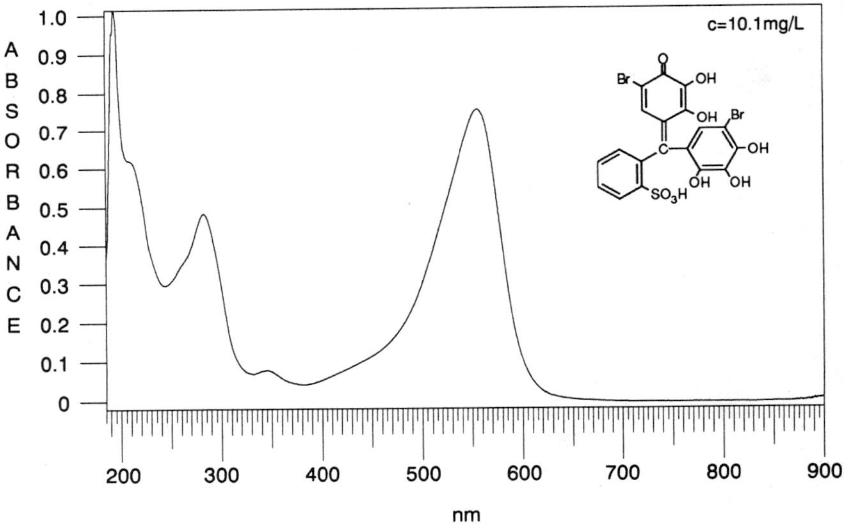

Bromopyrogallol Red is a member of the sulfonephthalein group of indicators. It is prepared by condensing two moles of pyrogallol with one mole of *o*-sulfobenzoic acid anhydride to form pyrogallolsulfonephthalein which is isolated, purified and dibrominated to yield the desired product. It is soluble in methanol, acetone, acetic acid and aqueous alkali solutions and slightly soluble in ethanol and water. Although it has acid-base indicator properties, it is used primarily as a metallochromic indicator.[1]

Reference:

1) Jenickova, A.; Suk, V.; Malat, M. *Collect. Czech. Chem. Commun.* **1956,** *21*, 1257.

General references:

Aldrich FT-IR Library	**1**(2), 1023C
Sigma FT-IR Library	**1**(2), 110C
IR Library	**3**, 1480C

BROMOTHYMOL BLUE

3',3"-Dibromothymolsulfonephthalein

Aldrich Cat. No.: 11,441-3
Sigma Prod. No.: B7146
Grade: A.C.S. reagent

CAS No.: [*76-59-5*]
Mol. Form.: $C_{27}H_{28}Br_2O_5S$
F.W.: 624.40
Appearance: Light-pink powder
λ_{max}: 420nm in methanol
Solubility: H_2O 1mg/ml
EGME 80mg/ml
EtOH 20mg/ml

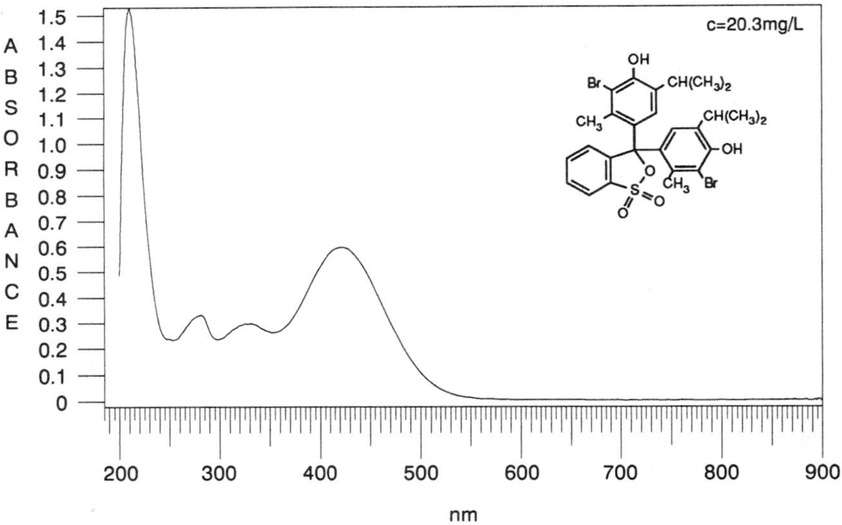

Bromothymol Blue (BTB) is a member of the sulfonephthalein series of indicators (see Bromochlorophenol Blue on pages 166 and 167). Like other members of the series, the sultone is soluble in ethanol, moderately soluble in ethyl acetate and acetone, slightly soluble in acetic acid and benzene and practically insoluble in water. Conversely, the sodium salt is soluble in water and practically insoluble in nonpolar organic solvents.

Reagent Chemicals[1] lists the visual-transition interval for BTB from pH 6.0 (yellow) to pH 7.6 (blue). The pK is 7.0. This pH range, which encompasses the neutral point, makes BTB the indicator of choice in many biological and biochemical studies and applications.

Information regarding the preparation of solutions of BTB for use as an acid-base indicator is included on the pH Indicator Chart in the back of this book.

Reference:

1) *Reagent Chemicals,* 7th ed.; Americal Chemical Society: Washington, DC, 1981; p 162.

General references:

Aldrich FT-IR Library	**1**(2), 1026D
Sigma FT-IR Library	**1**(2), 112B
IR Library	**3**, 1481G
Merck Index	**11**, 1435

BROMOTHYMOL BLUE, water soluble

3',3"-Dibromothymolsulfonephthalein, sodium salt

Aldrich Cat. No.:	11,442-1
Sigma Prod. No.:	B7271
Grade:	A.C.S. reagent

CAS No.:	*[34722-90-2]*
Mol. Form.:	$C_{27}H_{27}Br_2NaO_5S$
F.W.:	646.38
Appearance:	Orange-brown powder
λ_{max}:	615(392)nm in 0.1*N* sodium hydroxide
Solubility:	H_2O 60mg/ml
	EGME 30mg/ml
	EtOH 60mg/ml

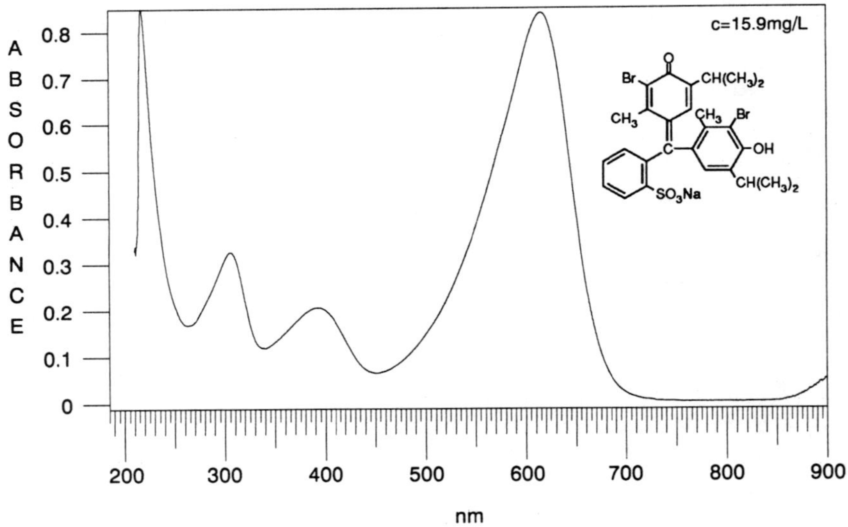

See description on pages 180 and 181 under Bromothymol Blue.

General references:

Aldrich FT-IR Library	**1**(2), 1027A
Sigma FT-IR Library	**1**(2), 112C
IR Library	**3**, 1481H

BROMOXYLENOL BLUE

3',3''-Dibromo-*m*-xylenolsulfonephthalein

Aldrich Cat. No.:	20,547-8
Sigma Prod. No.:	B1763
Grade:	Indicator

CAS No.:	[*40070-59-5*]
Mol. Form.:	$C_{23}H_{20}Br_2O_5S$
F.W.:	568.29
m.p.:	218°C (dec.)
Appearance:	Violet powder
λ_{max}:	417nm in methanol
Solubility:	H_2O 4mg/ml
	EGME 20mg/ml
	EtOH 4mg/ml

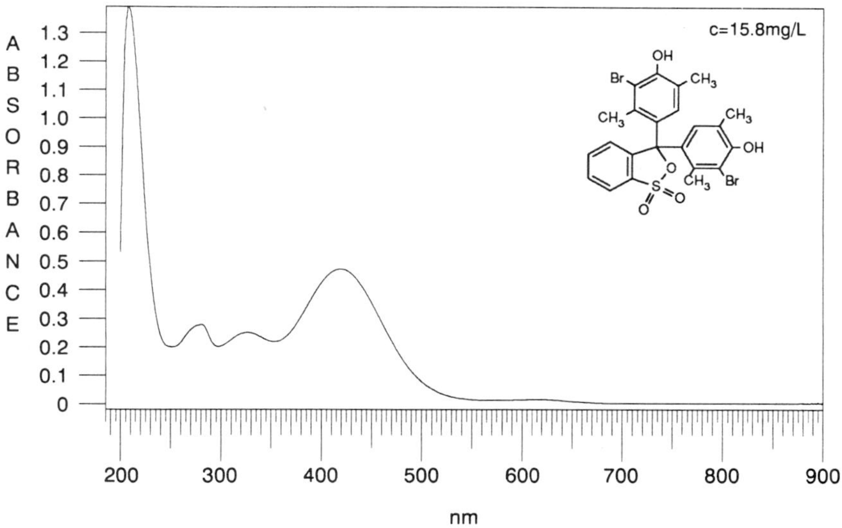

Bromoxylenol Blue (BXB) is probably the least used member of the sulfonephthalein series of indicators (see Bromochlorophenol Blue on pages 166 and 167). It was first made by Cohen[1,2] by condensing two moles of 2,5-dimethylphenol with one mole of *o*-sulfobenzoic acid anhydride and then dibrominating the resultant product. Since that time, very little has appeared in the literature about this item.

The sultone variety is the only form on the market. This product is soluble in ethyl acetate and acetone, slightly soluble in ethanol and very slightly soluble in acetic acid and benzene. It is only slightly soluble in water unless it is converted to a salt form first, by the addition of an appropriate base.

The visual-transition interval, which closely resembles that of Bromothymol Blue, occurs from pH 6.0 (yellow) to pH 7.6 (blue). The pK is 7.0. Information regarding the preparation of solutions of BXB for use as an acid-base indicator is included on the pH Indicator Chart in the back of this book.

References:

1) Cohen, A. *Biochem. J.* **1922,** *16*, 31.

2) *Idem ibid*. **1923,** *17*, 535.

General references:

Aldrich FT-IR Library **1**(2), 1026B

Sigma FT-IR Library **1**(2), 111A

IR Library **3**, 1481D

6'-BUTOXY-2,6-DIAMINO-3,3'-AZODIPYRIDINE

Aldrich Cat. No.: 20,677-6
Purity: 99%

CAS No.: [*617-19-6*]
Mol. Form.: $C_{14}H_{18}N_6O$
F.W.: 286.34
m.p.: 127-129°C
Appearance: Yellow powder
λ_{max}: 435nm in methanol
Solubility: H_2O 0.6mg/ml
EGME 8mg/ml
EtOH 8mg/ml

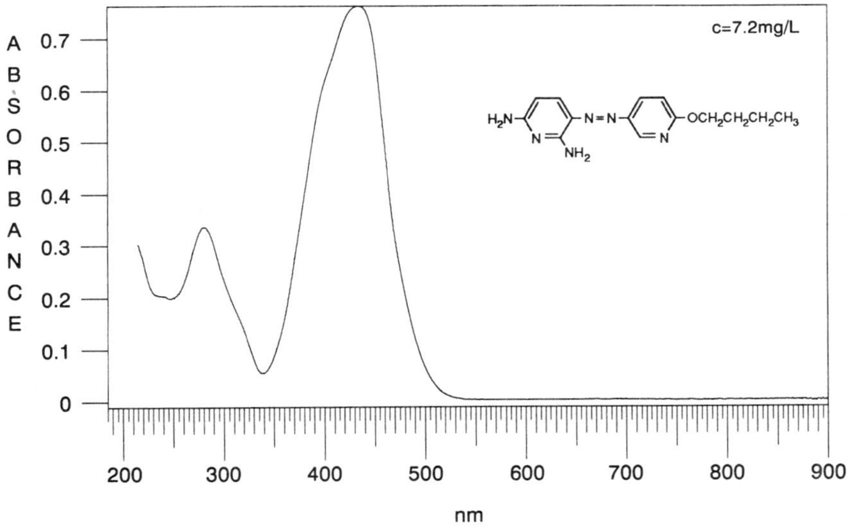

This orange, solvent monoazo dye is virtually insoluble in water, slightly soluble in ethanol and methyl Cellosolve and freely soluble in most other common organic solvents.

There is no evidence in the literature that it has been used industrially nor as a biological stain; however, *Merck*[1] describes it as a urinary antiseptic.

Reference:

1) *The Merck Index,* 11th ed.; Budavari, S. *et al.*, Eds.; Merck: Rahway, 1989; Monograph 2956.

General references:

Aldrich FT-IR Library **1**(2), 997B
IR Library **3**, 1467A

CALMAGITE

3-Hydroxy-4-(2-hydroxy-5-methylphenylazo)-1-naphthalenesulfonic acid

Aldrich Cat. No.: C20-4
Sigma Prod. No.: C3259
Grade: Indicator

CAS No.: [*3147-14-6*]
Mol. Form.: $C_{17}H_{14}N_2O_5S$
F.W.: 358.37
m.p.: 300°C
Appearance: Black powder with chunks
λ_{max}: 602nm in 0.01*N* sodium hydroxide
Solubility: H_2O 30mg/ml
EGME 60mg/ml
EtOH 20mg/ml

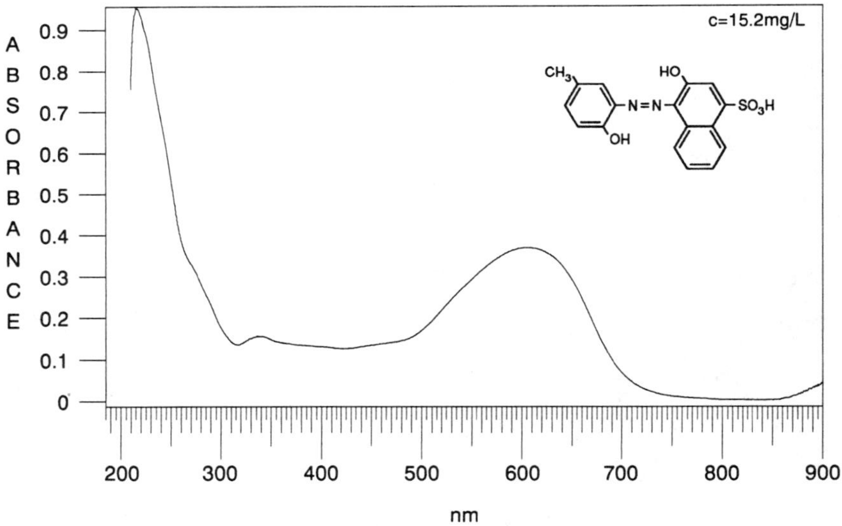

Calmagite is an anionic monoazo dye that Lindstrom[1] prepared by coupling diazotized 1-amino-2-naphthol-4-sulfonic acid to *p*-cresol. It is soluble in water and forms red crystals from acetone.

Calmagite is not listed in the *Colour Index* and apparently has not been used to any degree in industrial applications. However, it is ideally structured to form chelates with each of the two hydroxyl groups in the molecule that are positioned *ortho* to an azo-linkage nitrogen atom. The Calmagite structure was used to advantage by Chauhan and Ray Sarkar[2] as a reagent for the determination of calcium and magnesium. Possessing acid-base indicator properties, this dye has a visual-transition interval which extends from pH 7.1 (red) to pH 9.1 (blue).

There is no evidence that it has been used as a biological stain.

References:

1) Lindstrom, F.; Diehl, H. *Anal. Chem.* **1960,** *32*, 1123.

2) Chauhan, U.P.S.; Ray Sarkar, B.C.R. *Anal. Biochem.* **1969,** *32*, 70.

General references:

Aldrich FT-IR Library	**1**(2), 985D
Sigma FT-IR Library	**1**(2), 113B
IR Library	**3**, 1461E
Merck Index	**11**, 1725

CARBOL FUCHSIN

Aldrich Cat. No.: 20,693-8
Sigma Prod. No.: C4165
Grade: Biological stain

CAS No.: [*4197-24-4*]
Appearance: Dark-green crystalline solid
λ_{max}: 547nm in 50% ethanol
Solubility: H_2O 2mg/ml
EGME 40mg/ml
EtOH 30mg/ml

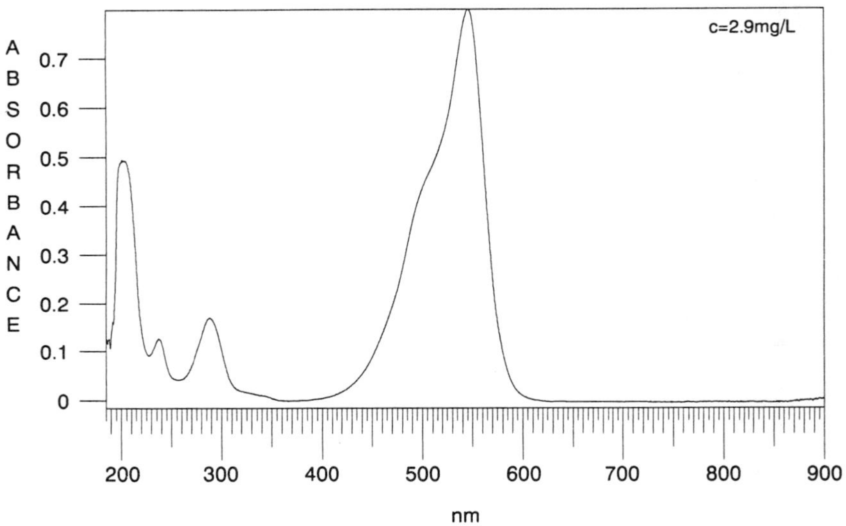

The use of dry powder Carbol Fuchsin allows Carbol Fuchsin solution to be prepared more simply and safely, obviating the bothersome and hazardous procedure that involves the melting and handling of phenol. On a dry-weight basis, this product contains approximately 90% Pararosaniline to 10% phenol, and a working solution can be prepared from the powder by first dissolving 1.5g of the powder in 100ml of 95% ethanol, then adding an equal volume of deionized water.

Emmel and Cowdry[1] describe methods for staining acid-fast bacilli and Nissl bodies. *Staining Procedures*[2] provides procedures for staining and a bibliography for demonstrating chromosomes and interphase chromatin, Ziehl-Neelsen technique, acid-fast staining at room temperature and a Lillie-modified Ziehl-Neelsen technique.

No industrial applications for Carbol Fuchsin have been found in the literature.

References:

1) Emmel, V.M.; Cowdry, E.V. *Laboratory Technique in Biology and Medicine*, 4th ed.; Williams and Wilkins: Baltimore, 1964; p 65.

2) Clark, G. *Staining Procedures,* 4th ed.; Williams and Wilkins: Baltimore, 1981; pp 349, 380, 384, 446.

General references:

Sigma FT-IR Library **1**(2), 113C
Safety **2**(1), 683A

CARMINE

Alum lake of carminic acid

Aldrich Cat. No.: 19,970-2
Sigma Prod. No.: C6152
Grade: Certified

CAS No.: [*1390-65-4*]
C.I. No.: 75470
Appearance: Red powder
λ_{max}**:** 531(563)nm in 0.1*N* sodium hydroxide
Solubility: H_2O 8mg/ml
EGME 6mg/ml
EtOH 0.8mg/ml

Structure according to Meloan et al.[5]

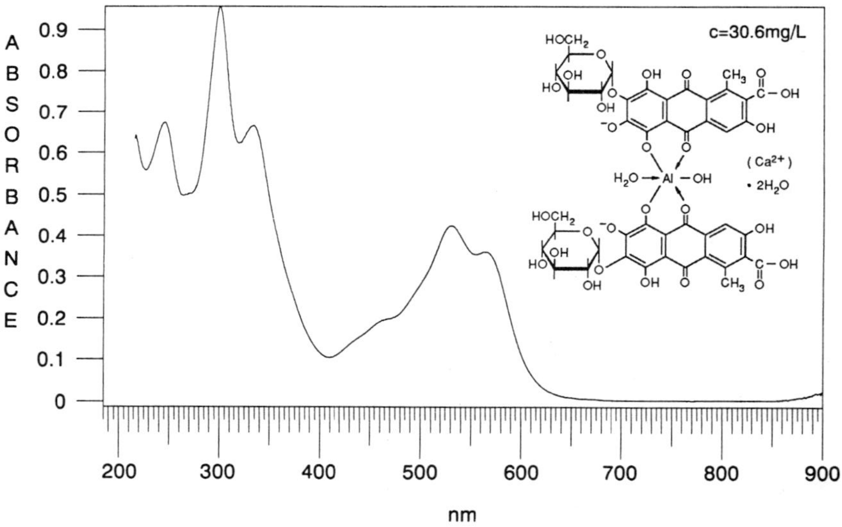

Carminic acid is a natural red dye that is responsible for the tinctorial properties of Cochineal powder and constitutes about 10% of its weight. Cochineal powder is produced by grinding the dried bodies of the female cochineal insect (*Coccus cacti* L.). The Aztecs cultivated the insect for its color value, and it is believed that Cortez observed the natives using it when he arrived in Mexico in 1518.[1] It was first believed to be Kermes, an ancient dye used in Europe at that time. Thereafter, it became an important commercial dye in Europe and its importance in commercial applications worldwide did not diminish until years after Perkin's discovery of the first synthetic dye, Mauve, in 1856. Cohn[2] reported that *Coccus cacti* was one of the first *Hemiptera* classified. The insect lives on the cactus, *Nopalea coccinelliferna,* indigenous to southern Mexico, Honduras and Guatemala. It was later introduced in South

America, the West Indies, East Indies, Spain, Algeria, Canary Islands, Malta, Palestine and Jordan.

To obtain optimum tinctorial yield, the fecund female insect is collected and then killed by either immersion in boiling water or by dehydration in a hot-air oven or on a hot plate. Cohn, in 1899, claimed the dehydration method yielded the "finest" product. Cochineal insects treated in this manner were described as "silver cochineal". However, the "inferior" black variety obtained by the immersion process was often treated with talc, calcium sulfate, lead carbonate, or tristearin to mimic the premium product.

Carmine is the aluminum or calcium-aluminum lake of carminic acid. The calcium-aluminum lake was first isolated from a Cochineal extract by Pelletier and Caventou in 1818. In 1847, Warren de la Rue succeeded in obtaining pure carminic acid. Structurally, carminic acid utilizes the anthraquinone chromophore as its nucleus. A sugar-like moiety and six other less bulky auxochromes, including a methyl, a carboxyl, and four hydroxyl functions, make up the remainder of this structure. These groups serve, in concert, to elaborate and enhance the potential of the basic molecule in its new-found role as a dye. The λ_{max} is shifted, bathochromically, approximately 90nm and the solubility in water and other aprotic solvents is greatly increased. The bond between the carbohydrate moiety and the anthraquinone nucleus is very heat labile, which makes the isolation of carminic acid from Cochineal very difficult and expensive. This, of course, explains why the less desirable Carmine, which can be isolated more readily, is used much more widely than carminic acid.

Cochineal has been used to dye wool and silk mordanted with Aluminum Crimson; or, when mordanted with tin or a single bath containing stannous chloride and Oxalic Acid Scarlet, alcoholic solutions have been used as an indicator for carbonates.[3] In the form of a lake, it has been used as a watercolor. Marmion[1] states that "Carmine is useful for producing pink shades in retorted protein products, candy, confections, rouge, eye shadow and pill coating."

Carmine is only slightly soluble in water. Therefore, in biological-stain applications, it must be used in a recipe that includes either an acid or a base to increase its solubility and concomitant staining ability. *Staining Procedures*[4] lists a number of applications, recipes and a bibliography for Carmine. The following are included: Orth's Lithium Carmine (for demonstrating nuclei in animal tissue); Best's stain for staining glycogen; Mayer's Mucicarmine in conjunction with Hematoxylin and Metanil Yellow (mucin stain); stain for trematode whole mounts, Iron Acetocarmine (nuclear stain for temporary preparations); precipitated Borax Carmine (nuclear stain for permanent preparations) and alcoholic-hydrochloric acid Carmine (for chromosomes at various stages of nuclear division).

References:

1) Marmion, D.M. *Handbook of U.S. Colorants for Foods, Drugs and Cosmetics;* Wiley: New York, 1979; p 80.
2) Cohn, A.I. *Indicators and Test Papers;* Wiley: New York, 1902.
3) *Colour Index,* 1st ed.; Society of Dyers and Colourists: Bradford, Yorkshire, 1924.
4) Clark, G. *Staining Procedures,* 4th ed.; Williams and Wilkins: Baltimore, 1981; pp 111,195, 212, 217, 258, 289, 290, 342, 344.
5) Meloan, S.N.; Puchtler, H.; Valentine, H. *Histochemie* **1975**, *27*, 87-95.

General reference:

Merck Index **11,** 1850

CARMINIC ACID

Natural Red 4

Aldrich Cat. No.:	22,925-3
Sigma Prod. No.:	C3522
CAS No.:	[*1260-17-9*]
C.I. No.:	75470
Mol. Form.:	$C_{22}H_{20}O_{13}$
F.W.:	492.40
Appearance:	Dark-red powder
λ_{max}**:**	495nm in methanol
Solubility:	H_2O 4mg/ml
	EGME 4mg/ml
	EtOH 2mg/ml

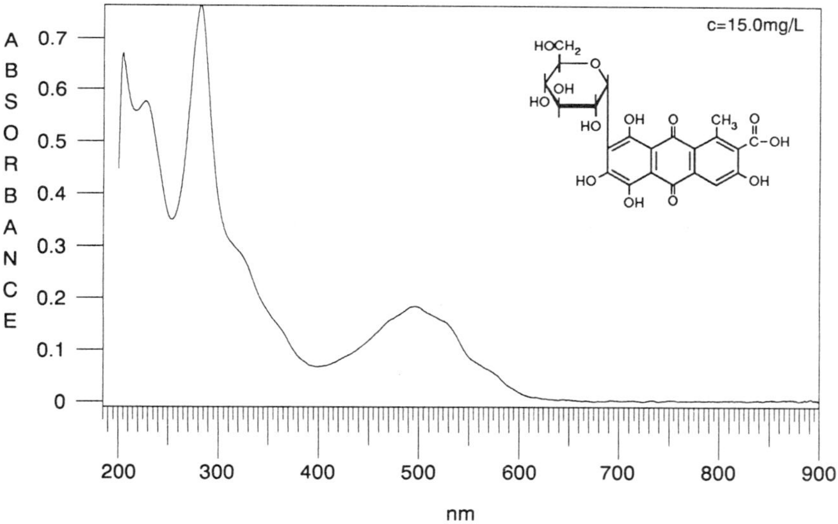

See description on pages 190-192 under Carmine.

General references:

Aldrich FT-IR Library	**1**(2), 246D
Sigma FT-IR Library	**1**(2), 114A
Merck Index	**11**, 1850

trans-β-CAROTENE
Food Orange 5

Aldrich Cat. No.:	85,555-3
Sigma Prod. No.:	C9750
CAS No.:	[*7235-40-7*]
C.I. No.:	40800
Mol. Form.:	$C_{40}H_{56}$
F.W.:	536.89
m.p.:	178-179°C
Appearance:	Red-orange powder
λ_{max}:	450(478)nm in hexane
Solubility:	H_2O 0.6mg/ml
	EGME 3mg/ml
	EtOH 2mg/ml

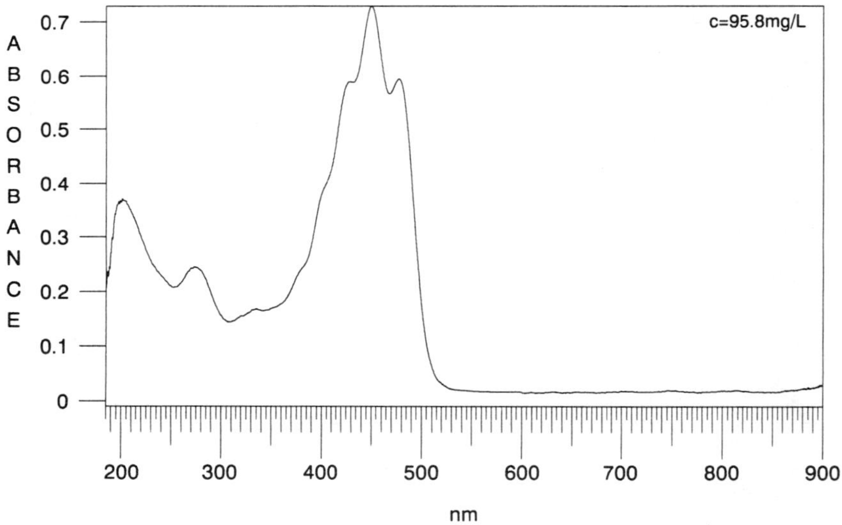

trans-β-Carotene is a carotenoid coloring matter belonging to the terpene group. This group of colors is tinctorially very intense, possessing a polyene chromophore containing not less than 18 carbon atoms in a conjugated double-bond system. The β-carotene of commerce is synthesized from acetone utilizing a process developed by Hoffman-LaRoche in the 1950s. It is, however, abundant in nature, and is found in certain plants and vegetables, as well as in animals that use these plants as food. The natural product is included in the *Colour Index* under C.I. 75130 (C.I. Natural Yellow 26 and C.I. Natural Brown 5). To a large degree, β-carotene is responsible for the color in butter, cheese, alfalfa and certain cereal grains, since it normally occurs in these substances as the dominant isomer in an isomeric mixture. This mixture has an average composition of 15% α, 85% β and 0.1% γ, varying considerably with the plant species from which it is obtained.[1]

The two main advantages gained from using β-carotene as a colorant for foods are: 1) nutritional (as provitamin A, it is metabolized to vitamin A) and 2) the ability to duplicate natural shades.

β-Carotene is virtually insoluble in water and very slightly soluble in ethanol and other polar solvents and only slightly soluble in aprotic solvents. Solubility in edible oils is about 0.08% at room temperature, 0.2% at 60°C and 0.8% at 100°C. It is used as a semisolid suspension in edible oils and as a water-dispersible oil beadlet.

Biologically, it is used as a probe to estimate lipid distribution in lipoproteins.[2]

References:

1) Marmion, D.M. *Handbook of U.S. Colorants for Foods, Drugs and Cosmetics;* Wiley: New York, 1979; pp 76-79.

2) Chen, G.C.; Krieger, M.; Kane, J.P.; Wu, C-S.C.; Brown, S.; Goldstein, J.L. *Biochemistry* **1980,** *19*, 4706.

General references:

Aldrich FT-IR Library	**1**(1), 54A
Sigma FT-IR Library	**1**(1), 630D
IR Library	**3**, 39C
Merck Index	**11**, 1860
Colour Index	3rd ed., 1971

CELESTINE BLUE

Mordant Blue 14

Aldrich Cat. No.:	20,634-2
Sigma Prod. No.:	C7143
CAS No.:	[*1562-90-9*]
C.I. No.:	51050
Mol. Form.:	$C_{17}H_{18}ClN_3O_4$
F.W.:	363.80
m.p.:	227-230°C
Appearance:	Black powder
λ_{max}:	642nm in water
Solubility:	H_2O 10mg/ml
	EGME 3mg/ml
	EtOH 2mg/ml

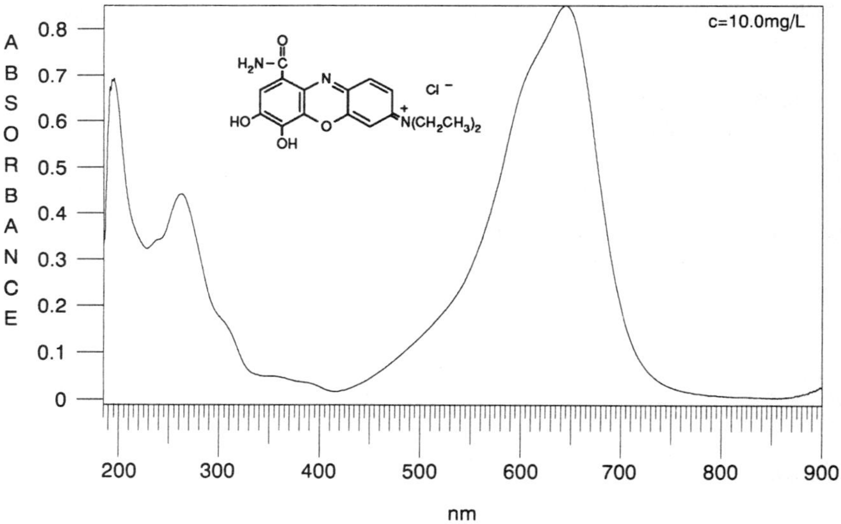

Celestine Blue is a cationic oxazine dye that is moderately soluble in water and slightly soluble in alcohol. It is prepared by condensing 4-nitroso-*N,N*-diethylaniline hydrochloride with gallamide.

Industrially, it is used to dye wool, cotton padded with chromium acetate and to print cotton, wool and silk mordanted with chromium.

Celestine Blue is a nuclear stain that has been used as a substitute for Hematoxylin. Lillie *et al.*[1] evaluated Celestine Blue, along with a number of other mordant dyes, as a substitute for Hematoxylin during a shortage of the latter in the early seventies. They considered it to be an adequate replacement for Hematoxylin in some applications. *Staining Procedures*[2] provides a Van Gieson-stain recipe and a general method for staining tissue embedded in glycol methacrylate that utilizes Celestine Blue along with Hematoxylin and Acid Fuchsin. With this method, nuclei are reportedly stained black; muscle, yellow and collagen, red.

References:

1) Lillie, R.D.; Pizzolato, P.; Donaldson, P.T. *Stain Technol.* **1976,** *51*, 25-41.

2) Clark, G. *Staining Procedures,* 4th ed.; Williams and Wilkins: Baltimore, 1981; p 37.

General references:

Aldrich FT-IR Library	**1**(2), 1041B
Sigma FT-IR Library	**1**(2), 114B
IR Library	**3**, 1488B
Merck Index	**11**, 1955
Colour Index	3rd ed., 1971

CHICAGO SKY BLUE 6B

Direct Blue 1

Aldrich Cat. No.:	20,138-3
Sigma Prod. No.:	C0392

CAS No.:	*[2610-05-1]*
C.I. No.:	24410
Mol. Form.:	$C_{34}H_{24}N_6Na_4O_{16}S_4$
F.W.:	992.82
Appearance:	Blue powder
λ_{max}:	618nm in water
Solubility:	H_2O 40mg/ml
	EGME 10mg/ml
	EtOH 0.6mg/ml

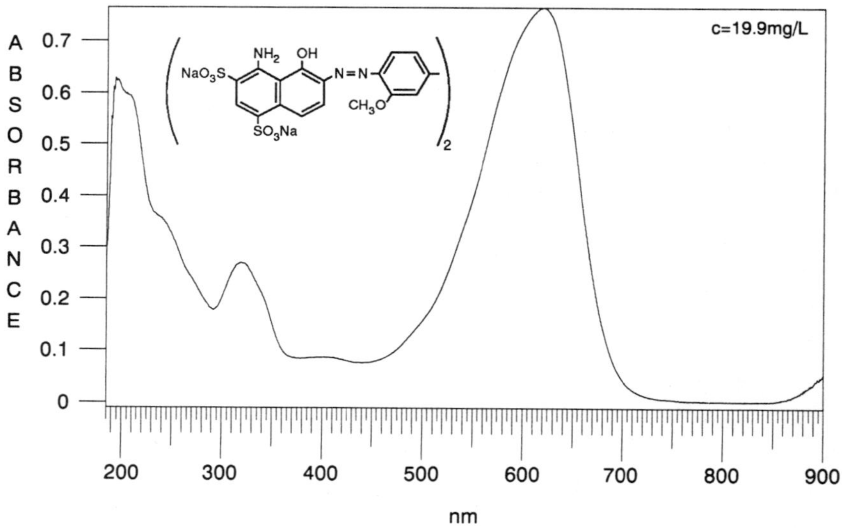

This direct disazo dye is made by coupling two moles of Chicago Acid (1-amino-8-naphthol-2,4-disulfonic acid) to one mole of tetraazotized *o*-dianisidine, which is then converted to the tetrasodium salt. It is soluble in water with a bright-blue color, slightly soluble in methyl Cellosolve and virtually insoluble in other organic solvents. It has a good affinity for cotton and other cellulosic materials, while it is somewhat less attracted to silk and wool (in the cold, the latter two are practically unstained, as are acetates and nylon). Nylon can be dyed with Chicago Sky Blue 6B by padding with ammonium tartrate. Other industrial applications include the production of aqueous inks, the production of lakes for pigments and the dyeing of paper and leather.

Lillie[1] described Chicago Sky Blue 6B as a fairly good, selective collagen stain when it is used in the Masson trichrome or Van Gieson techniques.

Reference:

1) Lillie, R.D. *J. Tech. Methods* **1945**, *25*, 1-47.

General references:

Aldrich FT-IR Library	**1**(2), 996D
Sigma FT-IR Library	**1**(2), 114C
IR Library	**3**, 1466G
Colour Index	3rd ed., 1971

CHLOROPHENOL RED

3',3"-Dichlorophenolsulfonephthalein

Aldrich Cat. No.:	19,952-4
Sigma Prod. No.:	C1268
Grade:	Indicator

CAS No.:	[*4430-20-0*]
Mol. Form.:	$C_{19}H_{12}Cl_2O_5S$
F.W.:	423.28
Appearance:	Green powder
λ_{max}:	572nm in 0.1*N* sodium hydroxide
Solubility:	H_2O 30mg/ml
	EGME 50mg/ml
	EtOH 80mg/ml

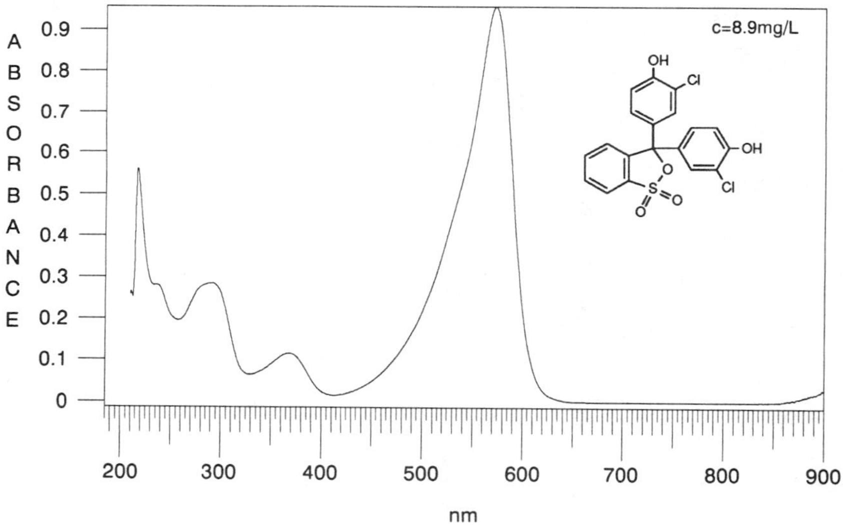

Chlorophenol Red is a member of the sulfonephthalein group of acid-base indicators. It is made by condensing two moles of 2-chlorophenol with one mole of 2-sulfobenzoic acid anhydride. Both the free-acid and the sodium-salt (water-soluble) forms are available. The free acid forms glistening green crystals from acetic acid. It is soluble in methanol, ethanol, ethyl acetate and acetic acid. The water-soluble form, as the description implies, is soluble in water and essentially insoluble in nonpolar solvents.

The visual-transition interval occurs from pH 4.8 (yellow) to pH 6.4 (red) and the pK is 6.0.[1] Information on the preparation of Chlorophenol Red solutions for use as a pH indicator is included in the pH Indicator Chart in the back of this book.

Apparently, Chlorophenol Red has no industrial application.

Reference:

1) Kolthoff, I.M.; Rosenblum, C. *Acid-Base Indicators;* Macmillan: New York, 1937.

General references:

Aldrich FT-IR Library	**1**(2), 1023A
Sigma FT-IR Library	**1**(2), 115A
IR Library	**3**, 1479H

CHLOROPHENOL RED, water soluble
3',3"-Dichlorophenolsulfonephthalein, sodium salt

Aldrich Cat. No.: 23,548-2

Mol. Form.: $C_{19}H_{11}Cl_2NaO_5S$
F.W.: 445.26
Appearance: Brown crystals
λ_{max}: 575nm in water
Solubility: H_2O 60mg/ml
EGME 30mg/ml
EtOH 10mg/ml

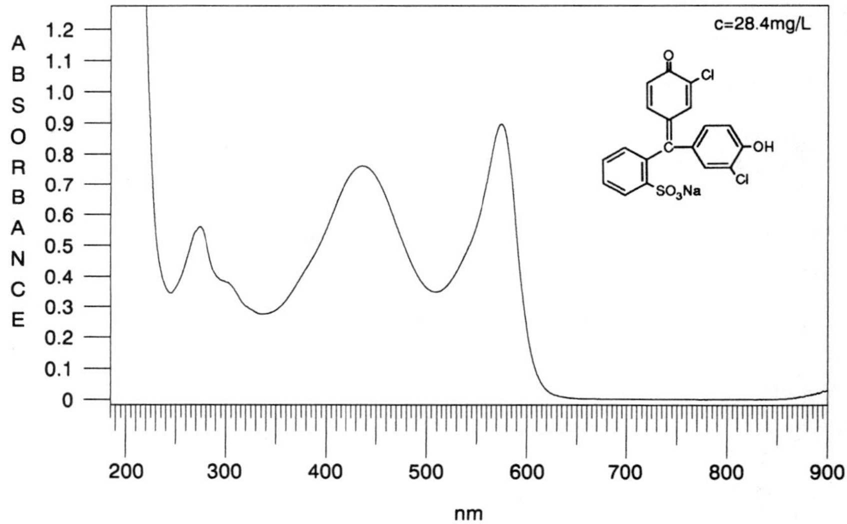

See description on Pages 200 and 201 under Chlorophenol Red.

General reference:

IR Library **3,** 1479H

CHROME AZUROL S
Mordant Blue 29

Aldrich Cat. No.: 19,953-2
Sigma Prod. No.: C1018

CAS No.: [*1667-99-8*]
C.I. No.: 43825
Mol. Form.: $C_{23}H_{13}Cl_2Na_3O_9S$
F.W.: 605.29
Appearance: Black crystalline powder
λ_{max}: 429nm in water
Solubility: H_2O 100mg/ml
EGME 100mg/ml
EtOH 4mg/ml

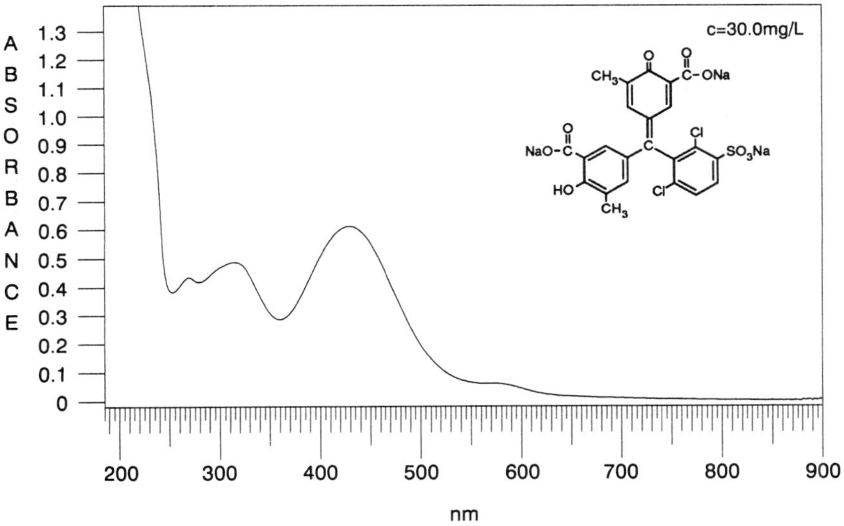

Chrome Azurol S is a blue triphenylmethane dye with a *para*-quinone chromophore. It is very soluble in water with a brownish-yellow color and slightly soluble in alcohol yielding a reddish-brown solution. When chromium is used as a mordant, a reddish-blue color is produced.

Industrial applications are confined to textile dyeing and printing. Chrome Azurol S dyes chrome-mordanted wool a reddish-blue, silk afterchromed a reddish-blue and chromate-padded cotton a reddish-blue. Printing on wool is accomplished with chromium acetate, on silk with chromate and cotton with vigoureux printing.

There is no evidence that it has been used as a biological stain.

General references:

Aldrich FT-IR Library **1**(2), 1032B
Sigma FT-IR Library **1**(2), 115C
IR Library **3**, 1484H
Colour Index 3rd ed., 1971

CHROMOTROPE FB

Acid Red 14 • Azorubin • Mordant Blue 79

Aldrich Cat. No.: 21,451-5

CAS No.: *[3567-69-9]*
C.I. No.: 14720
Mol. Form.: $C_{20}H_{12}N_2Na_2O_7S_2$
F.W.: 502.44
Appearance: Brown powder
λ_{max}: 515nm in water
Solubility: H_2O 30mg/ml
EGME 30mg/ml
EtOH 0.9mg/ml

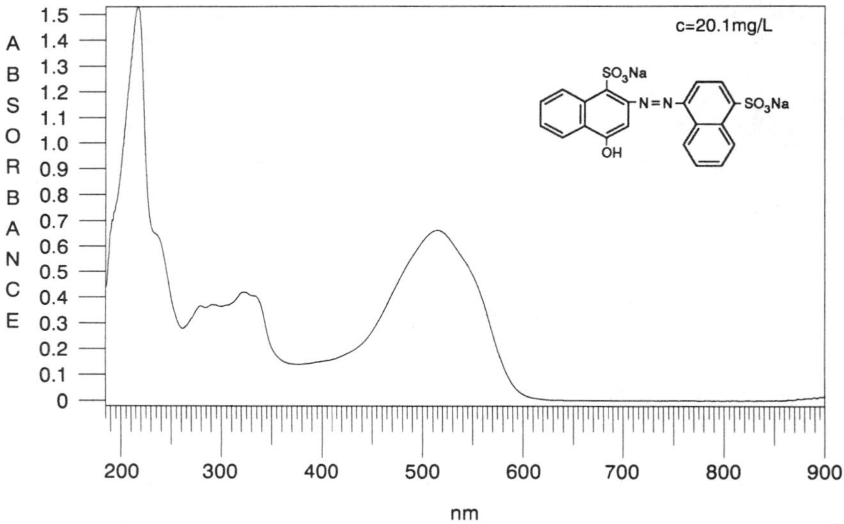

Chromotrope FB is a bluish-red, anionic monoazo dye that is made by coupling diazotized naphthionic acid to Nevile and Winther acid (1-naphthol-4-sulfonic acid) and converting the coupling product to the disodium salt.

Soluble in water and very slightly soluble in ethanol and acetone, it is used to dye and print wool, silk and nylon (lightfastness, 3), to dye chromed leather, to color paper and anodized aluminum, to stain wood and to prepare inks and pigments (heavy-metal salts). In Great Britain, it has been used as a food color (Food Red 3).

Conn's[1] cites the fact that Curtis, in 1905, used this dye under the name of Azorubin S and in combination with picric acid, to stain collagen fibers.

Reference:

1) *Conn's Biological Stains,* 9th ed.; Lillie, R.D., Ed.; Williams and Wilkins: Baltimore, 1977; p 109.

General references:

Aldrich FT-IR Library	**1**(2), 988C
IR Library	**3**, 1463B
Colour Index	3rd ed., 1971

CHROMOTROPE 2B

Acid Red 176 • Chromotrope Red 4B

Aldrich Cat. No.:	11,039-6
Sigma Prod. No.:	C1024
CAS No.:	[*548-80-1*]
C.I. No.:	16575
Mol. Form.:	$C_{16}H_9N_3Na_2O_{10}S_2$
F.W.:	513.37
m.p.:	>300°C
Appearance:	Greenish-brown powder
λ_{max}**:**	514nm in water
Solubility:	H_2O 20mg/ml
	EGME 30mg/ml
	EtOH 1mg/ml

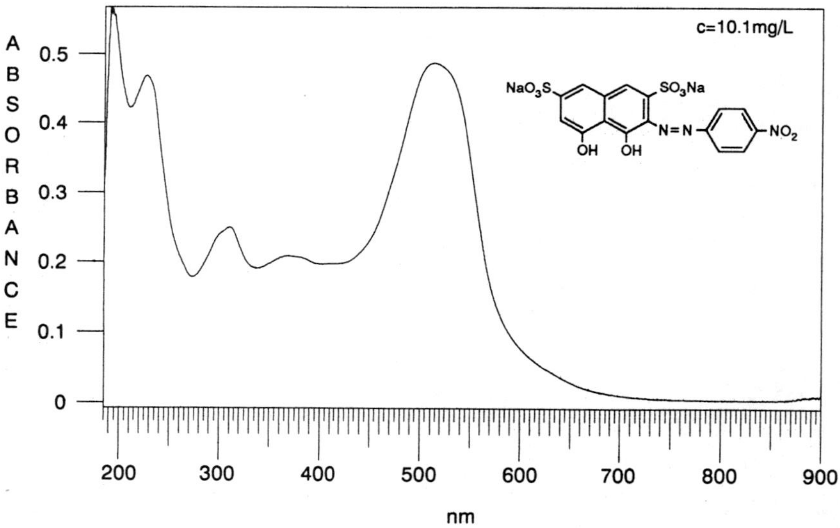

Chromotrope 2B is a bluish-red, acid monoazo dye that is made by coupling diazotized 4-nitroaniline to chromotropic acid. The coupling product is isolated and converted to the disodium salt. It is soluble in water and very slightly soluble in ethanol. Wool is dyed red from an acid bath, turning dark blue, with good lightfastness when afterchromed.

There is no evidence that it has been used as a biological stain.

General references:

Aldrich FT-IR Library	**1**(2), 986A
Sigma FT-IR Library	**1**(2), 115D
IR Library	**3**, 1462A
Colour Index	3rd ed., 1971

CHROMOTROPE 2R
Acid Red 29

Aldrich Cat. No.: 20,208-8
Sigma Prod. No.: C3143

CAS No.: [*4197-07-3*]
C.I. No.: 16570
Mol. Form.: $C_{16}H_{10}N_2Na_2O_8S_2$
F.W.: 468.37
Appearance: Reddish-brown powder
λ_{max}: 510(530)nm in methanol
Solubility: H_2O 200mg/ml
EGME 20mg/ml
EtOH 3mg/ml

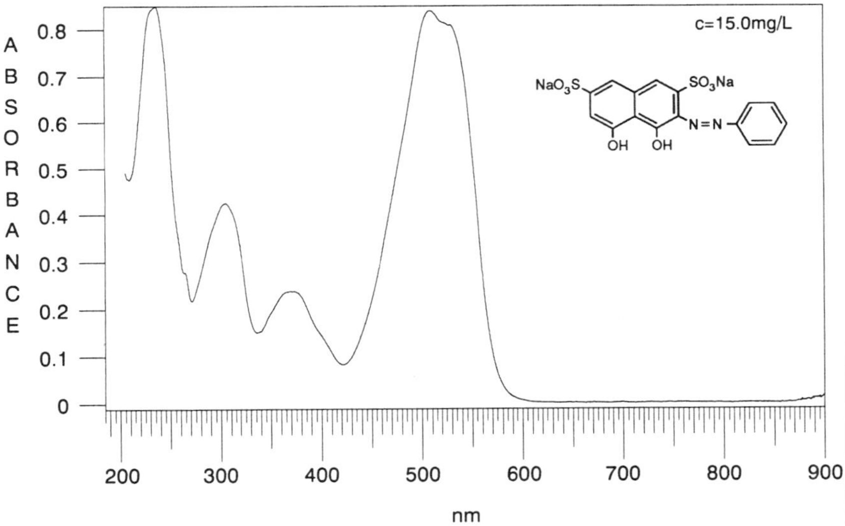

Chromotrope 2R is a red, anionic monoazo dye which is very water soluble. It has affinity for proteinaceous substrates and no attraction for cellulosic materials. It is made by coupling diazotized aniline to chromotropic acid (1,8-dihydroxynaphthalene-3,6-disulfonic acid). The coupling product is isolated and converted to the disodium salt. It dyes afterchromed wool a dark blue, silk out of an acetic acid bath, nylon out of an acetic acid or formic acid bath and chromed leather. It is used, also, for coloring plastics (*e.g.*, melamine, urea and nitrocellulose).

Lendrum[1] used Chromotrope 2R as a substitute for Eosin and Hematoxylin in a counterstain to Celestine Blue in animal histology. Ewen[2] included Chromotrope 2R as one of four dyes used in his recipe for Aldehyde Fuchsin to demonstrate neurosecretory products in insects.

References:

1) Lendrum, A.C. *J. Pathol. Bacteriol.* **1935,** *40*, 415-416.

2) Ewen, A.B. *Trans. Amer. Microsc. Soc.* **1962,** *81,* 94-96.

General references:

Aldrich FT-IR Library	**1**(2), 982D
Sigma FT-IR Library	**1**(2), 116A
IR Library	**3**, 1460B
Colour Index	3rd ed., 1971

CHROMOXANE CYANINE R

Mordant Blue 3 • Solochrome Cyanin R

Aldrich Cat. No.:	23,406-0
Sigma Prod. No.:	E2502
CAS No.:	[*3564-18-9*]
C.I. No.:	43820
Mol. Form.:	$C_{23}H_{15}Na_3O_9S$
F.W.:	536.40
Appearance:	Red powder
λ_{max}:	512nm in methanol
Solubility:	H_2O 50mg/ml
	EGME 10mg/ml
	EtOH 20mg/ml

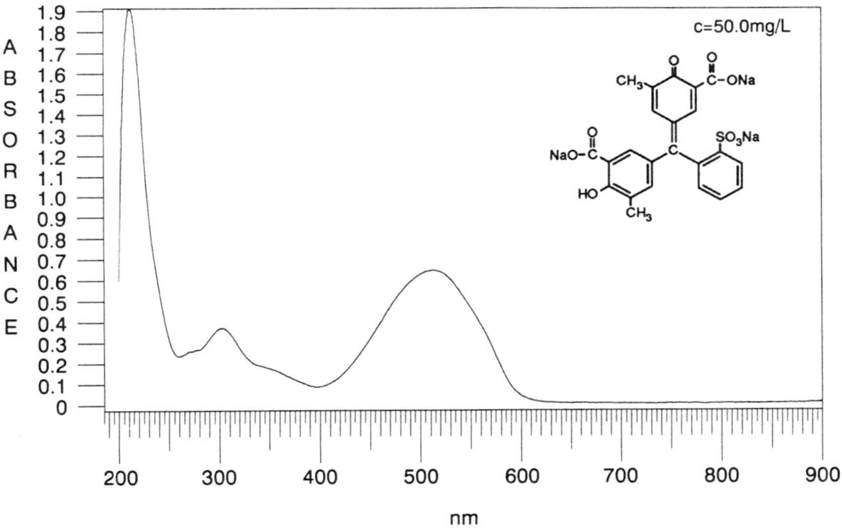

This anionic triphenylmethane dye is made by condensing one mole of *o*-formylbenzenesulfonic acid with two moles of 3-methylsalicylic acid, oxidizing the product with nitrosylsulfuric acid and converting to the trisodium salt.

Soluble in water (red) and somewhat less soluble in ethanol (orange-yellow), it is used to dye wool, nylon (afterchromed) and chromed leather and to print chromium-mordanted wool, silk and cotton.

Staining Procedures[1] provides a recipe and method, after Pearse (1957) amended by Clark (1979), for its use with ferric chloride to stain nuclei and myelin sheath. When counterstained with Eosin, it is claimed to closely resemble H&E stain. As a myelin-sheath stain, it is reported to closely resemble Luxol Fast Blue in results but differentiation is accomplished much more easily, suggesting that Chromoxane Cyanine R should replace the latter in this application.

During the Hematoxylin shortage, Llewellyn[2] suggested Mordant Blue 3 as a readily available substitute for the former in routine H&E applications.

References:

1) Clark, G. *Staining Procedures,* 4th ed.; Williams and Wilkins: Baltimore, 1981; p 109.

2) Llewellyn, B.D. *Stain Technol.* **1974,** *49*, 347-349.

General references:

Aldrich FT-IR Library **1**(2), 1031C
Sigma FT-IR Library **1**(2), 135D
Colour Index 3rd ed., 1971

CHRYSOIDIN
Basic Orange 2

Aldrich Cat. No.:	21,449-3
CAS No.:	*[532-82-1]*
C.I. No.:	11270
Mol. Form.:	$C_{12}H_{12}N_4 \cdot HCl$
F.W.:	248.72
m.p.:	235°C (dec.)
Appearance:	Burgundy crystalline powder
λ_{max}:	449nm in water
Solubility:	H_2O 20mg/ml
	EGME 20mg/ml
	EtOH 9mg/ml

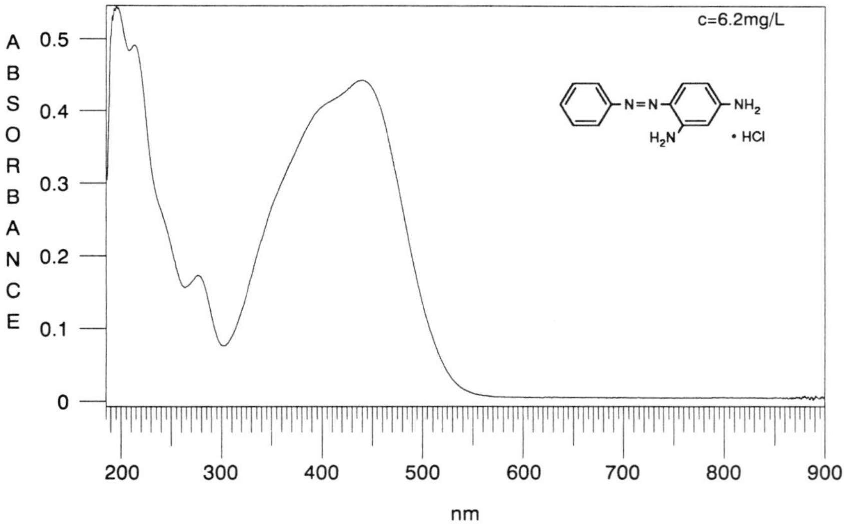

Chrysoidin is a cationic, orange dye with a monoazo chromophore. The product of commerce is the hydrochloride salt.

Soluble in water, methyl Cellosolve and acetone, slightly soluble in ethanol and practically insoluble in benzene, it is used to dye cotton mordanted with tannin, silk and chromed leather, to color cellulose-ester plastics and to stain wood.

Chrysoidin possesses acid-base-indicator properties, displaying a visual-transition interval from pH 4.0 (orange) to pH 7.0 (yellow). See the pH Indicator Chart in the back of this book.

The free-base form (Solvent Orange 3, C.I. 11270:1) is used to color fats, waxes, oils and polishes, while the hydrochloride, citrate and thiocyanate salt forms have been used as antiseptics.

Conn's[1] states that it is a good substitute for Bismarck Brown in some biological methods and, like Bismarck Brown, is unstable to excessive heat in solution preparation. The same author reports that it has been used as a vital stain for pollen tubes, a vital stain for insects, a stain to demonstrate Golgi apparatus and that it was included among a series of dyes for studying gastric secretions.

Reference:

1) *Conn's Biological Stains,* 9th ed.; Lillie, R.D., Ed.; Williams and Wilkins: Baltimore, 1977; pp 87-88.

General references:

Aldrich FT-IR Library	**1**(2), 966B
IR Library	**3**, 1450E
Safety	**2**(1), 844D
Colour Index	3rd ed., 1971

CHRYSOPHENINE

Direct Yellow 12

Aldrich Cat. No.: 20,190-1

CAS No.: [*2870-32-8*]
C.I. No.: 24895
Mol. Form.: $C_{30}H_{26}N_4Na_2O_8S_2$
F.W.: 680.67
Appearance: Dark-orange powder
λ_{max}: 389nm in water

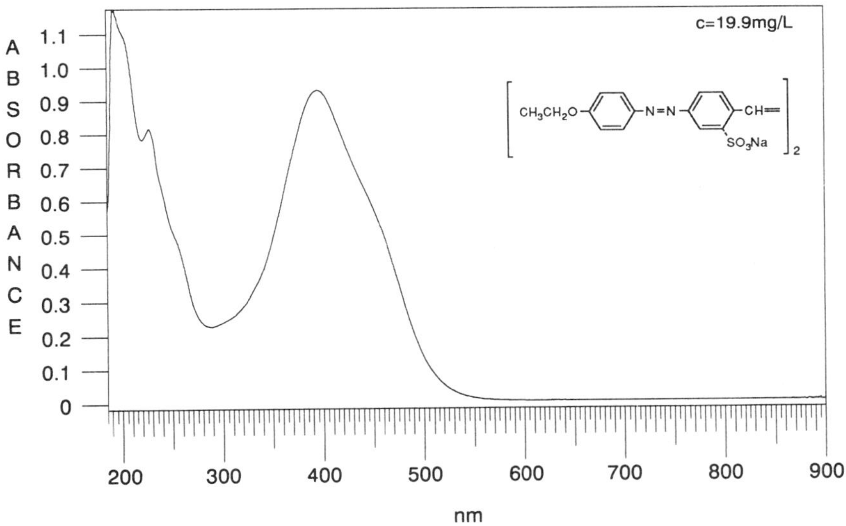

Chrysophenine is a yellow, anionic disazo dye. It has affinity for both proteinaceous and cellulosic substrates dyeing paper out of a neutral bath, wool with Glauber's salt or acetic acid, silk with a neutral or an acetic acid bath, nylon with acetic acid or ammonium tartrate and fully chromed leather, a bright yellow. Pigments prepared from heavy-metal salts are used for paper coatings, particularly in the manufacture of wallpaper.

Conn's[1] states that Chrysophenine has been used occasionally as a plasma stain in pathology.

Reference:

1) *Conn's Biological Stains,* 9th ed.; Lillie, R.D., Ed.; Williams and Wilkins: Baltimore, 1977; p 164.

General references:

Aldrich FT-IR Library	**1**(2), 980C
IR Library	**3**, 1458F
Safety	**2**(1), 885B
Colour Index	3rd ed., 1971

CIBACRON BRILLIANT RED 3B-A

Procion Brilliant Red H-7B • Reactive Red 4

Aldrich Cat. No.: 22,845-1
Sigma Prod. No.: R7878

CAS No.: [*17681-50-4*]
C.I. No.: 18105
Mol. Form.: $C_{32}H_{19}ClN_8Na_4O_{14}S_4$
F.W.: 995.23
Appearance: Maroon-purple powder
λ_{max}: 517nm in water
Solubility: H_2O 50mg/ml
EGME 30mg/ml
EtOH 1mg/ml

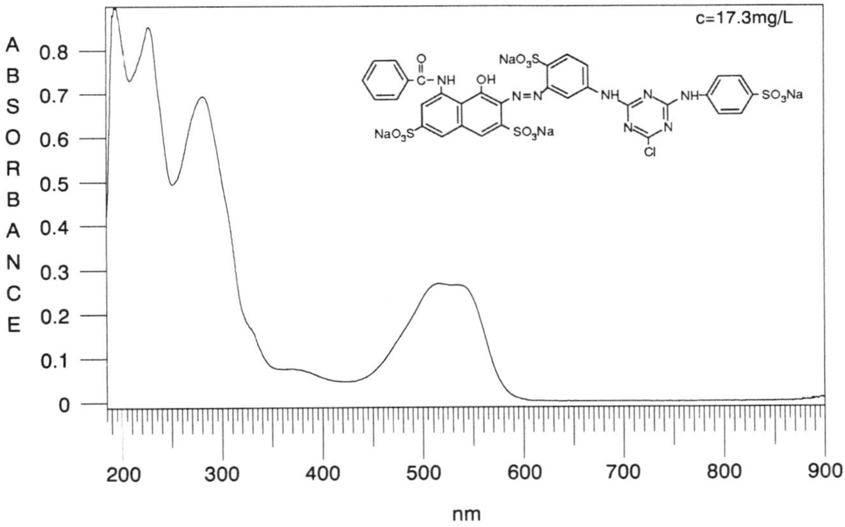

This red, monoazo monochlorotriazinyl dye is readily soluble in water and very slightly soluble in ethanol and methanol. A synthesis is described by Panchartek *et al.*[1]

Industrial applications include dyeing and direct printing of cellulose, nylon, silk and wool, and printing on the same materials padded with azoic coupling components No. 2 (3-hydroxy-2-naphthanilide) and No. 5 (4-nitro-*o*-anisidine).

Goland and Grand[2] used closely related dyes as markers and fixatives for the study of growth in teeth and bones. However, they preferred Procion M (dichlorotriazinyl) type dyes to the Procion H (monochlorotriazinyl) variety because of their greater reactivity at room temperature.

References:

1) Panchartek, J.; Allan, Z.J.; Muzik, F. *Collect. Czech. Chem. Commun.* **1960,** *25*, 2783-2799.

2) Goland, P.; Grand, N. *Amer. J. Phys. Anthropol.* **1968,** *29*, 201-218.

General references:

Aldrich FT-IR Library	**1**(2), 1005C
Sigma FT-IR Library	**1**(2), 207D
Safety	**2**(1), 3030D
Colour Index	3rd ed., 1971

CIBACRON BRILLIANT YELLOW 3G-P

Procion Brilliant Yellow H5G • Reactive Yellow 2

Aldrich Cat. No.:	22,847-8
Sigma Prod. No.:	R8003
CAS No.:	*[50662-99-2]*
C.I. No.:	18972
Mol. Form.:	$C_{25}H_{15}Cl_3N_9Na_3O_{10}S_3$
F.W.:	872.97
Appearance:	Gold-orange powder
λ_{max}:	404nm in water
Solubility:	H_2O 50mg/ml
	EGME 30mg/ml
	EtOH 2mg/ml

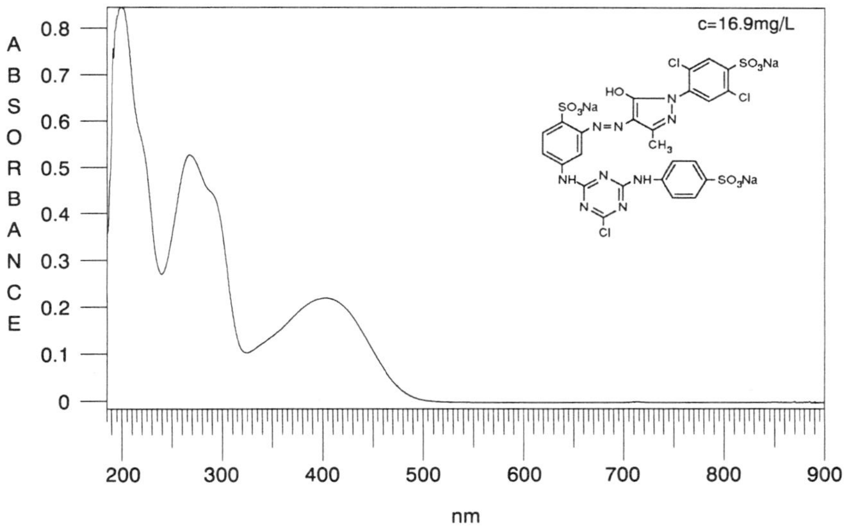

This yellow, monoazo monochlorotriazinyl dye is readily soluble in water and very slightly soluble in ethanol and methanol. Panchartek *et al.*[1] describe a preparative method for this dye, along with other closely related reactive dyes.

Industrial applications include dyeing and printing cotton, wool, silk, and nylon and printing on fabric padded with azoic coupling component No. 5 (4-nitro-*o*-anisidine).

There is no evidence that it has been used as a biological stain.

Reference:

1) Panchartek, J.; Allan, Z.J.; Muzik, F. *Collect. Czech. Chem. Commun.* **1960,** *25,* 2783-2799.

General references:

Aldrich FT-IR Library	**1**(2), 1005D
Sigma FT-IR Library	**1**(2), 208B
Safety	**2**(1), 3031B
Colour Index	3rd ed., 1971

COCHINEAL

Natural Red 4

Aldrich Cat. No.: 19,971-0

C.I. No.: 75470
Appearance: Reddish-black chunks
λ_{max}: 498nm in methanol
Solubility: H_2O 20mg/ml
EGME 20mg/ml
EtOH 0.9mg/ml

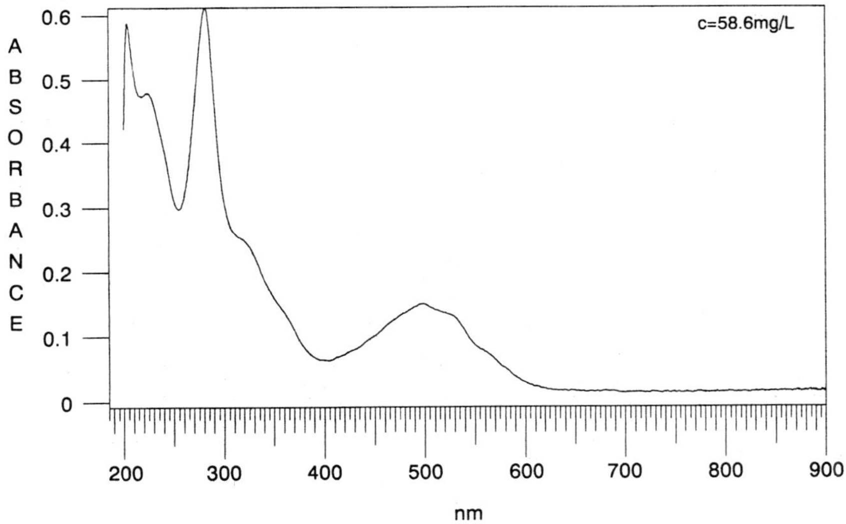

See description on pages 190-192 under Carmine.

General references:

Aldrich FT-IR Library	1(2), 1021A
IR Library	**3**, 1478E
Merck Index	**11**, 2453

CONGO RED
Direct Red 28

Aldrich Cat. No.:	86,095-6	C8,445-3
Sigma Prod. No.:	C6277	C6767
Grade:	Certified	Indicator
Dye content:	~85%	~65%

CAS No.:	*[573-58-0]*
C.I. No.:	22120
Mol. Form.:	$C_{32}H_{22}N_6Na_2O_6S_2$
F.W.:	696.67
Appearance:	Dark, red-brown powder
λ_{max}:	497nm in water + 1ml 1% Na_2CO_3
Solubility:	H_2O 40mg/ml
	EGME 10mg/ml
	EtOH 1mg/ml

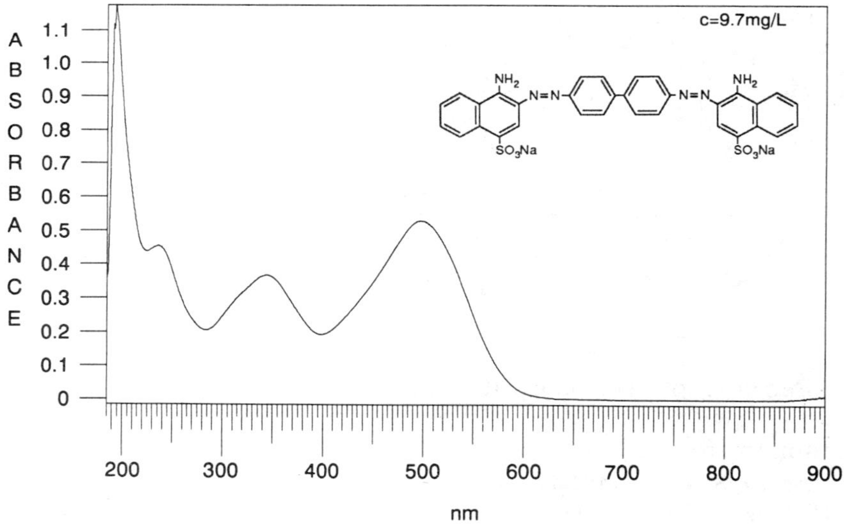

This red, anionic disazo dye was initially made by coupling two moles of naphthionic acid to one of tetraazotized benzidine. Currently, due to health and environmental considerations a hydrazobenzene rearrangement, in which benzidine is not isolated, is substituted for the carcinogenic benzidine. Congo Red is soluble in water, moderately soluble in methyl Cellosolve and ethylene glycol and very slightly soluble in ethanol.

Congo Red is an indicator with a visual-transition interval of pH 3.0 (blue) to pH 5.2 (red). While it has been used widely as an indicator in areas as diverse as test paper and estimation of free mineral acids, particularly in the presence of organic acids and culture media, Kolthoff and Rosenblum[1] did not consider it a very desirable indicator due to high salt and protein errors.

This dye has affinity for both proteinaceous and cellulosic substrates, dyeing cotton directly and wool out of a neutral bath. Other industrial applications include calico printing and dyeing paper.

Kelly[2] noted that Congo Red is the only known example of an anionic dye with metachromatic properties. He reported observing metachromasia in tissue sections and in solutions with certain basic materials stained with Congo Red. Baker[3] cited Indigo Carmine, Orange G and other anionic dyes, in addition to Congo Red, as being metachromatic. He indicated that the wavelength shift is bathochromic and not hypsochromic, as demonstrated by metachromatic cationic dyes. Congo Red is a good contrast stain or counterstain. Clark[4] included it in his list of substitutes for Eosin, as a counterstain for Hematoxylin, as a dry negative capsule stain and in the Dyar cell-wall stain. Bennhold,[5] Highman,[6] and Puchtler *et al.*[7] described Congo Red as a specific stain for amyloids in pathology.

References:

1) Kolthoff, I.M.; Rosenblum, C. *Acid-Base Indicators;* Macmillan: New York, 1937; p 150.
2) Kelly, W.J. *Stain Technol.* **1958,** *33*, 89-94.
3) Baker, J.R. *Principles of Biological Microtechnique*; Meuther: London, Wiley: New York, 1958; pp 260-261.
4) Clark, G. *Staining Procedures,* 4th ed.; Williams and Wilkins: Baltimore, 1981; pp 107, 387, 395.
5) Bennhold, H. *Munch. Med. Wochenshr.* **1922,** *2*, 1537-1538.
6) Highman, B. *Arch. Pathol.* **1946,** *41*, 559-562.
7) Puchtler, H.; Sweat, F.; Levine, M. *J. Histochem. Cytochem.* **1962,** *10*, 355-357.

General references:

Aldrich FT-IR Library	**1**(2), 993C	——
Sigma FT-IR Library	**1**(2), 117B	**1**(2), 117A
IR Library	**3**, 1465F	——
Safety	**2**(1), 908C	
Merck Index	**11**, 2493	
Colour Index	3rd ed., 1971	

COPPER PHTHALOCYANINE

Pigment Blue 15

Aldrich Cat. No.:	25,298-0
Sigma Prod. No.:	C1030
CAS No.:	[*147-14-8*]
C.I. No.:	74160
Mol. Form.:	$C_{32}H_{16}CuN_8$
F.W.:	576.08
λ_{max}:	795nm in conc. sulfuric acid
Appearance:	Dark-blue powder

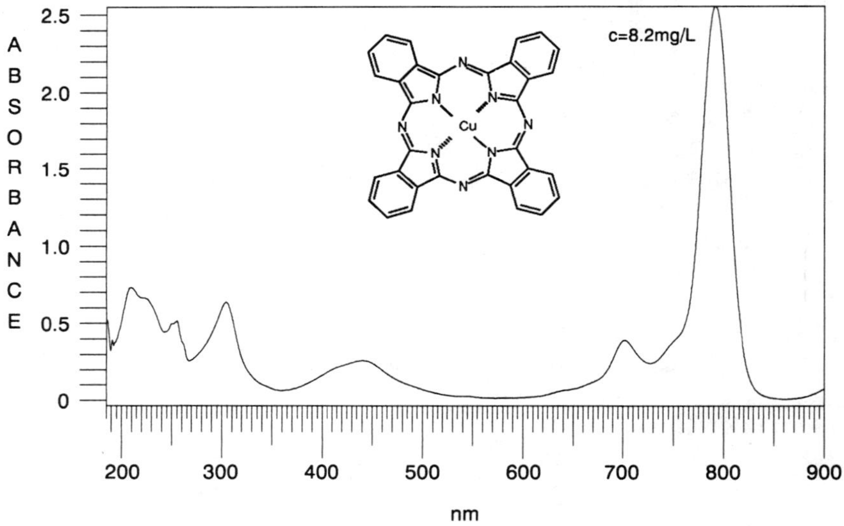

This phthalocyanine pigment is made by condensing phthalonitrile in the presence of cuprous chloride at 180-200°C. The product displays an overall good-to-excellent fastness in all categories. This high fastness coupled with its bright-blue color is responsible for its great demand as a pigment for paints, alkyd-resin enamels, printing inks, lacquers, emulsion paints and plastics. Its other industrial applications include its use as an intermediate for direct dyes such as Heligon Blue SBL (Direct Blue 86, C.I. 74180) and ingrain dyes, including Alcian Blue (Ingrain Blue 1, C.I. 74240).

There is no evidence that it has been used as a biological stain .

General references:

Aldrich FT-IR Library	**1**(2), 578D
Sigma FT-IR Library	**1**(2), 117C
Safety	**2**(2), 2473B
Merck Index	**11**, 2515
Colour Index	3rd ed., 1971

COPPER PHTHALOCYANINETETRASULFONIC ACID, tetrasodium salt

Acid Blue 249 • Heligon Blue S4

Aldrich Cat. No.:	27,401-1
Sigma Prod. No.:	C2029
CAS No.:	[*27360-85-6*]
C.I. No.:	74220
Mol. Form.:	$C_{32}H_{12}CuN_8Na_4O_{12}S_4$
F.W.:	984.26
Appearance:	Blue-violet powder
λ_{max}**:**	610(694)nm in water
Solubility:	H_2O 7mg/ml
	EGME 2mg/ml
	EtOH 0.2mg/ml

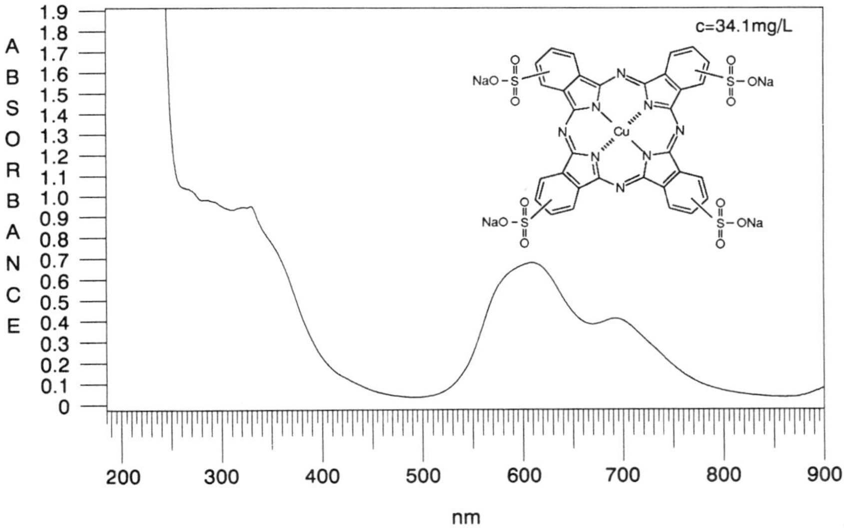

This anionic phthalocyanine dye is made by the direct sulfonation of Pigment Blue 15 (C.I. 74160) with oleum at 100°C. It may differ slightly from C.I. 74220, which is made by condensing 4-sulfophthalic acid and cyanoguanidine in the presence of cuprous chloride. The latter synthesis insures that the four sulfonic acid groups reside in the 4, 4',4'',4'''-positions, while the former tends to yield mixed levels and positioned sulfonic-acid substitution, although the tetrasubstituted 4-isomer variety just described accounts for 80% of the yield.

Slightly soluble in water with a bright blue color, it is used to dye wool and silk and to color cosmetics.

There is no evidence that it has been used as a biological stain.

General references:

Sigma FT-IR Library	**1**(2), 404C
Safety	**2**(1), 915D
Colour Index	3rd ed., 1971

COUMARIN

Aldrich Cat. No.:	C8,555-7
Sigma Prod. No.:	C4261
CAS No.:	*[91-64-5]*
Mol. Form.:	$C_9H_6O_2$
F.W.:	146.15
m.p.:	68.5-69.5°C
b.p.:	298°C/760mm Hg
Appearance:	White crystalline powder with a fibrous texture
λ_{max}:	275nm in ethanol
Solubility:	H_2O <0.1mg/ml
	EGME <0.1mg/ml
	EtOH <0.1mg/ml

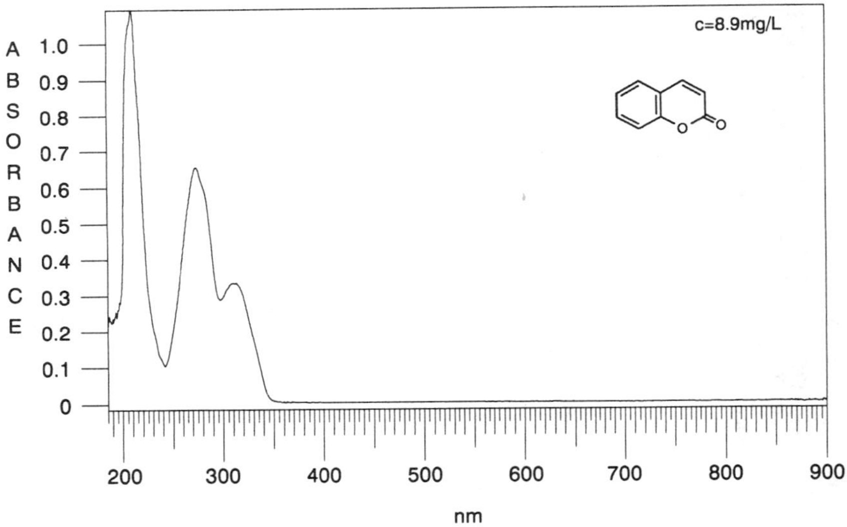

Coumarin is not a dye, but it serves as the nucleus for a number of laser dyes. Meuly[1] reports that one gram is soluble in 400ml of cold water and 50ml of hot water, and the compound is freely soluble in warm ethanol and in chloroform, ether and alkali-hydroxide solutions. Although there are no reported biological-stain applications, it has been used as a fluorescent indicator displaying a fluorescent-transition interval of pH 8.0 (weak green) to pH 9.5 (green).

Reference:

1) Meuly, W.C. In *Kirk-Othmer Encyclopedia of Chemical Technology*, 3rd edition; Wiley: New York, 1977; Vol. 7, pp 196-206.

General references:

Aldrich FT-IR Library	**1**(2), 322B
Sigma FT-IR Library	**1**(2), 117D
IR Library	**3**, 1044B
NMR Library	**2**(2), 309A
Safety	**2**(1), 921A
Merck Index	**11**, 2563

o-CRESOLPHTHALEIN

3,3'-Dimethylphenolphthalein

Aldrich Cat. No.: C8,577-8
Sigma Prod. No.: C2768
Grade: Indicator

CAS No.: [*596-27-0*]
Mol. Form.: $C_{22}H_{18}O_4$
F.W.: 346.38
m.p.: 223-225°C
Appearance: Beige powder
λ_{max}: 566(381)nm in 0.1*N* sodium hydroxide
Solubility: H_2O 1mg/ml
EGME 90mg/ml
EtOH 50mg/ml

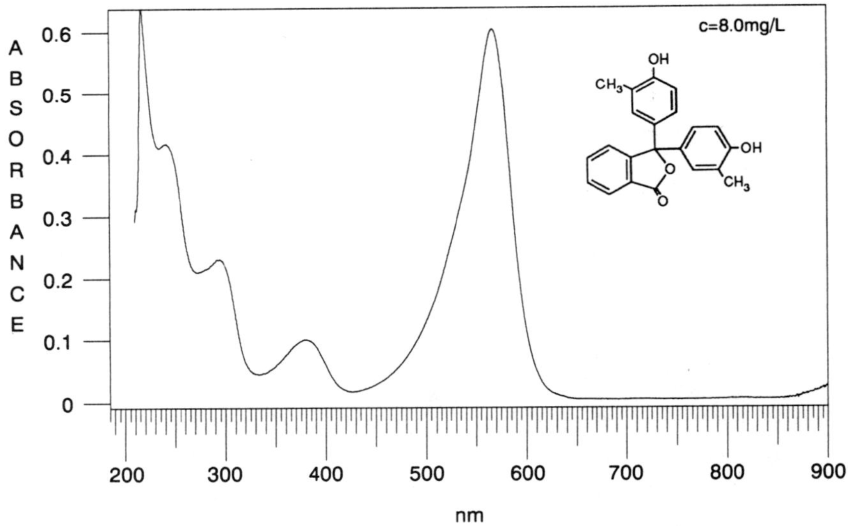

o-Cresolphthalein is a member of the phthalein-indicator series and is made by condensing one mole of phthalic anhydride with two moles of *o*-cresol.

This white, crystalline-powder product is soluble in alcohol and aqueous alkali solutions and very slightly soluble in water. An acid-base indicator,[1] it has a visual-transition interval extending from pH 8.2 (colorless) to pH 9.8 (red) and a pK of 9.4. For more complete information on the preparation of indicator solutions, see the pH Indicator Chart in the back of this book.

Welcher[2] lists its analytical reagent applications.

References:

1) Kolthoff, I.M.; Rosenblum, C. *Acid-Base Indicators*; Macmillan: New York, 1937; p 114.
2) Welcher, F.J. *Organic Analytical Reagents;* Van Nostrand: New York, 1949; Vol. 4, p 482.

General references:

Aldrich FT-IR Library	**1**(2), 1006D
Sigma FT-IR Library	**1**(2), 119B
IR Library	**3**, 1471E
Merck Index	**11**, 2582

o-CRESOLPHTHALEIN COMPLEXONE

o-Cresolphthalexon • Phthalein Purple

Aldrich Cat. No.:	18,819-0
Sigma Prod. No.:	P5631
CAS No.:	*[2411-89-4]*
Mol. Form.:	$C_{32}H_{32}N_2O_{12}$
F.W.:	636.62
Appearance:	White powder
λ_{max}:	575(377)nm in 0.1*N* sodium hydroxide
Solubility:	H_2O 2mg/ml
	EGME 20mg/ml
	EtOH 4mg/ml

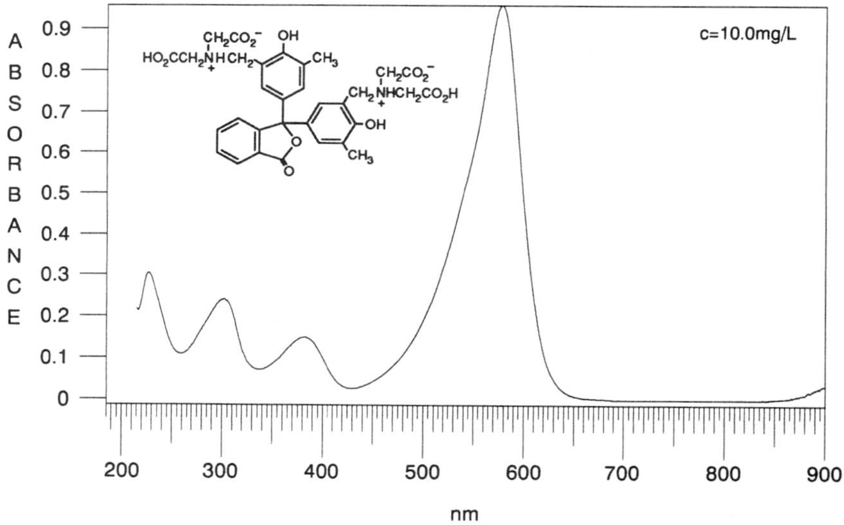

o-Cresolphthalein Complexone is a substituted phenolphthalein that is used as a reagent for manual and automated determination of calcium in serum. Information on the preparation of reagent solutions and the analytical procedures utilizing these solutions is provided in the references listed below.[1,2]

References:

1) Kessler, G.; Wolfman, M. *Clin. Chem.* **1964,** *10*, 686-703.

2) Ray Sakar, B.C.; Chauhan, U.P.S. *Anal. Biochem.* **1967,** *20*(1), 155-166.

General references:

Aldrich FT-IR Library	**1**(2), 1007D
Sigma FT-IR Library	**1**(2), 198A
IR Library	**3,** 1472A

m-CRESOL PURPLE

m-Cresolsulfonephthalein

Aldrich Cat. No.: 85,789-0
Sigma Prod. No.: C1393
Grade: Indicator

CAS No.: [*2303-01-7*]
Mol. Form.: $C_{21}H_{18}O_5S$
F.W.: 382.43
Appearance: Olive-green powder
λ_{max}: 579(371)nm in 0.005*N* sodium hydroxide in methanol
Solubility: H_2O 0.5mg/ml
EGME 3mg/ml
EtOH 0.4mg/ml

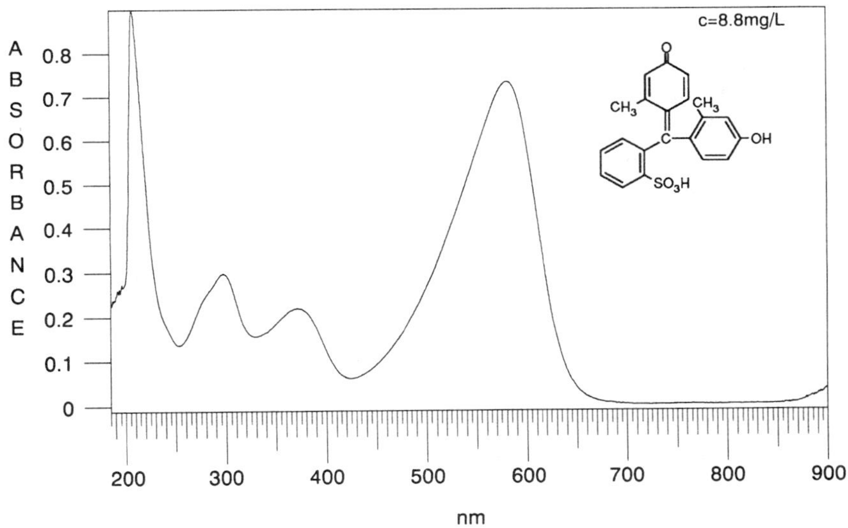

m-Cresol Purple is a member of the sulfonephthalein series of acid-base indicators and is made by condensing one mole of *o*-sulfobenzoic acid anhydride with two moles of *m*-cresol. The indicator grade is the free-acid form rather than the sultone and is very slightly soluble in water and ethanol and soluble in methanol and acetic acid. The monosodium salt is soluble in the lower aliphatic alcohols and slightly soluble in water. Both forms are dark, indicating the existence of quinoidal structures. See the pH Indicator Chart in the back of this book for more complete information on the preparation of indicator solutions. *m*-Cresol Purple has two visual-transition intervals: pH 1.2 (red) to pH 2.8 (yellow) and pH 7.4 (yellow) to pH 9.0 (purple).[1] Probably its greatest value is that it is the precursor of Bromocresol Green

(3',3",5',5"-tetrabromo-*m*-cresolsulfonephthalein) which is much more widely used.

Reference:

1) Kolthoff, I.M.; Rosenblum, C. *Acid-Base Indicators;* Macmillan: New York, 1937; p 131.

General references:

Aldrich FT-IR Library **1**(2), 1021C
Sigma FT-IR Library **1**(2), 119C
IR Library **3**, 1479A

m-CRESOL PURPLE, sodium salt

Aldrich Cat. No.: 21,176-1
Sigma Prod. No.: C0521

CAS No.: *[62625-31-4]*
Mol. Form.: $C_{21}H_{17}NaO_5S$
F.W.: 404.42
Appearance: Black metallic powder
λ_{max}: 436nm in water
Solubility: H_2O 2mg/ml
EGME 10mg/ml
EtOH 40mg/ml

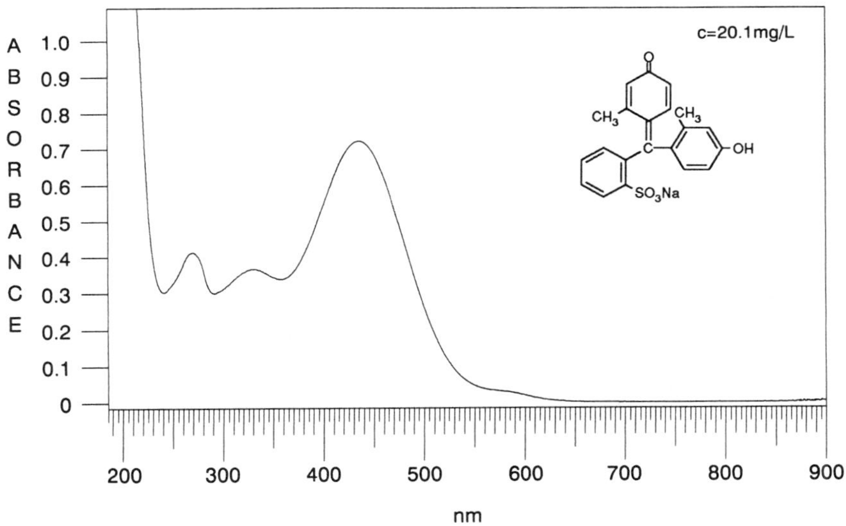

See description on pages 229 and 230 under *m*-Cresol Purple.

General references:

Aldrich FT-IR Library	**1**(2), 1021D
Sigma FT-IR Library	**1**(2), 119D
IR Library	**3**, 1479B

CRESOL RED
o-Cresolsulfonephthalein

Aldrich Cat. No.:	11,447-2
Sigma Prod. No.:	C7524
Grade:	Indicator

CAS No.:	[*1733-12-6*]
Mol. Form.:	$C_{21}H_{18}O_5S$
F.W.:	382.43
Appearance:	Metallic green powder
λ_{max}:	570nm in water
Solubility:	H_2O 1mg/ml
	EGME 10mg/ml
	EtOH 3mg/ml

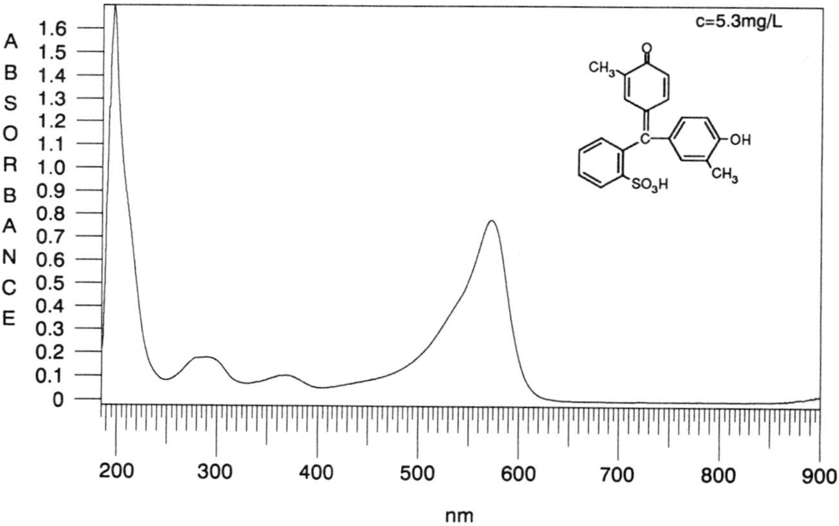

Cresol Red is a member of the sulfonephthalein series of indicators. It is made by condensing two moles of *o*-cresol with one of *o*-sulfobenzoic acid anhydride. There are two forms, both of which are dark, indicating that they both contain a chromophore. One is the free acid instead of the sultone and the other is the monosodium salt. The free acid is very slightly soluble in water and slightly soluble in methanol and ethanol, while the sodium salt is soluble in water and moderately soluble in methanol and ethanol. It has two visual-transition intervals, pH 0.2 (orange) to pH 1.8 (yellow) and pH 7.2 (yellow) to pH 8.8 (reddish-purple), with a pK of 8.3.

Please see the pH Indicator Chart in the back of this book for information pertaining to solution preparation and use as acid-base indicator.[1]

Cresol Red is the precursor of Bromocresol Purple, dibromo-*o*-cresolsulfonephthalein.

Reference:

1) Kolthoff, I.M.; Rosenblum, C. *Acid-Base Indicators;* Macmillan: New York, 1937; pp 108, 109, 118, 120, 279, 386.

General references:

Aldrich FT-IR Library	**1**(2), 1016B
Sigma FT-IR Library	**1**(2), 210A
IR Library	**3**, 1476C
Safety	**2**(1), 925B
Merck Index	**11**, 2583

CRESOL RED, water soluble

Aldrich Cat. No.:	11,448-0
Sigma Prod. No.:	C9877
Grade:	Indicator
CAS No.:	*[62625-29-0]*
Mol. Form.:	$C_{21}H_{17}NaO_5S$
F.W.:	404.42
Appearance:	Brown powder
λ_{max}**:**	425nm in ethanol
Solubility:	H_2O 30mg/ml
	EGME 20mg/ml
	EtOH 20mg/ml

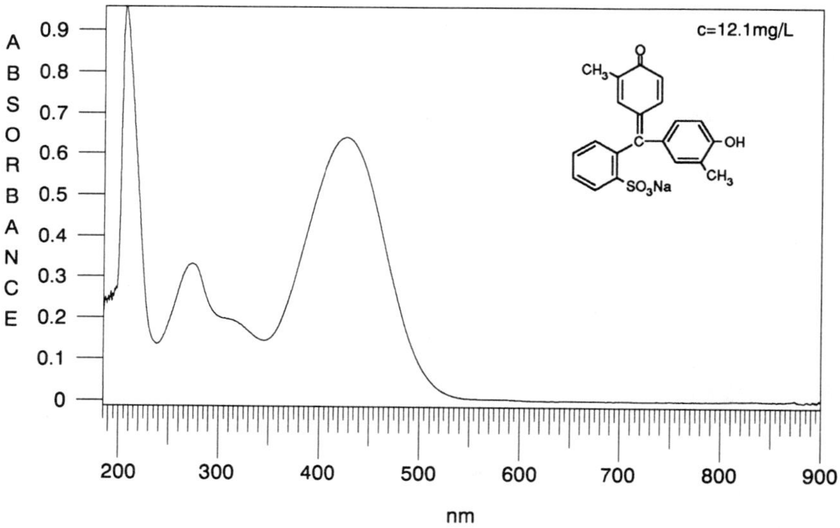

See description on pages 232 and 233 under Cresol Red.

General references:

Aldrich FT-IR Library	**1**(2), 1016C
Sigma FT-IR Library	**1**(2), 120B
IR Library	**3**, 1476C

CRESYL VIOLET acetate

5,9-Diaminobenzo[*a*]phenoxazonium acetate

Aldrich Cat. No.:	86,098-0
Sigma Prod. No.:	C6402
Grade:	Certified

CAS No.:	[*10510-54-0*]
Mol. Form.:	$C_{18}H_{15}N_3O_3$
F.W.:	321.34
m.p.:	140-143°C
Appearance:	Dark-green powder
λ_{max}:	596nm in 50% ethanol
Solubility:	H_2O 20mg/ml
	EGME 10mg/ml
	EtOH 10mg/ml

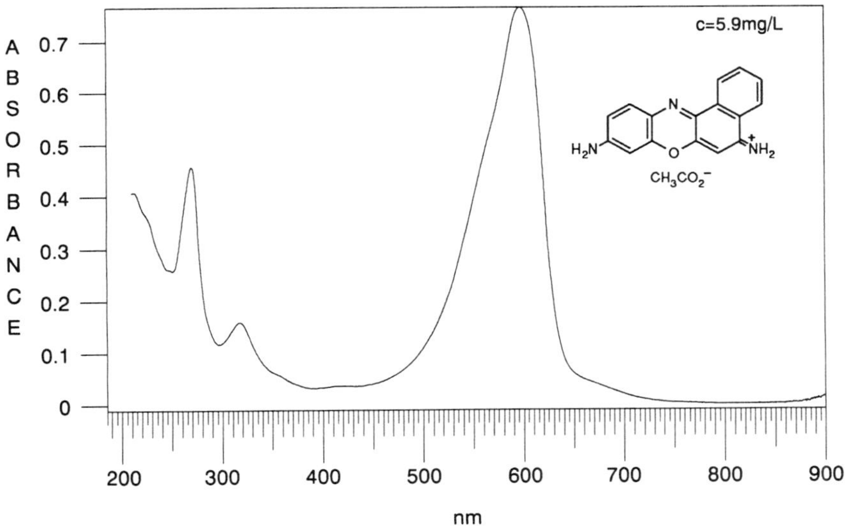

Cresyl Violet acetate is a metachromatic, cationic oxazine dye whose aqueous and alcoholic solutions are violet when observed in transmitted light and brilliant red in reflected light. Green[1] reported synthesizing this material in 1949 in response to the demand for a source for the Grubler Cresyl Echt Violet (Cresyl Fast Violet), a German product which had disappeared from commerce with the advent of World War II. Apparently, the structure or structures of the German product or products were closely held trade secrets. To further complicate identification of the German product, *Conn's*[2] reported the composition varied from lot to lot. Green[3] found one prewar sample to be a mixture of 5,9-diamino-10-methylbenzo[*a*]phenoxazonium chloride and

Pyronin Y. The same study evaluated two postwar-German Cresyl Echt Violet samples, described the synthesis and provided spectra for a number of dyes closely related in structure to the 5,9-diamino-10-methylbenzo[*a*]phenoxazonium chloride.

While it is not a cresyl derivative, Cresyl Violet acetate was found by Banny and Clark[4] to compare favorably with the formerly available German product or products. The BSC certified the first lot of this material in 1954 (for end-point buffered, nuclear metachromatic and Nissl staining) and a product of consistent quality has been available since that time.

Staining Procedures[5] provides staining methods and a bibliography for Nissl granules (tigroid bodies, Nissl bodies and nuclei of neuroglia cells) and the demonstration of basophilia of various tissue and tissue elements.

Cresyl Violet perchlorate (5,9-diaminobenzo[*a*]phenoxazonium perchlorate) has been used as a laser dye since the late 1960s. In this application, the fundamental dye molecule must be purified to remove 9-aminobenzo-[*a*]phenoxazin-5-one before the perchlorate salt is made. The oxazone is a by-product of the Cresyl Violet synthesis[1,3] and it tends to quench the lasing capability of the latter. Schafer[6] reported the optimum lasing λ_{max} (in ethanol) to be 650nm and the fluorescence λ_{max} (in ethanol) to be 630nm.

References:

1) Green, F.J. *Stain Technol.* **1966,** *41*(2), 115-120.

2) *Conn's Biological Stains,* 7th ed.; Lillie, R.D., Ed.; Williams and Wilkins: Baltimore, 1961.

3) Green, F.J. *Stain Technol.* **1966,** *41*(4), 223-228.

4) Banny, T.M.; Clark, G. *ibid.* **1950,** *25*, 195-196.

5) Clark, G. *Staining Procedures,* 4th ed.; Williams and Wilkins: Baltimore, 1981; pp 142-144, 206-207.

6) *Dye Lasers*; Schafer, F.P., Ed.; Springer: Berlin, 1973.

General references:

Aldrich FT-IR Library **1**(2)1036A
Sigma FT-IR Library **1**(2), 120D
IR Library **3**, 1486D

CRESYL VIOLET perchlorate

Aldrich Cat. No.:	25,524-6
Sigma Prod. No.:	C0664
Grade:	Laser dye
CAS No.:	[*41830-80-2*]
Mol. Form.:	$C_{16}H_{12}ClN_3O_5$
F.W.:	361.74
Appearance:	Green powder
λ_{max}:	602nm in ethanol
Solubility:	H_2O <0.1mg/ml
	EGME 5mg/ml
	EtOH 0.5mg/ml

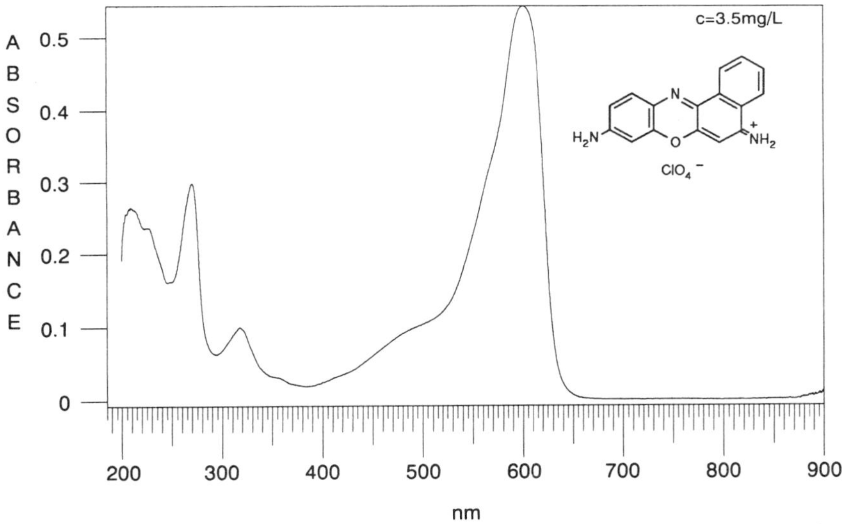

See description on pages 235 and 236 under Cresyl Violet acetate.

General references:

Aldrich FT-IR Library	**1**(2), 1033D
Safety	**2**(1), 925C

CROCEIN ORANGE G

Acid Orange 12

Aldrich Cat. No.: 19,813-7
Sigma Prod. No.: C3278
Grade: FIAT

CAS No.: [*1934-20-9*]
C.I. No.: 15970
Mol. Form.: $C_{16}H_{11}N_2NaO_4S$
F.W.: 350.33
Appearance: Red-orange powder
λ_{max}: 482nm in water
Solubility: H_2O 40mg/ml
EGME 20mg/ml
EtOH 20mg/ml

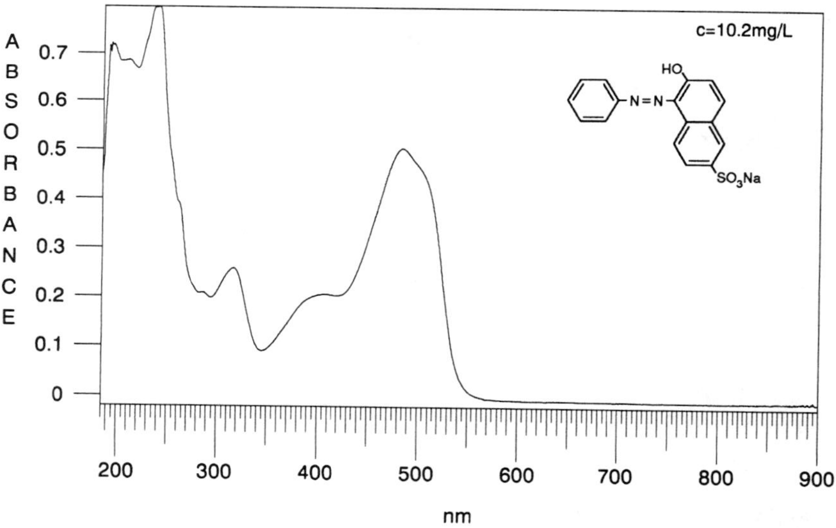

Crocein Orange G is an anionic monoazo dye that is soluble in water and moderately soluble in acetone, methyl Cellosolve and ethanol. It has been used to both dye and print wool and silk directly. Other industrial applications include dyeing paper and leather, and in the preparation of lakes. The calcium lake is used as a pigment (C.I. Pigment Orange 18, C.I. No. 15970:1). While the *Colour Index* indicates that it has been used as a biological stain, no verification has been found of that use.

General references:

Aldrich FT-IR Library	**1**(2), 981B
Sigma FT-IR Library	**1**(2), 121A
IR Library	**3**, 1459B
Colour Index	3rd ed., 1971

CRYSTAL VIOLET

Basic Violet 3 • Gentian Violet • Hexamethylpararosaniline chloride

Aldrich Cat. No.:	86,099-9	22,928-8
Sigma Prod. No.:	C0775	C6158
Grade:	Certified	A.C.S. reagent

CAS No.:	[*548-62-9*]
C.I. No.:	42555
Mol. Form:	$C_{25}H_{30}ClN_3$
F.W.:	407.99
m.p.:	205-215°C (dec.)
Appearance:	Green powder
λ_{max}:	590nm in water
Solubility:	H_2O 4mg/ml
	EGME 30mg/ml
	EtOH 30mg/ml

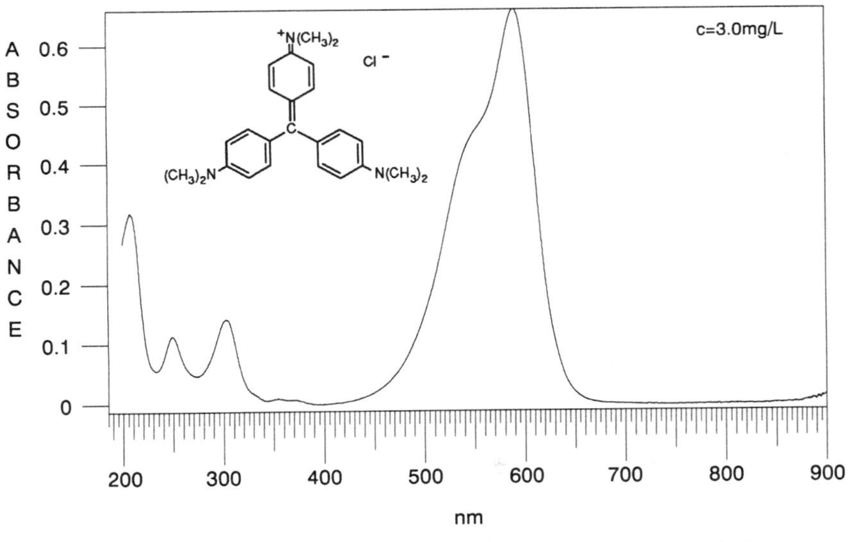

Crystal Violet is a cationic triarylmethane dye. The product of commerce is the chloride-salt form which is slightly soluble in water and soluble in a wide variety of organic solvents, not including aromatic and aliphatic hydrocarbons (see Gurr[1] for more detailed solubility information).

Crystal Violet has an affinity for both cellulosic and proteinaceous substrates and is used industrially to prepare inks, and to both dye and surface-coat paper. This dye has indicator properties with a range at the extreme lower end of the pH scale. The visual-transition interval is pH 0.0 (yellow) to pH 2.0 (blue-violet) using a 0.02% aqueous solution. *Reagent Chemicals*[2] lists specifications for using Crystal Violet as an indicator for nonaqueous titrations.

USP XX[3] states that Gentian Violet (a name frequently used synonymously with Crystal Violet) is usually the hexamethylpararosaniline chloride admixed with pentamethylpararosaniline chloride and tetramethylpararosaniline chloride. The *USP XX* assay for Gentian Violet can range from 96% to 100.5%, based on the formula weight of anhydrous Crystal Violet (hexamethylpararosaniline chloride). This, of course, does not allow for much of the pentamethyl and tetramethyl homologs to be present. In Europe, and formerly in the USA, Gentian Violet was considered to be a variable admixture of the Crystal Violet (hexamethylpararosaniline chloride) with the pentamethyl and tetramethyl homologs. The name Gentian Violet usually implied medicinal use and carried the connotation of having a higher purity than Methyl Violet or Crystal Violet. In fact, the hexamethyl compound (Crystal Violet) can be more highly purified since it is a single molecular entity and not a variable mixture of homologs. Medicinal Crystal Violet (Gentian Violet) is used as a topical antiinfective.[3] *Merck*[4] lists it as an anthelmintic.

Conn's[5] indicates that Crystal Violet can be satisfactorily substituted for both Methyl Violet and Gentian Violet in all of their biological staining applications, while the converse is not true. It is one of the more important biological stains available and is used extensively as a Gram-positive stain in bacteriology. In addition, it is an exceptionally good nuclear or chromatin stain. Clark[6] lists the following biological applications: stain for amyloid; used with Erythrosin B to demonstrate woody tissue in plants; stain for chromosomes in mitotic or meiotic divisions; wet and dry procedures for the Gram staining of microorganisms in smears; the Anthony modification of the Hiss capsule stain; positive-capsule stain; differential-spore stain; spirochetes stain and Brown and Hopps Gram stain for sections.

Leuco Crystal Violet is the completely reduced form of Crystal Violet. It is colorless, since the reduction destroys the quinoidal structure. This is a reversible reaction, however, and the leuco form rapidly reverts to the dye on irradiation, as a result of photoionization.[2]

References:

1) Gurr, E. *Synthetic Dyes*; Academic: London, 1971; p 107.
2) *Reagent Chemicals,* 7th ed.; American Chemical Society: Washington, DC, 1986; pp 233-234.
3) *The United States Pharmacopeia,* twentieth version; United States Pharmacopeial Convention; Mack: Easton, 1979; pp 348-349.
4) *The Merck Index,* 11th ed.; Budavari S. *et al.*, Eds.; Merck: Rahway, 1989; Monograph 4287.
5) *Conn's Biological Stains,* 9th ed.; Lillie, R.D., Ed.; Williams and Wilkins: Baltimore, 1977; pp 274-275.
6) Clark, G. *Staining Procedures,* 4th ed.; Williams and Wilkins: Baltimore, 1981; pp 214, 323, 347, 377, 388, 392, 411, 443, 444.

General references:

Aldrich FT-IR Library	**1**(2), 1029D
Sigma FT-IR Library	**1**(2), 121D
IR Library	**3**, 1483E
Safety	**2**(1), 929B
Colour Index	3rd ed., 1971

CUPROMERONIC BLUE

Aldrich Cat. No.: 30,904-4

Mol. Form.: $C_{36}H_{36}CuN_{12}O_{16}S_4$
F.W.: 1084.55
Appearance: Black powder
λ_{max}: 678nm, pH 5.7 acetate buffer, aliquot to water

Cupromeronic Blue (CB) is a highly colored, basic phthalocyanine dye that is used for the visualization of proteoglycans and other polyanions by light and electron microscopy. It has a molecular absorption of >100,000.

The dye dissociates from DNA and RNA complexes at relatively low electrolyte concentrations (<0.2*M* $MgCl_2$). CB is one of a group of dyes (like Alcec Blue) developed by Scott[1-3] on a molecular-biological approach to substrate-dye relationships.

Quaternary ammonium dyes such as CB are displaced from polyanionic substrates by, for example, $MgCl_2$ or NaCl, at critical electrolyte concentrations (CECs), determined mainly by the type of charged group (*i.e.*, polycarboxylates and polyphosphates at low CECs, polysulfate esters at much higher CECs). Thus, sulfated proteoglycans are stained at 0.3*M* $MgCl_2$, whereas polycarboxylates (*e.g.*, hyaluronate), sialoproteins and most double-stranded polynucleotides are not.[2,3]

The complexes are electron-dense because of copper, and therefore are visible by electron microscopy. The polyvalent nature of CB makes it possible to intensify the electron density by treating the polyanion-CB complex with sodium tungstate[3,4] to give pictures of such high contrast that the counterstaining of sections with UO_2 is very advantageous in locating the polyanion *vis-à-vis* tissue elements.

The use of CB has been outstandingly successful in demonstrating sulfated proteoglycans in association with collagen fibrils by both electron microscopy[5] and X-ray diffraction.

CB is usually used at 0.5% w/v in 0.025*M* sodium acetate buffer with $MgCl_2$ to the desired concentration. CB is unstable in alkali, in strong sunlight at high dilution and in contact with metals other than gold. In particular, contact with copper, nickel and iron (*e.g.*, stainless steel racks) should be avoided.

References:

1) Scott, J.E. *Trans. Biochem. Soc.* **1973,** *1*, 787.
2) *Idem Collagen Rel. Res.* **1985,** *5*, 541.
3) *Idem Biochem. J.* **1980,** *187*, 887.
4) Scott, J.E.; Haigh, M. *Biosci. Rep.* **1985,** *5*, 765.
5) Meek, K.M.; Scott, J.E.; Nave, C. *J. Microsc.* **1984,** *139*, 201.

DARROW RED

9-Acetylamino-5-aminobenzo[*a*]phenoxazonium chloride

Aldrich Cat. No.: 21,188-5
Grade: Certified

CAS No.: [*15391-59-0*]
Mol. Form.: $C_{18}H_{14}ClN_3O_2$
F.W.: 339.79
Appearance: Brown powder
λ_{max}: 502nm in 50% ethanol
Solubility: H_2O 1mg/ml
EGME 8mg/ml
EtOH 5mg/ml

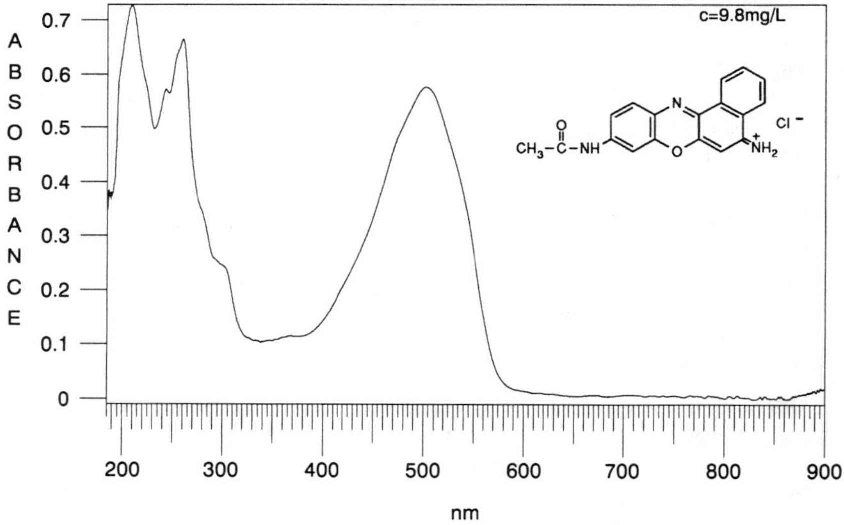

Darrow Red* is a metachromatic, cationic, oxazine dye whose aqueous and alcoholic solutions are red when observed in transmitted light and muddy yellow in reflected light.

Green[1] synthesized this material in 1949 as a precursor for Cresyl Violet acetate (CVA). Later, after Powers *et al.*[2] observed it as an impurity and separated it by paper chromatography from CVA, interest was aroused. It was readily isolated in quantity from a short-stopped CVA reaction.

Powers used it as a Nissl substance stain for nerve cells. *Staining*

Procedures[3] provides a procedure for using Darrow Red as a Nissl stain, especially recommended as a counterstain for the Weil or Kluver-Barrera myelin-sheath stains which are "designed to show RNA- and DNA-containing structures."

* Darrow Red was named in honor of Mary A. Darrow who served as an assistant and laboratory technologist under Dr. Harold J. Conn. She served in the same capacity under Dr. Victor Emmel at the Biological Stain Commission from 1937-1959.

References:

1) Green, F.J. *Stain Technol.* **1966,** *41* (2), 115-120.
2) Powers, M.M.; Clark, G.; Darrow, M.A.; Emmel, V.M. *ibid.* **1960,** *35*, 19-21.
3) Clark, G. *Staining Procedures,* 4th ed.; Williams and Wilkins: Baltimore, 1981; pp 144.

General reference:

Aldrich FT-IR Library **1** (2), 1027C

4',6-DIAMIDINO-2-PHENYLINDOLE dihydrochloride hydrate

DAPI

Aldrich Cat. No.:	21,708-5
Sigma Prod. No.:	D1388
Purity:	98%

CAS No.:	[*28718-90-3*]
Mol. Form.:	$C_{16}H_{15}N_5 \cdot 2HCl \cdot xH_2O$
F.W.:	350.25
Appearance:	Yellow powder
λ_{max}:	349(263)nm in water

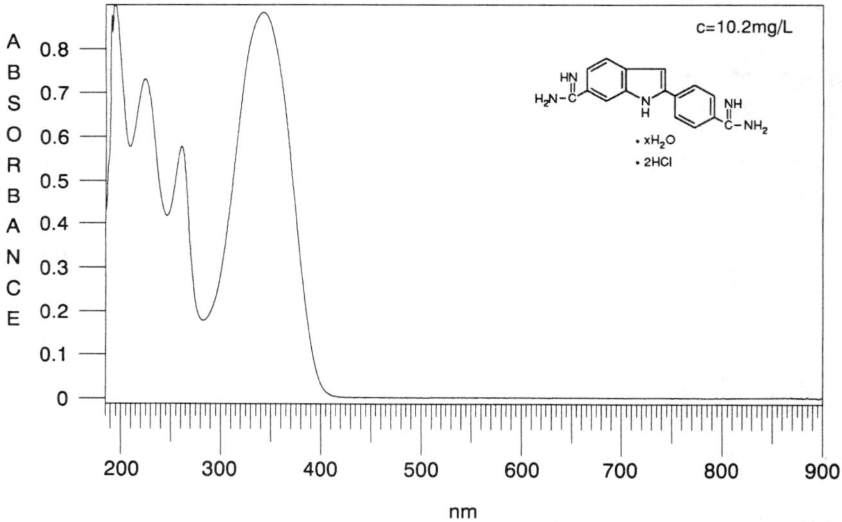

DAPI is a light-sensitive, yellow, water-soluble reagent that is used in combination with Evans Blue (Aldrich Cat. No. 20,633-4) and Primulin (Aldrich Cat. No. 20,686-5) in a simple double-labeling procedure for studying axonal branching.[1] It is also known to form highly fluorescent complexes with DNA from various sources.[2,3]

This reagent should be protected from light and stored in a freezer.

References:

1) van der Kooy, D; Kuypers, H.G.J.M. *Science* **1979,** *204*, 873.

2) Russell, W.C.; Newman, C.; Williamson, D.H. *Nature* **1975,** *253*, 461.

3) Brunk, C.F.; Jones, K.C.; James, T.W. *Anal. Biochem.* **1979,** *92*, 497-500.

General references:

Aldrich FT-IR Library	**1**(2), 678A
Sigma FT-IR Library	**1**(2), 427C
IR Library	**3,** 1261E
NMR Library	**2**(2), 547D
Safety	**2**(1), 1044D

3,6-DIAMINOACRIDINE hemisulfate
Proflavine

Aldrich Cat. No.: 19,822-6
Sigma Prod. No.: P2508

CAS No.: [*1811-28-5*]
Mol. Form.: $C_{13}H_{11}N_3 \cdot {}^1/_2\ H_2SO_4$
F.W.: 258.29
Appearance: Dull-orange powder
λ_{max}: 456nm in methanol
Solubility: H_2O 10mg/ml
EGME 1mg/ml
EtOH 0.2mg/ml

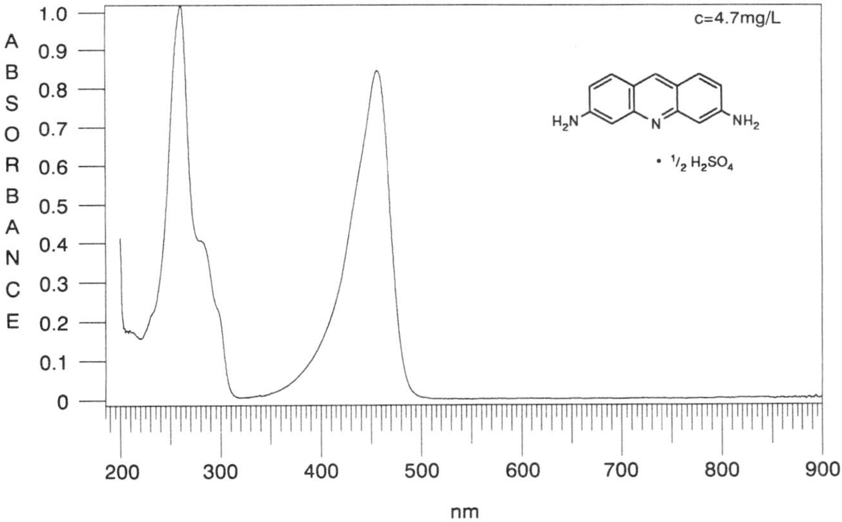

The Proflavines are yellow, fluorescent, cationic acridine dyes which are slightly soluble in water and very slightly to slightly soluble in ethanol. Depending on their concentration, their solutions are yellow to brown when observed by transmitted light, displaying a yellow-green fluorescence.

Remington[1] states that a 1.0% aqueous solution of 3,6-diaminoacridine hemisulfate should display a pH between 6.0 and 8.0, and lists its use as a topical antiseptic. It is used with Gentian Violet (Crystal Violet) and Brilliant Green to prepare a broad-spectrum "triple-dye" antiseptic solution for topical application to the umbilical-cord area of a newborn infant (in the neonatal nursery).[2] Rogers[3] listed it as an effective dye for inhibiting sulfate-reducing bacteria.

3,6-Diaminoacridine hydrochloride is an intercalating dye[4,5] listed by Remington[1] as a topical antiseptic.

References:

1) Remington, J.R. *Pharmaceutical Sciences*, 13th ed.; Mack: Easton, 1965; p 1257.

2) *Tests and Standards; New and Unofficial Remedies*; American Medical Association; Lippincott: Philadelphia, 1953; p 4.

3) Rogers, T.H. *J. Soc. Chem. Ind.* **1940,** *59*, 34-39.

4) Chan, L.M.; McCarter, J.A. *Biochim. Biophys. Acta* **1970,** *204,* 252.

5) Ramstein, J.; Leng, M. *ibid.* **1972,** *281*, 18.

General references:

Aldrich FT-IR Library	**1**(2), 876B
Sigma FT-IR Library	**1**(1), 704A
IR Library	**3**, 1400E
Safety	**2**(1), 1045A

3,6-DIAMINOACRIDINE hydrochloride
Proflavine hydrochloride

Aldrich Cat. No.:	13,110-5
Sigma Prod. No.:	P4646
CAS No.:	*[952-23-8]*
Mol. Form.:	$C_{13}H_{11}N_3 \cdot HCl$
F.W.:	245.71
m.p.:	270°C (dec.)
Appearance:	Orange to dark-red powder
λ_{max}:	456nm in methanol
Solubility:	H_2O 4mg/ml
	EGME 7mg/ml
	EtOH 7mg/ml

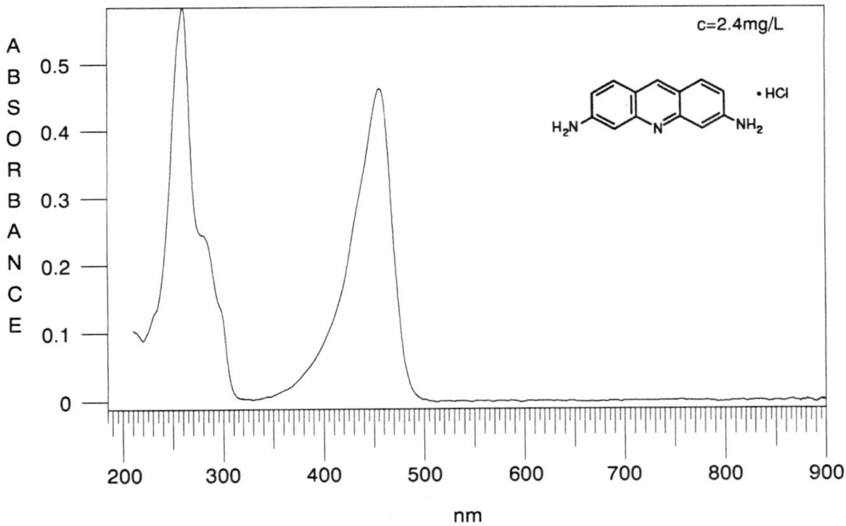

See description on pages 245 and 246 under 3,6-diaminoacridine hemisulfate.

General references:

Aldrich FT-IR Library	**1**(2), 875B
Sigma FT-IR Library	**1**(2), 200C
IR Library	**3**, 1400C
NMR Library	**2**(2), 751A
Safety	**2**(1), 1045B

DIAZO RED RC

Azoic Diazo No.10 • Fast Red RC salt

Aldrich Cat. No.:	20,147-2
Sigma Prod. No.:	F2256

CAS No.:	*[85252-22-8]*
C.I. No.:	37120
Mol. Form.:	$C_7H_6Cl_2N_2O \cdot {}^1/_2\ ZnCl_2$
F.W.:	273.18
Appearance:	Yellow powder
λ_{max}:	372nm in water
Solubility:	H_2O 80mg/ml
	EGME 20mg/ml
	EtOH 2mg/ml

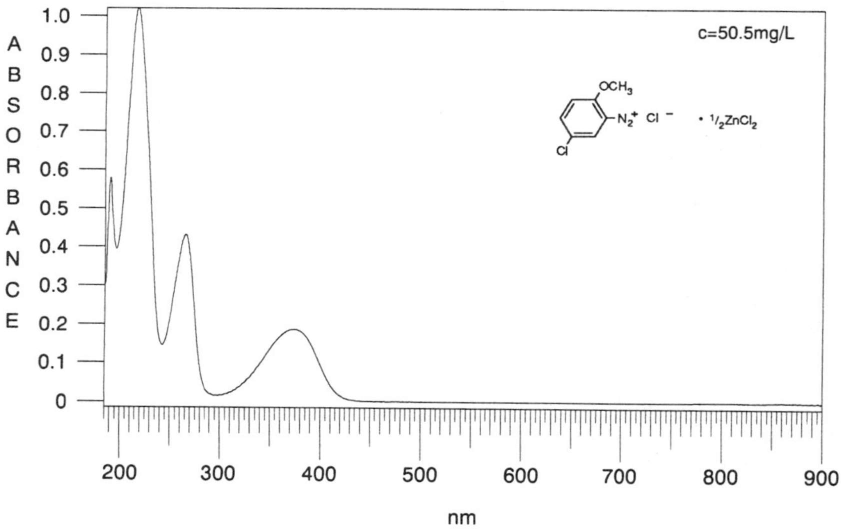

This stable diazonium salt is prepared by diazotizing 5-chloro-*o*-anisidine followed by precipitation as the zinc chloride double salt with the addition of a concentrated zinc chloride solution. Industrially, it is used in the *in situ* development of various dyes in cotton, silk and nylon which have been impregnated previously with a coupling component (padded).

Diazo Red RC is used as a coupling agent in histochemical reactions for hydrolases. Burstone[1,2] considered it useful in various phosphatase methods.

References:

1) Burstone, M.S. *J. Natl. Cancer Inst.* **1957,** *18*, 167-173.

2) *Idem ibid.* **1958,** *21*, 123-129.

General references:

Aldrich FT-IR Library	**1**(2), 957C
Sigma FT-IR Library	**1**(2), 143A
IR Library	**3**, 1445H
Colour Index	3rd ed., 1971

4',5'-DIBROMOFLUORESCEIN
Solvent Red 72

Aldrich Cat. No.:	21,672-0
Sigma Prod. No.:	D3017
CAS No.:	[*596-03-2*]
C.I. No.:	45370:1
Mol. Form.:	$C_{20}H_{10}Br_2O_5$
F.W.:	490.12
m.p.:	270-273°C
Appearance:	Orange powder
λ_{max}:	450nm in methanol + 1 drop 1*N* HCl
Solubility:	H_2O 0.3mg/ml
	EGME 80mg/ml
	EtOH 30mg/ml

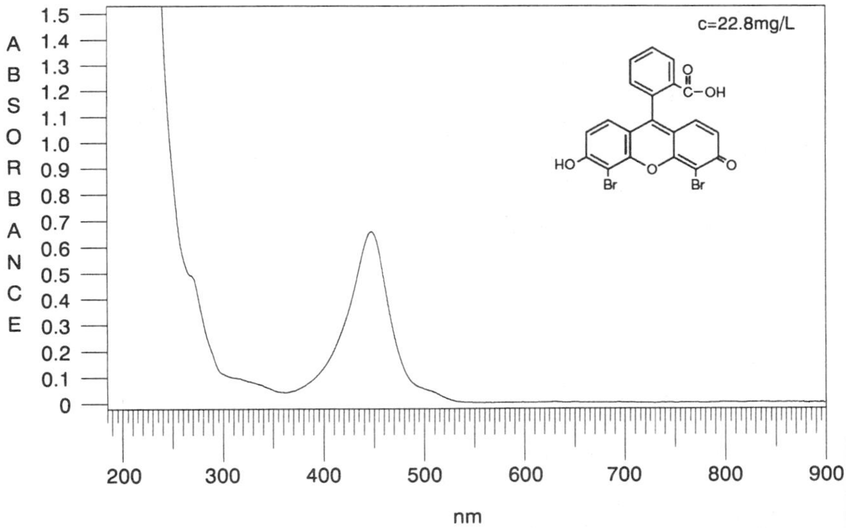

This bright, yellowish-red, xanthene solvent dye is prepared by dibrominating fluorescein in aqueous sodium hydroxide, acidifying with hydrochloric acid and washing free of inorganic salts.

Very soluble in acetone, and soluble in ethanol, butyl acetate and ethyl acetate, it is used to prepare Eosin Bluish (C.I. 45400) and other related dyes.

A highly purified material is certified by the U.S. Food and Drug Administration for use in cosmetics (D&C Orange 5).[1]

There is no evidence that it has been used as a biological stain.

Reference:

1) Marmion, D.M. *Handbook of U.S. Colorants for Foods, Drugs and Cosmetics;* Wiley: New York, 1979; p 61.

General references:

Aldrich FT-IR Library	**1**(2), 1011A
Sigma FT-IR Library	**1**(2), 124D
IR Library	**3**, 1473G
Merck Index	**11**, 3005
Colour Index	3rd ed., 1971

2',7'-DICHLOROFLUORESCEIN

Aldrich Cat. No.: 20,148-0
Sigma Prod. No.: D6132

CAS No.: *[76-54-0]*
Mol. Form.: $C_{20}H_{10}Cl_2O_5$
F.W.: 401.20
Appearance: Light-orange powder
λ_{max}: 509nm in ethanol + 1drop 1*N* NaOH
Solubility: H_2O <0.1mg/ml
EGME 80mg/ml
EtOH 30mg/ml

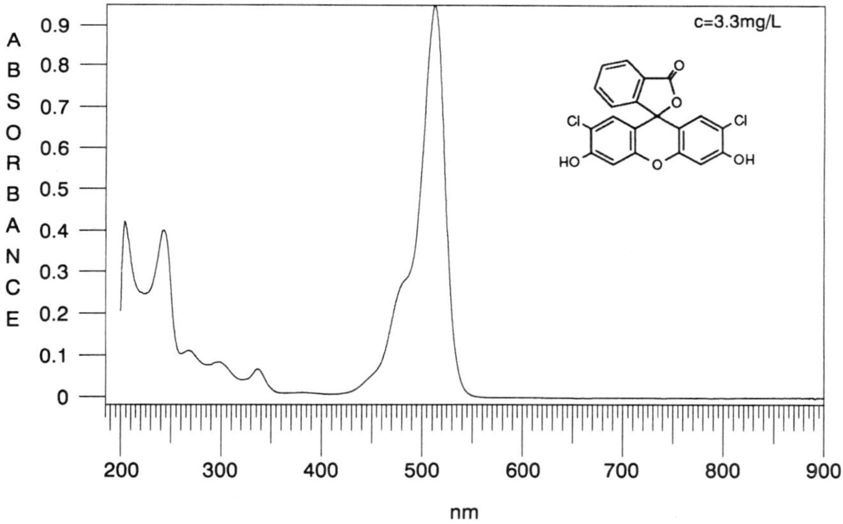

This red xanthene dye is insoluble in water, slightly soluble in methanol and ether and soluble in ethanol and aqueous alkali solutions. Produced on a laboratory scale, it apparently has no industrial application.

2',7'-Dichlorofluorescein is used as a fluorescent indicator with a fluorescent-transition interval extending from pH 4.0 (weakly green) to pH 6.0 (more intense green). In addition, it is used as an indicator for the argentometric titration of chloride ion and bromide ion. Tochikubo and Kaneko[1] used it as a reagent for the determination of the sodium chloride concentration in urine. The method is simple and reproducible.

This dye has limited utility as a laser dye with an absorption maximum of ~512nm and an optimum lasing wavelength of 570nm in basic ethanol. The molar absorptivity is 11.00 x 10^4 L $mole^{-1}$ cm^{-1} at 512nm.

Reference:

1) Tochikubo, O.; Kaneko, Y. *Igaku no Ayumi* **1984,** *131* (8), 545-550 (*Biochemical* Methods, **1984**).

General references:

Aldrich FT-IR Library	**1**(2), 1010D
Sigma FT-IR Library	**1**(2), 125A
IR Library	**3**, 1473D

5,7-DICHLORO-8-HYDROXYQUINOLINE

Chloroxine • 5,7-Dichloro-8-quinolinol

Aldrich Cat. No.:	D6,460-0
Sigma Prod. No.:	D2516
Purity:	99%
CAS No.:	*[773-76-2]*
Mol. Form.:	$C_9H_5Cl_2NO$
F.W.:	214.05
m.p.:	180-182°C
Appearance:	Off-white powder
λ_{max}:	253(335)nm in chloroform
Solubility:	H_2O <0.1mg/ml
	EGME 10mg/ml
	EtOH 3mg/ml

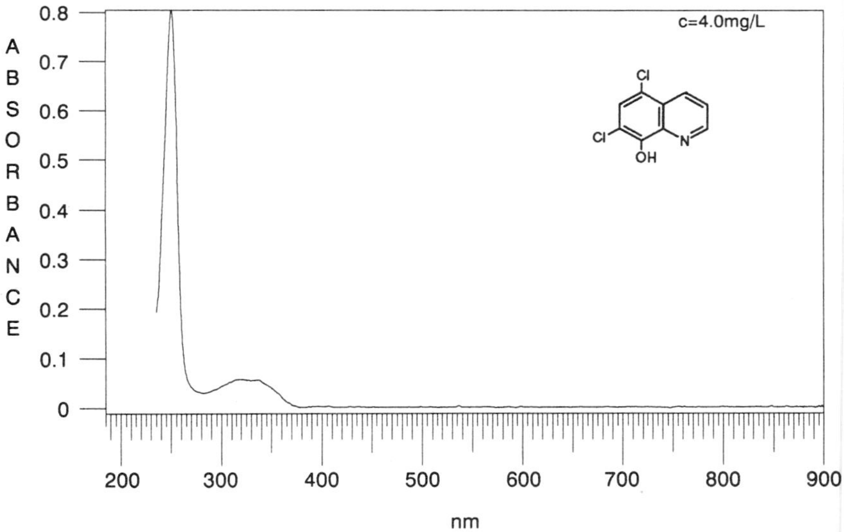

This chromogenic analytical reagent is prepared by chlorinating 8-hydroxyquinoline. It is soluble in benzene, acetone, alkali hydroxide solutions and acids and slightly soluble in ethanol.

The complexometric ability of the reagent is used to advantage in the determination of microgram quantities of vanadium.[1] Welcher[2] describes its physical characteristics and analytical uses.

References:

1) Chau, Y.K.; Lum-Shue-Chan, K. *Anal. Chim. Acta* **1970,** *50*, 201.

2) Welcher, F.J. *Organic Analytical Reagents,* Vol. 1; Van Nostrand: New York, 1949; pp 328-329.

General references:

Aldrich FT-IR Library	**1**(2), 861C
Sigma FT-IR Library	**1**(2), 125B
IR Library	**3**, 1389F
NMR Library	**2**(2), 738C
Safety	**2**(1), 1138B
Merck Index	**11**, 2175

2,6-DICHLOROINDOPHENOL, sodium salt hydrate

Aldrich Cat. No.:	11,981-4
Sigma Prod. No.:	D2908
Grade:	A.C.S. reagent

CAS No.:	[*620-45-1*]
Mol. Form.:	$C_{12}H_6Cl_2NNaO_2 \cdot xH_2O$
F.W.:	290.08
Appearance:	Dark-green powder
λ_{max}:	605nm in water + 2ml 1*N* NaOH
Solubility:	H_2O 30mg/ml
	EGME 40mg/ml
	EtOH 20mg/ml

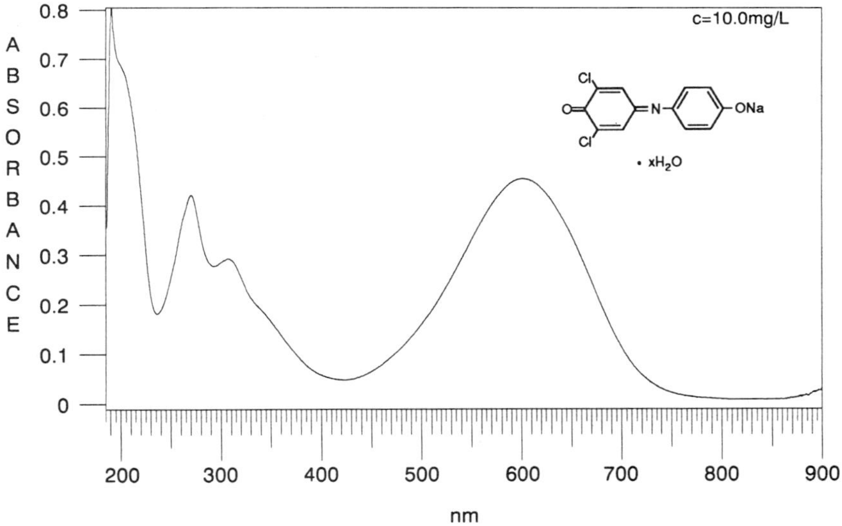

2,6-Dichloroindophenol, sodium salt, is a blue, water-soluble indophenol dye that is prepared by condensing an equal number of moles each of 2,6-dichloroquinone-4-chloroimide and phenol in an alkaline medium.

Reagent Chemicals[1] lists specifications and analytical methods for its use as a reagent for the determination of ascorbic acid (Vitamin C). In the presence of an excess of ascorbic acid, a typical blue dye solution is rendered colorless. This is the result of the complete reduction and concomitant elimination of the chromophore.

Reference:

1) *Reagent Chemicals,* 7th ed.; American Chemical Society: Washington, DC, 1986; p 254.

General references:

Aldrich FT-IR Library	**1**(2), 963A
Sigma FT-IR Library	**1**(2), 125C
IR Library	**3**, 1448B
Safety	**2**(1), 1138D

1,1'-DIETHYL-4,4'-CARBOCYANINE IODIDE

Cryptocyanine • Kryptocyanine

Aldrich Cat. No.:	D9,153-5
Purity:	99%
CAS No.:	*[4727-50-8]*
Mol. Form.:	$C_{25}H_{25}IN_2$
F.W.:	480.39
m.p.:	250.5-253°C
Appearance:	Dark-green crystalline powder
λ_{max}:	703(648)nm in methanol
Solubility:	H_2O 1mg/ml
	EGME 9mg/ml
	EtOH 2mg/ml

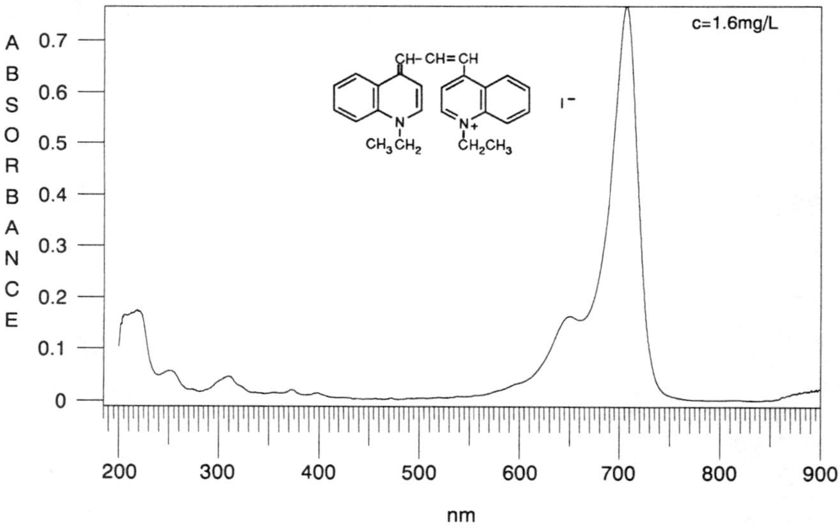

This basic carbocyanine dye is very slightly soluble in water and ethanol and soluble in ethylene glycol.

While there are no reported biological stain applications, it finds utility as a laser dye, both as a mode-locking dye for ruby lasers and as a dye for pulsed lasers. The λ_{max} in ethanol is 710nm with a molar absorptivity of 25.80 x 10^4 L mole^{-1} cm^{-1}. The optimum lasing wavelength in ethylene glycol is 745nm, and the fluorescence λ_{max} in ethanol is 720nm.[1]

Reference:

1) Spaeth, M.L.; Scoy, W.R. *J. Chem. Phys.* **1968,** *48*, 2315.

General references:

Aldrich FT-IR Library	**1**(2), 855A
IR Library	**3**, 1384H

DIIODOFLUORESCEIN

Solvent Red 73

Aldrich Cat. No.:	20,680-6
CAS No.:	[*31395-16-1*]
C.I. No.:	45425:1
Mol. Form.:	$C_{20}H_{10}I_2O_5$
F.W.:	584.10
m.p.:	240°C (dec.)
Appearance:	Orange powder
λ_{max}:	522nm in methanol
Solubility:	H_2O 1mg/ml
	EGME 200mg/ml
	EtOH 30mg/ml

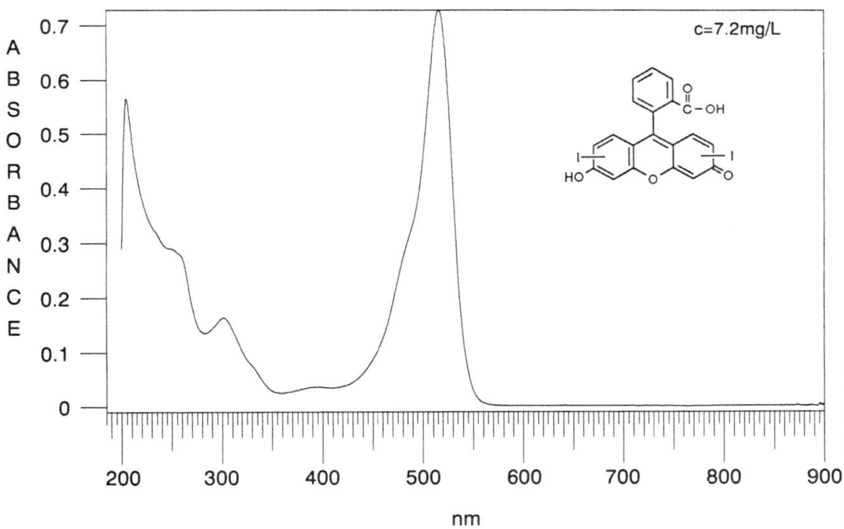

This fluorescent, orange, anionic xanthene dye is very slightly soluble in water and soluble in ethanol, methanol and alkali hydroxide solutions. It is prepared by the controlled iodination of fluorescein with either iodine and iodic acid or iodine chloride and alkali.

A highly purified grade, D&C Orange 10, has been used in cosmetics. Marmion,[1] showing the lactone form, listed the status as provisional.

Merck[2] states that it is used as an adsorption indicator for the determination of iodides in the presence of chlorides and bromides.

There is no evidence that it has been used as a biological stain.

References:

1) Marmion, D.M. *Handbook of U.S. Colorants for Foods, Drugs and Cosmetics;* Wiley: New York, 1979; p 12.

2) *The Merck Index,* 11th ed.; Budavari, S. *et al.*, Eds.; Merck: Rahway, 1989; Monograph 3174.

General references:

Aldrich FT-IR Library	**1**(2), 1011B
IR Library	**3**, 1473H
Colour Index	3rd ed., 1971

4-(DIMETHYLAMINO)BENZALDEHYDE

DMAB • Ehrlich's Reagent

Aldrich Cat. No.:	15,647-7	10,976-2
Sigma Prod. No.:	D8904	——
Grade:	A.C.S. reagent	——
Purity:	99%	98%
CAS No.:	[*100-10-7*]	
Mol. Form.:	$C_9H_{11}NO$	
F.W.:	149.19	
m.p.:	73-75°C	72-75°C
Appearance:	Off-white crystalline powder	Light greenish-tan powder and chunks
λ_{max}:	250nm in 10% hydrochloric acid	
Solubility:	H_2O <0.2mg/ml	
	EGME 0.2mg/ml	
	EtOH <0.2mg/ml	

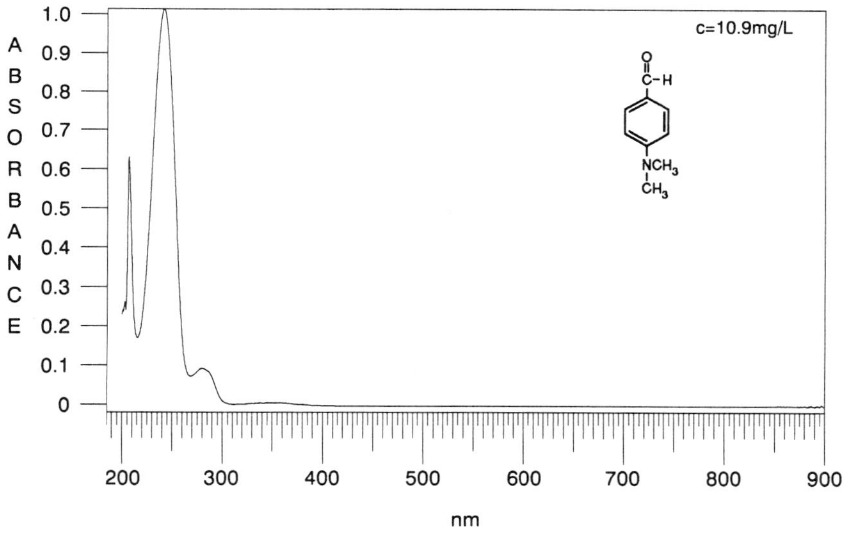

DMAB was first used as an intermediate for the manufacture of dyes and is currently used to produce certain valuable chrome colors.[1] If it is prepared by the formylation of dimethylaniline with dimethylformamide,[2] it can be purified readily and isolated as white, crystalline leaflets. However, this product must be protected from the light and atmosphere to remain in this state. It is practically insoluble in water and soluble in dilute hydrochloric acid (1:10), ether, chloroform and acetic acid.

Fieser and Fieser[3] reported it as a spot-test reagent for pyrrole and substances capable of forming pyrrole. Only pyrroles with one intact CH group (including indole, skatole and tryptophan) test positive, developing a violet color with DMAB. DMAB condenses with compounds containing the $NHCONH_2$ group to form yellow pigments and it develops yellow Schiff bases with primary amines; however, only pyrroles form violet pigments. Nixon[4] studied the colorimetric response to Ehrlich's reagent in plasma from patients with and without cancer. He noted that plasma from untreated cancer patients consistently developed a darker blue-purple color than did specimens from other subjects and he concluded that the procedure showed promise as a screening test for cancer.

References:

1) Abrahart, E.N. *Dyes and Their Intermediates;* Chemical: New York, 1977; p 63.
2) Archer, W. In *Organic Synthesis;* Rabjohn, N., Ed.; Wiley: New York, 1963; Coll. Vol. IV, p 331.
3) Fieser, L.F.; Fieser, M. *Reagents for Organic Synthesis;* Wiley: New York, 1967; Vol. 1, pp 273-274.
4) Nixon, D.W. *Cancer* **1973**, *51* (8), 596.

General references:

Aldrich FT-IR Library	**1**(2), 112B
IR Library	**3**, 915E
NMR Library	**2**(2), 111A
Safety	**2**(1), 1331C
Merck Index	**11**, 3219

5-(4-DIMETHYLAMINOBENZYLIDENE)RHODANINE

p-Dimethylaminobenzalrhodanine

Aldrich Cat. No.:	11,458-8
Purity:	97%
CAS No.:	[*536-17-4*]
Mol. Form.:	$C_{12}H_{12}N_2OS_2$
F.W.:	264.37
m.p.:	285-288°C
Appearance:	Red crystalline powder
λ_{max}:	451(321)nm in methanol
Solubility:	H_2O 0.5mg/ml
	EGME 2mg/ml
	EtOH 0.3mg/ml

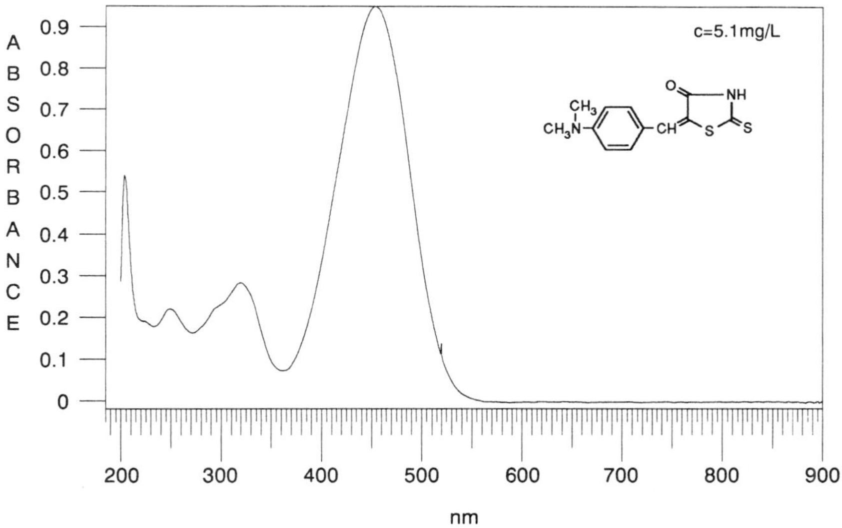

This red, crystalline (needles from xylene) chromogenic reagent is reported by *The Merck Index*[1] to be practically insoluble in water, very slightly soluble in chloroform, ether and benzene, slightly soluble in boiling alcohol, moderately soluble in acetone and soluble in strong acids, with a yellow color. The same source reported that a 0.03% acetone solution is used to detect platinum, gold, silver, mercury and copper ions. Feigl[2] describes its use as a spot-test reagent for copper(II).

References:

1) *The Merck Index,* 11th ed.; Budavari, S. *et al.*, Eds.; Merck: Rahway, 1989; Monograph 3220.

2) Feigl, F. *Spot Tests in Inorganic Analysis;* Elsevier: New York, 1958; p 59.

General references:

Aldrich FT-IR Library	**1**(2), 419C
IR Library	**3**, 1110A

4-DIMETHYLAMINO-2-METHYLAZOBENZENE

Aldrich Cat. No.: 13,043-5
Sigma Prod. No.: D3265

CAS No.: [*54-88-6*]
Mol. Form.: $C_{15}H_{17}N_3$
F.W.: 239.32
m.p.: 67-69°C
Appearance: Red-orange crystals

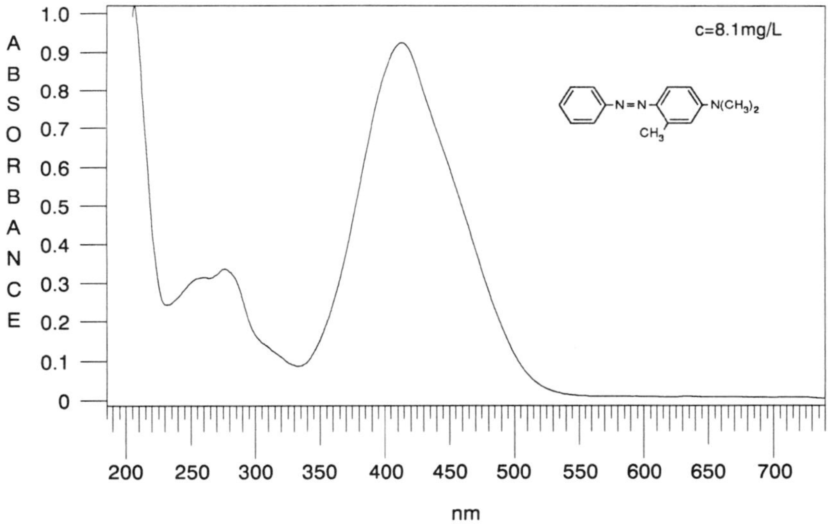

This basic, monoazo solvent dye is made by coupling diazotized aniline to *N,N*-dimethyl-*m*-toluidine. It is practically insoluble in water, soluble in ethanol, acetone and most aromatic solvents and fairly soluble in most aliphatic hydrocarbon solvents. Structure and solubility indicate that this dye has potential as a fat stain. In addition, it possesses indicator properties very close to Oil Yellow II (C.I. 11200). The visual-transition interval is pH 2.8 (red) to pH 4.4 (yellow).

General references:

IR Library	**3**, 1453A
Safety	**2**(1), 1338A

4-(4-DIMETHYLAMINOPHENYLAZO)BENZENEARSONIC ACID hydrochloride

Aldrich Cat. No.:	20,681-4
CAS No.:	*[73688-85-4]*
Mol. Form.:	$C_{14}H_{16}AsN_3O_3 \cdot HCl$
F.W.:	385.68
m.p.:	203°C (dec.)
Appearance:	Purple powder
λ_{max}:	452nm in methanol
Solubility:	H_2O 3mg/ml
	EGME 7mg/ml
	EtOH 5mg/ml

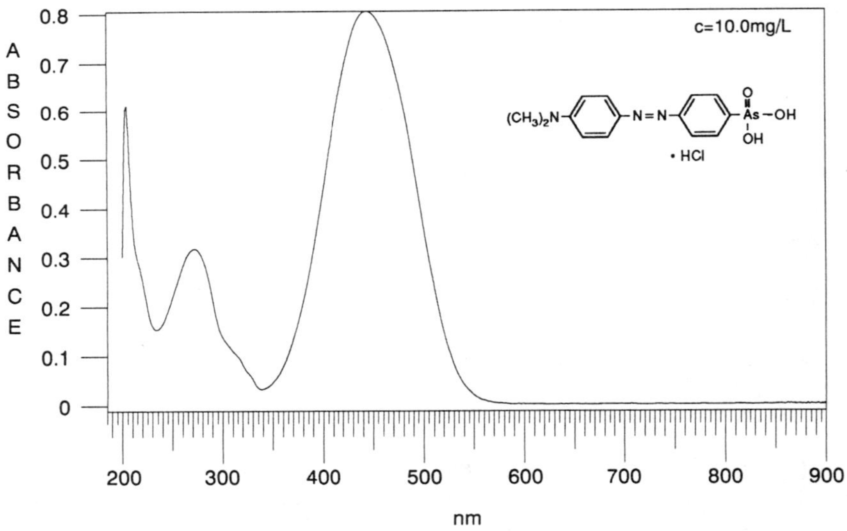

This slightly water-soluble, yellowish-red monoazo dye is prepared by coupling diazotized *p*-arsanilic acid to dimethylaniline, followed by conversion to the hydrochloride salt. Apparently it is not used in industrial applications; however, it has been used as a reagent for the colorimetric determination of zirconium.[1]

Reference:

1) Adler, S.; Haskelberg, L.; Bergmann, F. *J. Chem. Soc.* **1940,** 576-578.

General references:

Aldrich FT-IR Library	**1**(2), 998C
IR Library	**3**, 1467G

4,7-DIPHENYL-1,10-PHENANTHROLINE

Bathophenanthroline

Aldrich Cat. No.:	13,315-9
Sigma Prod. No.:	B1250
Purity:	97%

CAS No.:	[*1662-01-7*]
Mol. Form.:	$C_{24}H_{16}N_2$
F.W.:	332.41
m.p.:	218-220°C
Appearance:	White powder
λ_{max}:	288nm in chloroform
Solubility:	H_2O 0.2mg/ml
	EGME 10mg/ml
	EtOH 8mg/ml

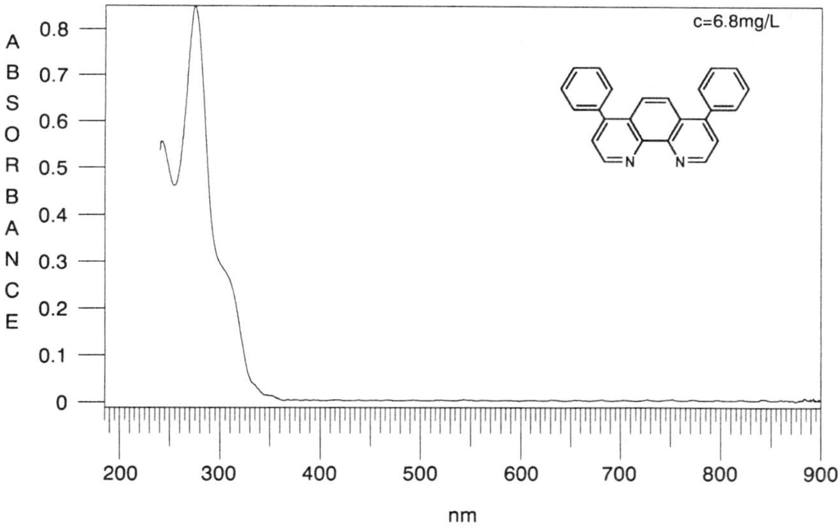

This chromogenic reagent was first synthesized by Case[1] and the trivial name Bathophenanthroline was derived from the fact that the deep-red complex formed with iron(II) absorbs at 533nm as compared to the 510nm of its iron 1,10-phenanthroline counterpart (a bathochromic shift of 23nm).

A colorless or light-yellow crystalline powder, it is virtually insoluble in water, slightly soluble in ethanol, other alcohols, acetone, benzene and aqueous acid solutions. The iron(II) complex forms in the 2-9 pH range and can be extracted with chloroform or a 1:1 mixture of isopentyl alcohol and isopropyl ether. Interference is encountered from copper, nickel and cobalt, but not from lithium, sodium, potassium, beryllium, magnesium, calcium, strontium, barium, cerium, rare earths, thorium, titanium, zirconium, uranium, chromium, tungsten, ruthenium, osmium, palladium, platinum, silver, zinc, cadmium, mercury, boron, aluminum, gallium, tellurium, tin and lead.

Smith *et al.*[2] introduced Bathophenanthroline as a reagent for iron in 1951 and since that time it has been generally accepted as a sensitive, colorimetric reagent for this purpose. Indeed, today it is used to determine iron in systems as diverse as serum, urine, wine, water, alloys, ingots, glass, petroleum, plants and other organic materials.[3]

References:

1) Case, F.J. *J. Org. Chem.* **1951,** *16*, 1541-1545.

2) Smith, G.F.; McCurdy, W.H., Jr.; Diehl, H. *Analyst* **1952,** *77*, 418-422.

3) Schilt, A.A.; *Analytical Application of 1,10-Phenanthroline and Related Compounds;* Pergamon: New York, 1969.

General references:

Aldrich FT-IR Library	**1**(2), 891C
Sigma FT-IR Library	**1**(2), 299D
IR Library	**3**, 1411F

DIRECT BLACK 22

Aldrich Cat. No.: 21,262-8

CAS No.: *[6473-13-8]*
C.I. No.: 35435
Mol. Form.: $C_{44}H_{32}N_{13}Na_3O_{11}S_3$
F.W.: 1083.99
Appearance: Black powder
λ_{max}: 481nm in water

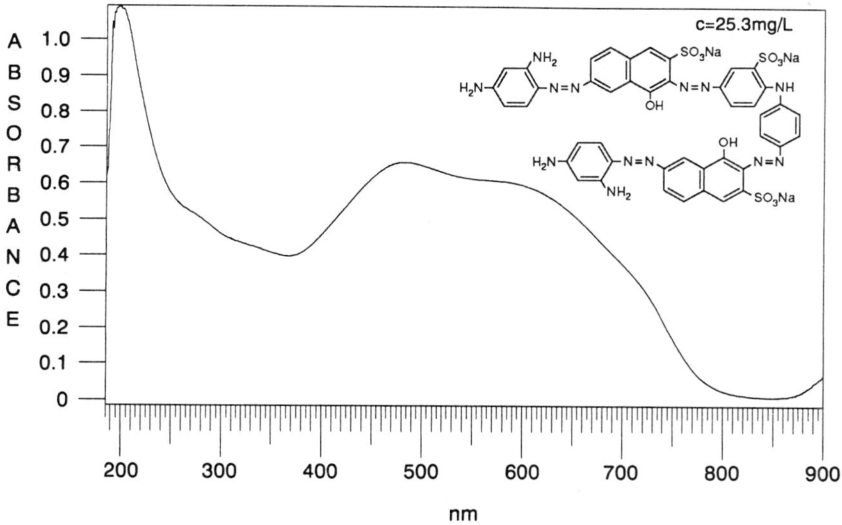

This water-soluble, polyazo dye dyes cotton greenish black from a direct bath; silk and bast fibers treated in the same manner are dyed less intensely than cotton. Other industrial applications include paper coating and dyeing beater paper and chromed leather.

There is no evidence that it has been used as a biological stain.

General references:

Aldrich FT-IR Library	**1**(2), 995C
IR Library	**3**, 1466B
Colour Index	3rd ed., 1971

DIRECT BLUE 71

Benzo Light Blue FFL • Sirius Light Blue BRR

Aldrich Cat. No.: 21,240-7

CAS No.: [*4399-55-7*]
C.I. No.: 34140
Mol. Form.: $C_{40}H_{23}N_7Na_4O_{13}S_4$
F.W.: 1029.88
Appearance: Black powder
λ_{max}: 594nm in water
Solubility: H_2O 40mg/ml
EGME 9mg/ml
EtOH 0.5mg/ml

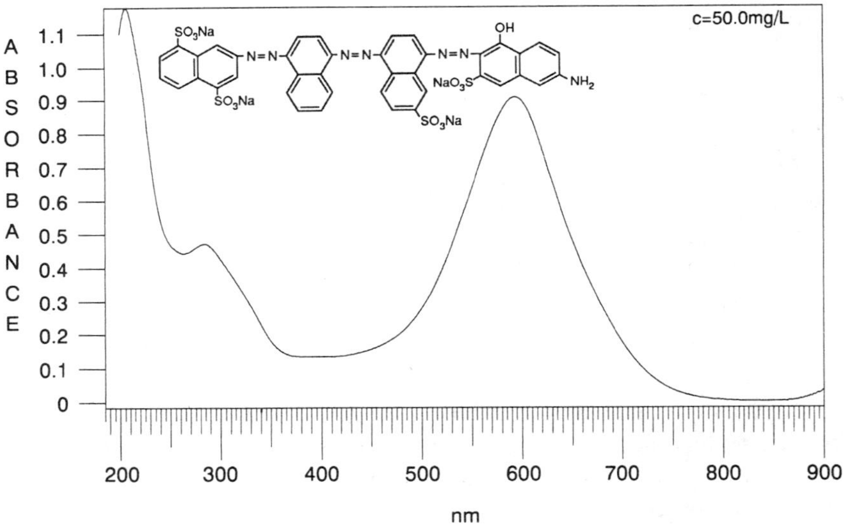

This blue, anionic trisazo dye is prepared by coupling diazotized 3-amino-1,5-naphthalenedisulfonic acid to 1-naphthylamine, diazotizing the resulting product and coupling it to 1,7-Cleve's acid (1-aminonaphthalene-7-sulfonic acid), then diazotizing this product and coupling it to J acid (6-amino-1-naphthol-3-sulfonic acid) in an alkaline medium.

Soluble in water, slightly soluble in methyl Cellosolve and very slightly soluble in ethanol, it is used to dye cellulose, wool, silk and paper. The wool and silk are dyed a much lighter hue than the cellulosic substrates.

There is no evidence that it has been used as a biological stain.

General references:

Aldrich FT-IR Library **1**(2), 992C
IR Library **3**, 1465C
Colour Index 3rd ed., 1971

DIRECT ORANGE 31

Aldrich Cat. No.: 19,527-8
Sigma Prod. No.: D4265

CAS No.: *[6420-03-7]*
C.I. No.: 23655
Mol. Form.: $C_{31}H_{24}N_6Na_2O_7S$
F.W.: 670.62
Appearance: Burnt-orange powder
λ_{max}: 428nm in water
Solubility: H_2O 30mg/ml
EGME 20mg/ml
EtOH 2mg/ml

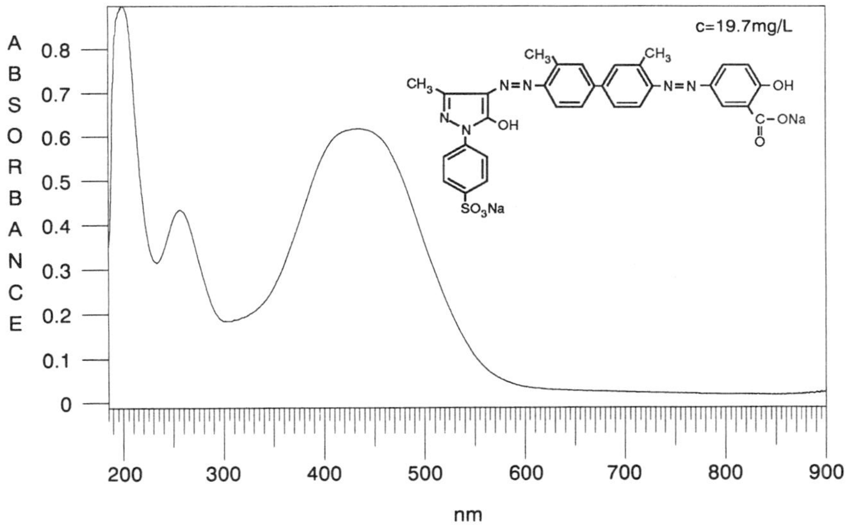

This reddish-orange, acid disazo dye is made by coupling one mole of salicylic acid and one mole of 3-methyl-1-(*p*-sulfophenyl)-5-pyrazolone to one mole of tetraazotized *o*-tolidine, followed by conversion to the disodium salt. It is soluble in water and methyl Cellosolve and slightly soluble in ethanol. Industrially it is used to dye cotton, wool and silk.

There is no evidence that it has been used as a biological stain.

General references:

Aldrich FT-IR Library **1**(2), 1003D
Sigma FT-IR Library **1**(2), 129C
IR Library **3**, 1470B
Colour Index 3rd ed., 1971

DIRECT RED 23

Benzo Fast Scarlet 4BS

Aldrich Cat. No.:	21,249-0
CAS No.:	[*3441-14-3*]
C.I. No.:	29160
Mol. Form.:	$C_{35}H_{25}N_7Na_2O_{10}S_2$
F.W.:	813.74
Appearance:	Dark-brown powder
λ_{max}:	507nm in water
Solubility:	H_2O 40mg/ml
	EGME 20mg/ml
	EtOH 0.7mg/ml

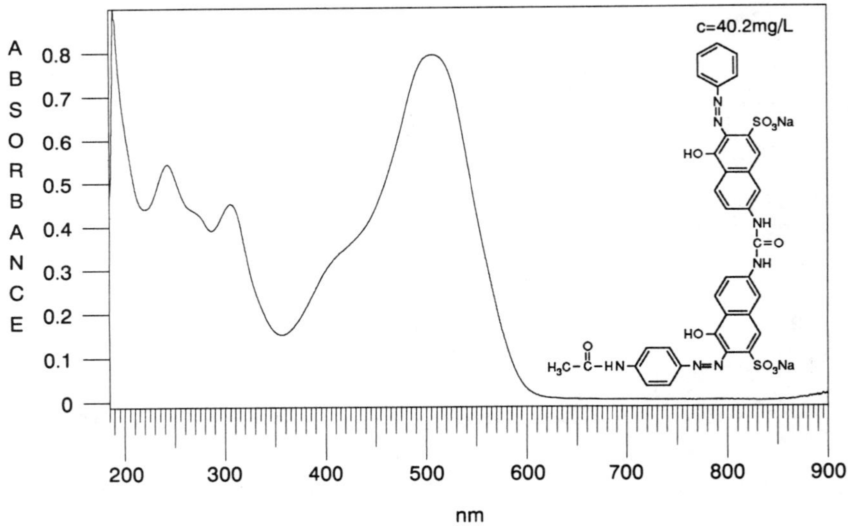

This anionic disazo dye is prepared by adding diazotized aniline and diazotized *p*-aminoacetanilide simultaneously to 6,6'-ureylenebis(1-naphthol-3-sulfonic acid), followed by conversion to the disodium salt.

Soluble in water (red), very slightly soluble in ethanol (orange) and insoluble in acetone, Direct Red 23 is used to dye cotton, wool and silk (the last two dye lighter than the first), to print cellulose and silk (urea process), to dye chromed leather and beater paper, to prepare paper-coating pigments from heavy-metal salts and to color cellophane and match heads.

Gurr[1] concluded that this is the Benzo Fast Scarlet that Emig[2] used as a general stain, instead of Carmine, for both plant and animal tissue. Apart from this reference, there is no evidence that it has been used as a biological stain.

References:

1) Gurr, E. *Encyclopedia of Microscopic Stains;* Hill: London, 1960; p 72.

2) Emig, W.H. *Stain Techniques;* Science Press: Pittsburgh, 1941.

General references:

Aldrich FT-IR Library	1(2), 996B
IR Library	**3**, 1466E
Colour Index	3rd ed., 1971

DIRECT RED 75

Benzo Fast Pink 2BL • Sirius Rose BB

Aldrich Cat. No.: 21,235-0

CAS No.: *[2829-43-8]*
C.I. No.: 25380
Mol. Form.: $C_{33}H_{22}N_8Na_4O_{15}S_4$
F.W.: 990.80
Appearance: Maroon crystalline powder
λ_{max}: 522nm in water
Solubility: H_2O 40mg/ml
EGME 10mg/ml
EtOH 0.8mg/ml

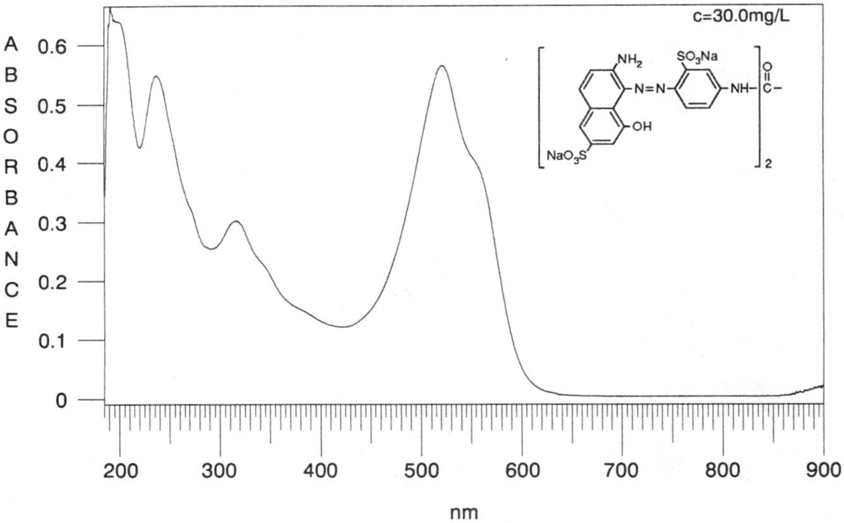

This bluish-pink, anionic disazo dye is prepared by coupling diazotized 2-amino-5-nitrobenzenesulfonic acid to gamma acid (7-amino-1-naphthol-3-sulfonic acid), reducing the nitro group and phosgenating the resulting amino function, followed by conversion to the tetrasodium salt.

Soluble in water, slightly soluble in methyl Cellosolve and very slightly soluble in ethanol, it dyes cotton, silk, wool, beater paper and chromed leather and prints cotton (urea process), wool, silk and nylon.

Conn's[1] states that it meets the linear configuration of the direct dyes that Puchtler and Sweat[2] recommended as lightfast substitutes for Acid Fuchsin in Van Gieson stains.

References:

1) *Conn's Biological Stains,* 9th ed.; Lillie, R.D., Ed.; Williams and Wilkins: Baltimore, 1977; p 165.

2) Puchtler, H.; Sweat, F. *Histochemie* **1964,** *4*, 24-34.

General references:

Aldrich FT-IR Library	**1**(2), 995D
IR Library	**3**, 1466A
Colour Index	3rd ed., 1971

DIRECT RED 81
Sirius Red 4B

Aldrich Cat. No.: 19,525-1
Sigma Prod. No.: D4140

CAS No.: [*2610-11-9*]
C.I. No.: 28160
Mol. Form.: $C_{29}H_{19}N_5Na_2O_8S_2$
F.W.: 675.61
m.p.: 240°C (dec.)
Appearance: Dark red-brown powder
λ_{max}: 508(397)nm in water
Solubility: H_2O 1mg/ml
EGME 10mg/ml
EtOH 1mg/ml

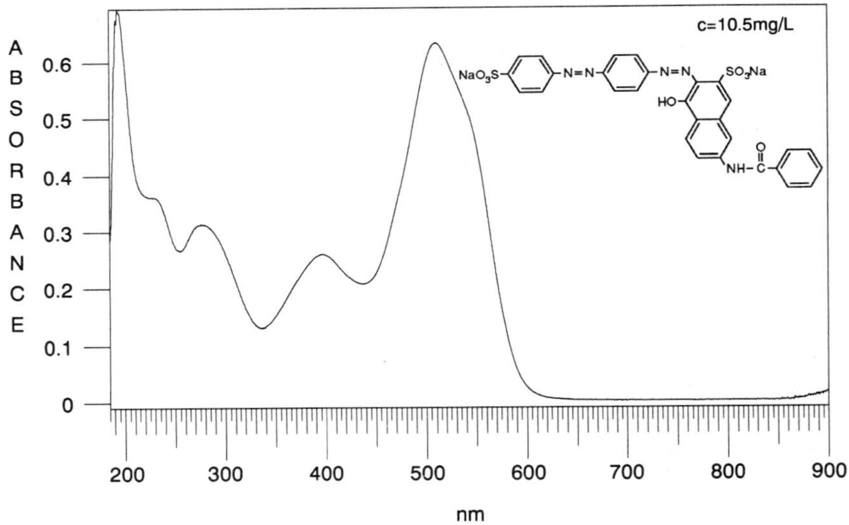

This acid disazo dye is soluble in methyl Cellosolve, very slightly soluble in water and ethanol and insoluble in all other organic solvents. It is made by first coupling diazotized sulfanilic acid to aniline. The product of this coupling is then diazotized and coupled to N-benzoylated J acid (2-amino-5-hydroxy-naphthalene-7-sulfonic acid), followed by conversion to the disodium salt. Industrial applications include the dyeing of cotton, silk, beater paper and fully chromed leather.

As a biological stain, it has been used primarily as a contrast stain.

Staining Procedures[1] substituted it for Eosin Y as a counterstain for Hematoxylin. Puchtler and Sweat[2] used it in place of Acid Fuchsin in the Van Gieson procedure.

References:

1) Clark, G. *Staining Procedures,* 4th ed.; Williams and Wilkins: Baltimore, 1981; p 107.

2) Puchtler, H.; Sweat, F. *Histochemie* **1964,** *4*, 24-34.

General references:

Aldrich FT-IR Library	1(2), 994D
Sigma FT-IR Library	1(2), 129D
IR Library	**3**,1465H
Colour Index	3rd ed., 1971

DIRECT VIOLET 51
Sirius Violet BB

Aldrich Cat. No.:	21,238-5
CAS No.:	[*5489-77-0*]
C.I. No.:	27905
Mol. Form.:	$C_{32}H_{27}N_5Na_2O_8S_2$
F.W.:	719.71
Appearance:	Black powder
λ_{max}:	549nm in water
Solubility:	H_2O 20mg/ml
	EGME 30mg/ml
	EtOH 10mg/ml

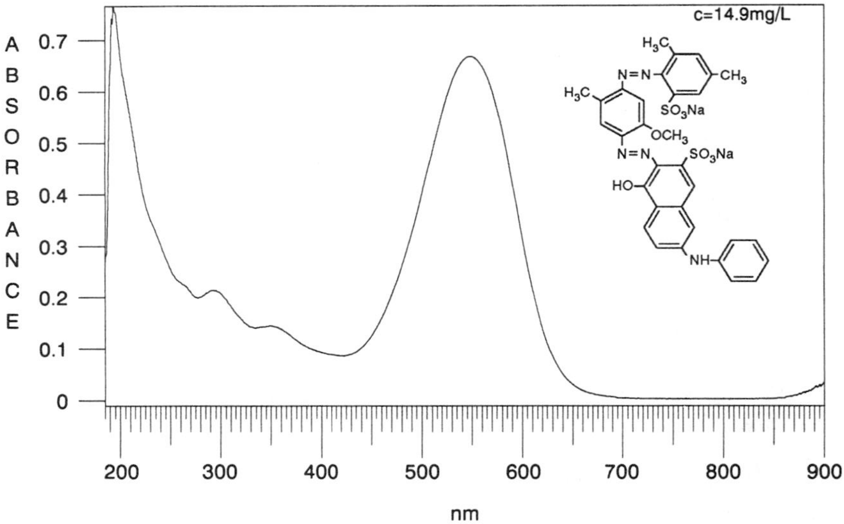

This bluish-violet, anionic disazo dye is prepared by first coupling diazotized 2-amino-3,5-xylenesulfonic acid to *m*-cresidine; then the product of this coupling is diazotized and coupled to *N*-phenyl J acid (6-amino-1-naphthol-3-sulfonic acid), at pH 8-9, and converted to the disodium salt.

Soluble in water and methyl Cellosolve and slightly soluble in ethanol, it is used to dye and print cotton, wool, silk and nylon, to dye chromed leather and beater paper and to color aqueous inks.

There is no evidence that it has been used as a biological stain.

General references:

Aldrich FT-IR Library	**1**(2), 995B
IR Library	**3**, 1466A
Colour Index	3rd ed., 1971

DIRECT YELLOW 8

Dianil Yellow 5G

Aldrich Cat. No.:	20,203-7
Sigma Prod. No.:	D4015
CAS No.:	[*10130-29-7*]
C.I. No.:	13920
Mol. Form.:	$C_{23}H_{19}N_4NaO_5S_2$
F.W.:	518.55
Appearance:	Orange powder
λ_{max}:	390nm in water
Solubility:	H_2O 30mg/ml
	EGME 50mg/ml
	EtOH 1mg/ml

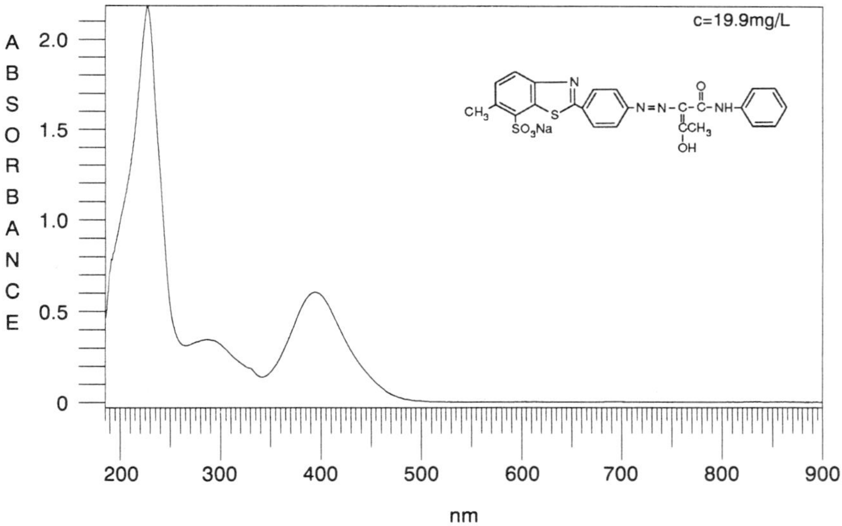

This bright, greenish-yellow, anionic monoazo dye is soluble in water and methyl Cellosolve and very slightly soluble in ethanol. It is prepared by coupling diazotized dehydrothio-*p*-toluidinesulfonic acid to acetoacetanilide, followed by conversion to the sodium salt.

Industrial uses include dyeing wool, cellulose, silk and beater paper, printing cellulose and silk (urea process with disodium phosphate) and coloring phenoplastic molding powder.

There is no evidence that it has been used as a biological stain.

General references:

Aldrich FT-IR Library	**1**(2), 1005A
Sigma FT-IR Library	**1**(2), 130A
IR Library	**3**, 1470G
Colour Index	3rd ed., 1971

DIRECT YELLOW 27

Sirius Light Yellow 5G

Aldrich Cat. No.:	20,187-1
Sigma Prod. No.:	D3890
CAS No.:	[*10190-68-8*]
C.I. No.:	13950
Mol. Form.:	$C_{24}H_{20}N_4Na_2O_9S_3$
F.W.:	650.62
Appearance:	Gold powder
λ_{max}:	393nm in water
Solubility:	H_2O 40mg/ml
	EGME 20mg/ml
	EtOH 2mg/ml

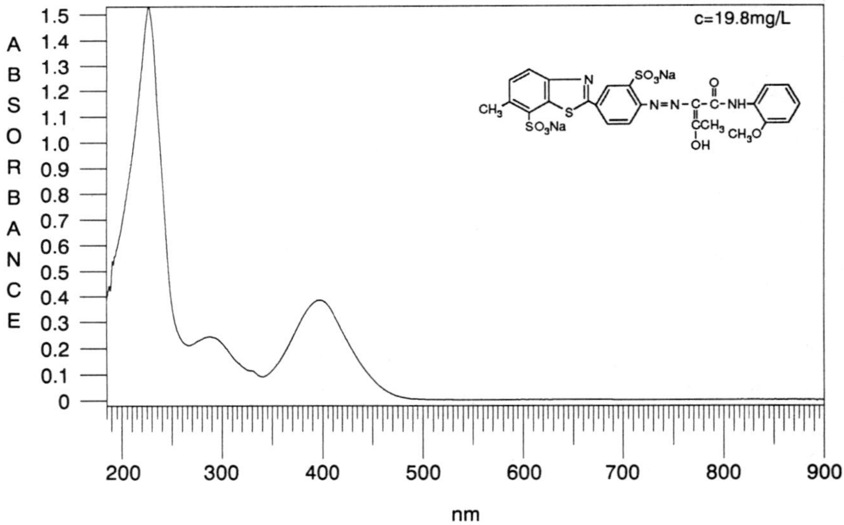

This yellow, anionic monoazo dye is made by coupling diazotized dehydrothio-*p*-toluidinedisulfonic acid to *o*-acetoacetanisidide, followed by conversion to the disodium salt. It is soluble in water and methyl Cellosolve and slightly soluble in ethanol. Industrial uses include dyeing cellulose, wool, silk, nylon, irregular quality viscose yarn and beater paper.

There is no evidence in the literature that it has been used as a biological stain.

General references:

Aldrich FT-IR Library	**1**(2), 1005B
Sigma FT-IR Library	**1**(2), 130B
IR Library	**3**, 1470H
Colour Index	3rd ed., 1971

DIRECT YELLOW 29

Sirius Light Yellow RT

Aldrich Cat. No.: 20,205-3
Sigma Prod. No.: D3765

CAS No.: [*6537-66-2*]
C.I. No.: 19556
Mol. Form.: $C_{42}H_{24}N_6Na_2O_6S_6$
F.W.: 947.06
Appearance: Orange powder
λ_{max}: 402nm in water
Solubility:

H_2O	50mg/ml
EGME	30mg/ml
EtOH	2mg/ml

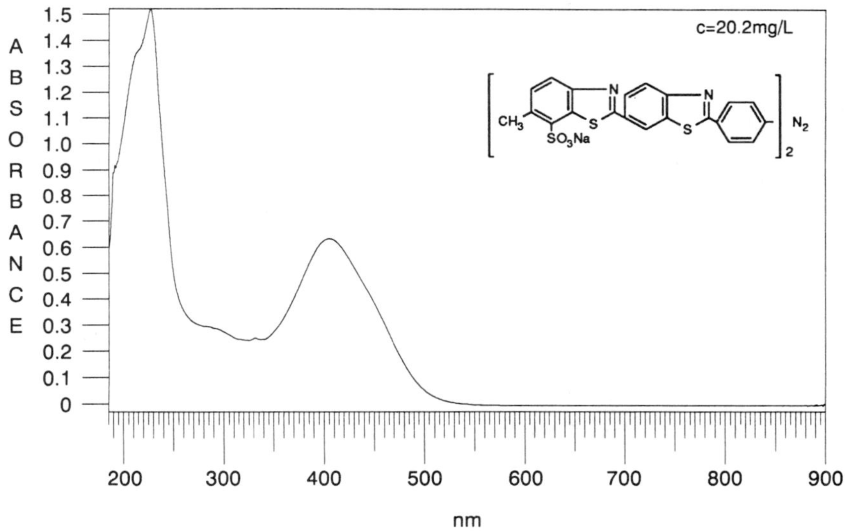

This yellow, water-soluble, anionic, monoazo thiazole dye is prepared by treating an aqueous solution of Primulin (Direct Yellow 59, C.I. 49000) with sodium hypochlorite.

Industrial applications include dyeing cotton, silk, wool, beater paper and chromed leather.

There is no evidence that it has been used as a biological stain.

General references:

Aldrich FT-IR Library	**1**(2), 1004B
Sigma FT-IR Library	**1**(2), 130C
IR Library	**3**, 1470E
NMR Library	**2**(2), 579A
Colour Index	3rd ed., 1971

DIRECT YELLOW 50

Sirius Light Yellow R

Aldrich Cat. No.:	20,189-8
Sigma Prod. No.:	D3640
CAS No.:	[*3214-47-9*]
C.I. No.:	29025
Mol. Form.:	$C_{35}H_{24}N_6Na_4O_{13}S_4$
F.W.:	956.83
Appearance:	Rust powder
λ_{max}:	390nm in water
Solubility:	H_2O 30mg/ml
	EGME 20mg/ml
	EtOH 1mg/ml

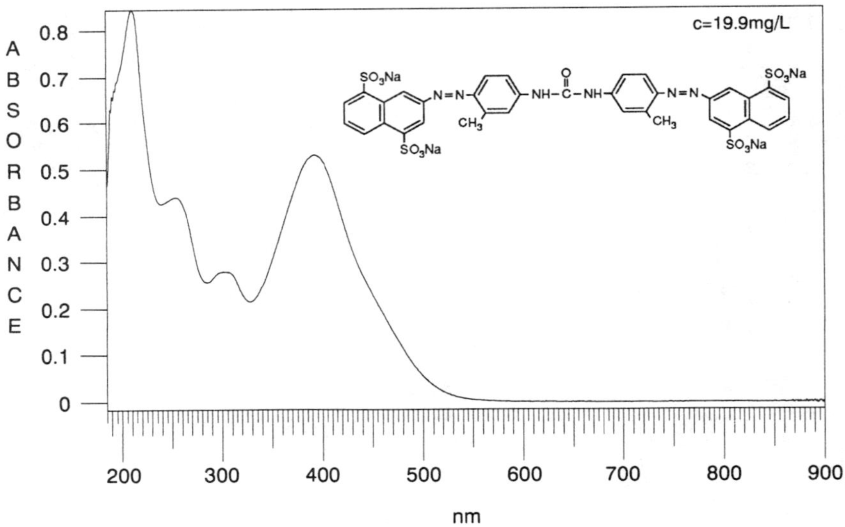

This orange-yellow, anionic disazo dye is made by coupling diazotized 3-amino-1,5-naphthalenedisulfonic acid with *m*-toluidine and phosgenating the reaction product, followed by conversion to the tetrasodium salt. It is soluble in water and very slightly soluble in ethanol. Industrial uses include dyeing cellulose, silk, nylon, paper, cellophane, chromed leather and anodized aluminum.

There is no evidence that it has been used as a biological stain.

General references:

Aldrich FT-IR Library	**1**(2), 996A
Sigma FT-IR Library	**1**(2), 130D
IR Library	**3**, 1466D
Colour Index	3rd ed., 1971

DIRECT YELLOW 62

Diazo Light Yellow GG

Aldrich Cat. No.:	20,206-1
Sigma Prod. No.:	D3515
CAS No.:	*[6409-90-1]*
C.I. No.:	36900
Mol. Form.:	$C_{35}H_{27}N_5Na_2O_9S_2$
F.W.:	771.74
Appearance:	Tan powder
λ_{max}:	341nm in water
Solubility:	H_2O 90mg/ml
	EGME 20mg/ml
	EtOH 0.6mg/ml

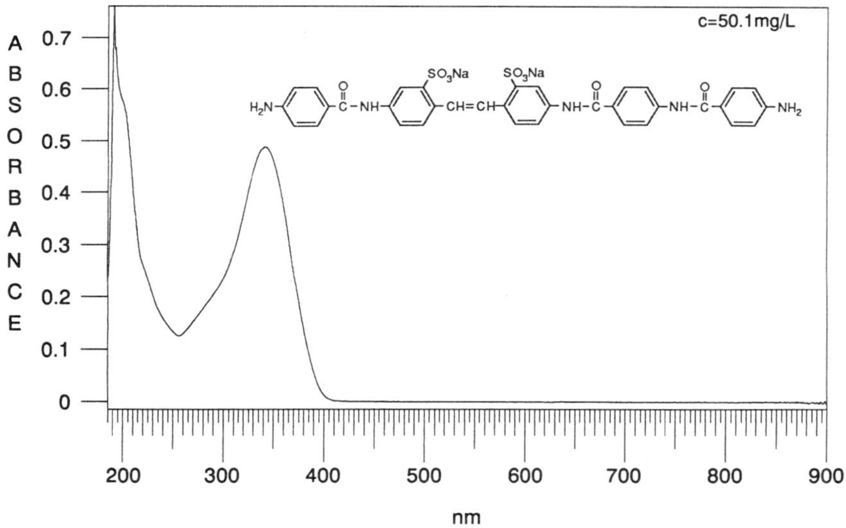

Direct Yellow 62 is a dye intermediate that is diazotized *in situ* on the fabric and then developed with methylphenylpyrazolone to form a dye with a greenish-yellow hue. It is soluble in water and only moderately soluble in methyl Cellosolve.

Industrially, it is used to dye cotton, silk and wool.

The *Colour Index* lists it as a biological stain with no specific application sited.

General references:

Aldrich FT-IR Library	**1**(2), 1039D
Sigma FT-IR Library	**1**(2), 131A
IR Library	**3**, 1487E
Colour Index	3rd ed., 1971

DISPERSE BLUE 1

Celliton Blue Extra • Solvent Blue 18

Aldrich Cat. No.:	21,564-3
Sigma Prod. No.:	D0892

CAS No.:	*[2475-45-8]*
C.I. No.:	64500
Mol. Form.:	$C_{14}H_{12}N_4O_2$
F.W.:	268.28
Appearance:	Black powder
λ_{max}:	620nm in 50% ethanol + 1 drop 1*N* NaOH
Solubility:	H_2O 30mg/ml
	EGME 30mg/ml
	EtOH 2mg/ml

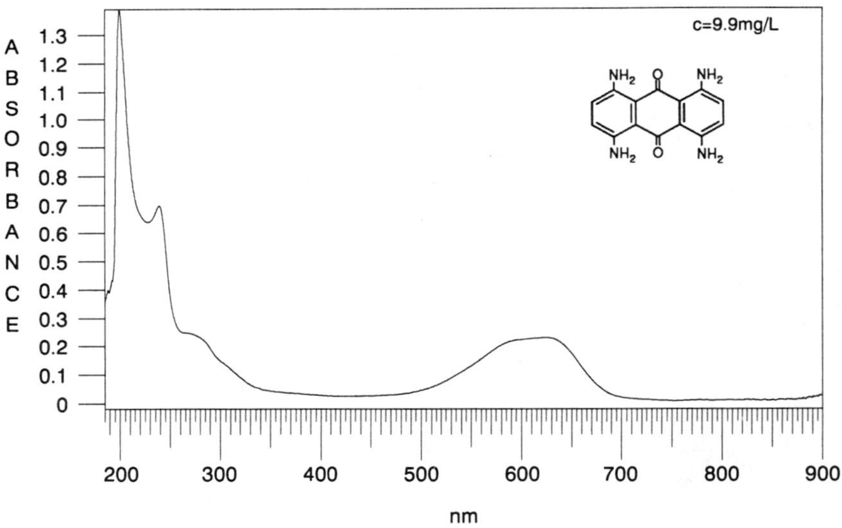

This solvent anthraquinone dye is prepared by acylating 1,5-diaminoanthraquinone with oxalic acid, nitrating in mixed acids, hydrolyzing and finally reducing to produce the 1,4,5,8-tetraaminoanthraquinone.

Soluble in acetone and methyl Cellosolve and slightly soluble in benzene, alcohol and linseed oil, Disperse Blue 1 is used in the textile industry to dye acetate, nylon, triacetate, polyester and acrylic. Other industrial applications include dyeing sheepskins and furs, surface coloring thermoplastics and coloring alcoholic and ester solvents, vinyl and alkyd resins and cellulose ester lacquers.

There is no evidence in the literature that it has been used as a biological stain.

General references:

Aldrich FT-IR Library	**1**(2), 1018A
Sigma FT-IR Library	**1**(2), 131B
IR Library	**3**, 1477B
Safety	**2**(1), 1476B
Colour Index	3rd ed., 1971

DISPERSE BLUE 3

Celliton Fast Blue FFR

Aldrich Cat. No.: 21,565-1

CAS No.: [*2475-46-9*]
C.I. No.: 61505
Mol. Form.: $C_{17}H_{16}N_2O_3$
F.W.: 296.33
Appearance: Black powder
λ_{max}: 640(594)nm in 50% ethanol
Solubility: H_2O 30mg/ml
EGME 20mg/ml
EtOH 8mg/ml

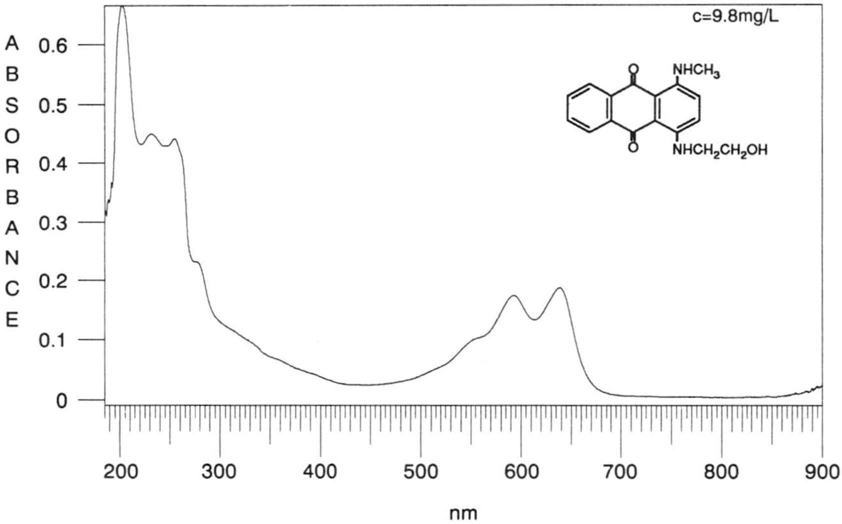

This anthraquinone solvent dyestuff is prepared by condensing methylamine and ethanolamine with quinizarin and leucoquinizarin in isobutanol, and oxidizing the condensation product.

Soluble in acetone, benzene and methyl Cellosolve and slightly soluble in alcohol and carbon tetrachloride, Disperse Blue 3 is used to dye acetate, nylon, triacetate and acrylic fibers.

There is no evidence that it has been used as a biological stain.

General references:

Aldrich FT-IR Library **1**(2), 1017C
IR Library **3**, 1476H
NMR Library **2**(2), 93A
Colour Index 3rd ed., 1971

DISPERSE BLUE 14

Celliton Fast Blue B

Aldrich Cat. No.:	21,567-8
Sigma Prod. No.:	D1017
CAS No.:	*[2475-44-7]*
C.I. No.:	61500
Mol. Form.:	$C_{16}H_{14}N_2O_2$
F.W.:	266.30
Appearance:	Green powder
λ_{max}:	640(594)nm in 50% ethanol
Solubility:	H_2O 50mg/ml
	EGME 30mg/ml
	EtOH 3mg/ml

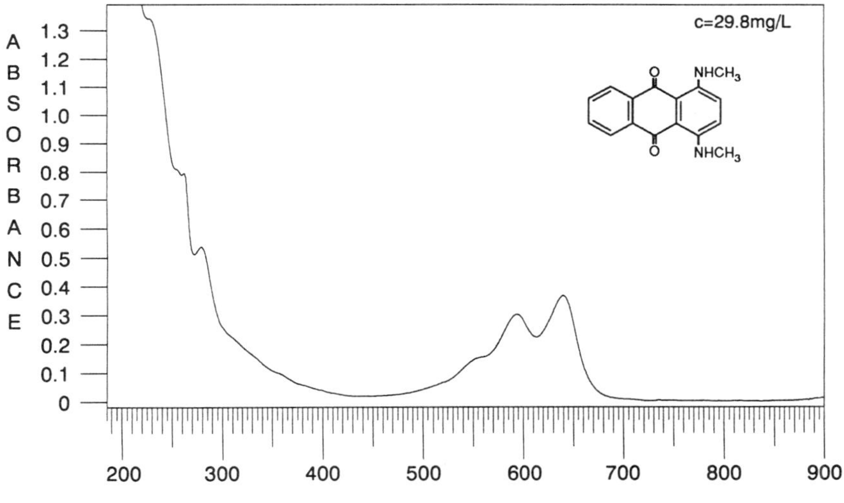

This anthraquinone solvent dye is prepared by treating quinizarin with aqueous monomethylamine and sodium hydrosulfite or zinc dust under pressure, and oxidizing the resulting leuco dye with nitrobenzene, or directly in the presence of hot sulfuric acid to yield the desired product.

Soluble in water, acetone, glacial acetic acid, nitrobenzene, pyridine and toluene and slightly soluble in ethanol, it is used to dye acetate, nylon, sheepskins and fur and to surface dye thermoplastics.

There is no evidence that it has been used as a biological stain.

General references:

Aldrich FT-IR Library	**1**(2), 1016D
Sigma FT-IR Library	**1**(2), 131C
IR Library	**3**, 1476E
Colour Index	3rd ed., 1971

DISPERSE ORANGE 1

Celliton Fast Orange • 4-(4-Nitrophenylazo)diphenylamine

Aldrich Cat. No.:	21,228-8
Sigma Prod. No.:	D3517

CAS No.:	*[2581-69-3]*
C.I. No.:	11080
Mol. Form.:	$C_{18}H_{14}N_4O_4$
F.W.:	318.34
Appearance:	Maroon powder
λ_{max}:	483nm in 50% ethanol
Solubility:	H_2O 60mg/ml
	EGME 40mg/ml
	EtOH 8mg/ml

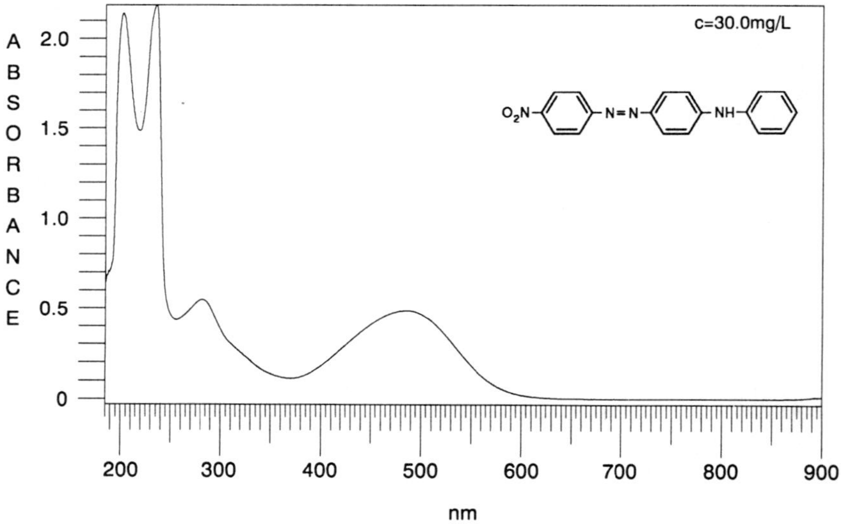

This orange, solvent monoazo dye is prepared by coupling diazotized 4-nitroaniline with diphenylamine.

Disperse Orange 1 is used to dye and print acetates, triacetates, polyesters and nylon, to dye poly(vinyl chloride) fibers, sheepskins and furs and to surface dye thermoplastics.

There is no evidence that it has been used as a biological stain.

General references:

Aldrich FT-IR Library	**1**(2), 974A
Sigma FT-IR Library	**1**(2), 131D
IR Library	**3,** 1455G
Safety	**2**(1), 1476C
Colour Index	3rd ed., 1971

DISPERSE ORANGE 3

Celliton Fast Orange • Solvent Orange 9

Aldrich Cat. No.: 21,571-6
Sigma Prod. No.: D1142

CAS No.: [*730-40-5*]
C.I. No.: 11005
Mol. Form.: $C_{12}H_{10}N_4O_2$
F.W.: 242.24
Appearance: Brown powder
λ_{max}: 443nm in 50% ethanol
Solubility: H_2O 40mg/ml
EGME 40mg/ml
EtOH 6mg/ml

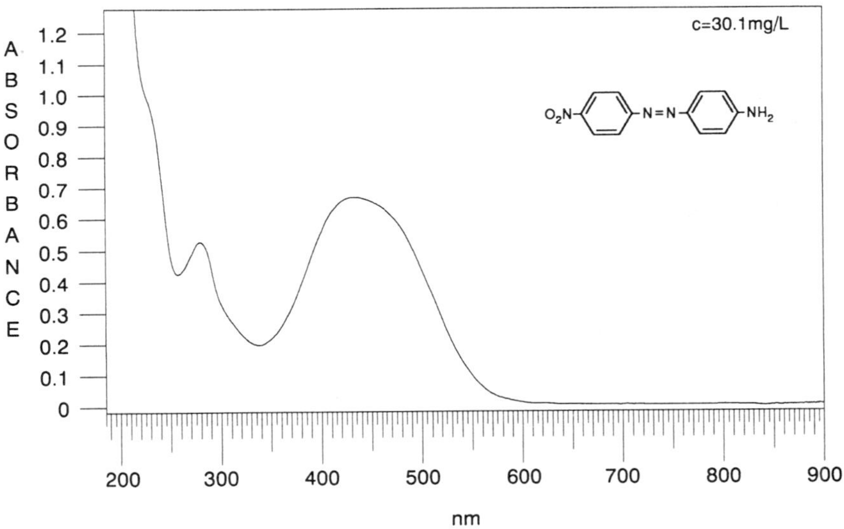

This orange, solvent monoazo dye is prepared by coupling diazotized 4-nitroaniline to anilinomethanesulfonic acid, followed by alkaline hydrolysis at the boiling point to remove the methanesulfonic acid group.

Soluble in water, acetone, methyl Cellosolve and toluene and slightly soluble in ethanol, it is used to dye acetate, nylon, triacetate, polyester, acrylic, sheepskin and furs, to surface dye thermoplastics and to color alcohols, esters, ketones and lacquers (both vinyl and synthetic resin).

There is no evidence that it has been used as a biological stain.

General references:

Aldrich FT-IR Library	**1**(2), 966C
Sigma FT-IR Library	**1**(2), 132A
IR Library	**3**, 1450F
Safety	**2**(1), 1476D
Colour Index	3rd ed., 1971

DISPERSE ORANGE 13
Solvent Orange 52

Aldrich Cat. No.: 21,229-6

CAS No.: [*6253-10-7*]
C.I. No.: 26080
Mol. Form.: $C_{22}H_{16}N_4O$
F.W.: 352.40
Appearance: Dark-brown powder with gold specks
λ_{max}: 426nm in 50% ethanol
Solubility: H_2O 60mg/ml
EGME 40mg/ml
EtOH 8mg/ml

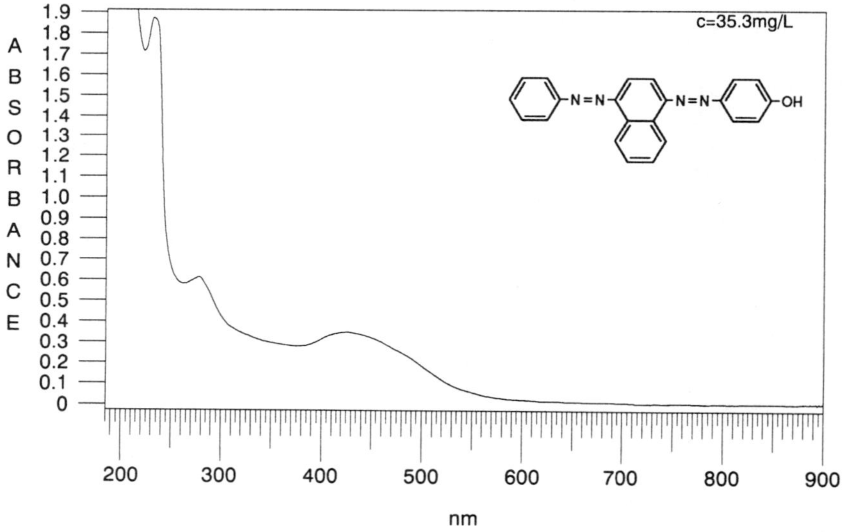

This orange, solvent disazo dye is prepared by coupling diazotized aniline to 1-naphthylamine, then isolating the product, diazotizing and coupling to phenol.

Solvent Orange 52 is used to dye and print acetate, triacetate, nylon, polyester and acrylic, to dye poly(vinyl chloride) fibers, sheepskins and fur, and to surface dye cellulose acetate, poly(styrene) and other thermoplastics.

There is no evidence that it has been used as a biological stain.

General references:

Aldrich FT-IR Library **1**(2), 975D
IR Library **3**, 1456E
Colour Index 3rd ed., 1971

DISPERSE ORANGE 25

Aldrich Cat. No.: 21,573-2

CAS No.: [*31482-56-1*]
Mol. Form.: $C_{17}H_{17}N_5O_2$
F.W.: 323.36
Appearance: Maroon powder
λ_{max}: 455nm in methanol
Solubility: H_2O 40mg/ml
EGME 30mg/ml
EtOH 2mg/ml

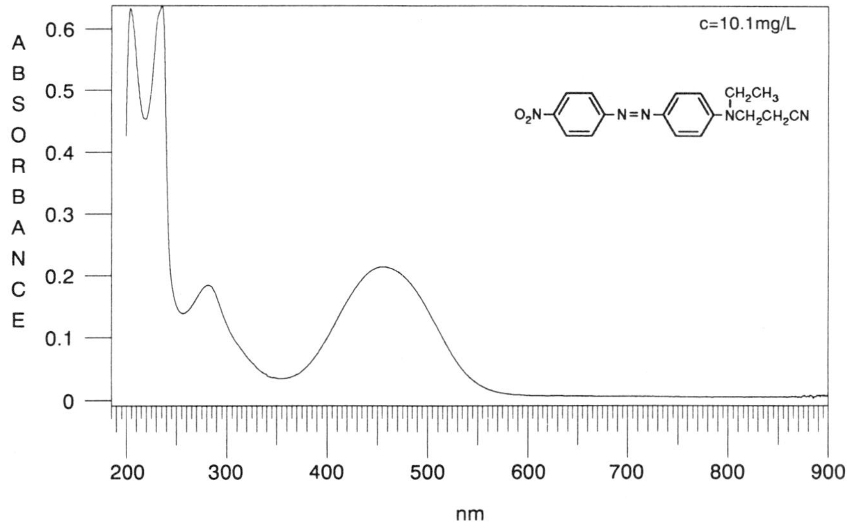

This monoazo dye is soluble in water, acetone and methyl Cellosolve and slightly soluble in ethanol.

In the textile industry, it is used to dye and print acetate, triacetate and polyester.

There is no evidence in the literature that it has been used as a biological stain.

General references:

Aldrich FT-IR Library	**1**(2), 967A
IR Library	**3**, 1450H
Safety	**2**(1), 1477B
Colour Index	3rd ed., 1971

DISPERSE RED 1

Celliton Scarlet B

Aldrich Cat. No.:	21,574-0
Sigma Prod. No.:	D1392
CAS No.:	[*2872-52-8*]
C.I. No.:	11110
Mol. Form.:	$C_{16}H_{18}N_4O_3$
F.W.:	314.35
Appearance:	Very dark, brown powder
λ_{max}**:**	503nm in 50% ethanol
Solubility:	H_2O 30mg/ml
	EGME 20mg/ml
	EtOH 6mg/ml

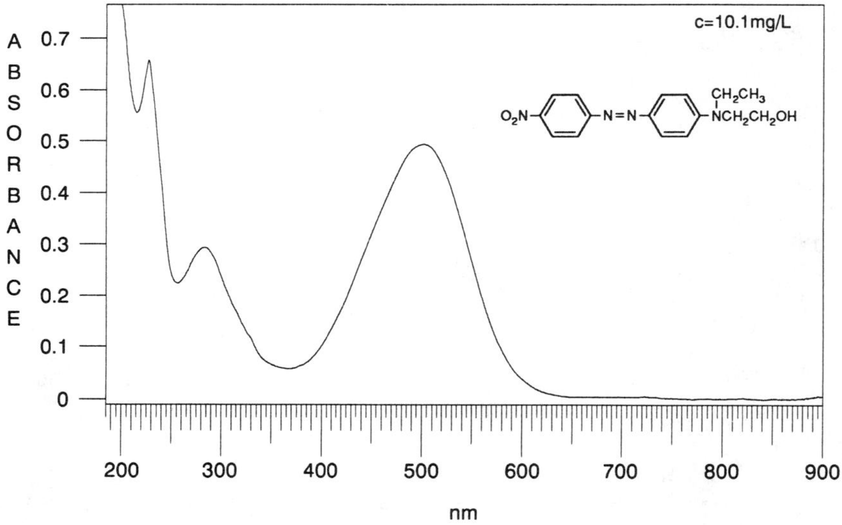

This red, monoazo solvent dye is prepared by coupling diazotized 4-nitroaniline to 2-(*N*-ethylanilino)ethanol.

Soluble in water, acetone and benzene and slightly soluble in ethanol, it is used to dye and print acetate, nylon, triacetate, polyester and acrylic, to dye sheepskins and furs, to surface dye thermoplastics (*e.g.*, methyl methacrylate) and to color smokes.

There is no evidence that it has been used as a biological stain.

General references:

Aldrich FT-IR Library	**1**(2), 966D
Sigma FT-IR Library	**1**(2), 132C
IR Library	**3**, 1450G
Colour Index	3rd ed., 1971

DISPERSE RED 13

Celliton Fast Ruby B

Aldrich Cat. No.: 21,575-9
Sigma Prod. No.: D1517

CAS No.: [*3180-81-2*]
C.I. No.: 11115
Mol. Form.: $C_{16}H_{17}ClN_4O_3$
F.W.: 348.79
Appearance: Black powder
λ_{max}: 517nm in 50% ethanol
Solubility: H_2O 30mg/ml
EGME 40mg/ml
EtOH 10mg/ml

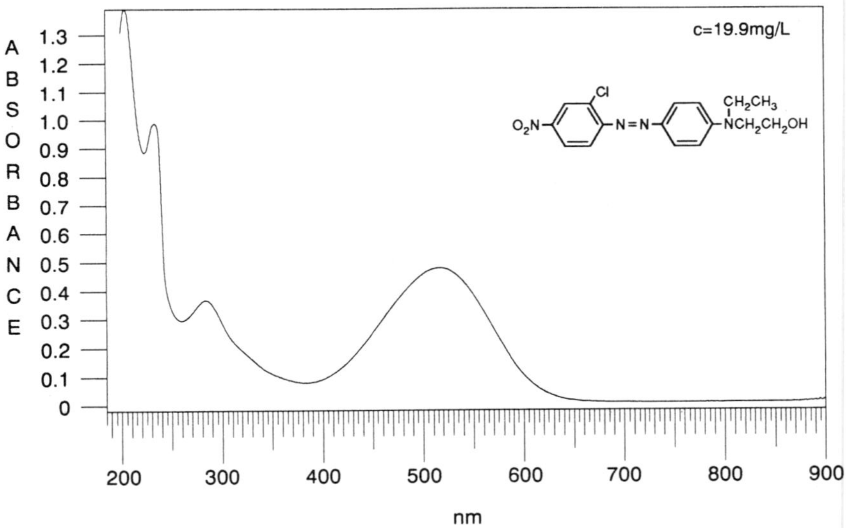

This monoazo solvent dye is prepared by coupling diazotized 2-chloro-4-nitroaniline to 2-(*N*-ethylanilino)ethanol.

Slightly soluble in ethanol and acetone, and soluble in benzene and most aliphatic hydrocarbon solvents, it is used to dye and print acetate, nylon, triacetate, polyester and acrylic, to dye sheepskins and furs and to surface dye thermoplastics.

There is no evidence that it has been used as a biological stain.

General references:

Aldrich FT-IR Library **1**(2), 967B
Sigma FT-IR Library **1**(2), 132D
IR Library **3**, 1451A
Colour Index 3rd ed., 1971

DISPERSE RED 19

Aldrich Cat. No.: 21,576-7
Sigma Prod. No.: D1642

CAS No.: [*2734-52-3*]
C.I. No.: 11130
Mol. Form.: $C_{16}H_{18}N_4O_4$
F.W.: 330.35
Appearance: Violet-brown powder
λ_{max}: 495nm in 50% ethanol

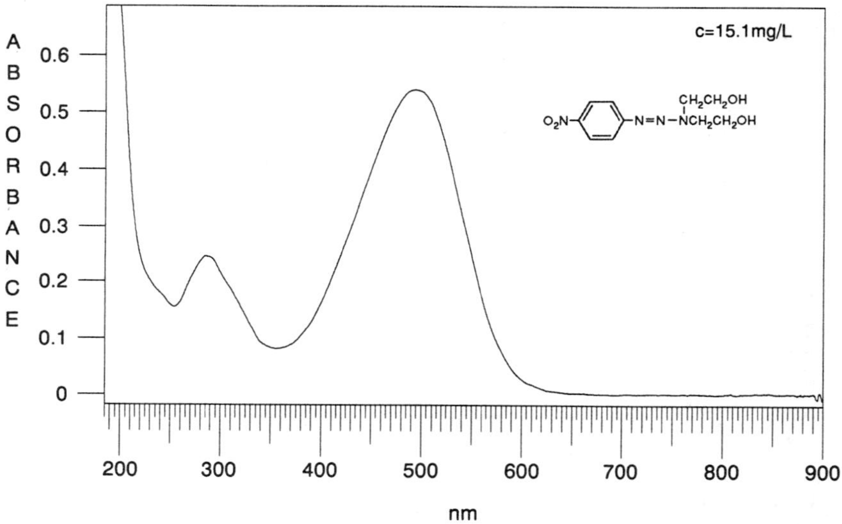

This red, monoazo solvent dye is prepared by coupling diazotized 4-nitroaniline to 2,2'-(phenylimino)diethanol.

Soluble in ethanol, acetone and benzene, it is used to dye and print acetate, nylon and triacetate, to dye polyester and acrylic (limited), to dye sheepskins and furs, to surface dye thermoplastics and color smokes.

There is no evidence that it has been used as a biological stain.

General references:

Aldrich FT-IR Library **1**(2), 966A
Sigma FT-IR Library **1**(2), 133A
IR Library **3**, 1450D
Colour Index 3rd ed., 1971

DISPERSE YELLOW 3

Celliton Fast Yellow G • Solvent Yellow 77

Aldrich Cat. No.: 21,568-6
Sigma Prod. No.: D1892

CAS No.: [*2832-40-8*]
C.I. No.: 11855
Mol. Form.: $C_{15}H_{15}N_3O_2$
F.W.: 269.31
m.p.: 268-270°C
Appearance: Brown powder
λ_{max}: 357nm in 50% ethanol
Solubility: H_2O 20mg/ml
EGME 50mg/ml
EtOH 10mg/ml

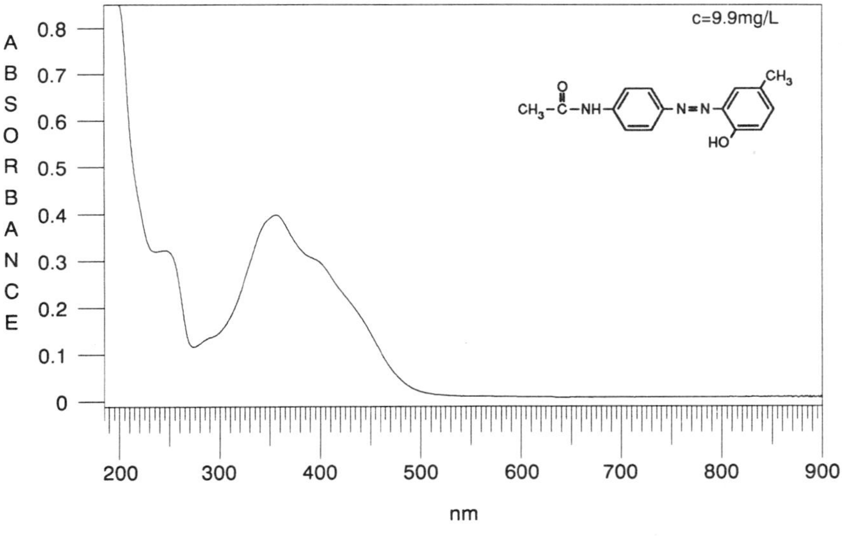

This solvent monoazo dye is prepared by coupling diazotized *p*-amino-acetanilide to *p*-cresol.

Soluble in acetone and benzene but less soluble in ethanol, Disperse Yellow 3 is used to dye acetate, nylon, triacetate, polyester, acrylic, poly(vinyl chloride) fibers, sheepskins and furs and to surface dye thermoplastics.

There is no evidence that it has been used as a biological stain.

General references:

Aldrich FT-IR Library	**1**(2), 969A
Sigma FT-IR Library	**1**(2), 133B
IR Library	**3**, 1452A
Safety	**2**(1), 1477D
Colour Index	3rd ed., 1971

DISPERSE YELLOW 5

Celliton Yellow 5G

Aldrich Cat. No.: 21,224-5

CAS No.: *[6439-53-8]*
C.I. No.: 12790
Mol. Form.: $C_{16}H_{12}N_4O_4$
F.W.: 324.30
Appearance: Brown powder
λ_{max}: 405nm in 50% ethanol
Solubility: H_2O 50mg/ml
EGME 30mg/ml
EtOH 4mg/ml

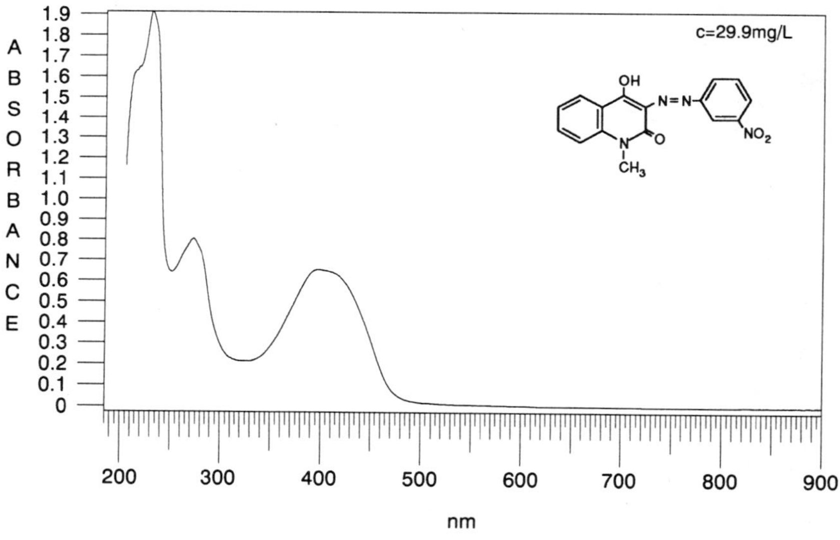

This monoazo solvent dye is prepared by coupling diazotized 3-nitroaniline with 4-hydroxy-1-methylcarbostyril.

Slightly soluble in ethanol and soluble in water, acetone, benzene, methyl Cellosolve and carbon tetrachloride, it is used to dye and print acetate, nylon, triacetate and polyester, to dye poly(vinyl chloride) fibers, wool, sheepskins and furs, to color transparent styrene resins and to surface dye thermoplastics.

There is no evidence that it has been used as a biological stain.

General references:

Aldrich FT-IR Library **1**(2), 985C
IR Library **3**, 1461C
Colour Index 3rd ed., 1971

DISPERSE YELLOW 7
Celliton Fast Yellow 5R

Aldrich Cat. No.: 21,569-4
Sigma Prod. No.: D2017

CAS No.: [*6300-37-4*]
C.I. No.: 26090
Mol. Form.: $C_{19}H_{16}N_4O$
F.W.: 316.37
Appearance: Brown powder
λ_{max}: 388nm in 50% ethanol
Solubility: H_2O 30mg/ml
EGME 80mg/ml
EtOH 10mg/ml

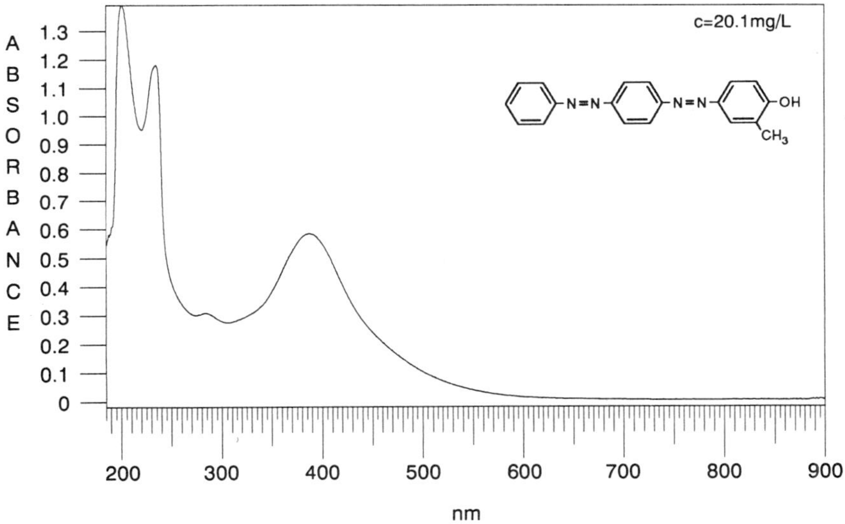

This neutral disazo dye is prepared by coupling diazotized *p*-phenylazoaniline to *o*-cresol.

Soluble in acetone and benzene and somewhat soluble in ethanol, it is used to dye acetate, nylon, triacetate, polyester, sheepskins and furs, and to surface dye thermoplastic resins, such as poly(methyl methacrylate).

There is no evidence that it has been used as a biological stain.

General references:

Aldrich FT-IR Library **1**(2), 973D
Sigma FT-IR Library **1**(2), 133C
IR Library **3**, 1455F
Colour Index 3rd ed., 1971

DISPERSE YELLOW 9

N-(2,4-Dinitrophenyl)-1,4-phenylenediamine • Solvent Orange 53

Aldrich Cat. No.:	21,225-3
Sigma Prod. No.:	D2142

CAS No.:	[*6373-73-5*]
C.I. No.:	10375
Mol. Form.:	$C_{12}H_{10}N_4O_4$
F.W.:	274.24
Appearance:	Red-brown powder
λ_{max}:	359nm in chloroform
Solubility:	H_2O 1mg/ml
	EGME 30mg/ml
	EtOH 2mg/ml

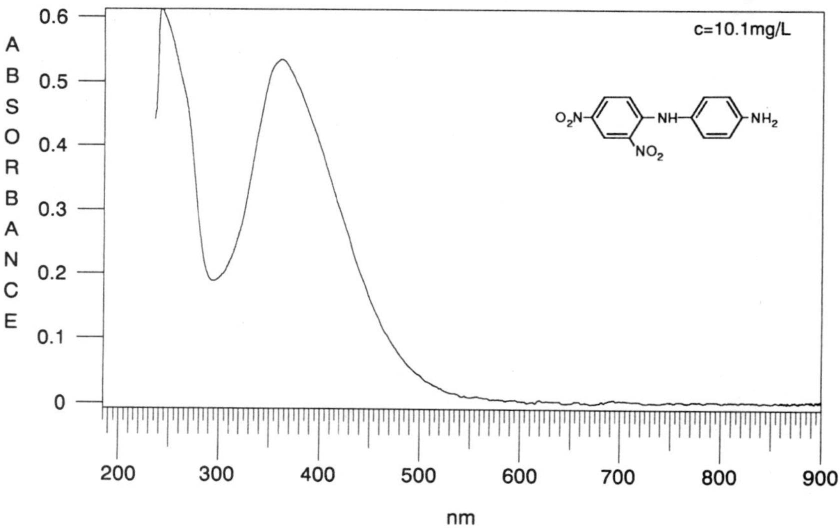

This yellow, dinitro functional dye is prepared by condensing 1,4-phenylenediamine with 1-chloro-2,4-dinitrobenzene.

Soluble in toluene, xylene, pyridine and very slightly soluble in ethanol, acetone and benzene, it is used both as a disperse dye and a solvent dye. Industrial applications include dyeing and printing acetates, triacetates, polyesters and acrylics, dyeing wool, sheepskins and furs, leveling pastel shades of plastic, surface dyeing poly(methyl methacrylate), and coloring cellulose acetate, poly(styrene) and other thermoplastics.

While no biological stain applications have been reported, its solubility characteristics indicate that it might have potential as a fat stain.

General references:

Aldrich FT-IR Library	**1**(2), 953D
Sigma FT-IR Library	**1**(2), 133D
IR Library	**3**, 1444D
Safety	**2**(1), 1478A
Colour Index	3rd ed., 1971

EMODIN

6-Methyl-1,3,8-trihydroxyanthraquinone

Aldrich Cat. No.: 23,051-0
Sigma Prod. No.: E7881
Purity: 97%

CAS No.: [*518-82-1*]
Mol. Form.: $C_{15}H_{10}O_5$
F.W.: 270.24
m.p.: 255°C (dec.)
Appearance: Orange powder
λ_{max}: 521(309)nm in 0.1*N* sodium hydroxide

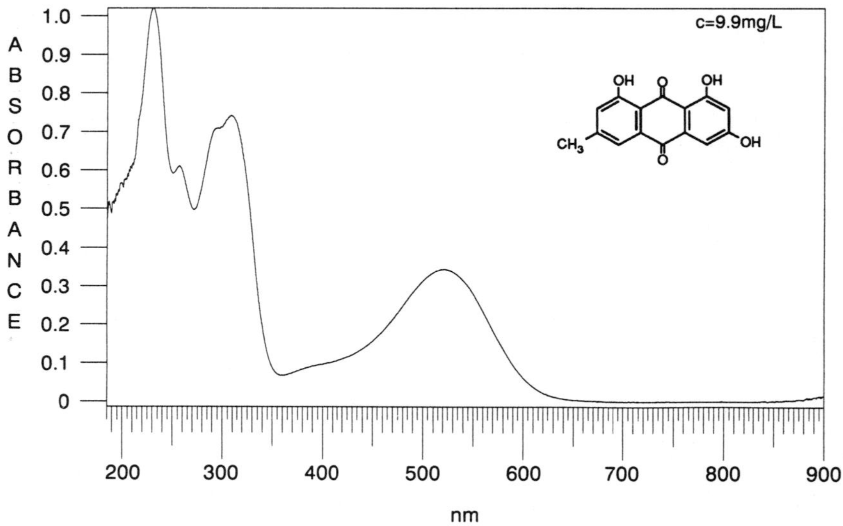

This natural, anthraquinone solvent dye is practically insoluble in water, soluble in ethanol and moderately soluble in aqueous alkali solutions.

While Emodin is naturally occurring as a rhamnoside in rhubarb roots, Ayyangar *et al.*[1] have synthesized it from 2-methylanthraquinone. This is not to be confused with an isomer, 2-methyl-1,3,8-trihydroxyanthraquinone, which is listed in the *Colour Index* as Emodin (C.I. 75440, C.I. Natural Green 2 and C.I. Natural Yellow 14). The latter is now obsolete as a dye.

The Merck Index[2] lists Emodin (the *6-methyl*-form) as a cathartic. There is no evidence that it has been used as a biological stain.

References:

1) Ayyangar, N.R.; Bapat, D.S.; Joshi, B.S. *J. Sci. Ind. Res. (India)* **1961,** *20B*, 493; *Chem. Abstr.* **1962,** *57,* 8514b.

2) *The Merck Index,* 11th ed.; Budavari, S. *et al.*, Eds. Merck: Rahway, 1989; Monograph 3518.

General references:

Aldrich FT-IR Library	**1**(2), 88D
Sigma FT-IR Library	**1**(2), 655A
IR Library	**3**, 900F
NMR Library	**2**(2), 91D

EOSIN B

Acid Red 91

Aldrich Cat. No.:	86,100-6
Grade:	Certified
CAS No.:	*[548-24-3]*
C.I. No.:	45400
Mol. Form.:	$C_{20}H_6Br_2N_2Na_2O_9$
F.W.:	624.08
Appearance:	Brownish-green powder
λ_{max}**:**	514nm in water + 2ml 1% sodium carbonate
Solubility:	H_2O 50mg/ml
	EGME 60mg/ml
	EtOH 20mg/ml

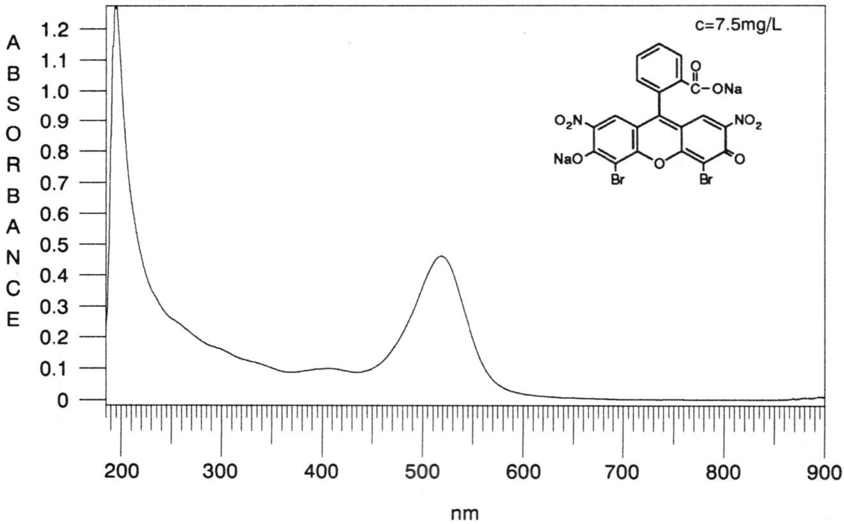

Eosin B, certified, is the disodium salt of 4',5'-dibromo-2',7'-dinitrofluorescein. It is an anionic xanthene dye that is made by the 2',7'-dinitration of 4',5'-dibromofluorescein, which then is converted to the disodium salt. This salt form is water and alcohol soluble.

Eosin B, spirit soluble, as the name indicates, is soluble in most aromatic and aliphatic solvents and virtually insoluble in water. It is predominantly the free-acid form with some of the lactone form present.

These products are distinctly bluer in shade than Eosin Y. However, this is not reflected in the λ_{max} since it closely coincides with the Eosin Y maximum.

Eosin B dyes silk from either a neutral or an acetic acid bath. It is used in inks and to color wax pencils.

Staining Procedures[1] includes the biological methods and bibliography for the following applications: counterstain for Hematoxylin used in Lillie- and Fullmer-modified Nocht's method with Azure A to demonstrate cell granules; stain for microorganisms and nuclei; and in the Glenner- and Lillie-modified Mann's Methyl Blue-Eosin B technique for demonstrating acidophils, basophils and chromophobes.

Reference:

1) Clark, G. *Staining Procedures,* 4th ed.; Williams and Wilkins: Baltimore, 1981; pp 209-210, 248.

General references:

Aldrich FT-IR Library	**1**(2), 1011D
IR Library	**3**, 1474B
Merck Index	**11**, 3554
Colour Index	3rd ed., 1971

EOSIN B, spirit soluble

Aldrich Cat. No.: 21,295-4
Sigma Prod. No.: E6880

CAS No.: [*56360-46-4*]
Mol. Form.: $C_{20}H_8Br_2N_2O_9$
F.W.: 580.11
m.p.: 275°C (dec.)
Appearance: Bright-orange powder
λ_{max}: 522nm in methanol
Solubility: H_2O 0.1mg/ml
EGME 30mg/ml
EtOH 4mg/ml

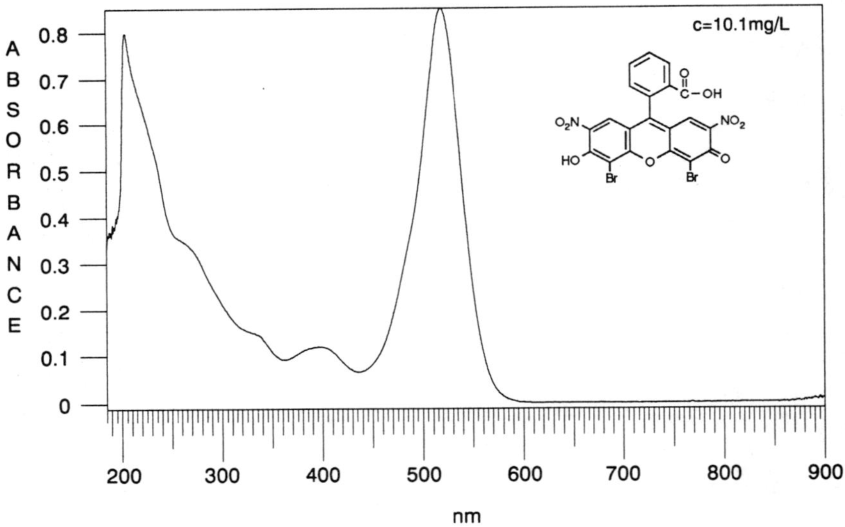

See description on pages 300 and 301 under Eosin B.

General references:

Aldrich FT-IR Library	**1**(2), 1012C
Sigma FT-IR Library	**1**(2), 134C
IR Library	**3**, 1474E

EOSIN BLUISH blend

Aldrich Cat. No.: 22,932-6

Appearance: Dark-red powder
λ_{max}: 517(301)nm in water
Solubility: H_2O 70mg/ml
EGME 10mg/ml
EtOH 40mg/ml

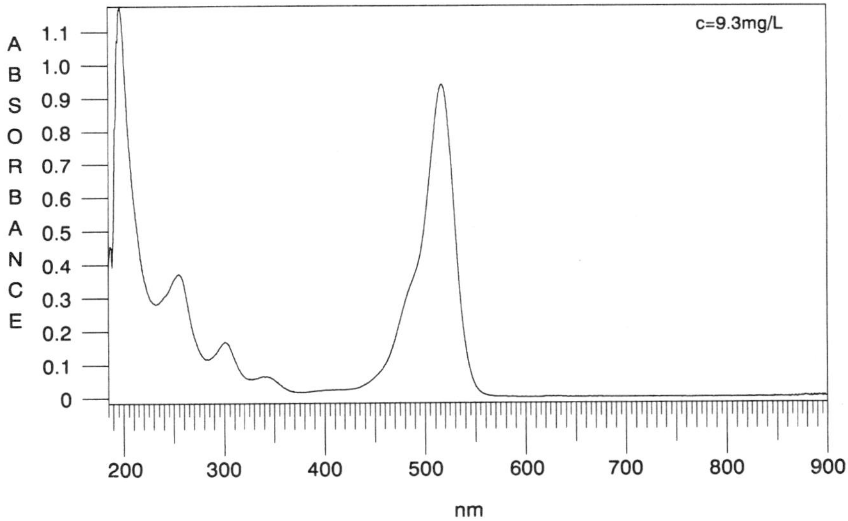

Eosin Bluish Blend is a deliberate blend of Eosin Y (C.I. 45380) and Erythrosin B (C.I. 45430) in an 80/20 ratio, w/w. It is readily soluble in water and somewhat less soluble in ethanol and methanol.

Conn's[1] indicates that it is very useful in certain staining applications where a stain bluer than Eosin Y is desired; however, it gives the color effect of Eosin Y when used as a component of a neutral stain.

Eosin Bluish Blend should not be confused with Eosin B (bluish) (C.I. 45400; Aldrich Cat. No. 86,100-6) which, ideally, is a pure molecular entity.

Reference:

1) *Conn's Biological Stains,* 9th ed.; Lillie, R.D., Ed.; Williams and Wilkins: Baltimore, 1977; p 346.

General references:

Aldrich FT-IR Library 1(2), 1034D
Colour Index 3rd ed., 1971

EOSIN Y

Acid Red 87

Aldrich Cat. No.:	11,983-0	21,612-7
Sigma Prod. No.:	E4382	——
Grade:	Certified	Indicator
Dye content:	~90%	~80%
CAS No.:	*[548-26-5]*	
C.I. No.:	45380	
Mol. Form.:	$C_{20}H_6Br_4Na_2O_5$	
F.W.:	691.88	
Appearance:	Rust powder	Dark red-brown powder
λ_{max}:	514nm in water + 1ml 1% sodium carbonate	517nm in water + 1ml 1% sodium carbonate
Solubility:	H_2O 40mg/ml	200mg/ml
	EGME 40mg/ml	50mg/ml
	EtOH 10mg/ml	20mg/ml

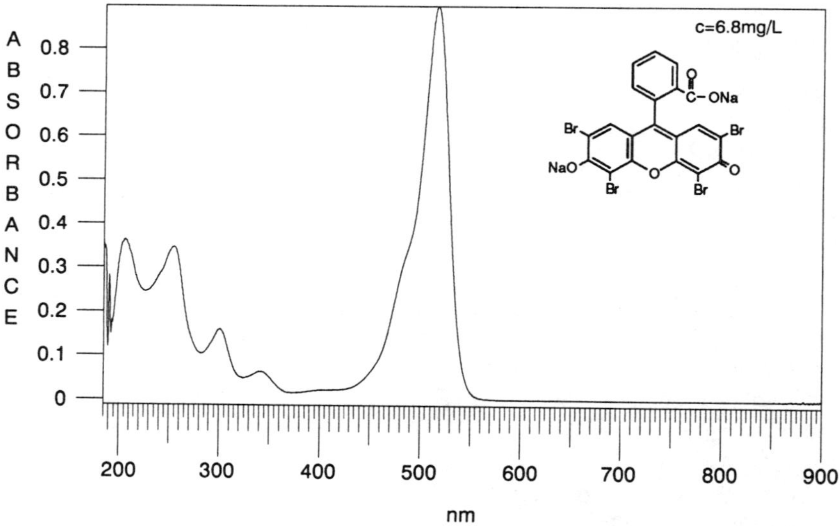

This fluorescent (yellow), red xanthene dye is made by 2',4',5',7'-tetrabrominating fluorescein. The free acid and lactone forms are isolated at this point. Closely related, these forms are essentially insoluble in water. The free-acid form is slightly soluble in methanol and ethanol, while the lactone is even less soluble in these solvents. Both are moderately soluble in methyl Cellosolve.

The Eosin Y certified- and indicator-grade products are made by converting tetrabromofluorescein to the disodium salt. They differ only in degree of purity. The former has a dye content of approximately 90% while the latter has a dye content of approximately 80%. This salt form is water soluble and slightly soluble in methanol and ethanol. It dyes nylon from a formic acid bath and silk from a neutral or acetic acid bath, while silk and wool are printed directly. Other industrial applications include coloring inks, paper, pencil leads, spirit stains, cosmetics, drugs and preparing pigments (heavy-metal salts) (*Colour Index*).

Eosin Y is an important component of hematological stains of the Romanowsky variety. Solutions made from the various Romanowsky-stain powder mixtures are used universally for running differential counts in peripheral blood and in the examination of bone-marrow smears. This, undoubtedly, is Eosin Y's most important application. Wittekind *et al.*,[1] using dyes purified by column chromatography, found that a balanced Romanowsky-type stain could be made from two pure components, Azure B and Eosin Y. Previously, the mixtures were much more complex.

Staining Procedures[2] provides methods and bibliography for the biological applications that follow: Hematoxylin and Eosin stain; counterstain for Hematoxylin; use in combination with Azure II to show morphological details of marrow cells; buffered Azure A/Eosin Y method for demonstrating cell granules, microorganisms and nuclei; for showing insect musculature; for demonstrating cells infected with mosquito-iridescent virus; use in combination with Alum Hematoxylin/Azure II to show mast cells, early necrobiotic changes in cells and protozoans, rickettsias and bacteria in tissues; to demonstrate volutin granules in any microorganism; Bond's modification of Mann's Eosin-Methyl Blue for Negri bodies in smears; Harris stain for Negri bodies in smears; Mann's method for staining Negri bodies in sections; Gridley's stain for *Endamoeba* in tissue sections; a modified methenamine-silver nitrate method for the demonstration of *Endamoeba* in sections.

References:

1) Wittekind, D.; Kretschmer, V.; Lohr, W. *Blut* **1976,** *32*, 71-78.

2) Clark, G. *Staining Procedures,* 4th ed.; Williams and Wilkins: Baltimore, 1981; pp 37, 180, 209-210, 260-261, 308, 407, 432-433, 462, 471-472.

General references:

Aldrich FT-IR Library	**1**(2), 1011C	——
Sigma FT-IR Library	**1**(2), 135B	——
IR Library	**3**, 1474A	**3**,1474F
Merck Index	**11**, 3555	
Colour Index	3rd ed., 1971	

EOSIN Y, free acid

Solvent Red 43

Aldrich Cat. No.: 23,025-1

CAS No.: [*15086-94-9*]
C.I. No.: 45380:2
Mol. Form: $C_{20}H_8Br_4O_5$
F.W.: 647.92
Appearance: Orange powder
λ_{max}: 521nm in water + 1ml 1% sodium carbonate
Solubility: H_2O 0.8mg/ml
EGME 40mg/ml
EtOH 10mg/ml

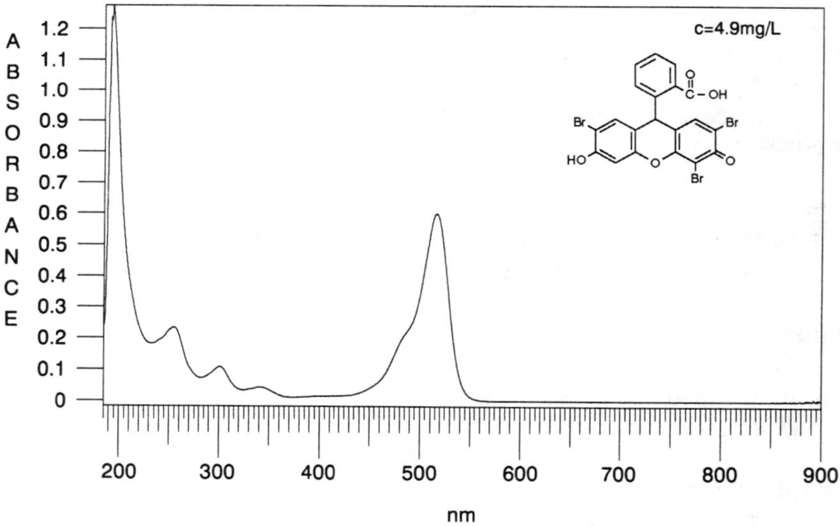

See description on pages 304 and 305 under Eosin Y.

General reference:

IR Library **3**, 1474A

EOSIN Y lactone

Aldrich Cat. No.: 29,395-4
Sigma Prod. No.: E4009
Grade: High purity

CAS No.: [*15086-94-9*]
Mol. Form.: $C_{20}H_8Br_4O_5$
F.W.: 647.92
Appearance: Pink powder
λ_{max}: 524nm in methanol
Solubility: H_2O <0.1mg/ml
EGME 30mg/ml
EtOH 5mg/ml

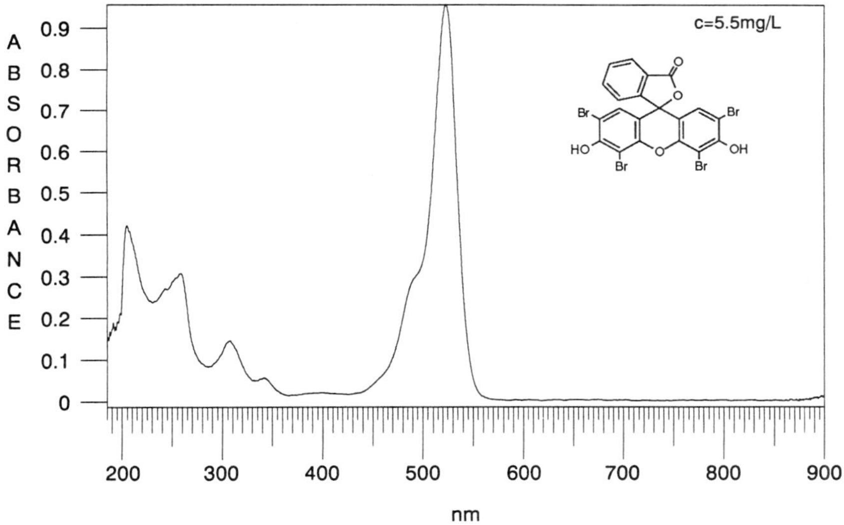

See description on pages 304 and 305 under Eosin Y.

General reference:

Safety **2**(1), 1512D

ERIOCHROME BLACK T

Mordant Black 11

Aldrich Cat. No.:	85,839-0	31,891-4
Sigma Prod. No.:	E2377	——
Grade:	Indicator	0.5 wt. % solution in triethanolamine
CAS No.:	[*1787-61-7*]	
C.I. No.:	14645	
Mol. Form.:	$C_{20}H_{12}N_3NaO_7S$	
F.W.:	461.39	
Appearance:	Black powder	Very dark-blue, viscous liquid
Solubility:	H_2O 80mg/ml EGME 50mg/ml EtOH 10mg/ml	

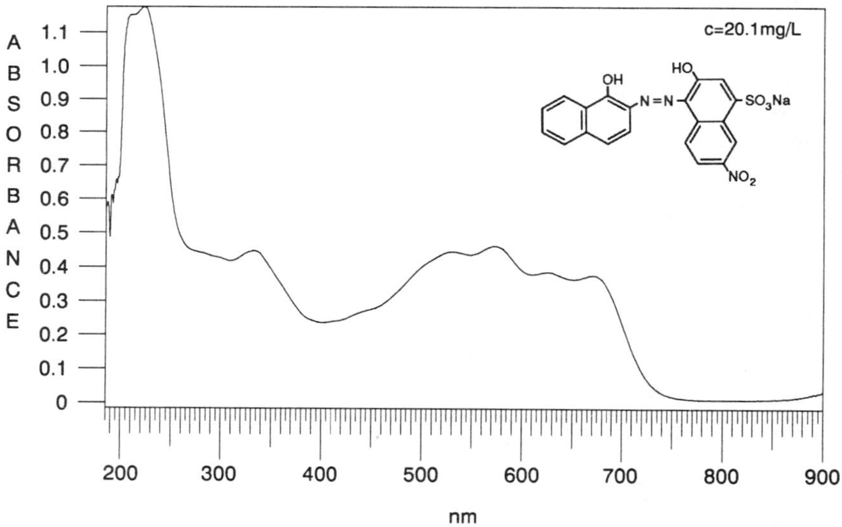

Eriochrome Black T is a red monoazo dye that is soluble in water and moderately soluble in the lower alcohols. It dyes wool, nylon and silk bluish black when afterchromed, and produces a print of the same color on wool mordanted with chromium.

Mordant Black 11 is used as an analytical indicator for the determination of calcium and magnesium or, more generally, as a complexometric indicator in EDTA titrations.[1]

Reference:

1) Young, A.; Sweet, T.R. *Anal. Chem.* **1955,** *27*, 418.

General references:

Aldrich FT-IR Library	1(2), 997D
Sigma FT-IR Library	1(2), 540C
IR Library	**3**, 1467C
Merck Index	**11**, 3615
Colour Index	3rd ed., 1971

ERIOCHROME BLUE BLACK 2B

Mordant Black 3

Aldrich Cat. No.:	20,149-9
Sigma Prod. No.:	E2252
CAS No.:	[*3564-14-5*]
C.I. No.:	14640
Mol. Form.:	$C_{20}H_{13}N_2NaO_5S$
F.W.:	416.39
Appearance:	Black powder
λ_{max}:	528nm in water + 1ml 1*N* hydrochloric acid
Solubility:	H_2O 30mg/ml
	EGME 60mg/ml
	EtOH 10mg/ml

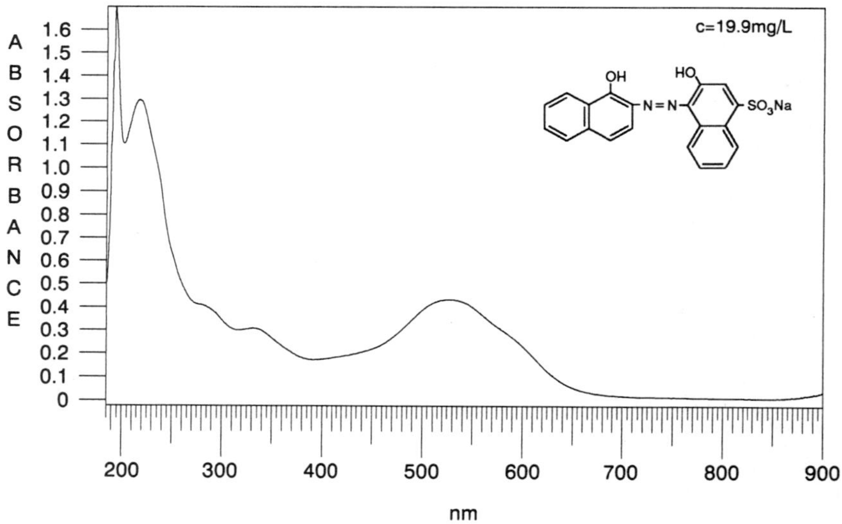

This monoazo dye is made by coupling diazotized 1-amino-2-naphthol-4-sulfonic acid (1,2,4-acid) to 1-naphthol, followed by conversion to the sodium salt. It is used as an indicator in chelometric applications.

Soluble in water (aqueous solutions are violet; alkaline solutions are a wine red) and moderately soluble in ethanol (red) and acetone, it is used to dye afterchromed wool, silk and nylon a bluish black. Wool is printed a bluish black utilizing chromium acetate and the vigoureux process, and leather mordanted with chromium displays the same color.

There is no evidence that it has been used as a biological stain.

General references:

Aldrich FT-IR Library	**1**(2), 987C
Sigma FT-IR Library	**1**(2), 135C
IR Library	**3,** 1462H
Colour Index	3rd ed., 1971

ERIOCHROME RED B

Mordant Red 7

Aldrich Cat. No.: 21,000-5

CAS No.: [*3618-63-1*]
C.I. No.: 18760
Mol. Form.: $C_{20}H_{15}N_4NaO_5S$
F.W.: 446.42
Appearance: Brick-red powder
λ_{max}: 483nm in methanol
Solubility: H_2O 30mg/ml
EGME 40mg/ml
EtOH 10mg/ml

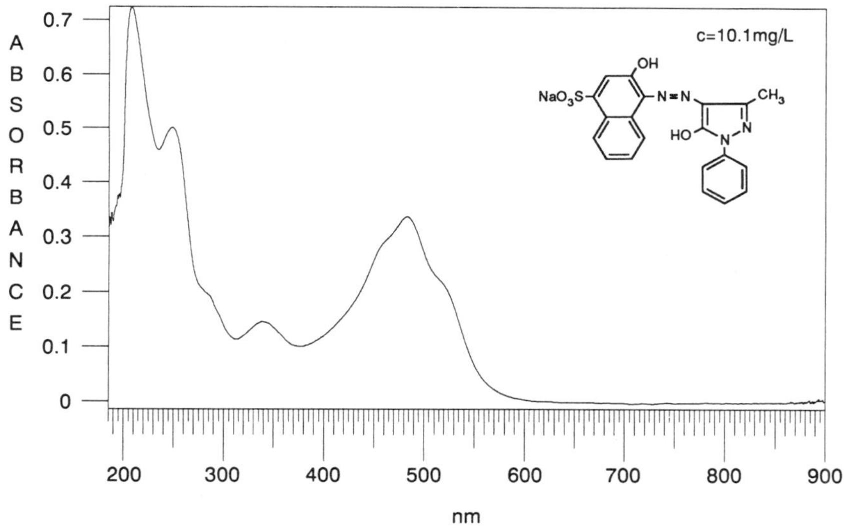

This red, anionic monoazo dye is soluble in water and moderately soluble in alcohol.

Eriochrome Red B dyes wool and silk a bluish red with afterchroming, and nylon red in the presence of a chromium mordant. Wool and silk are vigoureux printed using chromium as a mordant. Other industrial applications include dyeing fully chromed leather and loose wool and fur for hats, and anodizing aluminum.

There is no evidence that it has been used as a biological stain.

General references:

Aldrich FT-IR Library **1**(2), 1003A
IR Library **3,** 1469
Colour Index 3rd ed., 1971

ERIOGLAUCINE

Acid Blue 9 • Alphazurine FG

Aldrich Cat. No.:	86,114-6
Sigma Prod. No.:	E8504
CAS No.:	[*3844-45-9*]
C.I. No.:	42090
Mol. Form.:	$C_{37}H_{34}N_2Na_2O_9S_3$
F.W.:	792.86
m.p.:	283°C (dec.)
Appearance:	Purple powder
λ_{max}:	625(406)nm in water
Solubility:	H_2O 30mg/ml
	EGME 20mg/ml
	EtOH 3mg/ml

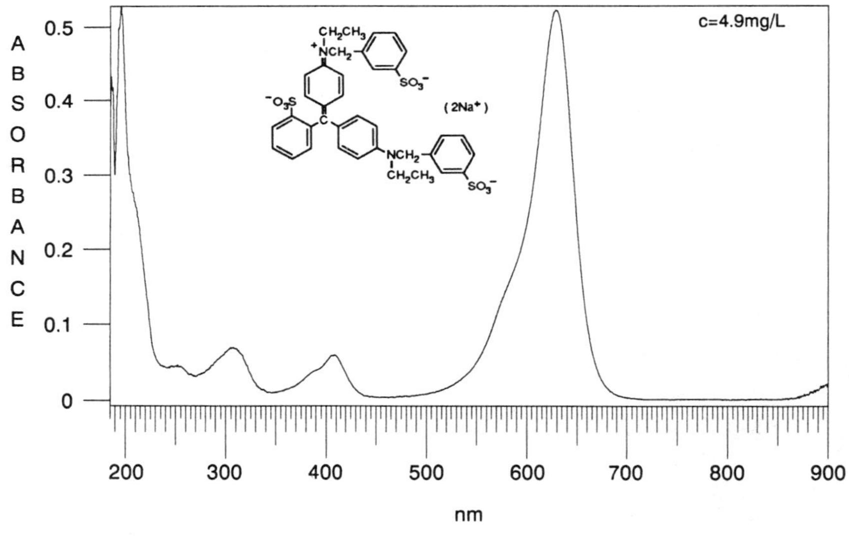

This blue, anionic triphenylmethane dye has a strong affinity for proteinaceous substrates but little attraction for cellulosic materials.

Soluble in water (greenish blue) and methyl Cellosolve and slightly soluble in ethanol, it dyes wool, silk and nylon greenish blue from an acid bath. Nylon is dyed less intensely than the other two fabrics. Other applications include dyeing leather, surface-coating paper and preparing heavy-metal lakes for use as pigments.

Gurr[1] states that it is occasionally called for in animal histology and cytology in Romania and in neighboring Balkan states. He did not provide details regarding its use.

Reference:

1) Gurr, E. *Encyclopedia of Microscopic Stains*; Leonard Hill: London, 1960; p 182.

General references:

Aldrich FT-IR Library	**1**(2), 1033B
Sigma FT-IR Library	**1**(2), 136A
IR Library	**3**, 1485C
Safety	**2**(1), 1530C
Merck Index	**11**, 1366
Colour Index	3rd ed., 1971

ERYTHROSIN B

Acid Red 51 • 2',4',5',7'-Tetraiodofluorescein, disodium salt

Aldrich Cat. No.:	19,826-9
Sigma Prod. No.:	E7379
Grade:	Certified

CAS No.:	*[568-63-8]*
C.I. No.:	45430
Mol. Form.:	$C_{20}H_6I_4Na_2O_5$
F.W.:	879.87
Appearance:	Maroon powder
λ_{max}:	525nm in water + 1ml 1% sodium carbonate
Solubility:	H_2O 20mg/ml
	EGME 20mg/ml
	EtOH 20mg/ml

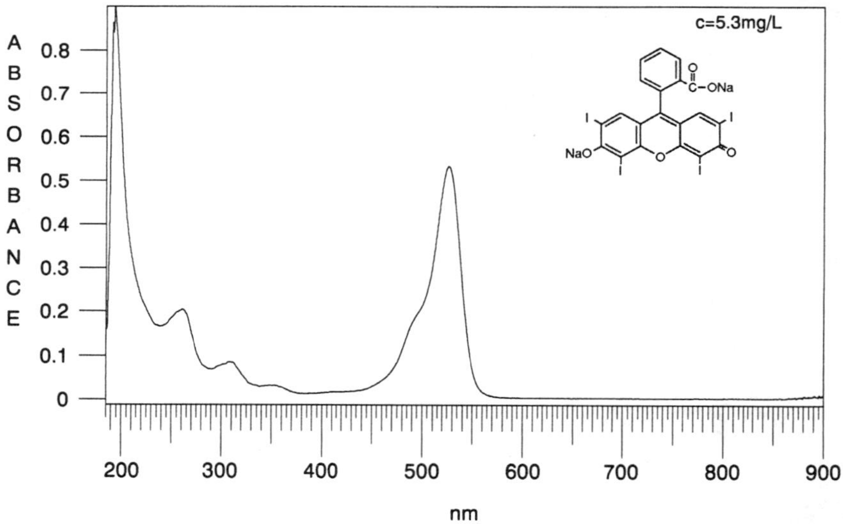

Erythrosin B is a red, anionic xanthene dye that is made by the tetra-iodination of fluorescein. The BSC certified product (Aldrich Cat. No. 19,826-9) is the disodium-salt form which is moderately soluble in both water and ethanol. A highly purified product of this salt form is certified by the FDA for use as a food color (C.I. Food Red 14 or FD&C Red 3).

Due to its light instability, Erythrosin B is not very practical as a textile dye. However, it is used as a dye for paper, in the preparation of inks and lacquers and in staining wood. Aluminum and lead lakes are used as pigments that find application in printing-ink formulations.

Erythrosin B has been used widely as a contrast or counterstain to

Hematoxylin, Crystal Violet and other nuclear stains, in both plant and animal tissue. *Staining Procedures*[1] lists methods and bibliography for the following applications: bacteria in plant tissue, stain for insect tissue, as a counterstain for Crystal Violet to demonstrate vascular bundles and differentiation of lignified and nonlignified cell walls in the woody tissue of vascular plants.

The free-acid form Erythrosin B, spirit soluble (Aldrich Cat. No. 20,096-4), is soluble in methyl Cellosolve, glycerol, ethylene glycol, slightly soluble in alcohol and ether and essentially insoluble in water. It has been used in the preparation of printing inks and lacquers.

Reference:

1) Clark, G. *Staining Procedures,* 4th ed.; Williams and Wilkins: Baltimore, 1981; pp 50, 268, 323.

General references:

Aldrich FT-IR Library	**1**(2), 1014D
Sigma FT-IR Library	**1**(2), 136C
IR Library	**3**, 1475F
Merck Index	**11**, 3640
Colour Index	3rd ed., 1971

ERYTHROSIN B, spirit soluble

2',4',5',7'-Tetraiodofluorescein

Aldrich Cat. No.:	20,096-4
Sigma Prod. No.:	E7505
Purity:	95+%
CAS No.:	[*15905-32-5*]
C.I. No.:	45430:2
Mol. Form.:	$C_{20}H_8I_4O_5$
F.W.:	835.90
m.p.:	303°C (dec.)
Appearance:	Red-orange powder
λ_{max}:	533nm in methanol
Solubility:	H_2O 0.7mg/ml
	EGME 20mg/ml
	EtOH 2mg/ml

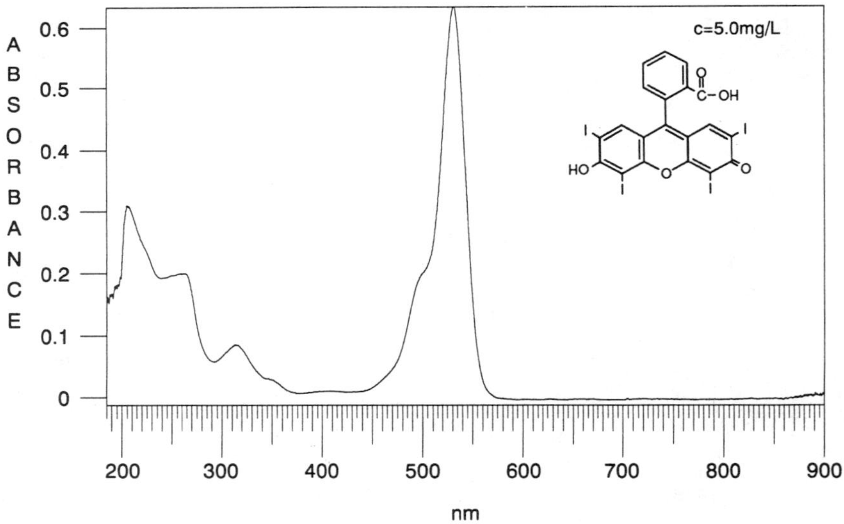

See description on pages 314 and 315 under Erythrosin B.

General references:

Aldrich FT-IR Library	**1**(2), 1014C
Sigma FT-IR Library	**1**(2), 136B
IR Library	**3**, 1475E

ERYTHROSIN YELLOWISH blend

Aldrich Cat. No.: 23,407-9

Appearance: Red powder
λ_{max}: 525(305)nm in water
Solubility: H_2O 30mg/ml
EGME 10mg/ml
EtOH 20mg/ml

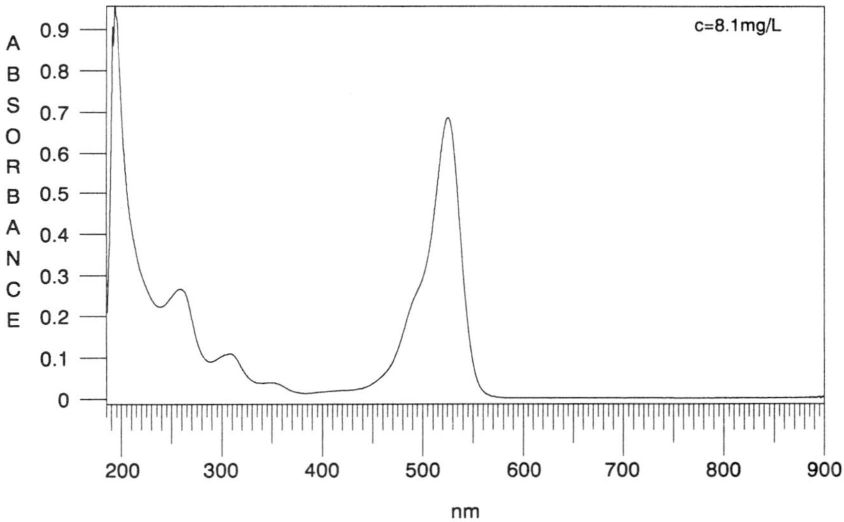

Erythrosin Yellowish blend is composed of 90% Erythrosin B (Acid Red 51, Aldrich Cat. No. 19,826-9) and 10% Eosin Y (Acid Red 87, Aldrich Cat. No. 11,983-0).

Soluble in water and moderately soluble in methanol and ethanol, it is used as a biological stain when a more yellowish contrast stain or counterstain than Erythrosin B is desired.

A more detailed description of Erythrosin B may be found on pages 314 and 315, and of Eosin Y on pages 304 and 305.

ETHIDIUM BROMIDE

Homidium bromide

Aldrich Cat. No.:	16,053-9
Sigma Prod. No.:	E8751
Purity:	95%

CAS No.:	[*1239-45-8*]
Mol. Form.:	$C_{21}H_{20}BrN_3$
F.W.:	394.32
m.p.:	260-262°C (dec.)
Appearance:	Maroon powder
λ_{max}:	525(296)nm in methanol
Solubility:	H_2O 20mg/ml
	EGME 20mg/ml
	EtOH 2mg/ml

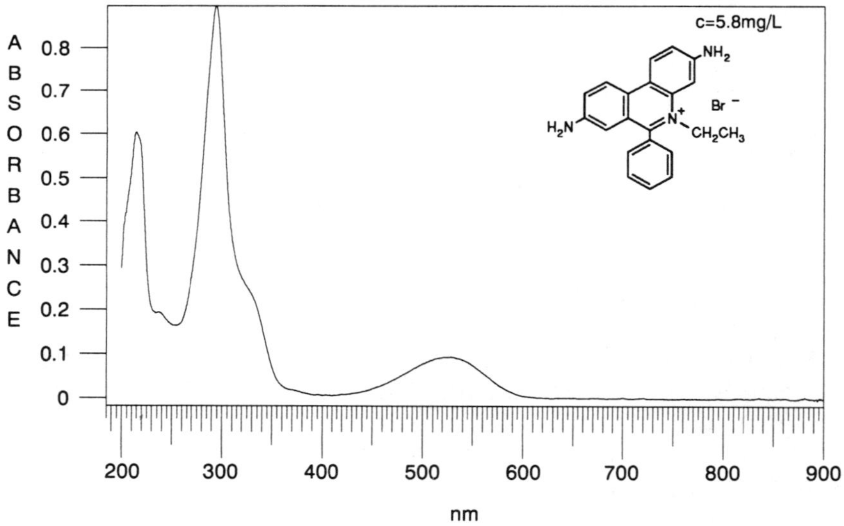

This red, cationic fluorochromic dye is moderately soluble in water and slightly soluble in chloroform. A preparative method is described by Watson.[1]

Ethidium bromide is a frameshift mutagen which intercalates double-stranded DNA[2] and RNA.[3] Haag *et al.*[4] reported its affinity for nuclear chromatin to be non-specific since reactivity with both keratin and collagen was also observed.

More recent studies have utilized it increasingly as a DNA probe. Atherton and Beaumont[5] used it to determine water accessibility to the interior of DNA. Pruetz[6] studied the effect of various radicals on the base-pair region of DNA. Kool *et al.*[7] utilized it for physiochemical studies characterizing mitochondrial DNA. Moore and Sutherland[8] used it to quantify DNA.

References:

1) Watkins, T.L. *J. Chem. Soc.* **1952,** 3059.
2) Le Pecq, J.B.; Paoletti, C. *J. Mol. Biol.* **1967,** *27*, 87.
3) Donthart, R.J.; Burnett, J.P.; Beasley, F.W.; Frank, B.H. *Biochemistry* **1973,** *12*, 214.
4) Haag, D.; Tschanargane, C.; Goerttler, K. *Histochemie* **1971,** *27*, 119-124.
5) Atherton, S.J.; Beaumont, P.C. *Photobiochem. Photobiophys.* **1984,** *8* (2), 103.
6) Pruetz, W.A. *Radiat. Environ. Biophys.* **1984,** *23*(1), 1,7.
7) Kool, A.J.; De Haas, J.M.; Mol, J.N.M.; Van Marrewijk, G.A.M. *Theor. Appl. Genet.* **1985,** *69* (3), 223.
8) Moore, S.P.; Sutherland, B.M. *Anal. Biochem.* **1985,** *144* (1), 15.

General references:

Aldrich FT-IR Library	**1**(2), 873B
Sigma FT-IR Library	**1**(2), 136D
IR Library	**3**, 1399A
Safety	**2**(1), 1549B
Merck Index	**11**, 4650

ETHYL EOSIN

Solvent Red 45

Aldrich Cat. No.: 19,954-0
Sigma Prod. No.: E4004
Grade: Certified

CAS No.: [*6359-05-3*]
C.I. No.: 45386
Mol. Form.: $C_{22}H_{11}Br_4KO_5$
F.W.: 714.07
Appearance: Red powder
λ_{max}: 532nm in 50% ethanol
Solubility: H_2O 1mg/ml
EGME 40mg/ml
EtOH 1mg/ml

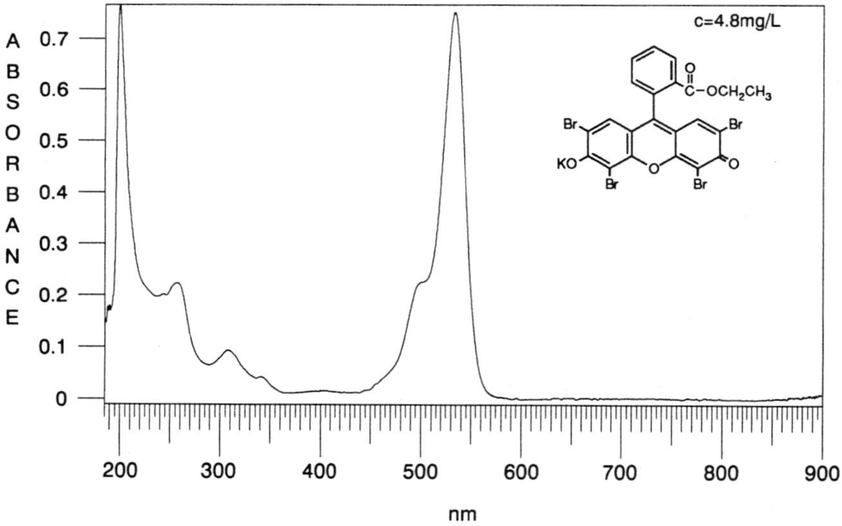

Ethyl Eosin is a fluorescent, red xanthene dye that is practically insoluble in cold water and moderately soluble in hot water, remaining in solution on cooling. It is very slightly soluble in ethyl alcohol at room temperature. Industrially, it is used to prepare spirit varnishes for metal foils.

Conn's[1] describes it as a valuable counterstain after Delafield's Hematoxylin. *Staining Procedures*[2] provides the Stoval and Black procedure for demonstrating Negri bodies in acetone-fixed tissue. This method utilizes acidified Ethyl Eosin as a substrate stain and acidified Methylene Blue as a counterstain which will not replace Ethyl Eosin. "The Negri bodies should be stained a reddish brown to bright red. Nucleoli and cytoplasm should be blue, and other material pink."

References:

1) *Conn's Biological Stains,* 9th ed.; Lillie, R.D., Ed.; Williams and Wilkins: Baltimore, 1977; p 344.

2) Clark, G. *Staining Procedures,* 4th ed.; Williams and Wilkins: Baltimore, 1981; p 461.

General references:

Aldrich FT-IR Library	**1**(2), 1012B
Sigma FT-IR Library	**1**(2), 137A
IR Library	**3**, 1474
Colour Index	3rd ed., 1971

ETHYL ORANGE, sodium salt

4-(4-Diethylaminophenylazo)benzenesulfonic acid, sodium salt

Aldrich Cat. No.:	19,956-7
Sigma Prod. No.:	E6505
Grade:	Indicator

CAS No.:	*[62758-12-7]*
Mol. Form.:	$C_{16}H_{18}N_3NaO_3S$
F.W.:	355.39
Appearance:	Orange powder
λ_{max}:	474nm in water
Solubility:	H_2O 100mg/ml
	EGME 30mg/ml
	EtOH 2mg/ml

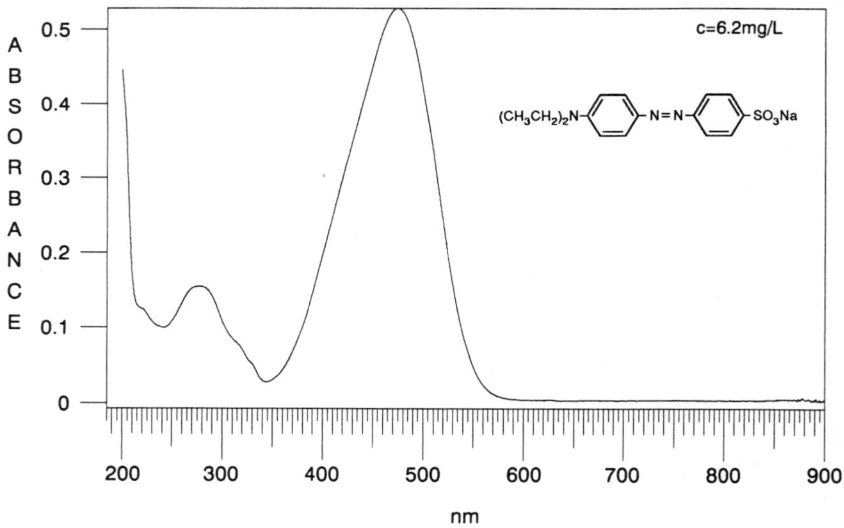

Ethyl Orange is an amphoteric monoazo dye that is used primarily as an acid-base indicator. It is the diethyl homolog of Methyl Orange and is made similarly. The synthesis is accomplished by coupling diazotized sulfanilic acid to diethylaniline acidified with acetic acid, and converting the resulting product to the sodium salt. This form is very soluble in water and very slightly soluble in ethanol. Theil and Springemann[1] reported Ethyl Orange with a visual-transition interval of pH 3.4 (rose-red) to pH 4.8 (yellow). Long[2] found the range to be much closer to Methyl Orange: pH 3.0 (red) to pH 4.5 (yellow). Our laboratory is in closer agreement with Theil and Springemann.

There is no evidence in the literature that Ethyl Orange has been used as a biological stain, and it probably offers no advantage over Methyl Orange for this purpose.

References:

1) Theil, A.; Springemann, W. *Z. Anorg. Allg. Chem.* **1927,** *176*, 112.

2) Long, C. *Biochemist's Handbook;* Van Nostrand: Princeton, 1961.

General references:

Aldrich FT-IR Library	**1**(2), 976D
Sigma FT-IR Library	**1**(2), 137B
IR Library	**3**, 1457A
Safety	**2**(1), 1643D

ETHYL RED

2-(4-Diethylaminophenylazo)benzoic acid

Aldrich Cat. No.:	19,957-5
Sigma Prod. No.:	E7629
Grade:	Indicator
CAS No.:	[*76058-33-8*]
Mol. Form.:	$C_{17}H_{19}N_3O_2$
F.W.:	297.36
m.p.:	135°C (dec.)
Appearance:	Red powder
λ_{max}:	447nm in 0.1N sodium hydroxide
Solubility:	H_2O 3mg/ml
	EGME 7mg/ml
	EtOH 3mg/ml

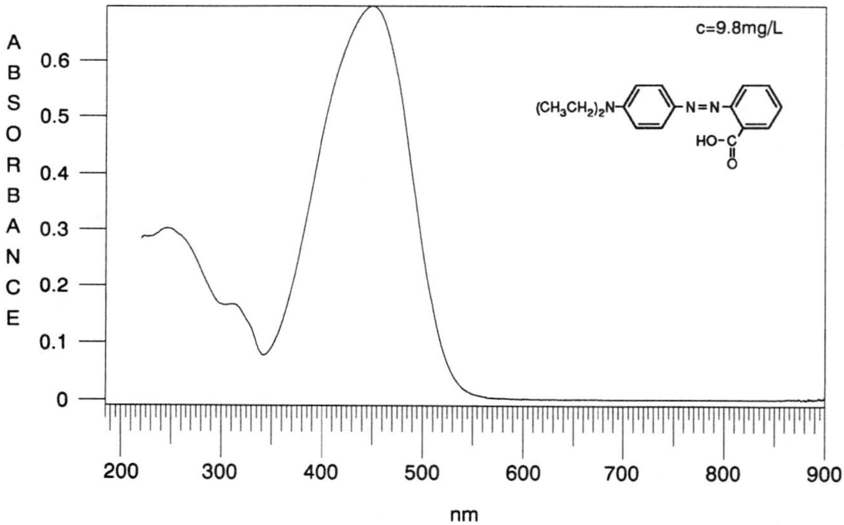

Ethyl Red is an amphoteric monoazo dye. Like its more prominent methyl homolog, it is used as a pH indicator with a visual-transition interval of pH 4.5 (red) to pH 6.5 (yellow). This crystalline, neutral form is soluble in benzene and slightly soluble in alcohol.

There is no evidence in the literature that Ethyl Red has been used as a biological stain or in any industrial applications.

General references:

Aldrich FT-IR Library	**1**(2), 968B
Sigma FT-IR Library	**1**(2), 137C
IR Library	**3**, 1451F

ETHYL VIOLET
Basic Violet 4

Aldrich Cat. No.: 22,884-2
Sigma Prod. No.: E4506

CAS No.: [*2390-59-2*]
C.I. No.: 42600
Mol. Form.: $C_{31}H_{42}ClN_3$
F.W.: 492.15
Appearance: Olive-green powder
λ_{max}: 596nm in water
Solubility: H_2O 9mg/ml
EGME 50mg/ml
EtOH 30mg/ml

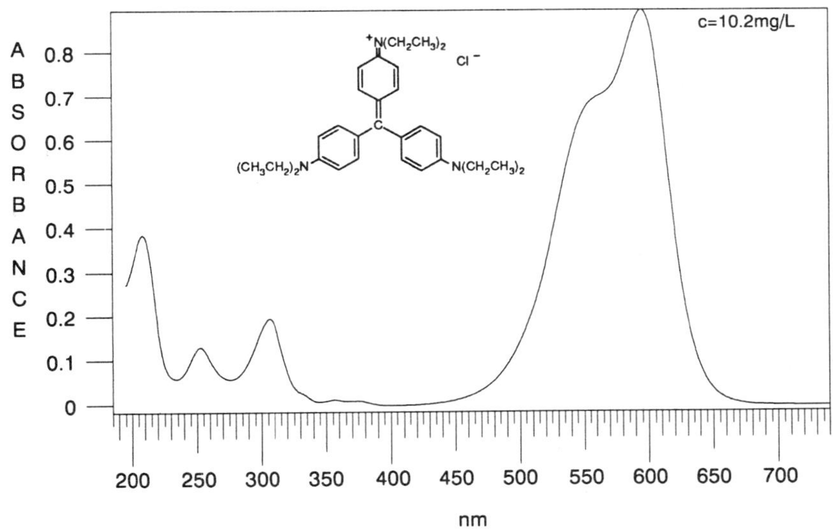

This violet, cationic triphenylmethane dye is prepared by reacting *N,N*-diethylaniline with phosgene in the presence of zinc chloride.

Slightly soluble in water and soluble in methanol and ethanol, Ethyl Violet has a visual-transition interval of pH 0.0 (yellow) to pH 3.5 (blue).

Industrial applications include dyeing cotton with a tannin mordant (poor lightfastness, 1), dyeing paper and chromed leather, compounding shoe polish and suede dressing and preparing the copper ferrocyanide lake (Fanal Blue) which is used as a pigment for printing inks.

Since Ethyl Violet is the ethyl homolog of Crystal Violet, it predictably demonstrates a bathochromic shift of 6nm in λ_{max} compared to the latter dye.

Ethyl Violet has been used successfully in animal histology in a number of the same applications as Crystal Violet; however, since it presents no obvious advantage, it probably has not been investigated nearly as thoroughly. *Conn's*[1] cites the following: Bowie (in 1934) used Ethyl Violet in combination with Biebrich Scarlet as a neutral stain for staining the islets of Langerhans; Kernohan (in 1931) and Proescher (in 1934) used it to stain nervous tissue; Ono (in 1934) utilized it to stain blood spirochetes. Litsky *et al.*[2] described it as a selective dye for the isolation of Gram-negative bacteria.

References:

1) *Conn's Biological Stains,* 9th ed.; Lillie, R.D., Ed.; Williams and Wilkins: Baltimore, 1977; pp 282-283.

2) Litsky, W., Mallman, W.L.; Fifield, C.W. *Stain Technol.* **1952,** *27*, 229-232.

General references:

Aldrich FT-IR Library	**1**(2), 964B
Sigma FT-IR Library	**1**(2), 137D
Safety	**2**(1), 1666C
Colour Index	3rd ed., 1971

EVANS BLUE

Direct Blue 53

Aldrich Cat. No.:	20,633-4
Sigma Prod. No.:	E2129
CAS No.:	[*314-13-6*]
C.I. No.:	23860
Mol. Form.:	$C_{34}H_{24}N_6Na_4O_{14}S_4$
F.W.:	960.82
Appearance:	Dark-brown powder
λ_{max}:	611nm in water
Solubility:	H_2O 60mg/ml
	EGME 40mg/ml
	EtOH 0.1mg/ml

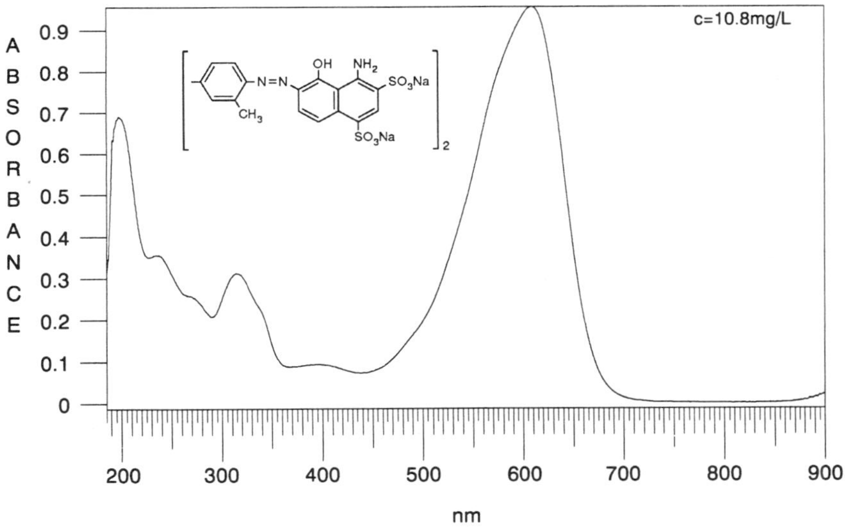

This blue, anionic disazo dye is soluble in water, ethylene glycol and methyl Cellosolve and practically insoluble in ethanol, benzene, carbon tetrachloride, chloroform and ether.

Preparation is accomplished by coupling one mole of tetraazotized *o*-tolidine to two moles of Chicago acid (1-amino-8-naphthol-2,4-disulfonic acid) in an alkaline solution.

Despite its poor lightfastness (1-2), Evans Blue has been used to dye cotton, paper and leather directly and to print silk and wool directly.

Like isomeric Trypan Blue, Evans Blue is used widely as a vital stain. Brunschwig, Schmitz and Jennings[1] used it in blood-uptake studies, and

Brunschwig, Schmitz and Clarke[2] utilized it as a vital stain for malignant tumors in man.

Crippen and Perrier[3] reported that dead cells of diatoms in sea water were stained blue, while living cells were unstained. *Science*[4] reported the dye useful, in combination with 4',6-diamidino-2-phenylindole dihydrochloride (DAPI) and Primulin (Aldrich Cat. No. 21,708-5 and 20,686-5, respectively), to study axonal branching by a simple, retrograde double-labeling procedure.

Gregersen and Shiro[5] noted that Evans Blue is not taken up by red cells, making it valuable in blood-plasma volume studies. *The USP XX*[6] notes that highly purified Evans Blue is used as a diagnostic aid to determine blood volumes. Specifications and test methods for determining the purity of Evans Blue and its suitability for this *in vivo* application are also given.

References:

1) Brunschwig, A.; Schmitz, R.L.; Jennings, S. *Proc. Soc. Exp. Biol. Med.* **1940,** *44,* 64-66.

2) Brunschwig, A.: Schmitz, R.L.; Clarke, T.H. *Arch. Pathol.* **1940,** *30,* 902-910.

3) Crippen, R.W.; Perrier, J.L. *Stain Technol.* **1974,** *49,* 97-104.

4) van der Kooy, D.; Kuypers, H.C.J.M. *Science* **1979**, *204*, 873.

5) Gregersen, M.I.; Shiro, H. *Amer. J. Physiol.* **1938,** *121,* 293-309; *Chem. Abstr.* **1938,** *32*, 1779.

6) *The United States Pharmacopeia,* twentieth version; United States Pharmacopeial Convention; Mack: Easton, 1979; p 315.

General references:

Aldrich FT-IR Library	**1**(2), 996C
Sigma FT-IR Library	**1**(2), 138A
IR Library	**3**, 1466F
NMR Library	**2**(1), 1673A
Merck Index	**11**, 3863
Colour Index	3rd ed., 1971

FAST BLACK K salt
Azoic Diazo No. 38

Aldrich Cat. No.: 20,151-0
Sigma Prod. No.: F7253

CAS No.: [*64071-86-9*]
C.I. No.: 37190
Mol. Form.: $C_{14}H_{12}ClN_5O_4 \cdot {}^1/_2 ZnCl_2$
F.W.: 417.87
m.p.: 150°C (dec.)
Appearance: Dark-brown powder
λ_{max}: 457nm in water
Solubility: H_2O 60mg/ml
EGME 20mg/ml
EtOH 3mg/ml

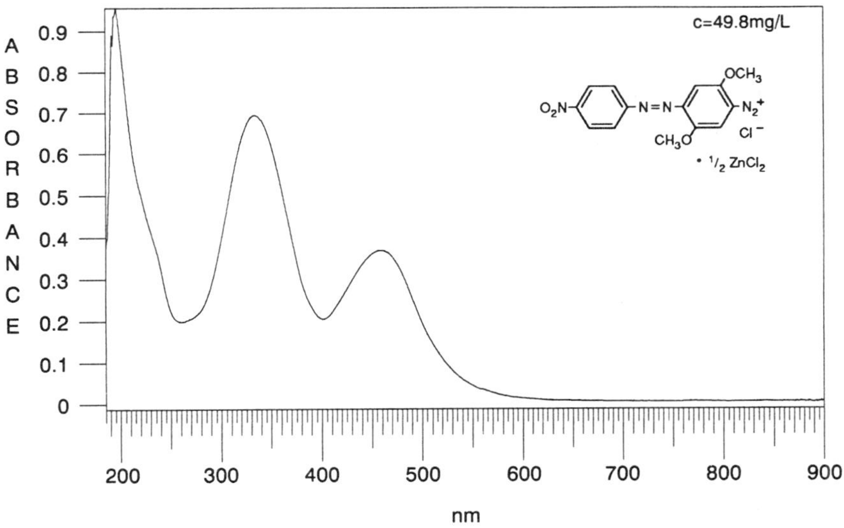

Fast Black K salt [4-(4-nitrophenylazo)-2,5-dimethoxybenzenediazonium chloride, zinc chloride double salt] is a water-soluble, stabilized diazonium salt which is prepared from the base by diazotization, followed by conversion to the zinc chloride double salt by the addition of zinc chloride to the reaction mixture.

It is one of a group of similar dye precursors which are used in the textile industry to produce insoluble azo dyes directly on the textile fiber. This is accomplished by first impregnating the fabric with a water-soluble coupling component such as the sodium or potassium salt of β-naphthol or one of its derivatives. The impregnated fabric is then passed through a solution containing the stabilized diazonium salt. Under favorable coupling conditions, the two smaller water-soluble molecules combine to form a larger water-insoluble dye molecule *in situ*. By varying the coupling component, different colors and hues can be developed on the fabric. Fast Black K salt has been used to dye both cotton and silk in this manner. Colors varying from black to Bordeaux Red with intermediary hues of bluish black, greenish black and reddish black have been attained by varying the coupling components used with this particular salt. The commercial product assays approximately 30%, although 95% material can be made readily.

Burstone[1] used Fast Black K salt for the histochemical demonstration of acid phosphatases with naphthol AS phosphates. Lillie *et al.*[2] found it useful for direct coupling with proteinaceous tissue components.

References:

1) Burstone, M.S. *J. Natl. Cancer Inst.* **1958**, *21*, 523-529.

2) Lillie, R.D.; Gilmer, P.R.; Welsh, R.A. *Stain Technol.* **1961**, *36*, 361-363.

General references:

Aldrich FT-IR Library **1**(2), 957B
Sigma FT-IR Library **1**(2), 138B
IR Library **3**, 1445G
Colour Index 3rd ed., 1971

FAST BLUE

Diamidino 253/50

Aldrich Cat. No.: 28,400-9
Sigma Prod. No.: F5756

CAS No.: *[74749-42-1]*
Appearance: Dark-yellow powder
λ_{max}: 382nm in methanol

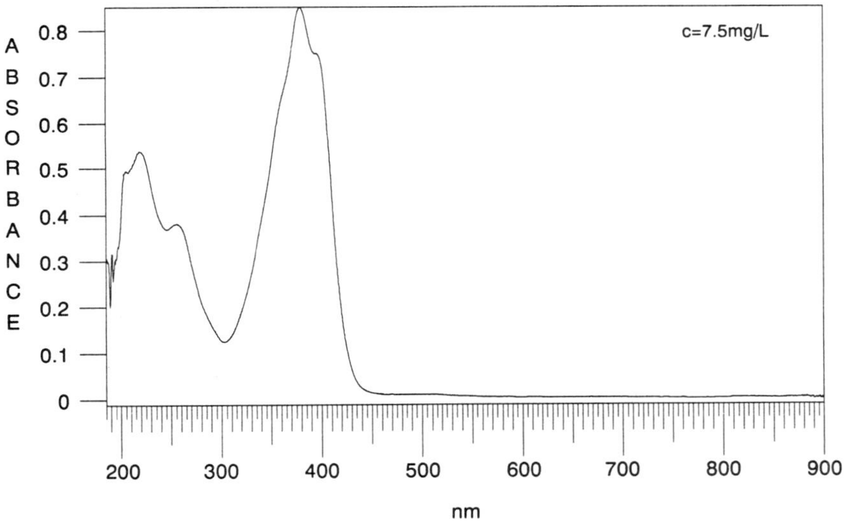

The fluorescent tracer, Fast Blue, was first synthesized by Bentivoglio *et al.*[1] Without disclosing the structure, they reported it to be related to True Blue, a retrograde neuronal tracer, also discovered by Dann and his co-workers.

Fast Blue is soluble in cold water, methanol and ethanol. Like True Blue, it is used as a fluorescent retrograde neuronal tracer, but is transported more rapidly and effectively over longer distances in the cat than the former. Labeling in the rat and cat occurs mainly in the neuronal cytoplasm while 2-(4-sulfamylphenyl)-6-[6-(4-methylpiperazino)-2-benzimidazolyl]benzimidazole (Nuclear Yellow) described in the same paper labels mainly the neuronal nuclei.

Reference:

1) Bentivoglio, M.; Kuypers, H.G.J.M.; Catsman-Berrevoets, C.E.; Lowe, H.; Dann, O. *Neuroscience Lett.* **1980,** *18,* 25-30.

FAST BLUE B salt
Azoic Diazo No. 48

Aldrich Cat. No.:	20,142-1	34,077-4
Sigma Prod. No.:	D3502	——
Dye content:	~30%	>90%

CAS No.:	[*14263-94-6*]	
C.I. No.:	37235	
Mol. Form.:	$C_{14}H_{12}Cl_2N_4O_2 \cdot ZnCl_2$	
F.W.:	475.46	
m.p.:	——	>300°C
Appearance:	Dark, greenish-yellow crystalline powder	
λ_{max}**:**	371(307)nm in water	
Solubility:	H_2O 70mg/ml	
	EGME 30mg/ml	
	EtOH 3mg/ml	

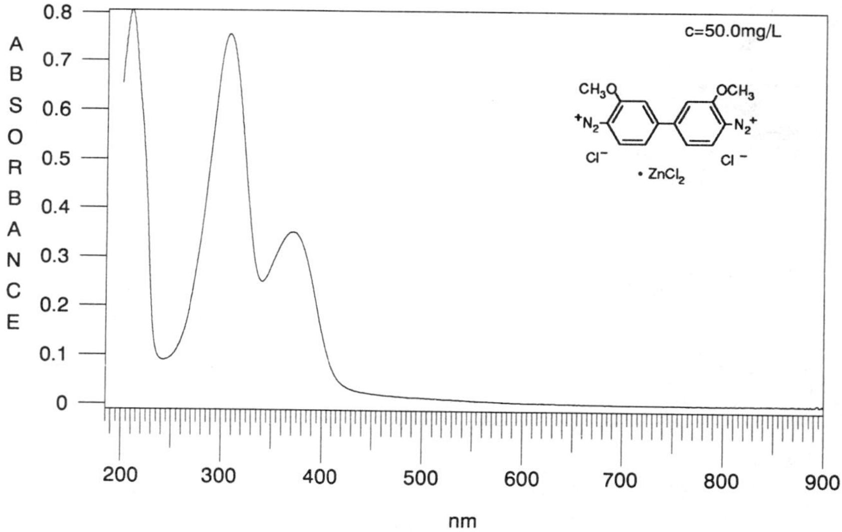

Fast Blue B salt [3,3'-dimethoxy-4,4'-biphenylbis(diazonium chloride), zinc chloride double salt] is a water-soluble, stabilized (zinc chloride) tetrazonium salt. It is a dye precursor used commercially to develop a dye directly on a fabric (see Fast Black K salt, pages 329 and 330). The standardized commercial product contains approximately 30% active material.

Conn's[1] states that it is used extensively in the "coupled-tetrazonium" procedures for the identification of aromatic residues in proteins and for the intensification of the 2-hydroxy-3-naphthoic acid hydrazide reaction for carbonyls.

Reference:

1) *Conn's Biological Stains,* 9th ed.; Lillie, R.D., Ed.; Williams and Wilkins: Baltimore, 1977; p 214.

General references:

Aldrich FT-IR Library	**1**(2), 958C
Sigma FT-IR Library	**1**(2), 124A
IR Library	**3**, 1446D
Safety	**2**(1), 1676B
Colour Index	3rd ed., 1971

FAST BLUE BB base

Azoic Diazo No. 20

Aldrich Cat. No.:	20,152-9
Sigma Prod. No.:	F0125
CAS No.:	[*120-00-3*]
C.I. No.:	37175
Mol. Form.:	$C_{17}H_{20}N_2O_3$
F.W.:	300.36
m.p.:	98-100°C
Appearance:	Grey powder
λ_{max}:	318nm in ethanol
Solubility:	H_2O 6mg/ml
	EGME 40mg/ml
	EtOH 20mg/ml

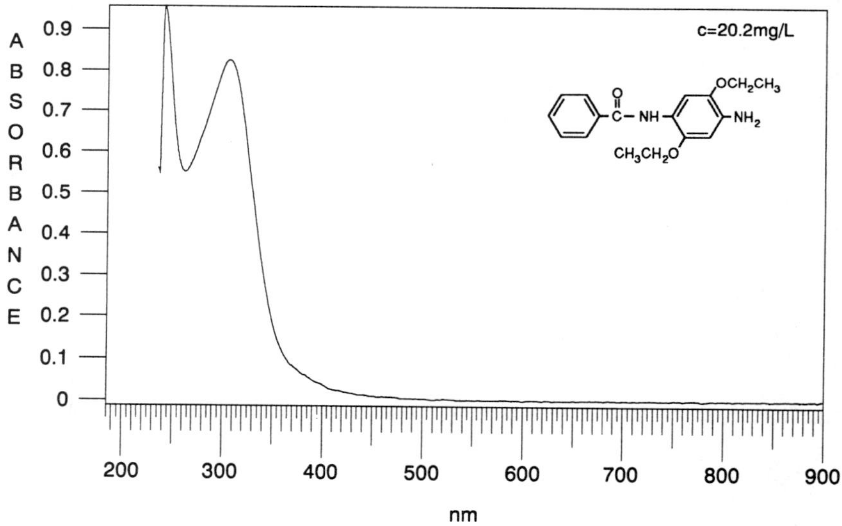

This base is prepared by condensing 2,5-diethoxyaniline with benzoyl chloride, and then isolating, nitrating and reducing the condensation product. Fast Blue BB salt (Aldrich Cat. No. 20,153-7, found on page 336) is prepared by first diazotizing the base, followed by conversion to the zinc chloride double salt by the addition of a concentrated zinc chloride solution to the reaction mixture. It is used in the textile industry to develop dyes, *in situ,* on variously padded cotton or silk fabrics.

Fast Blue BB salt is used as a coupling agent in histochemical reactions for hydrolases. Burstone[1,2] found it satisfactory for staining phosphatases.

References:

1) Burstone, M.S. *J. Natl. Cancer Inst.* **1957,** *18*, 167-173.

2) *Idem ibid.* **1958,** *21*, 123-129.

General references:

Aldrich FT-IR Library	**1**(2), 953B
Sigma FT-IR Library	**1**(2), 138C
IR Library	**3**, 1444B
Colour Index	3rd ed., 1971

FAST BLUE BB salt

Azoic Diazo No.20

Aldrich Cat. No.:	20,153-7
Sigma Prod. No.:	F3378
CAS No.:	*[5486-84-0]*
C.I. No.:	37175
Mol. Form.:	$C_{17}H_{18}ClN_3O_3 \cdot {}^1/_2\ ZnCl_2$
F.W.:	415.94
m.p.:	157°C (dec.)
Appearance:	Yellow powder with chunks
λ_{max}**:**	395nm in water
Solubility:	H_2O 40mg/ml
	EGME 60mg/ml
	EtOH 6mg/ml

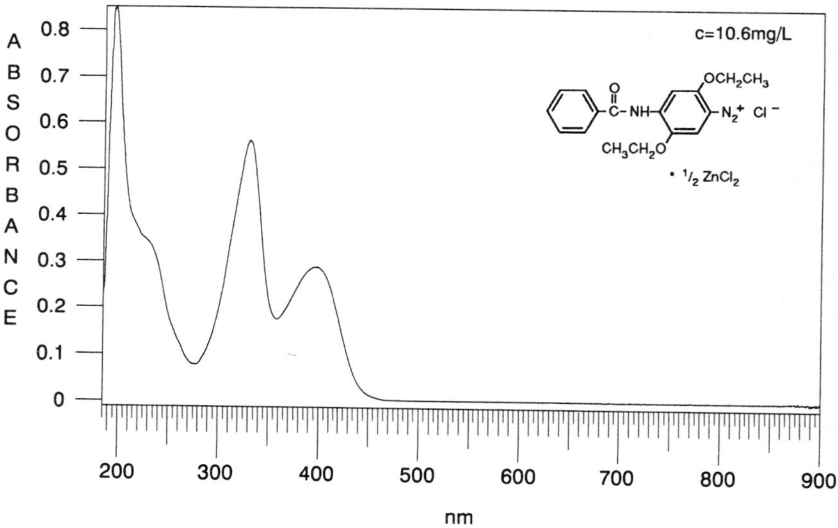

See description on pages 334 and 335 under Fast Blue BB.

General references:

Aldrich FT-IR Library	**1**(2), 959A
Sigma FT-IR Library	**1**(2), 138D
IR Library	**3**, 1446E
Safety	**2**(1), 1676C

FAST BLUE RR

Azoic Diazo No.24 • 4-Benzoylamino-2,5-dimethoxyaniline

Aldrich Cat. No.: 20,154-5
Sigma Prod. No.: F0375
Purity: 97%

CAS No.: [*6268-05-9*]
C.I. No.: 37155
Mol. Form.: $C_{15}H_{16}N_2O_3$
F.W.: 272.31
m.p.: 165-168°C
Appearance: Light brown-violet powder
λ_{max}: 319nm in methanol
Solubility: H_2O 7mg/ml
EGME 20mg/ml
EtOH 3mg/ml

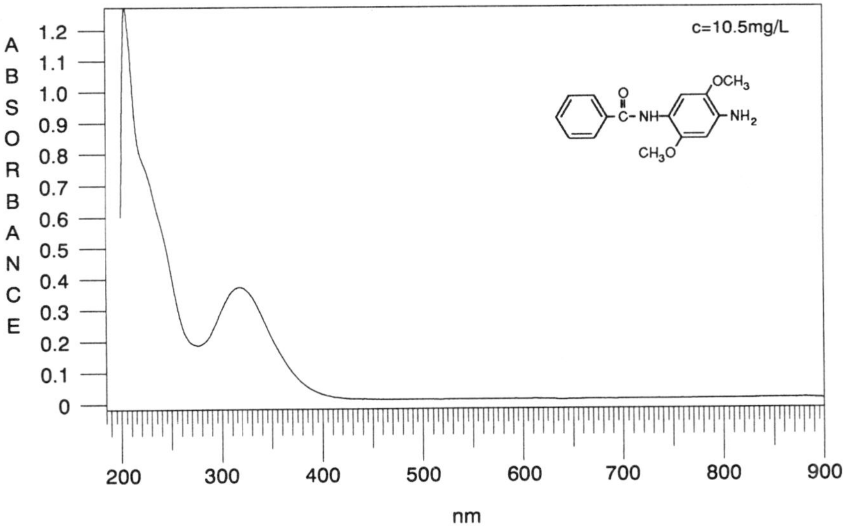

For description, see pages 338 and 339 under Fast Blue RR salt.

General references:

Aldrich FT-IR Library	**1**(2), 953C
Sigma FT-IR Library	**1**(2), 139B
IR Library	**3**, 1444C

FAST BLUE RR salt

Azoic Diazo No. 24 • 4-Benzoylamino-2,5-dimethoxybenzene-diazonium chloride, zinc chloride double salt

Aldrich Cat. No.: 20,122-7
Sigma Prod. No.: F0500
Purity: 85%

CAS No.: [*14726-29-5*]
C.I. No.: 37155
Mol. Form.: $C_{15}H_{14}ClN_3O_3 \cdot {}^1/_2\ ZnCl_2$
F.W.: 387.89
Appearance: Yellow-green powder
Solubility: H_2O 40mg/ml
EGME 50mg/ml
EtOH 4mg/ml

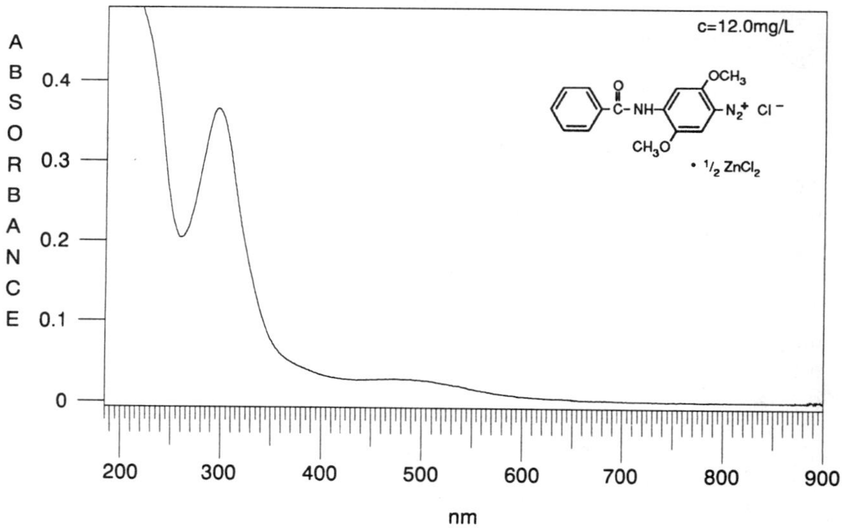

Fast Blue RR base (see page 337, Aldrich Cat. No. 20,154-5), a dye intermediate, is the precursor of Fast Blue RR salt, a stable diazonium salt. The latter is used in the textile industry for the *in situ* development of dyes on fabrics padded (impregnated) with suitable coupling components, frequently derivatives of 1- and 2-naphthols.

Fast Blue RR salt is used as a coupling agent in histochemical reactions for hydrolases (see Gomori[1] for acid phosphatase and esterases, Pearse[2] for phosphatases, Gossner[3] for esterases and Burstone[4] for phosphatases).

References:

1) Gomori, G. *Microscopic Histochemistry;* University of Chicago: Chicago, 1952.

2) Pearse, A.G.E. *Nature* **1958,** *181*, 1531-1532.

3) Gossner, W. *Histochemie* **1958,** *1*, 48-69.

4) Burstone, M.S. *J. Natl. Cancer Inst.* **1958,** *20*, 601-615.

General references:

Aldrich FT-IR Library	**1**(2), 959C
Sigma FT-IR Library	**1**(2), 139C
IR Library	**3**, 1446G
Colour Index	3rd ed., 1971

FAST CORINTH V salt
Azoic Diazo No. 39

Aldrich Cat. No.:	22,736-6
Sigma Prod. No.:	F6383
CAS No.:	*[61966-14-1]*
C.I. No.:	37220
Mol. Form.:	$C_{15}H_{14}ClN_5O_3 \cdot {}^1/_2\ ZnCl_2$
F.W.:	415.90
m.p.:	147°C (dec.)
Appearance:	Brown-gold powder
Solubility:	H_2O 20mg/ml
	EGME 70mg/ml
	EtOH 5mg/ml

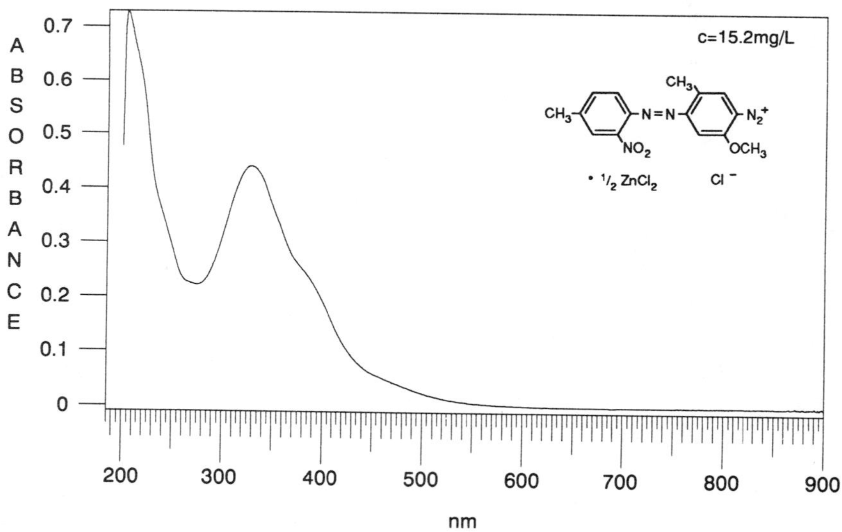

This stable diazonium salt is prepared by coupling diazotized 2-nitro-*p*-toluidine to cresidine (5-methyl-*o*-anisidine), diazotizing the coupling product and precipitating as the zinc chloride double salt by the addition of a concentrated zinc chloride solution.

Soluble in water, it is suitable for the *in situ* development of different disazo dyes on cotton and silk. The color and shade can be varied depending on the coupling component used to pad the textile being dyed.

Lillie[1] provides information and references regarding its use in enzyme-demonstration reactions.

Reference:

1) Lillie, R.D. *Histopathologic Technic and Practical Histochemistry*, 3rd edition; McGraw-Hill: New York, 1965; p 352.

General references:

Aldrich FT-IR Library	**1**(2), 961A
Sigma FT-IR Library	**1**(2), 139D
Safety	**2**(1), 1676D
Colour Index	3rd ed., 1971

FAST GARNET GBC base

4'-Amino-2,3-dimethylazobenzene • Solvent Yellow 3

Aldrich Cat. No.:	12,156-8
Sigma Prod. No.:	F8504
Purity:	97%

CAS No.:	[*97-56-3*]
C.I. No.:	11160
Mol. Form.:	$C_{14}H_{15}N_3$
F.W.:	225.30
m.p.:	101-102°C
Appearance:	Red-violet powder
λ_{max}:	328nm in 50% ethanol + 1ml 1*N* HCl
Solubility:	H_2O 0.1mg/ml
	EGME 50mg/ml
	EtOH 30mg/ml

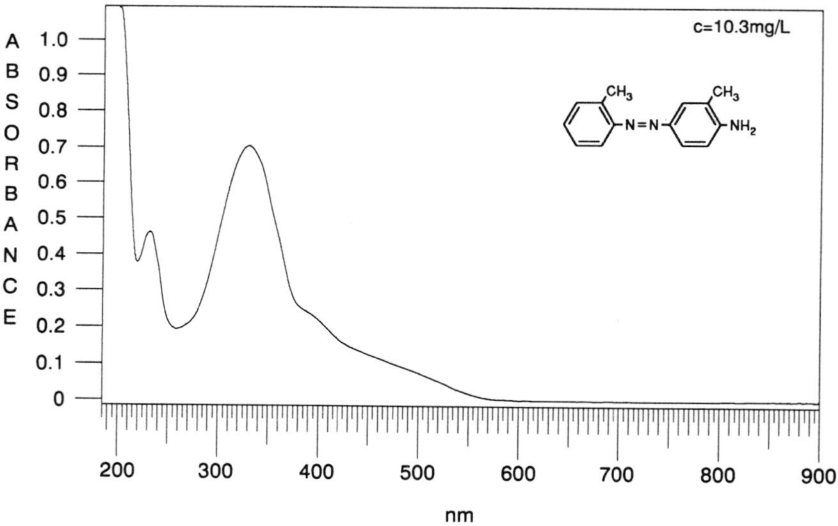

Fast Garnet GBC base is a yellow, monoazo solvent dye that has been used to color wax polishes, like shoe polish. It has indicator properties and is virtually insoluble in water, soluble in methyl Cellosolve and ethanol, chloroform, ether, acetone, ethyl acetate, linseed oil, stearic acid, oleic acid and benzene.

Biological applications of Fast Garnet GBC base have been limited. It has been used as a fat stain and in liver studies involving its inherent carcinogenicity.

Fast Garnet GBC salt (Azoic Diazo 4) was studied by Pugh and Walker[1] and Hayashi[2] for suitability as a reagent to demonstrate *N*-acetyl-β-glucosaminidase. Results from the former study were not very satisfactory due to low 1-naphthol substantivity. When compared to hexaazotized Pararosaniline in the latter study, Fast Garnet salt was found to be more convenient, but the overall second choice.

References:

1) Pugh, D.; Walker, P.G. *J. Histochem. Cytochem.* **1961,** *9*, 242.

2) Hayashi, M. *ibid.* **1965,** *13*, 355-360.

General references:

Aldrich FT-IR Library	**1**(2), 956C
Sigma FT-IR Library	**1**(2), 140B
IR Library	**3**,1445E
Safety	**2**(1), 1677B
Merck Index	**11**, 431
Colour Index	3rd ed., 1971

FAST GARNET GBC salt

Azoic Diazo No. 4

Aldrich Cat.. No.:	20,123-5	23,465-6
Sigma Prod. No.:	F0875	F8761
Grade:	——	Purified
CAS No.:	[*101-89-3*]	
C.I. No.:	37210	
Mol. Form.:	$C_{14}H_{14}N_4O_4S$	
F.W.:	334.36	
Appearance:	Rust-brown powder	Reddish-brown powder
λ_{max}:	360nm in water	
Solubility:	H_2O 80mg/ml	20mg/ml
	EGME 20mg/ml	40mg/ml
	EtOH 4mg/ml	3mg/ml

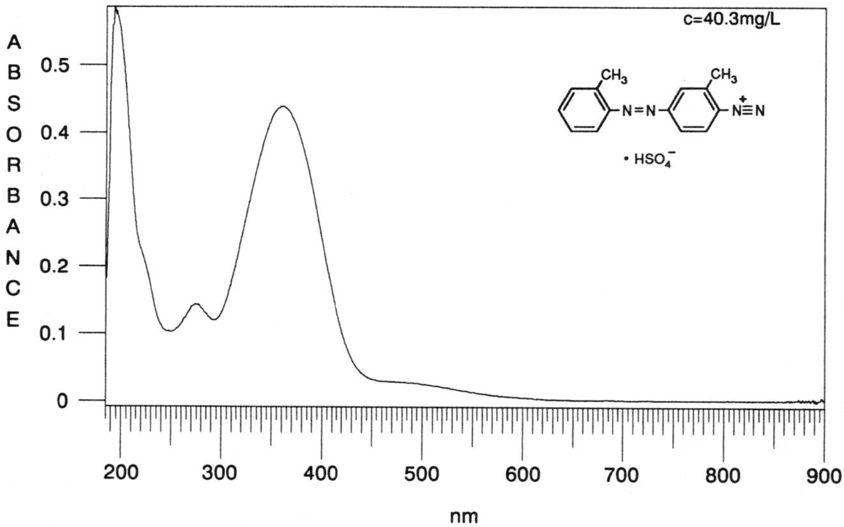

See description on pages 342 and 343 under Fast Garnet GBC base.

General references:

Aldrich FT-IR Library	**1**(2), 956D	——
Sigma FT-IR Library	**1**(2), 140D	1(2), 140C
IR Library	**3**, 1445E	**3**, 1445F
Safety	**2**(1), 1677C	

FAST GREEN FCF

Food Green 3

Aldrich Cat. No.:	86,115-4
Sigma Prod. No.:	F7258
Grade:	Certified
CAS No.:	*[2353-45-9]*
C.I. No.:	42053
Mol. Form.:	$C_{37}H_{34}N_2Na_2O_{10}S_3$
F.W.:	808.86
m.p.:	290°C (dec.)
Appearance:	Maroon powder
λ_{max}:	624nm in 50% ethanol
Solubility:	H_2O 60mg/ml
	EGME 60mg/ml
	EtOH 5mg/ml

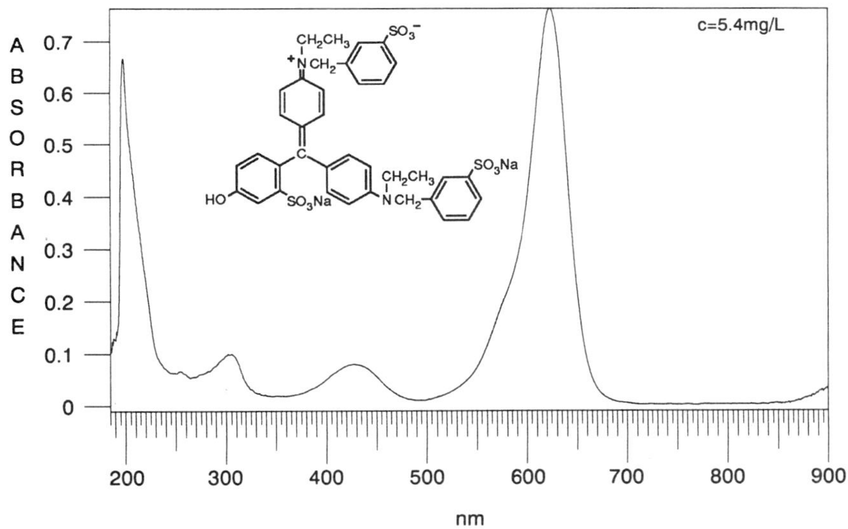

This bluish-green, anionic triphenylmethane dye is soluble in water and slightly soluble in alcohol. Although closely related structurally to Light Green SF Yellowish (LG), it is more lightfast than LG and has gained wide acceptance as a substitute for this material as a stain in plant histology and cytology. Previously, it was used as a general-purpose food dye. Recently, however, it has been subjected to scrutiny as a cancer-suspect agent.

Staining Procedures[1] lists methods and a bibliography for the following applications: Fast-Green modification of the Van Gieson stain; Lillie modification of Masson's trichrome stain; Kornhauser's "Quad" stain; Orcein and

Feulgen techniques; modified-Azan-staining technique for insects; Hematoxylin-Biebrich Scarlet-Fast Green (designed to show mantle-tissue glandular muscle, connective and nervous tissue); tannic acid-ferric chloride-Safranin-Fast Green for staining meristemic tissues; use with Safranin as a stain for plant tissue; stain for demonstrating nuclear histones; counterstain for Acid Fuchsin-lactophenol (designed to differentiate between fungal structures on leaf structures and those that are subcuticular or intracellular); Gimenez Stain for psittacosis and other rickettsial agents in yolk-sac cultures.

Reference:

1) Clark, G. *Staining Procedures,* 4th ed.; Williams and Wilkins: Baltimore, 1981; pp 114, 120-121, 125, 197, 262, 301, 319-320, 324-325, 336-337, 355, 368, 415.

General references:

Aldrich FT-IR Library **1**(2), 1033C
Sigma FT-IR Library **1**(2), 141B
IR Library **3**, 1485D
Safety **2**(1), 1678A
Merck Index **11**, 3886
Colour Index 3rd ed., 1971

FAST RED AL salt
Azoic Diazo No. 36

Aldrich Cat. No.: 20,124-3
Sigma Prod. No.: F5002

C.I. No.: 37275
Mol. Form.: $C_{14}H_7ClN_2O_2 \cdot {}^1/_2\, ZnCl_2$
F.W.: 270.68
Appearance: Straw-colored powder
Solubility: H_2O 70mg/ml
EGME 20mg/ml
EtOH 5mg/ml

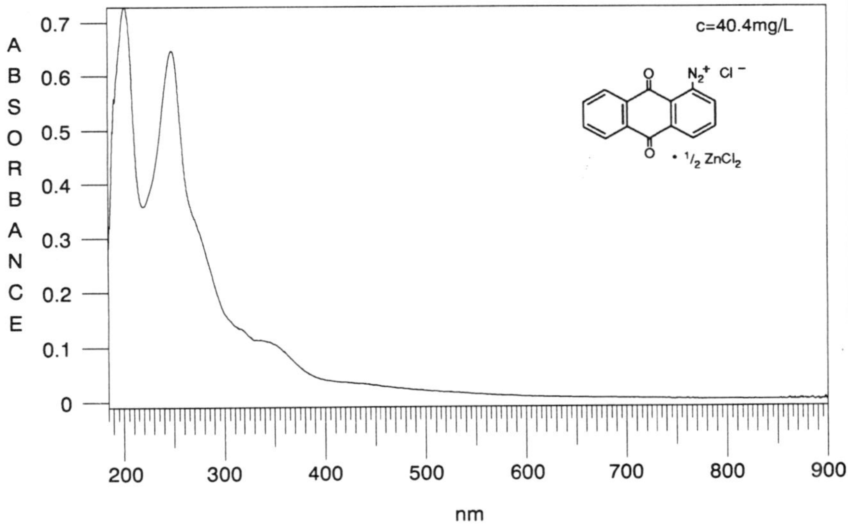

This stable diazonium salt is prepared by diazotizing 1-aminoanthraquinone (Aldrich Cat. No. A3,900-9) to form 1-anthraquinonediazonium chloride, which is isolated in its stable zinc chloride double-salt form by precipitation with zinc chloride. Industrially, it is used as a coupling agent for the *in situ* dyeing of padded cellulosic fabrics, primarily cotton.

Fast Red AL salt is used as a coupling agent in histochemical reactions for hydrolases. Seligman and Mannheimer[1] used it for acid phosphatases with α-naphthyl phosphate.

Reference:

1) Seligman, A.M.; Mannheimer, L.H. *J. Natl. Cancer Inst.* **1949,** *9*, 427-437.

General references:

Aldrich FT-IR Library	**1**(2), 960C
Sigma FT-IR Library	**1**(2), 141C
IR Library	**3**, 1477B
Colour Index	3rd ed., 1971

FAST RED B salt
Azoic Diazo No. 5

Aldrich Cat. No.:	20,125-1
Sigma Prod. No.:	F1125
CAS No.:	*[49735-71-9]*
C.I. No.:	37125
Mol. Form.:	$C_{17}H_{13}N_3O_9S_2$
F.W.:	467.44
Appearance:	Pale-orange powder
λ_{max}:	375nm in water
Solubility:	H_2O 80mg/ml
	EGME 40mg/ml
	EtOH 20mg/ml

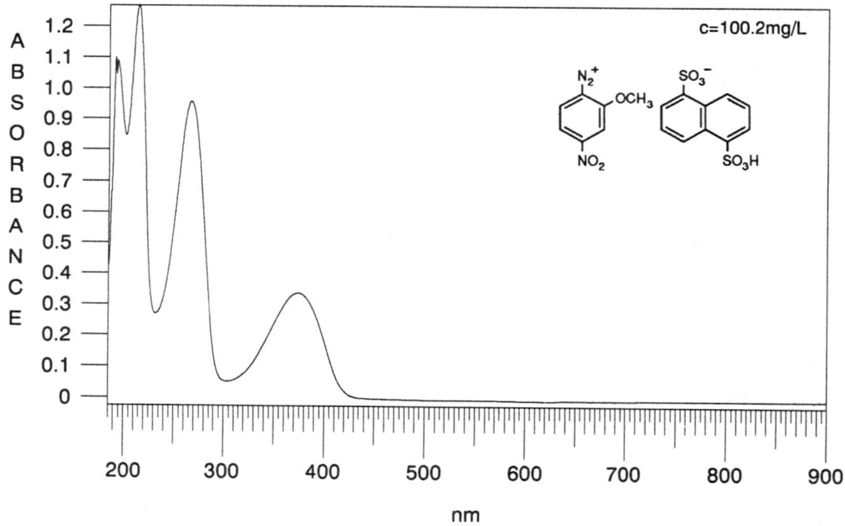

This stable diazonium salt is prepared by diazotizing 4-nitro-*o*-anisidine and treating the reaction product with 1,5-naphthalenedisulfonic acid to precipitate the desired product. It is used industrially as a coupling agent for the *in situ* development of dyes on variously padded (impregnated with one of numerous 1- and 2-naphthol derivatives selected to attain the desired color and/or shade) cotton, silk and polyester.

Fast Red B salt is used as a coupling agent in histochemical reactions for hydrolases. *Staining Procedures*[1] provides an alkaline phosphatase method from Lillie *et al.* (1961) and Gomori (1952) for enterochromaffin-cell granules.

The tetrafluoroborate salt (Aldrich Cat. No. 30,546-4) has similar uses, but it is less soluble in water and virtually insoluble in ethanol.

Reference:

1) Clark, G. *Staining Procedures,* 4th ed.; Williams and Wilkins: Baltimore, 1981; p 239.

General references:

Aldrich FT-IR Library	**1**(2), 960B
Sigma FT-IR Library	**1**(2), 141D
IR Library	**3**, 1447A
Colour Index	3rd ed., 1971

FAST RED B, tetrafluoroborate salt

Aldrich Cat No.: 30,546-4
Sigma Prod. No.: F1263

CAS No.: *[2357-51-9]*
Mol. Form.: $C_7H_6BF_4N_3O_3$
F.W.: 266.45
m.p.: 122-124°C
Appearance: Tan crystalline powder
λ_{max}: 375(266)nm in water
Solubility: H_2O 60mg/ml
EGME 70mg/ml
EtOH 0.9mg/ml

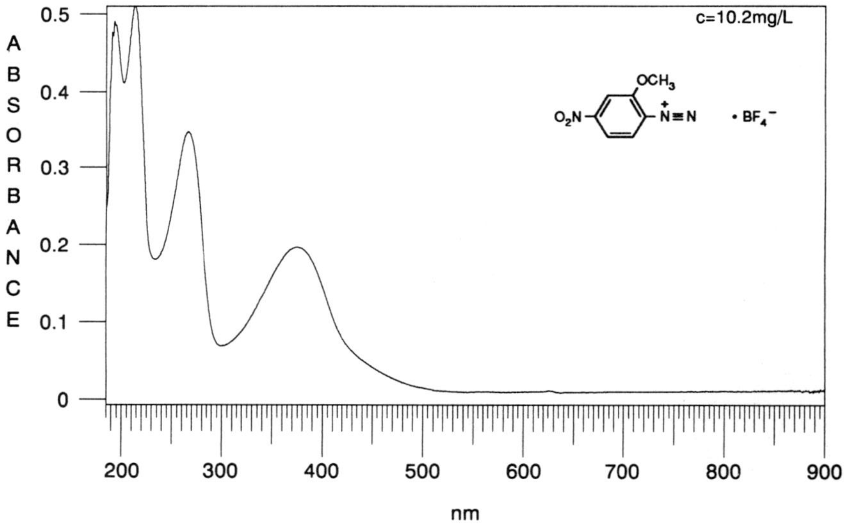

See pages 348 and 349 under Fast Red B salt for description.

General reference:

Safety **2**(1), 1678B

FAST RED ITR base

Azoic Diazo No. 42

Aldrich Cat. No.:	20,128-6
Sigma Prod. No.:	F1250
Purity:	98%

CAS No.:	[*97-35-8*]
C.I. No.:	37150
Mol. Form.:	$C_{11}H_{18}N_2O_3S$
F.W.:	258.34
Appearance:	Light-brown powder
λ_{max}:	256nm in dioxane
Solubility:	H_2O 0.5mg/ml
	EGME 100mg/ml
	EtOH 30mg/ml

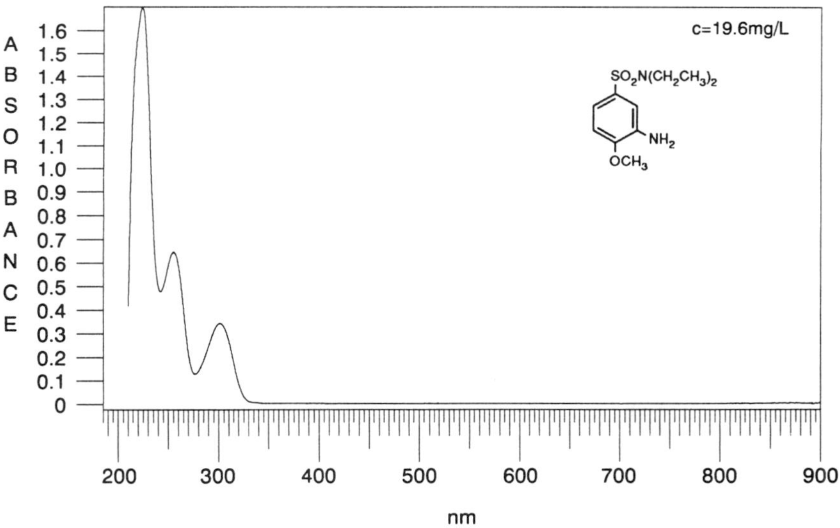

For description, see page 352 under Fast Red ITR salt.

General references:

Aldrich FT-IR Library	**1**(2), 953A
Sigma FT-IR Library	**1**(2), 142A
IR Library	**3**, 1444A

FAST RED ITR salt

Azoic Diazo No. 42

Aldrich Cat. No.:	20,127-8
Sigma Prod. No.:	F1375
C.I. No.:	37150
Mol. Form.:	$C_{11}H_{16}ClN_3O_3S \cdot {}^1/_2\ ZnCl_2$
F.W.:	373.92
Appearance:	Off-white powder
λ_{max}:	357nm in water
Solubility:	H_2O 80mg/ml
	EGME 20mg/ml
	EtOH 7mg/ml

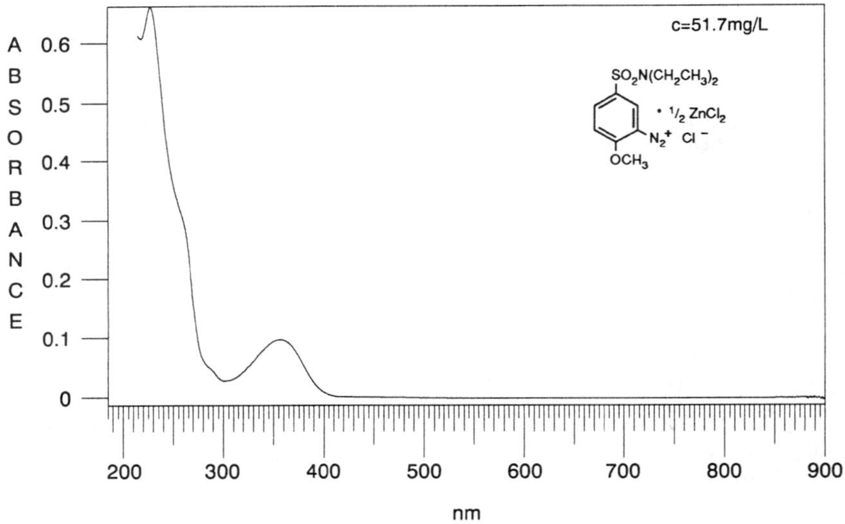

Fast Red ITR salt is made by first diazotizing Fast Red ITR base (see page 351) and then adding zinc chloride to the reaction product to precipitate the zinc chloride double salt end-product. Industrially, it is used as a coupling agent for the *in situ* development of various dyes on cotton, silk and nylon padded with one of a number of coupling components selected to attain the desired color and/or shade.

Gossner[1] reported good stability when this compound was used in acid phosphatase applications.

Reference:

1) Gossner, W. *Histochemie* **1958,** *1*, 48-69.

General references:

Aldrich FT-IR Library	**1**(2), 958B
Sigma FT-IR Library	**1**(2), 142B
IR Library	**3**, 1446C
Colour Index	3rd ed., 1971

FAST RED PDC salt

Azoic Diazo No. 14

Aldrich Cat. No.:	20,129-4
Sigma Prod. No.:	F6876

C.I. No.:	37151
Mol. Form.:	$C_{11}H_{16}ClN_3O_3S \cdot {}^1/_2\ ZnCl_2$
F.W.:	373.92
Appearance:	Cream powder
λ_{max}:	352nm in methanol
Solubility:	H_2O 40mg/ml
	EGME 50mg/ml
	EtOH 10mg/ml

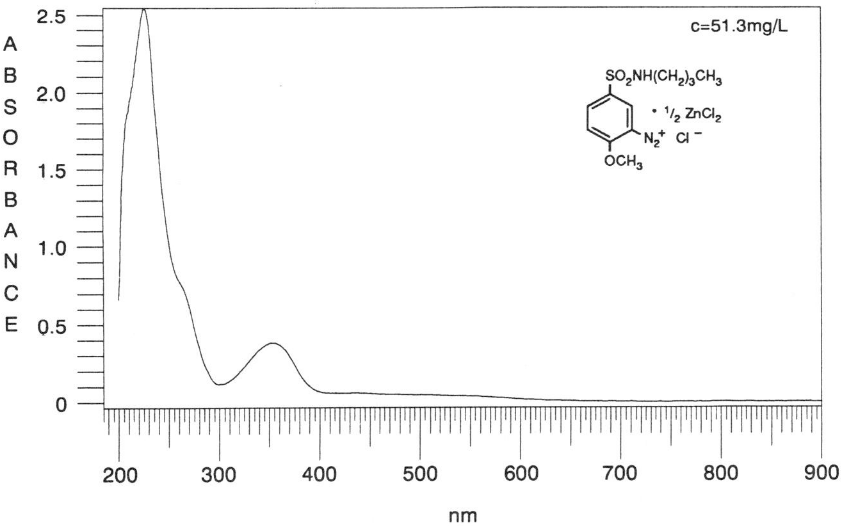

This stable diazonium salt is prepared by diazotizing *N*-butyl-4-methoxy-metanilamide, followed by precipitating as the zinc chloride double salt with the addition of a concentrated zinc chloride solution. It is used in the textile industry for the *in situ* development of various dyes in padded (previously treated with a coupling component) cotton, silk, acetate and nylon. The color and/or shade is varied within limits by the coupling component selected.

Fast Red PDC salt is used as a coupling agent in histochemical reactions for hydrolases.

General references:

Aldrich FT-IR Library	**1**(2), 958A
Sigma FT-IR Library	**1**(2), 142D
IR Library	**3**, 1446B
Colour Index	3rd ed., 1971

FAST RED TR salt

Azoic Diazo No. 11

Aldrich Cat. No.:	20,155-3
Sigma Prod. No.:	F1500

CAS No.:	*[89453-69-0]*
C.I. No.:	37085
Mol. Form.:	$C_7H_6Cl_2N_2 \cdot {}^1/_2\ ZnCl_2$
F.W.:	257.18
Appearance:	Light-beige powder
λ_{max}:	285nm in water
Solubility:	H_2O 50mg/ml
	EGME 20mg/ml
	EtOH 10mg/ml

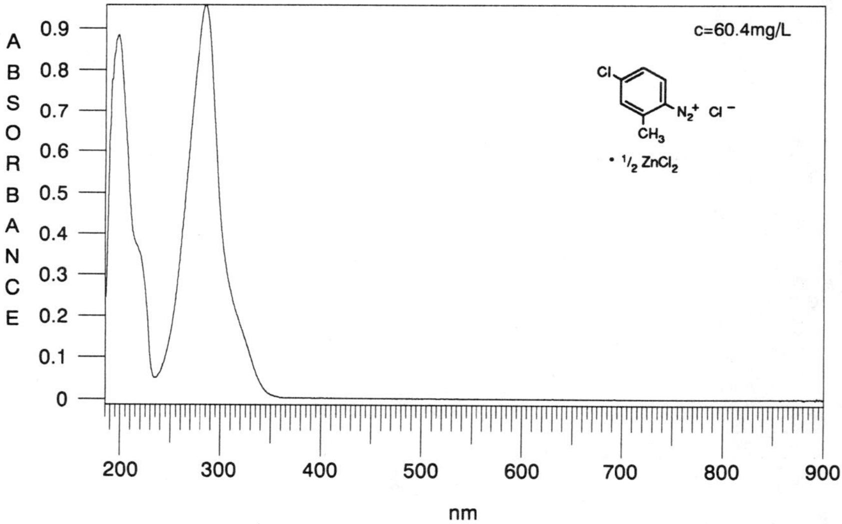

Fast Red TR salt is a stable diazonium salt that is made by diazotizing the base, 4-chloro-*o*-toluidine, followed by conversion to the zinc chloride double salt by the addition of zinc chloride to the reaction mixture. It is used in the textile industry for the *in situ* dyeing of variously padded cotton, silk, acetate and nylon.

Burstone[1] found it satisfactory for phosphatase methods and Gossner[2] found it very useful for both alkaline phosphatases and esterases.

References:

1) Burstone, M.S. *J. Natl. Cancer Inst.* **1958,** *21*, 167-173.
2) Gossner, W. *Histochemie* **1958,** *1*, 48-69.

General references:

Aldrich FT-IR Library	**1**(2), 960A
Sigma FT-IR Library	**1**(2), 143B
IR Library	**3**, 1446H
Colour Index	3rd ed., 1971

FAST RED VIOLET LB base

4'-Amino-2'-chloro-5'-methylbenzanilide

Aldrich Cat. No.: 27,405-4

CAS No.: [121-22-2]
C.I. No.: 37165
Mol. Form.: $C_{14}H_{13}ClN_2O$
F.W.: 260.73
m.p.: 194°C (dec.)
Appearance: Off-white powder
λ_{max}: 293nm in methanol
Solubility: H_2O 0.6mg/ml
EGME 40mg/ml
EtOH 3mg/ml

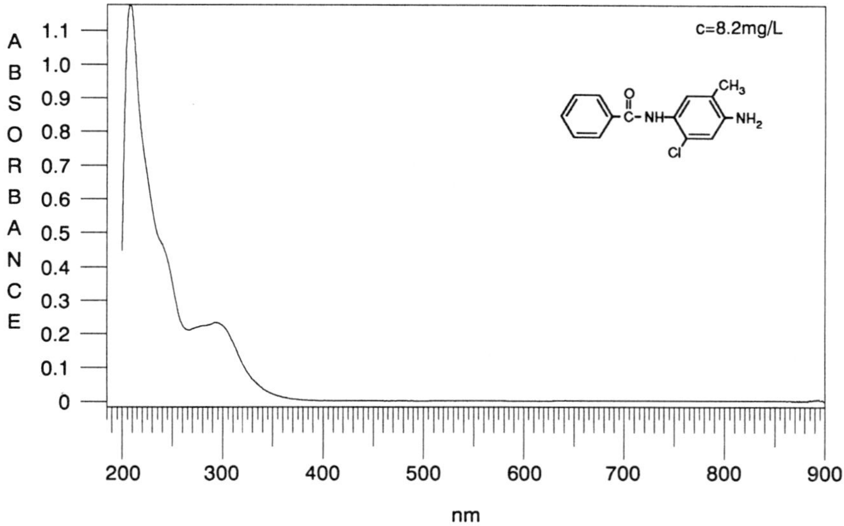

See description on pages 356 and 357 under Fast Red Violet LB salt.

General reference:

Safety **2**(1), 1678C

FAST RED VIOLET LB salt

Aldrich Cat. No.: 24,732-4
Sigma Prod. No.: F3381

Mol. Form.: $C_{14}H_{11}Cl_2N_3O \cdot {}^1/_2\, ZnCl_2$
F.W.: 376.31
Appearance: Tan powder
λ_{max}: 342nm in water
Solubility: H_2O 20mg/ml
EGME 60mg/ml
EtOH 2mg/ml

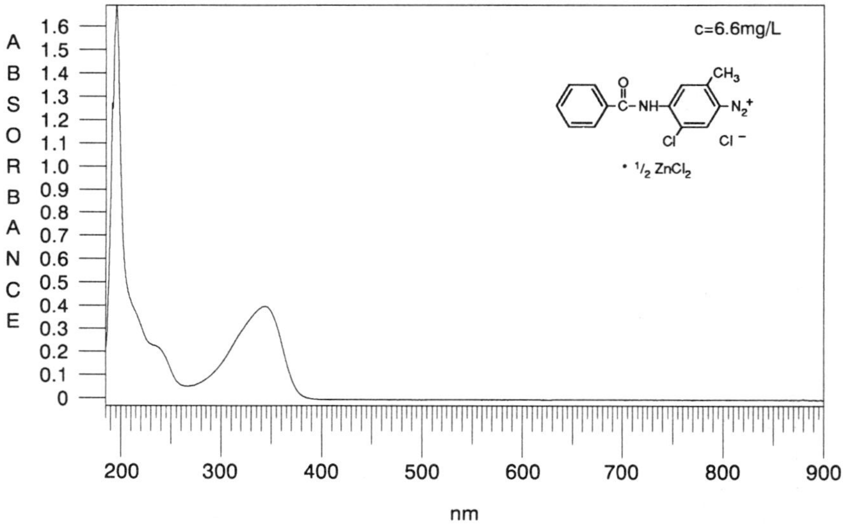

Fast Red Violet LB salt is a stable diazonium salt made by diazotizing the base and converting the resulting product to the zinc chloride double salt. Both products are available with a purity of approximately 95%.

The salt (Aldrich Cat. No. 24,732-4) couples with Naphthols AS, BI, TR, MX and BS to form red to purplish-red dyes. Burstone used it with excellent results with Naphthol AS phosphates in both acid[1] and alkaline[2,3] phosphatase couplings. Tanaka[4] proposed a procedure for using Fast Red Violet LB salt as a reagent for the semi-quantitative assay of leucocyte alkaline phosphatase. A criterion for scoring the assay and interpreting the results is also included in Tanaka's paper.

References:

1) Burstone, M.S. *J. Natl. Cancer Inst.* **1958,** *21*, 523-529.
2) *Idem ibid.* **1957,** *18*, 167-173.
3) *Idem ibid.* **1958,** *20*, 601-605.
4) Tanaka, K.R., chair; and the Subcommittee on Cellular Enzymology *National Committee for Clinical Laboratory Standards* **1984,** *4* (14), 346-354.

General references:

Aldrich FT-IR Library **1**(2), 958D
Sigma FT-IR Library **1**(2), 143C

FAST VIOLET B
Azoic Diazo No. 41

Aldrich Cat. No.:	20,158-8
Sigma Prod. No.:	F1875
CAS No.:	[*99-21-8*]
C.I. No.:	37165
Mol. Form.:	$C_{15}H_{16}N_2O_2$
F.W.:	256.31
m.p.:	185-188°C
Appearance:	Off-white powder
λ_{max}**:**	297nm in ethanol
Solubility:	H_2O 0.4mg/ml
	EGME 50mg/ml
	EtOH 9mg/ml

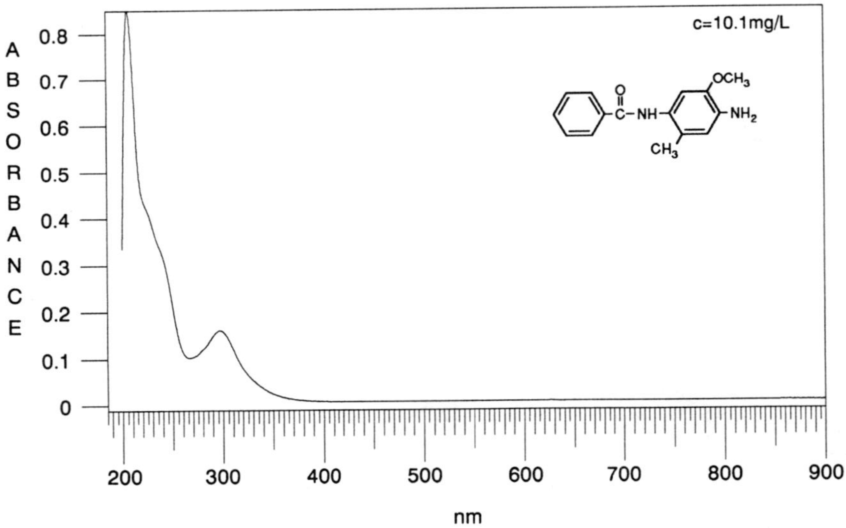

For description, see page 359 under Fast Violet B salt.

General references:

Aldrich FT-IR Library	**1**(2), 954C
Sigma FT-IR Library	**1**(2), 144B
IR Library	**3**, 1444G

FAST VIOLET B salt

Azoic Diazo No. 41

Aldrich Cat. No.: 20,159-6

CAS No.: [*14726-28-4*]
C.I. No.: 37165
Mol. Form.: $C_{15}H_{14}ClN_3O_2 \cdot {}^1/_2\, ZnCl_2$
F.W.: 371.89
Appearance: Yellow crystalline powder
λ_{max}: 325nm in water
Solubility: H_2O 50mg/ml
EGME 50mg/ml
EtOH 20mg/ml

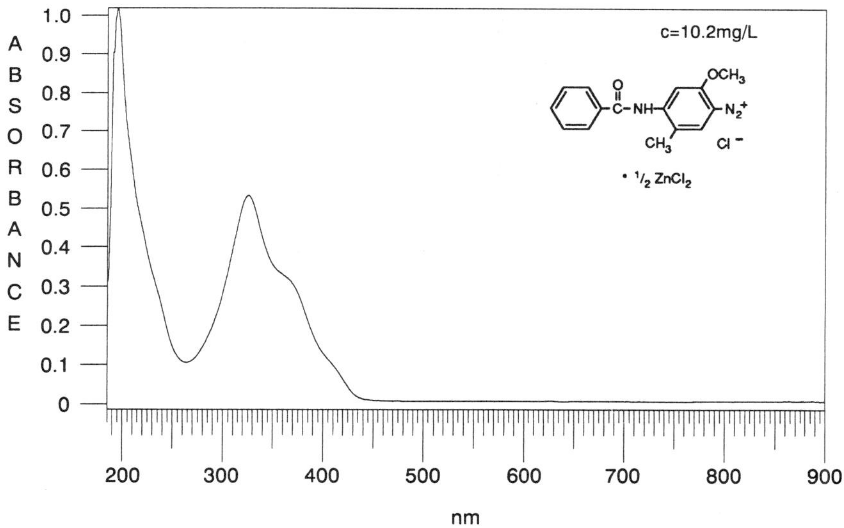

Fast Violet B salt is a typical, stable diazonium salt that is made by first diazotizing the base (see page 358, Aldrich Cat. No. 20,158-8), and then converting to the zinc chloride double salt by the addition of zinc chloride to the reaction mixture. It is used in the textile industry for the *in situ* dyeing of cotton padded with various coupling components.

It is used as a coupling agent, in histochemical reactions, for hydrolases. Burstone[1] considered it useful for phosphatases.

Reference:

1) Burstone, M.S. *J. Natl. Cancer Inst.* **1958,** *21*, 167-173.

General references:

Aldrich FT-IR Library **1**(2), 959B
Sigma FT-IR Library **1**(2), 144C
IR Library **3**, 1446F
Colour Index 3rd ed., 1971

FAST YELLOW

Acid Yellow 9 • 4-Amino-1,1'-azobenzene-3,4'-disulfonic acid, sodium salt

Aldrich Cat. No.: 25,030-9
Sigma Prod. No.: A9536
Purity: 97%

CAS No.: [*2706-28-7*]
C.I. No.: 13015
Mol. Form.: $C_{12}H_{10}N_3NaO_6S_2$
F.W.: 379.35
Appearance: Red-brown powder
λ_{max}: 480(380)nm in 0.1*N* hydrochloric acid
Solubility: H_2O 80mg/ml
EGME 0.9mg/ml
EtOH 0.1mg/ml

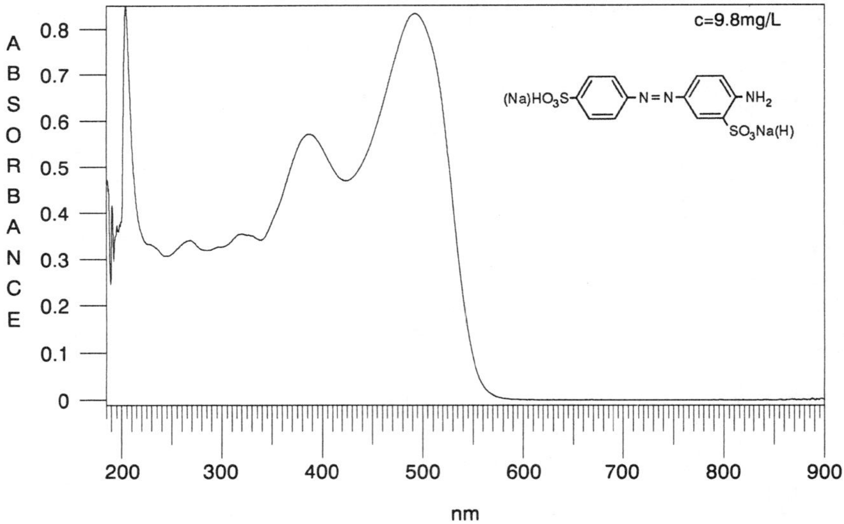

Fast Yellow is an anionic monoazo dye that is prepared by disulfonating 4-phenylazoaniline.

Soluble in water and very slightly soluble in ethanol and methyl Cellosolve, it is used to dye wool from a sulfuric acid bath, silk from an acetic acid bath, nylon from a formic acid bath and to dye beater paper; while the aluminum salt is used in inks for tinplate printing. A highly purified grade, Food Yellow 2, is used to color foods, drugs and cosmetics.

This dye has not been used widely as a biological stain. *Conn's*[1] cites

Unna (1921) who used it to characterize various basic proteins, and Wallart and Houette (1934) who used it in a trichrome stain mixture with Hematoxylin and Acid Fuchsin.

Reference:

1) *Conn's Biological Stains,* 9th ed.; Lillie, R.D., Ed.; Williams and Wilkins: Baltimore, 1977; p 97.

General references:

Aldrich FT-IR Library	**1**(2), 97C
Safety	**2**(1), 136A
Colour Index	3rd ed., 1971

FAST YELLOW GC salt

Azoic Diazo No. 44

Aldrich Cat. No.:	23,563-6
Sigma Prod. No.:	F9879

CAS No.:	[*14263-92-4*]
C.I. No.:	37000
Mol. Form.:	$C_6H_4Cl_2N_2 \cdot ZnCl_2$
F.W.:	243.16
Appearance:	Light tannish-orange powder
λ_{max}:	324(268)nm in water
Solubility:	H_2O 50mg/ml
	EGME 20mg/ml
	EtOH 3mg/ml

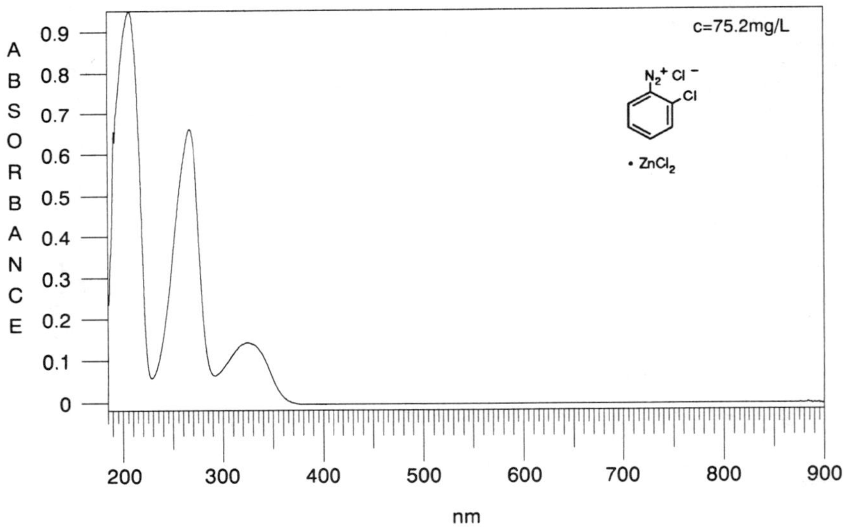

Fast Yellow GC salt is a stable diazonium salt that is made by diazotizing the base, 2-chloroaniline, and isolating the reaction product as the zinc chloride double salt.

This water-soluble dye intermediate must be coupled to an Azoic Coupling Component (ACC) to function as a dye. The ACC's are generally known as naphthols since 2-naphthol and its derivatives comprise the bulk of the list, which also includes the anilide of 3-hydroxy-2-naphthoic acid, acetoacetic acid esters, benzoylacetic acid esters, benzocarbazoles and dibenzofurans.

Cellulosic textiles usually are dyed in two steps. The material is impregnated with an alkaline solution of the ACC which is then passed through a bath containing a solution of the diazonium salt (Fast Yellow GC salt, in this case). As a result, an insoluble dye is formed *in situ* in the fabric. By varying the ACC, the color of dye can be varied greatly. Printing can be accomplished by following the ACC-impregnating step with a Fast Yellow GC salt printing step.

There is no evidence in the literature that it has been used as a biological stain.

General references:

Aldrich FT-IR Library	**1**(2), 956A
Sigma FT-IR Library	**1**(2), 145B
Colour Index	3rd ed., 1971

FAT BROWN B

Solvent Red 3

Aldrich Cat. No.: 23,602-0
Sigma Prod. No.: F3257

CAS No.: [*6535-42-8*]
C.I. No.: 12010
Mol. Form.: $C_{18}H_{16}N_2O_2$
F.W.: 292.34
m.p.: 152-155°C (dec.)
Appearance: Very dark-brown powder
λ_{max}: 408nm in ethanol
Solubility: H_2O 0.3mg/ml
EGME 40mg/ml
EtOH 20mg/ml

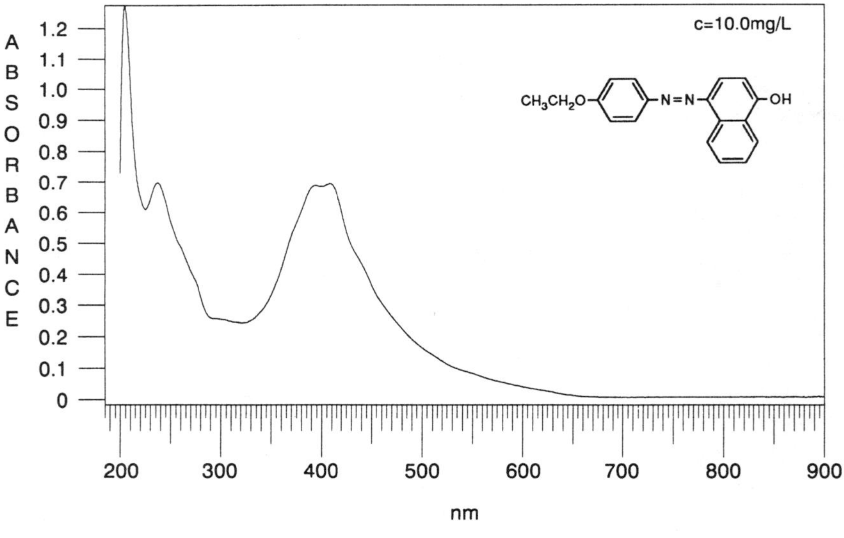

Fat Brown B is a monoazo dye that imparts a bordeaux to brown color to a wide spectrum of organic solvents. The color variation is dependent upon the solvent system.

The *Colour Index* gives the following solubilities: acetone, 2.5%; butyl acetate, 0.5%; ethyl acetate, very soluble; linseed oil, 1.4%; mineral oil, 0.1%; oleic acid, 1.2%; paraffin wax, slightly soluble; toluene, 0.7%; stearic acid, 0.6% and xylene, 0.1%. The same source states that Fat Brown B is used to color oils, fats, waxes, polishing waxes, creams, wood stains and molding powders.

There is no evidence in the literature that Fat Brown B has been used as a biological stain; however, it appears to have the requisite properties to function as a fat stain in histological staining techniques.

General references:

Aldrich FT-IR Library	**1**(2), 969C
Sigma FT-IR Library	**1**(2), 145C
Colour Index	3rd ed., 1971

FAT BROWN RR

Solvent Brown 1

Aldrich Cat. No.: 23,603-9
Sigma Prod. No.: F3382

CAS No.: [*6416-57-5*]
C.I. No.: 11285
Mol. Form.: $C_{16}H_{14}N_4$
F.W.: 262.32
m.p.: 147-158°C
Appearance: Red powder
λ_{max}: 451nm in ethanol
Solubility: H_2O 0.5mg/ml
EGME 20mg/ml
EtOH 8mg/ml

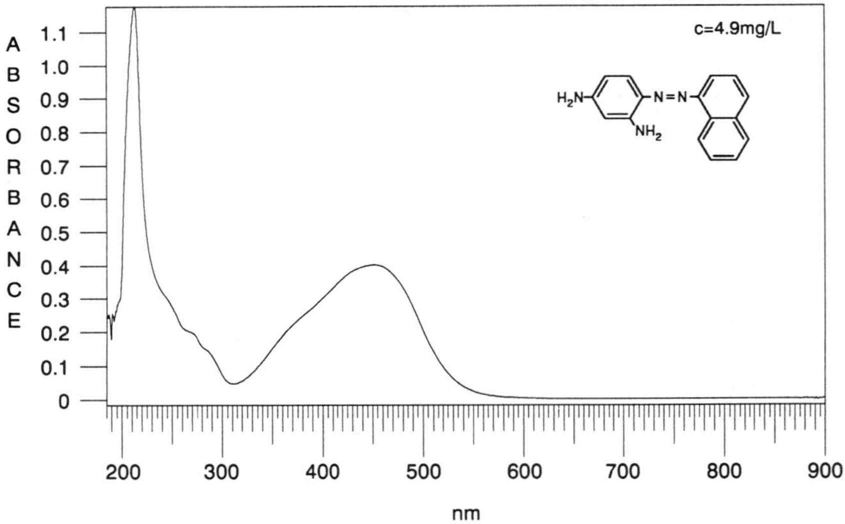

Fat Brown RR is a monoazo dye that imparts a reddish-orange to brown color to various solvent systems making it useful as a lithographic varnish.

The *Colour Index* lists Solvent Brown 1 as soluble in acetone, ethanol, oleic acid and stearic acid, and slightly soluble in benzene, ethyl acetate, linseed oil, mineral oil, paraffin wax and xylene. Industrial applications include coloring alcoholic solvents, oils, fats, waxes, double-tone- and photogravure-printing inks, spirit lacquers and polishing waxes.

While there is no evidence that it has been used as a biological stain, it appears that it may be useful as a fat stain in histological techniques.

General references:

Aldrich FT-IR Library **1**(2), 970C
Sigma FT-IR Library **1**(2), 145D
Colour Index 3rd ed., 1971

FERROCENE

Dicyclopentadienyliron

Aldrich Cat. No.:	F40-8
Sigma Prod. No.:	F3375
Purity:	~98%
CAS No.:	[*102-54-5*]
Mol. Form.:	$C_{10}H_{10}Fe$
F.W.:	186.04
m.p.:	174-176°C
b.p.:	249°C/760mm
Appearance:	Burnt-orange crystalline powder
Solubility:	H_2O <0.1mg/ml
	EGME 20mg/ml
	EtOH 7mg/ml

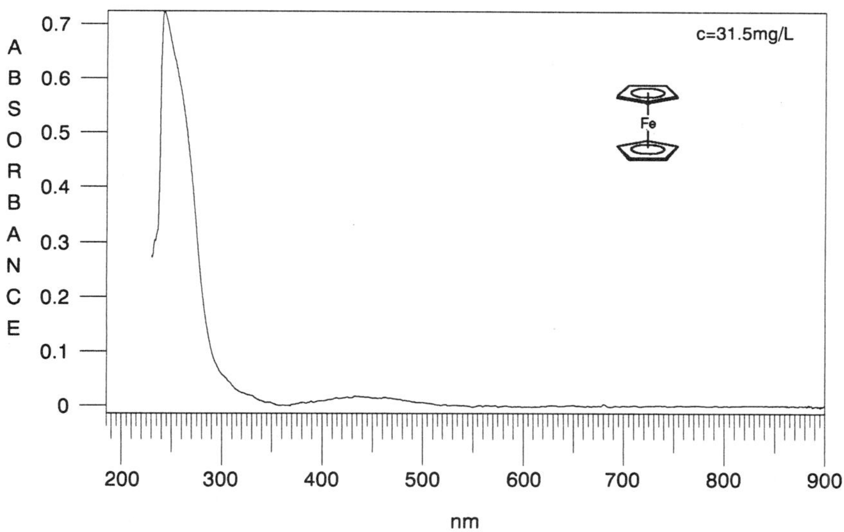

Ferrocene is classified as a nonbenzenoid aromatic compound (*i.e.*, it possesses some measure of aromatic character typical of benzene, but does not possess a benzenoid ring). The molecule is composed of a ferrous ion sandwiched between two cyclopentadienyl rings. Bonding between the metal and the rings involves the *pi* electrons of the two rings, with all carbons equally bonded to the central ferrous ion. Roberts and Caserio[1] state that the ferrous ion, in accepting a share of 12 *pi* electrons from two cyclopentadienyl anions, achieves the electronic configuration of the inert gas, krypton. The molecule is diamagnetic and the dipole moment is effectively zero.

Ferrocene is an orange solid (needles from methanol or ethanol) with a camphor-like odor. It is remarkably stable and resists thermal breakdown even at 400°C. It is quite resistant to most acids and bases, even concentrated sulfuric acid. However, it is readily oxidized with nitric acid to the less stable ferricinium ion. It is insoluble in water, 10% sodium hydroxide and concentrated boiling hydrochloric acid, and somewhat soluble in alcohol, ether and benzene. Solutions in dilute nitric acid and concentrated sulfuric acid are deep red with a blue fluorescence.

Commercially, it is used as an antiknock additive for gasoline and as a catalyst.

Although ferrocene is not a dye, it is used as an electron stain by virtue of its iron content (increased density) and finds application in the study of biological specimens by electron microscopy.

Ferrocene can also be used in histological techniques evaluated by light microscopy. When tissue sections are treated with ferrocene and a 2% silver nitrate solution and allowed to stand in the dark for about an hour, the ferrous ion is oxidized and converted to the ferricinium ion. Oxidation differentially stains the tissue components so that they can be effectively observed by light microscopy.

Reference:

1) Roberts, J.D.; Caserio, M.C. *Basic Principles of Organic Chemistry;* Benjamin: New York, 1984; p 830.

General references:

Aldrich FT-IR Library	**1**(2), 1150B
Sigma FT-IR Library	**1**(2), 560B
IR Library	**3**,1563C
Safety	**2**(1), 1681A
Merck Index	**11**, 3985

FLAVAZIN L

Acid Yellow 11

Aldrich Cat. No.:	20,199-5
Sigma Prod. No.:	F8879
CAS No.:	[*6359-82-6*]
C.I. No.:	18820
Mol. Form.:	$C_{16}H_{13}N_4NaO_4S$
F.W.:	380.36
Appearance:	Bright-yellow powder
λ_{max}**:**	407nm in water
Solubility:	H_2O 20mg/ml
	EGME 30mg/ml
	EtOH 6mg/ml

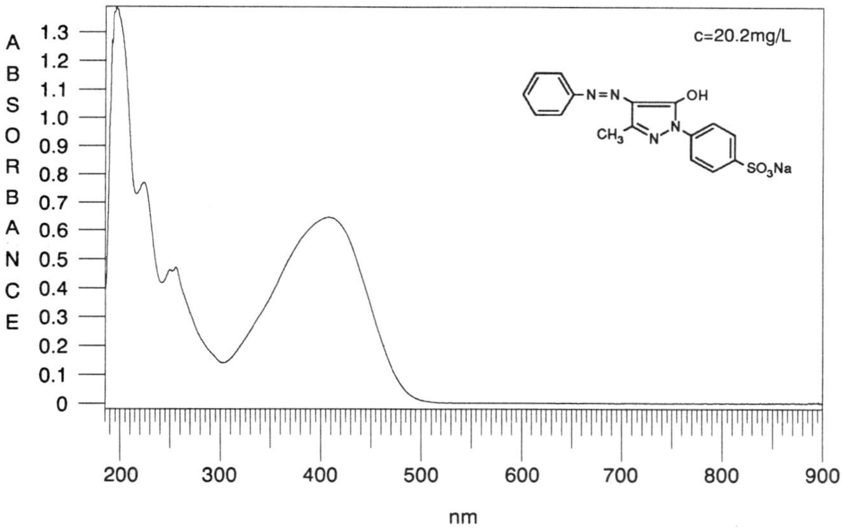

Flavazin L is a yellow, anionic monoazo pyrazolone dye that is soluble in water, methyl Cellosolve and acetone, slightly soluble in ethanol and benzene and insoluble in other organic solvents. It dyes wool, silk and nylon from an acid bath and prints directly on wool and silk. Leather is dyed a greenish yellow when fully chromed. Other industrial applications include paper coating and surface coloring and coloring resins, cellulose esters and spirit varnishes. Heavy-metal salts of Flavazin L absorbed on alumina are used as pigments in lacquers, paints, enamels and for textile printing.

Flavazin L has found limited application as a contrast or counterstain used in combination with a nuclear stain such as Hematoxylin.

General references:

Aldrich FT-IR Library	**1**(2), 1000A
Sigma FT-IR Library	**1**(2), 146B
IR Library	**3**, 1468C
Colour Index	3rd ed., 1971

FLAVIANIC ACID hydrate

8-Hydroxy-5,7-dinitro-2-naphthalenesulfonic acid

Aldrich Cat. No.:	22,501-0
Sigma Prod. No.:	F6500
CAS No.:	[*483-84-1*]
Mol. Form.:	$C_{10}H_6N_2O_8S \cdot xH_2O$
F.W.:	314.23
m.p.:	148-151°C
Appearance:	Yellow crystalline powder
λ_{max}:	428(390)nm in water
Solubility:	H_2O 20mg/ml
	EGME 10mg/ml
	EtOH 5mg/ml

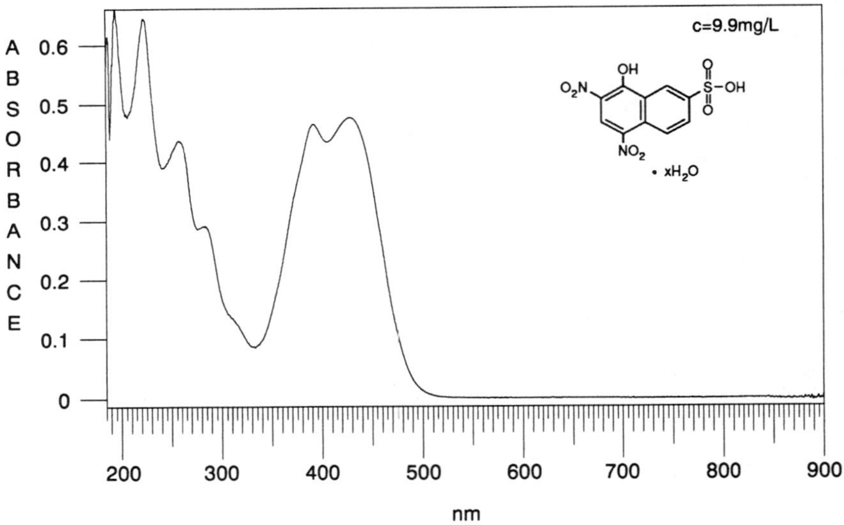

See description on pages 500 and 501 under Naphthol Yellow S.

General references:

Aldrich FT-IR Library	**1**(2), 500D
Sigma FT-IR Library	**1**(2), 561B
IR Library	**3**, 1159E
Safety	**2**(1), 1684D

FLUORESCAMINE

Aldrich Cat. No.:	20,165-0
Sigma Prod. No.:	F9878
CAS No.:	*[38183-12-9]*
Mol. Form.:	$C_{17}H_{10}O_4$
F.W.:	278.27
m.p.:	154-156°C
Appearance:	Off-white to light-yellow powder
Solubility:	H_2O 0.2mg/ml
	EGME 70mg/ml
	EtOH 40mg/ml

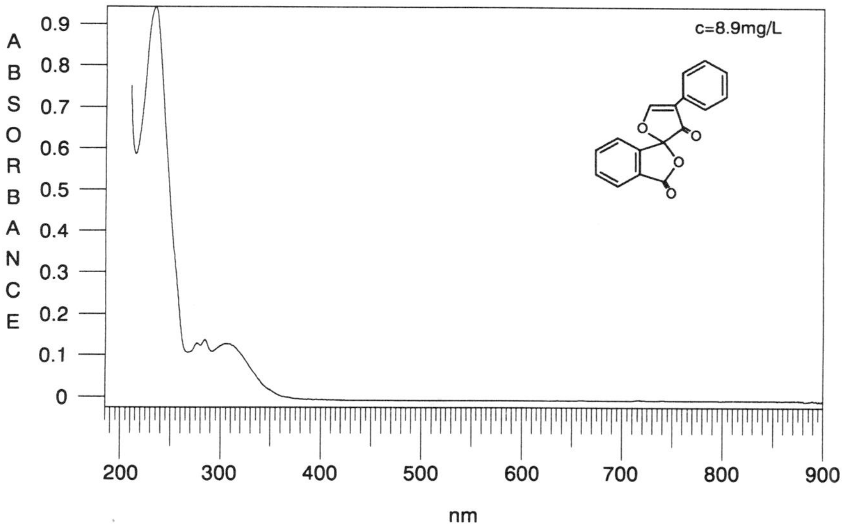

Fluorescamine is a colorless, fluorogenic reagent that is used for the fluorometric determination of primary amines, amino acids, peptides and proteins in picomole quantities.[1-3] Its reaction with amines is almost instantaneous at room temperature in aqueous media. The products are highly fluorescent, whereas the reagent and its degradation products are non-fluorescent.

Staining Procedures[4] describes a fluorescence cell-surface labeling, a slight modification of the method of Hawkes and Bartholomew (1977). This method is used to analyze tissue-culture cell surface by flow microfluorometry.

References:

1) Weigele, M; DeBernardo, J.P.; Tengi, J.P.; Leimgruber, W. *J. Am. Chem. Soc.* **1972,** *94*, 5927-5928.

2) Udenfriend, S.; Stein, S.; Böhlen, P.; Dairman, W.; Leimgruber, W.; Weigele, M. *Science* **1972,** *178*, 871-872.

3) Böhlen, P.; Stein, S.; Udenfriend, S. *Arch. Biochem. Biophys.* **1974,** *163*, 390-399.

4) Clark, G. *Staining Procedures,* 4th ed.; Williams and Wilkins: Baltimore, 1981; pp 96-97.

General references:

Aldrich FT-IR Library	**1**(2), 1009D
Sigma FT-IR Library	**1**(2), 146C
IR Library	**3**, 1472H
Merck Index	**11**, 4084

FLUORESCEIN
Solvent Yellow 94

Aldrich Cat. No.:	F245-6
Sigma Prod. No.:	F7505
CAS No.:	[*2321-07-5*]
C.I. No.:	45350:1
Mol. Form.:	$C_{20}H_{12}O_5$
F.W.:	332.31
m.p.:	320°C
Appearance:	Maroon powder
λ_{max}:	496nm in 0.1*N* sodium hydroxide
Solubility:	H_2O 0.3mg/ml
	EGME 40mg/ml
	EtOH 20mg/ml

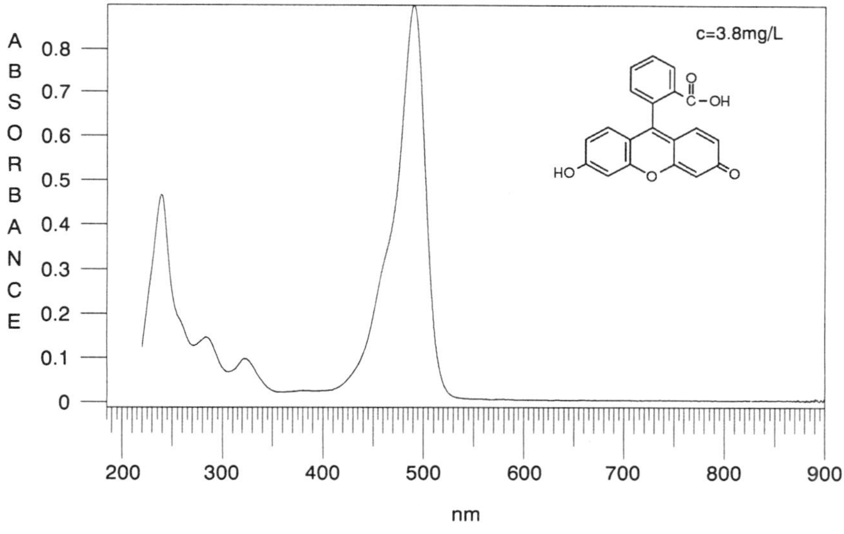

Fluorescein (Solvent Yellow 94) is a yellow fluorone dye (hydroxylated xanthene). It is made by condensing phthalic anhydride with resorcinol. Usually this is done with the aid of a condensing agent such as zinc chloride, sulfuric acid or phosphoric acid. It is soluble in the common solvents acetone, ethanol, methanol, ethyl acetate and methyl Cellosolve. It is insoluble in most other organic solvents and very slightly soluble in water. A highly purified form (D & C Yellow No. 7) is used in cosmetics.

The disodium salt of fluorescein is known as Uranine or fluorescein, water soluble. It has a very low tinctorial efficiency and an intense-green fluorescence. The latter persists even in extremely dilute, aqueous solutions (1 part in 40 million). Because of its low tinctorial efficiency, Uranine is not used as a paper, textile or leather dye. Conversely, because of its intense fluorescence and biodegradability, it is in great demand for tracing the course of underground streams and of factory effluents and sewage sources that contaminate underground water supplies. It is also used as a marker for aiding sea-rescue operations. A highly purified product (D & C Yellow No. 8) is used in solution as a diagnostic aid for evaluating corneal trauma and for treating intraocular inflammation.

General references:

Aldrich FT-IR Library	**1**(2), 1010A
Sigma FT-IR Library	**1**(2), 146D
IR Library	**3**, 1473A
Merck Index	**11**, 4085
Colour Index	3rd ed., 1971

FLUORESCEIN, water soluble

Acid Yellow 73 • Uranine

Aldrich Cat. No.: 16,630-8
Sigma Prod. No.: F6377

CAS No.: [*518-47-8*]
C.I. No.: 45350
Mol. Form.: $C_{20}H_{10}Na_2O_5$
F.W.: 376.28
Appearance: Orange-brown powder
λ_{max}: 491nm in water
Solubility: H_2O 40mg/ml
EGME 40mg/ml
EtOH 70mg/ml

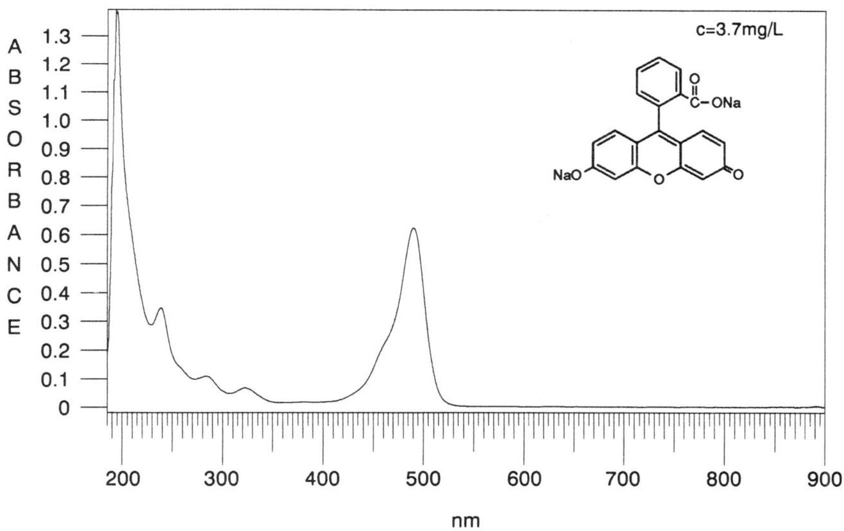

See description on pages 373 and 374 under fluorescein.

General references:

Aldrich FT-IR Library	**1**(2), 1010B
Sigma FT-IR Library	**1**(2), 147A
IR Library	**3**, 1473B
Safety	**2**(1), 1688C

FLUORESCEIN DIACETATE

Aldrich Cat. No.:	20,164-2
Sigma Prod. No.:	F7378
CAS No.:	[*596-09-8*]
Mol. Form.:	$C_{24}H_{16}O_7$
F.W.:	416.39
m.p.:	200-203°C
Appearance:	Light-yellow powder
λ_{max}:	490nm in ethanol
Solubility:	H_2O 0.1mg/ml
	EGME 10mg/ml
	EtOH 0.3mg/ml

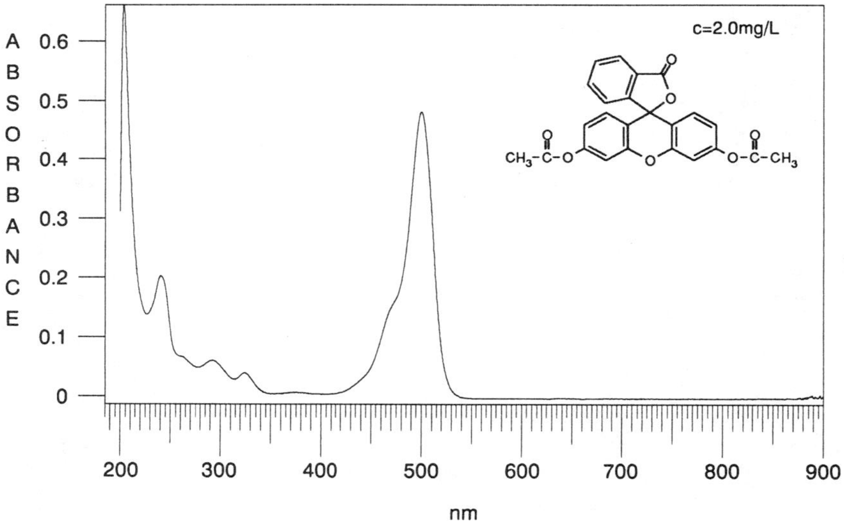

This fluorescent, yellow, solvent xanthene dye is made by the diesterification of fluorescein with acetic acid. It is soluble in methanol, acetic acid, hot acetone and alkali solutions and very slightly soluble in ethanol.

Kvach *et al.*[1] used it, along with ethidium bromide (Aldrich Cat. No. 16,053-9), for staining tissue-derived *Mycobacterium leprae*. Schnuerer and Rosswall[2] utilized the hydrolysis of fluorescein diacetate as a measure of total microbial activity in soil and litter.

References:

1) Kvach, J.T.; Munguia, G.; Strand, S.H. *Int. J. Lepr. Other Mycobact. Dis.* **1984,** *52*(2), 176-182.

2) Schnuerer, J.; Rosswall, T. *Appl. Environ. Microbiol.* **1982,** *43* (6), 1256-1261.

General references:

Aldrich FT-IR Library	**1**(2), 1010C
Sigma FT-IR Library	**1**(2), 147D
IR Library	**3**, 1473C

FLUORESCEIN ISOTHIOCYANATE, isomer I
FITC

Aldrich Cat. No.: F250-2
Sigma Prod. No.: F7250

CAS No.: [*3326-32-7*]
C.I. No.: 45350:1
Mol. Form.: $C_{21}H_{11}NO_5S$
F.W.: 389.39
m.p.: >360°C
Appearance: Orange powder
λ_{max}: 490nm in 0.1*N* sodium hydroxide
Solubility: H_2O <0.1mg/ml
EGME 9mg/ml
EtOH 20mg/ml

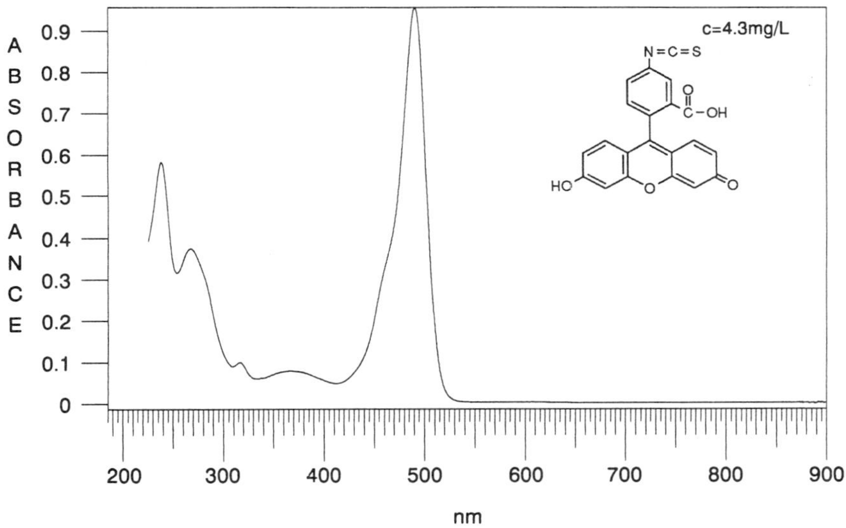

This fluorescent xanthene dye was introduced by Riggs and his coworkers for use as an antibody conjugate. Stable in solid form, it has largely replaced fluorescein isocyanate, which is so unstable it must be prepared immediately before it is used. FITC is insoluble in water and soluble in ethanol and acetone.

Biological applications include use as a fluorescent labelling reagent for proteins,[1-4] a fluorescent reagent for protein tracing,[5] and a reagent in a fluorescent antibody technique for the rapid identification of pathogens.[6,7]

Staining Procedures[8] describes the preparation of serum globulin/fluorescein isothiocyanate conjugates.

References:

1) Rinderknecht, H. *Nature* **1962,** *193*, 167.
2) Mann, K.G.; Fish, W.W. *Methods Enzymol.* **1972,** *26*, 28.
3) Sinsheimer, J.E.; Jagodic, V.; Burckhalter, J.H. *Anal. Biochem.* **1974,** *57*, 227.
4) Weber, K.; Pollack, R.; Bibring, T. *Proc. Natl. Acad. Sci. U.S.A.* **1975,** *72*, 459.
5) *Fluorescent Protein Tracing,* 3rd ed.; Nairn, R.C., Ed.; Williams and Wilkins: Baltimore, 1969.
6) Riggs, J.L.; Seoward, R.J.; Burckhalter, J.H.; Downs, C.M.; Metcalf, T.G. *Am. J. Pathol.* **1958,** *34*, 1081.
7) Cherry, W.B.; McKinney, R.M.; Emmel, V.M.; Spillane, J.T.; Hebert, G.A.; Pittman, B. *Stain Technol.* **1969,** *44*, 179.
8) Clark, G. *Staining Procedures,* 3rd ed.; Williams and Wilkins: Baltimore, 1973; p 185.

General references:

Aldrich FT-IR Library	**1**(2), 469C
Sigma FT-IR Library	**1**(2), 148B
IR Library	**3,** 1142D
NMR Library	**2**(2), 440C
Safety	**2**(1), 1688D

FLUORESCEINAMINE, isomer I

4-Aminofluorescein

Aldrich Cat. No.: 20,162-6
Sigma Prod. No.: F7000

CAS No.: [*3326-34-9*]
Mol. Form.: $C_{20}H_{13}NO_5$
F.W.: 347.33
m.p.: 223°C (dec.)
Appearance: Maroon powder
λ_{max}: 496nm in ethanol + 1ml 0.1*N* sodium hydroxide
Solubility: H_2O 1mg/ml
EGME 50mg/ml
EtOH 60mg/ml

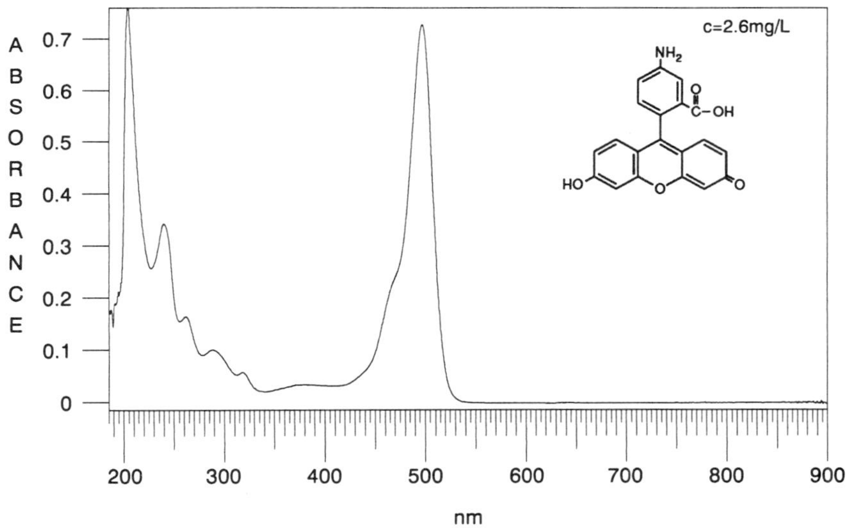

These aminofluorescein isomers are made by first condensing 4-nitrophthalic anhydride with two equivalents of resorcinol and converting the resulting isomeric mixture to the diacetate. The resulting mixed-nitro isomers are separated by fractional crystallization. Isomer I (4-nitro-) is isolated as pale-yellow crystals melting at 221-225.5°C, while isomer II (5-nitro-) is separated as pale-yellow, asbestos-like needles melting at 189-190°C. The amine isomers are made in turn from their isomeric nitrofluorescein diacetate counterparts by hydrogenation with Raney nickel followed by hydrolysis. Isomer I is a red-brown solid with a melting point of 223°C (dec.) that is soluble in methanol. Isomer II is also a red-brown solid, with a melting point of

285°C (dec.), that is soluble in methanol.

Fluorescein isothiocyanate (FITC, Aldrich Cat. No. F250-2), which is used widely as a fluorescent protein tag in immunofluorescent techniques, is made from 4-aminofluorescein diacetate by treating it with thiophosgene.[1]

Reference:

1) Coons, A.H.; Kaplan, M.H. *J. Exp. Med.* **1950,** *91*, 1-14.

General references:

Aldrich FT-IR Library	**1**(2), 1013B
Sigma FT-IR Library	**1**(2), 147B
IR Library	**3**, 1474H

FLUORESCEINAMINE, isomer II
5-Aminofluorescein

Aldrich Cat. No.:	20,163-4
Sigma Prod. No.:	F7125
CAS No.:	[*51649-83-3*]
Mol. Form.:	$C_{20}H_{13}NO_5$
F.W.:	347.33
m.p.:	285°C (dec.)
Appearance:	Orange powder
λ_{max}:	495nm in 0.1*N* sodium hydroxide
Solubility:	H_2O 10mg/ml
	EGME 100mg/ml
	EtOH 20mg/ml

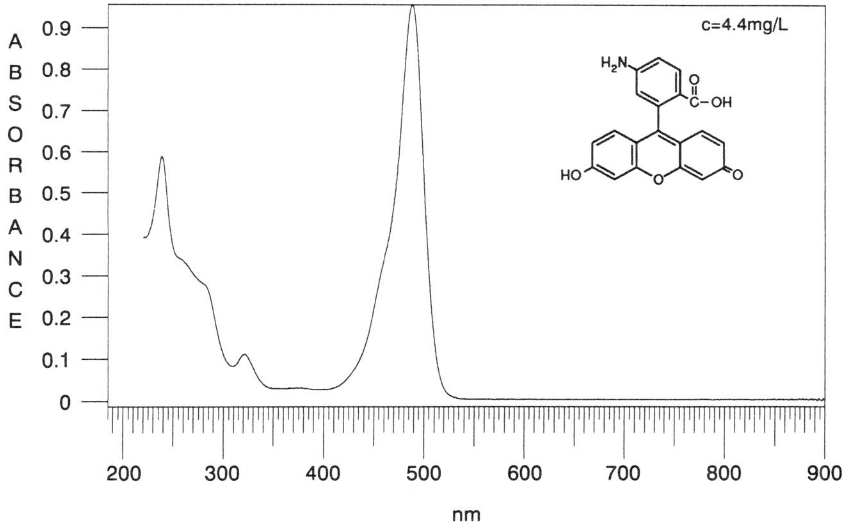

See description on pages 379 and 380 under fluoresceinamine, isomer I.

General references:

Aldrich FT-IR Library	**1**(2), 1013C
Sigma FT-IR Library	**1**(2), 147C
IR Library	**3**, 1475A

FLUORESCENT BRIGHTENER 28

Aldrich Cat. No.:	29,418-7
Sigma Prod. No.:	F6259
CAS No.:	*[4404-43-7]*
C.I. No.:	40622
Mol. Form.:	$C_{40}H_{42}N_{12}Na_2O_{10}S_2$
F.W.:	960.96
m.p.:	290°C (dec.)
Appearance:	Yellow powder
λ_{max}**:**	350nm in methanol
Solubility:	H_2O 80mg/ml
	EGME 9mg/ml
	EtOH 2mg/ml

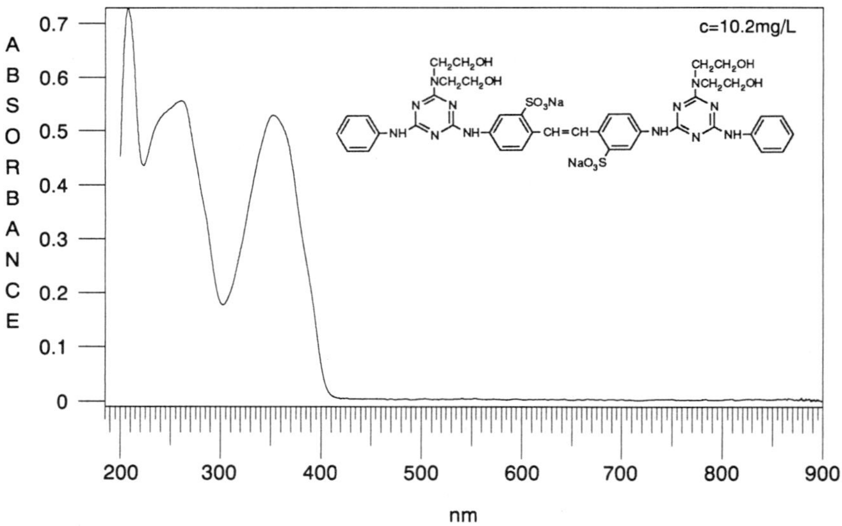

Fluorescent Brightener 28 (FB 28) is not a dye but a colorless organic compound that fluoresces (clear bluish) under ultraviolet radiation. It is made by condensing one mole of 4,4'-diamino-2,2'-stilbenedisulfonic acid with two moles of cyanuric acid and reacting the product, first with two moles of aniline followed by two moles of diethanolamine. The final product is soluble in water and slightly soluble in ethanol, and its fastness to light, chlorine and washing are good to very good. Industrially, it is used to make white cellulosic and silk textiles look whiter and is incorporated into detergents to accomplish the same end.

Apparently there are no specific citations in the literature regarding FB 28's use as a biological fluorochrome; however, it is related structurally to

Calcofluor White MR (Fluorescent Brightener 9) which was used by Jefferies and Belcher[1] and Maeda and Ischida[2] in plant tissue studies. Since Fluorescent Brightener 9 is no longer produced, FB 28 might prove to be a suitable substitute.

References:

1) Jefferies, C.J.; Belcher, A.R.A. *Stain Technol.* **1974,** *49*, 199-202.

2) Maeda, H.; Ischida, N. *J. Biochem.* **1967,** *62*, 276-278.

General references:

Safety **2**(1), 1690A
Colour Index 3rd ed., 1971

FLUOREXON

Bis[*N,N*-bis(carboxymethyl)aminomethyl]fluorescein • Calcein

Aldrich Cat. No.: 11,985-7
Sigma Prod. No.: C0875
Grade: Indicator

CAS No.: [*1461-15-0*]
Mol. Form.: $C_{30}H_{26}N_2O_{13}$
F.W.: 622.55
Appearance: Orange powder
λ_{max}: 499nm in 1*N* sodium hydroxide
Solubility: H_2O 4mg/ml
EGME 9mg/ml
EtOH 0.6mg/ml

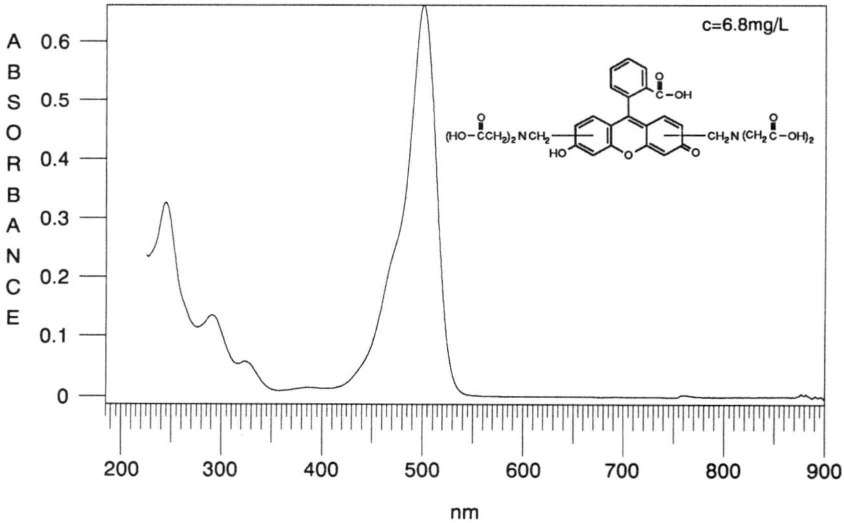

Fluorexon is not a dye but an acid fluorochrome that is made by introducing two methyliminodiacetic acid groups into a fluorescein molecule. It is composed of mixed isomers including the 2,7-disubstitution. Commercially, it is available as the disodium salt, which is soluble in water and ethanol. Diehl and Ellingboe[1] recognized its potential as a complexometric indicator and used it for the EDTA determination of calcium in the presence of magnesium. Wallach *et al.*[2] did further work in this area, separating and comparing the various isomers. They found the 2,4-isomer to be the best suited, of the isomers studied, for the foregoing determination.

Suzuki and Mathews[3] used it and tetracycline, intravitally, in a study demonstrating incremental bone growth. Calcium tagged with Fluorexon displayed a green fluorescence, while the calcium labeled with tetracycline presented a yellow fluorescence. Hahn von Dorsche and Opitz[4] compared it with 46 other dyes in a study designed to stain mast cells. Fluorexon was one of 23 stains that showed promise. Mast cells treated with this fluorochrome demonstrated a yellow fluorescence when subjected to ultraviolet or blue-violet light.

References:

1) Diehl, H.; Ellingboe, J.L. *Anal. Chem.* **1956,** *28*, 882.

2) Wallach, D.F.H.; Surgenor, D.M.; Soderberg, J.; Delano, E. *ibid.* **1966,** *31*, 456-466.

3) Suzuki, H.K.; Mathews, A. *Stain Technol.* **1966,** *41*, 57-60.

4) Hahn von Dorsche, H.; Opitz, M. *Acta Histochem.* **1970,** *36*, 74-86.

General references:

Aldrich FT-IR Library **1**(2), 1015A
Sigma FT-IR Library **1**(2), 112D
IR Library **3**, 1475G

GALLOCYANINE

1-Carboxy-7-(dimethylamino)-3,4-dihydroxyphenoxazin-5-ium chloride • Mordant Blue 10

Aldrich Cat. No.:	12,450-8
Sigma Prod. No.:	G8884
CAS No.:	[*1562-85-2*]
C.I. No.:	51030
Mol. Form.:	$C_{15}H_{13}ClN_2O_5$
F.W.:	336.73
Appearance:	Black powder
λ_{max}:	620nm in ethanol
Solubility:	H_2O 10mg/ml
	EGME 2mg/ml
	EtOH 0.7mg/ml

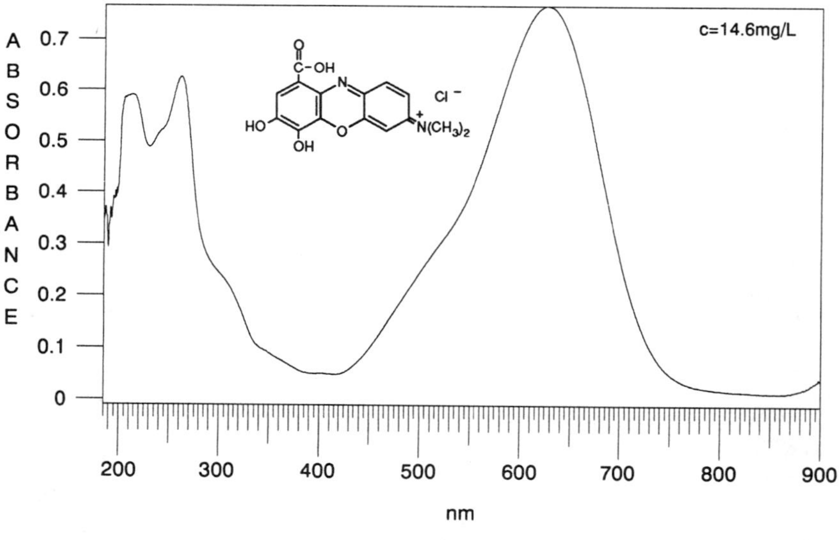

Gallocyanine is a blue oxazine dye that is made by condensing 4-nitrosodimethylaniline hydrochloride with gallic acid in boiling methanol. Mordant Blue 10 is used to dye wool, acetate and cellulose. Cotton is dyed utilizing a chromium acetate pad. Wool, cotton and silk are printed using a chromium mordant. *The Merck Index*[1] lists Gallocyanine as a reagent for lead with which it is reported to form a "deep-violet color."

The chloride-salt form is available for use as a stain in biological applications. This salt is practically insoluble in cold water and slightly soluble in hot water.

Gallocyanine was considered by Proescher and Arkush[2] to be a fine nuclear stain. Cole[3] used Gallocyanine as a nuclear stain for frozen sections. He suggested that, in specialized cases, it could be substituted for Hematoxylin where the latter is difficult to use. Clark and Powers,[4] in an evaluation of a series of aniline dyes for use as nuclear stains in trichrome staining procedures, found Gallocyanine gave excellent results. *Staining Procedures*[5] provides a method for using it as a Hematoxylin substitute in Van Gieson-type staining applications resulting in the blue-black staining of nuclei, Nissl granules and other sites containing cytoplasmic RNA. Allison,[6] in citing the need for a standardizable synthetic substitute for Hematoxylin in automated cervical-smear screening, stated "Gallocyanin is under intense scrutiny for this role."

References:

1) *The Merck Index,* 11th ed.; Budavari, S. *et al.*, Eds.; Merck: Rahway, 1989; Monograph 4256.

2) Proescher, F.; Arkush, A.S. *Stain Technol.* **1928,** *3*, 28.

3) Cole, W.V. *ibid.* **1947,** *22,* 5-7, 103-106.

4) Clark, G.; Powers, M.M. *ibid.* **1953,** *28*, 93-96.

5) Clark, G. *Staining Procedures,* 4th ed.; Williams and Wilkins: Baltimore, 1981; p 109.

6) Allison, R.T. *MLW* **1986,** 29-30.

General references:

Aldrich FT-IR Library	**1**(2), 888A
Sigma FT-IR Library	**1**(2), 149B
IR Library	**3,** 1410A
Colour Index	3rd ed., 1971

GIEMSA STAIN

Aldrich Cat. No.: 12,139-8
Sigma Prod. No.: G5637
Grade: Certified

CAS No.: [*51811-82-6*]
Appearance: Greenish-black powder
λ_{max}: 648(521)nm in methanol
Solubility: H_2O 7mg/ml
EGME 30mg/ml
EtOH 5mg/ml

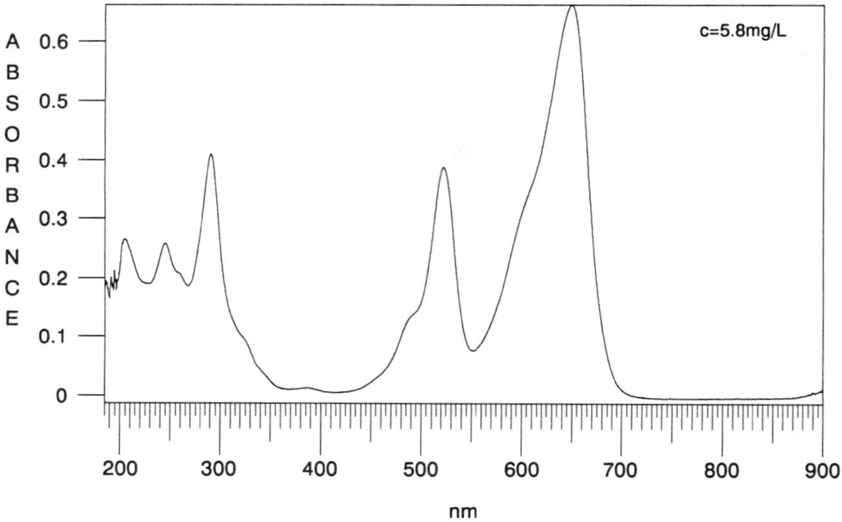

Giemsa Stain is a Romanowsky-type stain that was first compounded and used by Giemsa[1] in hematology and protozoology. He found it especially suited for staining *Plasmodium malariae*.

The original formulation was a closely held trade secret that has never been publicly disclosed. Apparently, it was sold to Grübler and passed on to his successor, Hollborn. Until World War II, these companies were considered by most investigators, worldwide, to be the only sources of reliable Giemsa Stain powder. With supplies from Germany cut off, the Allies had to find another quality source of Giemsa Stain, since numerous troops were operating in areas where malaria was highly endemic. In the United States, investigators at the U.S. Public Health Service[2] were assigned the task, and, through their efforts, the secret ingredient Azur I, claimed by Giemsa to be the critical component in the stain bearing his name, was concluded to be Azure B. Giemsa's[1] observation regarding the importance of Azur I (Azure B) in his stain was reinforced by Lillie[3] and Wittekind *et al.* (see also p 115).[4]

Simply described, Giemsa Stain is a blend of Eosin Y, Methylene Blue and oxidation products of the latter, including the methylene azures. Unfortunately the preparation is not that straightforward. Formulations may vary from supplier to supplier; and, in the past, individual manufacturers have experienced problems maintaining batch-to-batch consistency due to the difficulty involved in precisely controlling all of the many process variables. Even with precise spectrophotometric and HPLC monitoring, subtle solubility differences are observed from batch to batch, and these differences are often reflected in a reduced stock-solution shelf life. Equally unfortunately, the problems do not cease after the dry powder has been manufactured. Only too often, solutions are prepared by people with no formal laboratory training and little, if any, understanding of the staining process involved. Over the years, we have become aware of workers attempting to fortify purchased formulations, without knowing their composition, with their own "magic" ingredient. Erroneously, Giemsa Stain has been used to "fortify" other Romanowsky-type stains, such as Wright stain, apparently because in its classic application, the Giemsa Stain was observed to stain blood films more intensely than the formula it was used to fortify. This leads one to wonder if any consideration was given to differences in the classic staining protocols of these differently named but very similarly formulated powders. In our hands, for example, properly prepared and optimally color-balanced Wright Stain and Giemsa Stain display much the same staining quality, when examined side by side in the same application, utilizing the same staining protocol.

Giemsa Stain continues to be an important differential stain for smears taken from both peripheral blood and from bone marrow. In addition, it is used to study trypsan-induced banding of human chromosomes and, on a more limited scale, for staining cervical smears. *Staining Procedures*[5] lists methods and a bibliography for the following applications: C-banding of chromosomes; Giemsa Stain for thin films; Giemsa Stain for thick films; Wolbach's Giemsa Colophonium Stain; accelerated Giemsa Stain for thick films; Giemsa Stain on saponin-lysed thick films; Giemsa Stain for wet films or sections; Robinow Stain for bacterial nuclear structures; Klienberger-Nobel agar-fixation stain for colonies of *Mycoplasma* and L-phase variants; demonstration of malaria organisms, *Leishmania*, *Trypanosoma,* microfilariare, in blood smears; Goodburn and Marmion Stain for *Mycoplasma* in frozen sections.

References:

1) Giemsa, G. *Zentrabl. Bakteriol.* **1902,** *31,* 429-430.
2) Roe, M.A.; Lillie, R.D.; Wilcox, A. *Public Health Rep.* **1940,** *55*, 1272-1278.
3) Lillie, R.D. *ibid.* **1943,** *58*, 304-317.
4) Wittekind, D.; Kretschmer, V.; Lohr, W. *Blut* **1976,** *32*, 71-78.
5) Clark, G. *Staining Procedures,* 4th ed.; Williams and Wilkins: Baltimore, 1981; pp 173-174, 270, 305-307, 352, 398, 418, 439, 465-466.

General references:

Aldrich FT-IR Library	**1**(2), 1037C
Sigma FT-IR Library	**1**(2), 149D
IR Library	**3**, 1486F
Safety	**2**(1), 1769A

GUINEA GREEN B
Acid Green 3

Aldrich Cat. No.: 20,772-1

CAS No.: [*4680-78-8*]
C.I. No.: 42085
Mol. Form.: $C_{37}H_{35}N_2NaO_6S_2$
F.W.: 690.82
m.p.: 255°C (dec.)
Appearance: Maroon-purple powder
λ_{max}: 618nm in water
Solubility: H_2O 30mg/ml
EGME 20mg/ml
EtOH 9mg/ml

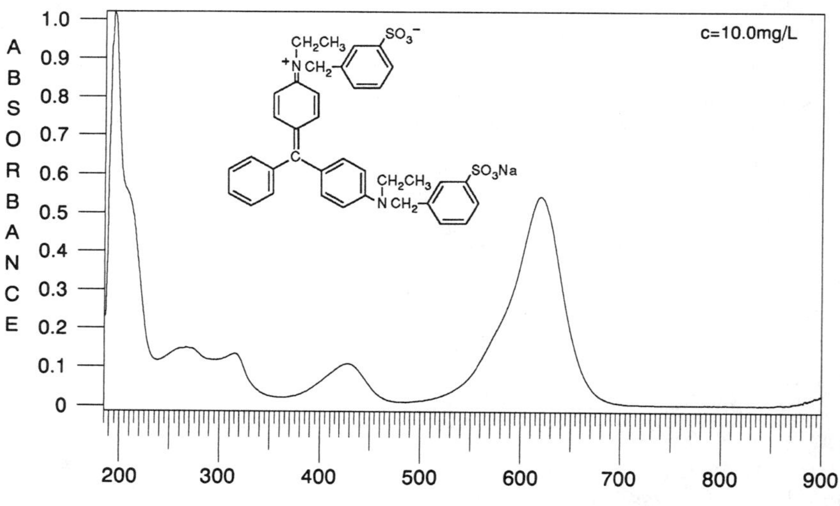

Guinea Green B is a bright-green, anionic aminotriphenylmethane dye which is a less sulfonated variation of Light Green SF. It is soluble in water, methyl Cellosolve and ethylene glycol and slightly soluble in ethanol. It is prepared by condensing one mole of benzaldehyde with two moles of α-(*N*-ethylanilino)-3-toluenesulfonic acid, oxidizing the reaction product and converting it to the sodium salt.

Because of its low lightfastness (2), it has limited application both in dyeing and printing wool and silk (from an acid bath). Although it dyes fully chromed leather a brilliant green, this application, too, is flawed by a low lightfastness (1). Other industrial applications include use in barium and aluminum lakes that are used as pigments, in wood stain and in paper dyes.

Lillie[1] used Guinea Green, with moderately good results, as a fiber and plasma stain in variations of Masson's trichrome procedure.

The 1,3-di-*o*-tolylguanidine salt of Guinea Green (Luxol Brilliant Green BL, C.I. Solvent Green 11, Aldrich Cat. No. 27,726-6) is used as a stain to demonstrate virus inclusions in plant-pathology studies.[2]

References:

1) Lillie, R.D. *J. Technol. Methods* **1945,** *25*, 1-47.

2) Christie, R.G. *Virology* **1967,** *31*, 268-271.

General references:

Aldrich FT-IR Library	**1**(2), 1034B
IR Library	**3**, 1485F
Safety	**2**(1), 1805C
Merck Index	**11**, 4487
Colour Index	3rd ed., 1971

HABA
2-(4-Hydroxyphenylazo)benzoic acid

Aldrich Cat. No.:	14,803-2
Sigma Prod. No.:	H5126
CAS No.:	[*1634-82-8*]
Mol. Form.:	$C_{13}H_{10}N_2O_3$
F.W.:	242.23
m.p.:	205-207°C
Appearance:	Dark-orange powder
λ_{max}:	348nm in 1% phosphate buffer, pH 6.2
Solubility:	H_2O 0.2mg/ml
	EGME 200mg/ml
	EtOH 30mg/ml

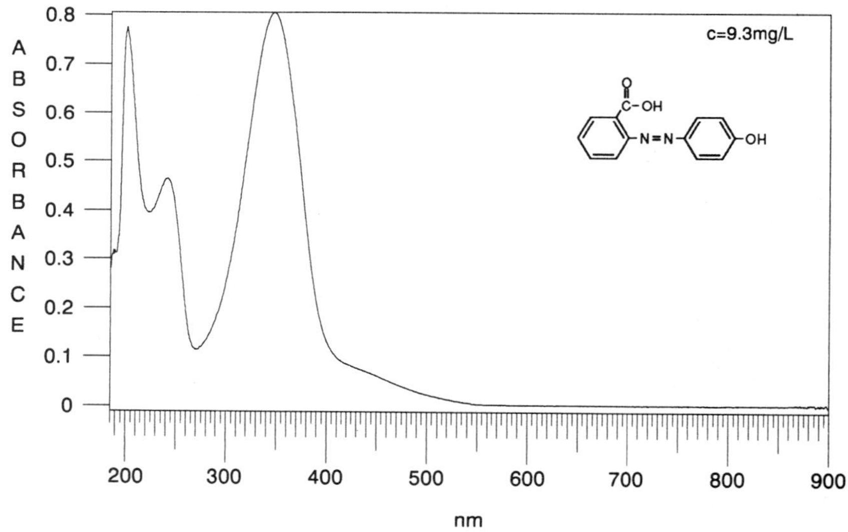

This yellow, anionic monoazo dye is virtually insoluble in water and soluble in ethanol and dimethylformamide.

HABA is prepared by coupling diazotized anthranilic acid to phenol. It is not included in the *Colour Index* and apparently is not used in industrial applications.

It is a valuable reagent for the determination of albumin in serum, being used both in manual kits and with automated systems for this test.[1-4] This reagent was also utilized in a study investigating the binding of L-thyroxine to serum albumins.[5]

References:

1) Rutstein, D.D.; Ingenito, E.F.; Reynolds, W.E.; Burke, J.M. *J. Clin. Invest.* **1954,** *33*, 211.

2) Ness, A.T.; Dickerson, H.C.; Pastewka, J.V. *Clin. Chim. Acta* **1965,** *12*, 532.

3) Martinek, R.G. *Clin. Chem.* **1965,** *11*, 441.

4) Porter, E.G.; Waters, W.J. *J. Lab. Clin. Med.* **1966,** *67*, 660.

5) Nazareth, R.I.; Sokoloski, T.D.; Witiak, D.T.; Hopper, A.T. *J. Pharm. Sci.* **1974,** *63*, 199.

General references:

Aldrich FT-IR Library	**1**(2), 971B
Sigma FT-IR Library	**1**(2), 151A
IR Library	**3**, 1454E
NMR Library	**2**(2), 980A

HEMATOXYLIN

Natural Black 1

Aldrich Cat. No.:	86,118-9	85,842-0
Sigma Prod. No.:	H3136	H9627
Grade:	Certified	Indicator, trihydrate
CAS No.:	[*517-28-2* (anhydrous)]	
C.I. No.:	75290	
Mol. Form.:	$C_{16}H_{14}O_6$	$C_{16}H_{14}O_6 \cdot 3H_2O$
F.W.:	302.29	356.33
m.p.:	200°C (dec.)	229-231°C (dec.)
Appearance:	Tan powder	
λ_{max}:	292nm in methanol	
Solubility:	H_2O 30mg/ml	——
	EGME 50mg/ml	——
	EtOH 30mg/ml	20mg/ml

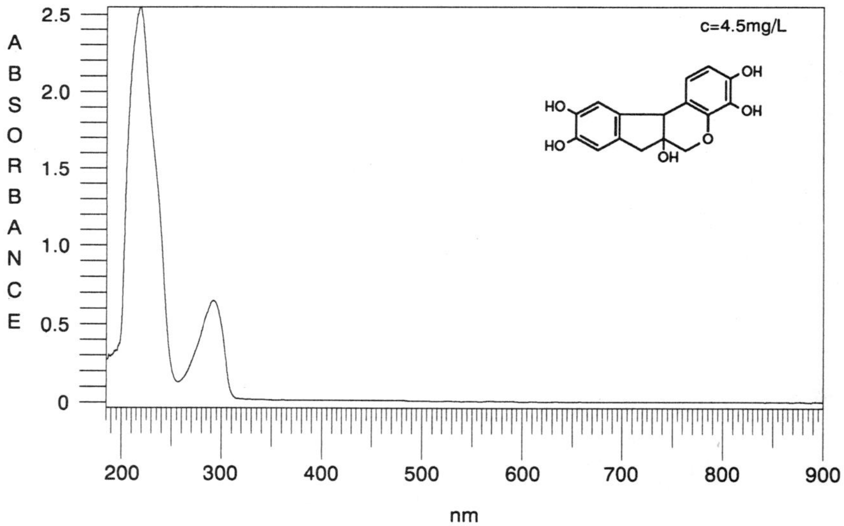

Hematoxylin is a natural leuco compound (extracted from chipped logwood, *Haematoxylon campechianum*, indigenous to Central America) that is rapidly converted to the dye, Hematein (which contains a paraquinoidal chromophore), by oxidation. Even exposure to atmospheric oxygen in the presence of ultraviolet light will elicit the formation of Hematein and cause solid Hematoxylin, which in a pure state is white to pure yellow, to darken. In fact, many of the color-related properties attributed to Hematoxylin are due to the presence of Hematein as an impurity. Notably, the acid-base indicator properties listed by Long[1] for Hematoxylin [visual-transition intervals pH 0.0 (red) to pH 1.0 (yellow) and pH 5.0 (pale yellow) to pH 6.0 (violet)] are not observed when fresh test solutions are prepared with pure Hematoxylin. Ideally, a pure form of solid Hematein could shortcut the preparation of many of the Hematoxylin-stain-solution recipes used in cytology and histology. However, the oxidative process does not stop cleanly after Hematein has been formed. Less desirable oxidation products are formed that are insoluble in water and have little tinctorial strength. Isohematein, which was first prepared by Thorpe in 1922 and evaluated as a biological stain by Cole,[2] may be one of the more desirable ones.

Hematoxylin is soluble in water, ethanol, methyl Cellosolve and ethylene glycol. Industrial use has declined in recent years; however, it is still used to dye textiles (particularly woolens) and furs and to prepare inks. It forms chelates with various metals and different colors can be developed (reddish-violet, blues to black) depending on the metal used as a mordant. This property is used to advantage when Hematoxylin is used as a reagent for the detection and determination of various metals in fluids, extracts and tissue. Dalal[3] describes a colorimetric method for the determination of aluminum in soil extracts.

Hematoxylin is an excellent nuclear stain that is used widely in cytology and histology, two of the most important uses being as a nuclear stain in the Papanicolaou procedure for the diagnosis of vaginal cancer and in combination with the contrast stain Eosin (H & E stain). Several thousand pounds of this dye are used worldwide each year in biological-stain applications, ranking it as the leader, or among the leaders, in volume sold for this use. *Staining Procedures*[4] provides in-depth information regarding methods and recipes for the PAP smear, H & E stain and a number of other important uses.

References:

1) Long, C. *Biochemist's Handbook;* Van Nostrand: Princeton, 1961.
2) Cole, E.C. *Stain Technol.* **1931,** *6*, 93-96.
3) Dalal, R.C. *Plant Soil* **1972,** *36*, 223.
4) Clark, G. *Staining Procedures,* 4th ed.; Williams and Wilkins: Baltimore, 1981; pp 36, 208.

General references:

Aldrich FT-IR Library	**1**(2), 954D	——
Sigma FT-IR Library	**1**(2), 150C	**1**(2), 150B
IR Library	**3,** 1444H	——
Safety	**2**(1), 1812B	
Merck Index	**11,** 4556	
Colour Index	3rd ed., 1971	

HOECHST 33258

Bisbenzimide • 2'-(4-Hydroxyphenyl)-5-(4-methyl-1-piperazinyl)-2,5'-bi-1*H*-benzimidazole trihydrochloride pentahydrate

Aldrich Cat. No.: 86,140-5
Sigma Prod. No.: B2883

CAS No.: [*23491-44-3* (anhydrous, free base)]
Mol. Form.: $C_{25}H_{24}N_6O \cdot 3HCl \cdot 5H_2O$
F.W.: 623.97
m.p.: >300°C
Appearance: Yellowish-green powder
λ_{max}: 343nm in methanol

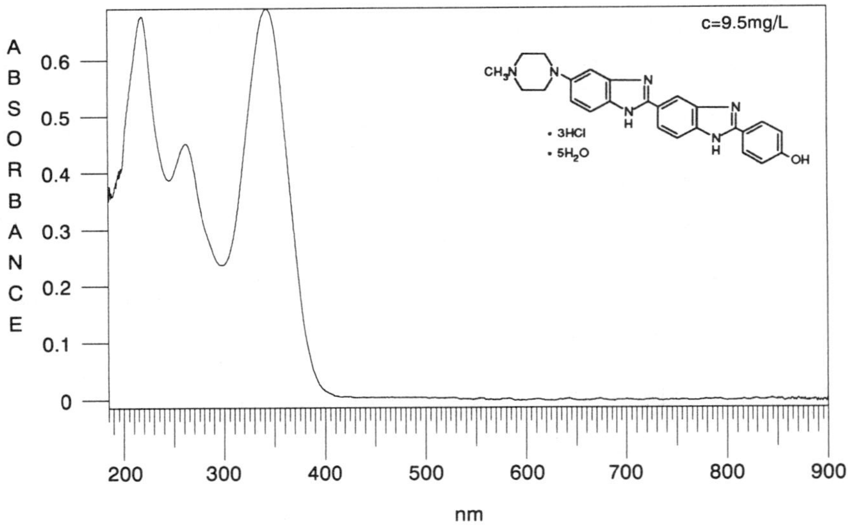

This cationic fluorochrome is completely soluble in water up to 2% (pH of the solution, 1.7). However, precipitation will occur when dissolution is attempted in phosphate-buffer solutions. If kept in the dark, aqueous solutions of up to one milligram per milliliter concentration are stable for up to two weeks.

This fluorochrome is generally nontoxic to cells, and the combination of dye plus light is also nontoxic to cells unless the cells have incorporated 5-bromodeoxyuridine. Hoechst 33258 has been found to enhance the photosensitivity of proliferating cells which had incorporated 5-bromodeoxyuridine into their DNA.[1] The modification increases the effectiveness of the method by at least 50 times. In addition, it is a DNA-binding fluorochrome that is useful for staining chromosomes.[2-5]

There are no published industrial applications.

References:

1) Maryanski, J.L.; Cerottini, J.-C.; Brunner, K.T. *J. Immunol.* **1980,** *124*, 839.

2) Hilwig, I.; Gropp, A. *Exp. Cell Res.* **1972,** *75*, 122.

3) *Idem ibid.* **1973,** *81*, 474.

4) Jalal, S.M.; Markvong, A.; Hsu, T.C. *ibid.* **1975,** *90*, 443.

5) Binder, M.; Kim, M.A. *Exp. Pathol.* **1975,** *10*, 220.

General references:

Sigma FT-IR Library **1**(2), 105C
Safety **2**(1), 1880A

3-HYDROXY-4-(2-HYDROXY-4-SULFO-1-NAPHTHYLAZO)-2-NAPHTHALENECARBOXYLIC ACID

Calconcarboxylic acid

Aldrich Cat. No.: 13,079-6
Sigma Prod. No.: H3128
Grade: Indicator

CAS No.: [*3737-95-9*]
Mol. Form.: $C_{21}H_{14}N_2O_7S$
F.W.: 438.41
m.p.: 300°C
Appearance: Very dark violet powder
λ_{max}: 570nm in methanol
Solubility: H_2O 8mg/ml
EGME 8mg/ml
EtOH 10mg/ml

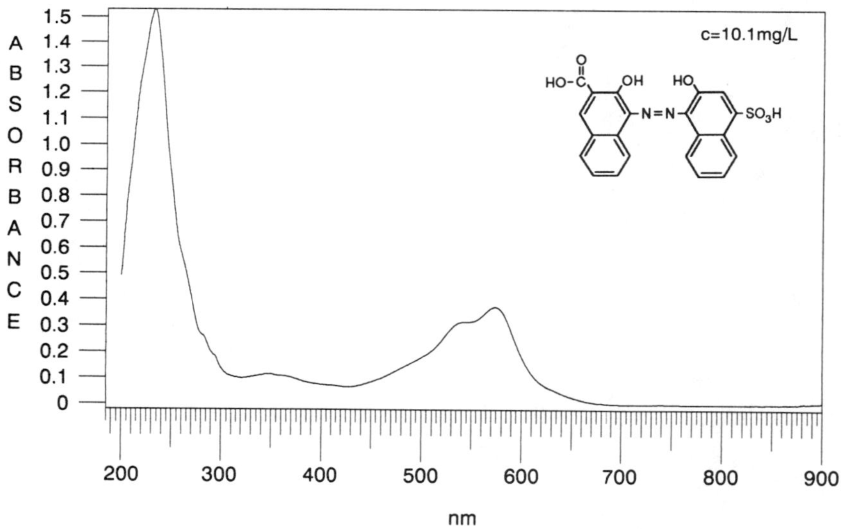

This somewhat water-soluble, red, acid monoazo dye is not listed in the *Colour Index* and apparently has not been of consequence industrially. However, it forms a complex with calcium and has been used to determine this ion in the presence of magnesium with ethylenediaminetetraacetic acid (EDTA).[1]

It is prepared by coupling diazotized 1-2-4 acid (1-amino-2-naphthol-4-sulfonic acid) to 3-hydroxy-2-naphthoic acid.

Reference:

1) Patton, J.; Reeder, W. *Anal. Chem.* **1956,** *28*, 1026.

General references:

Aldrich FT-IR Library **1**(2), 988A
Sigma FT-IR Library **1**(2), 151B

7-HYDROXY-4-METHYLCOUMARIN

4-Methylumbelliferone

Aldrich Cat. No.:	12,872-4
Sigma Prod. No.:	M1381
Purity:	97%
CAS No.:	[*90-33-5*]
Mol. Form.:	$C_{10}H_8O_3$
F.W.:	176.17
m.p.:	190-192°C
Appearance:	White powder
λ_{max}:	322nm in methanol
Solubility:	H_2O 0.1mg/ml
	EGME 40mg/ml
	EtOH 10mg/ml

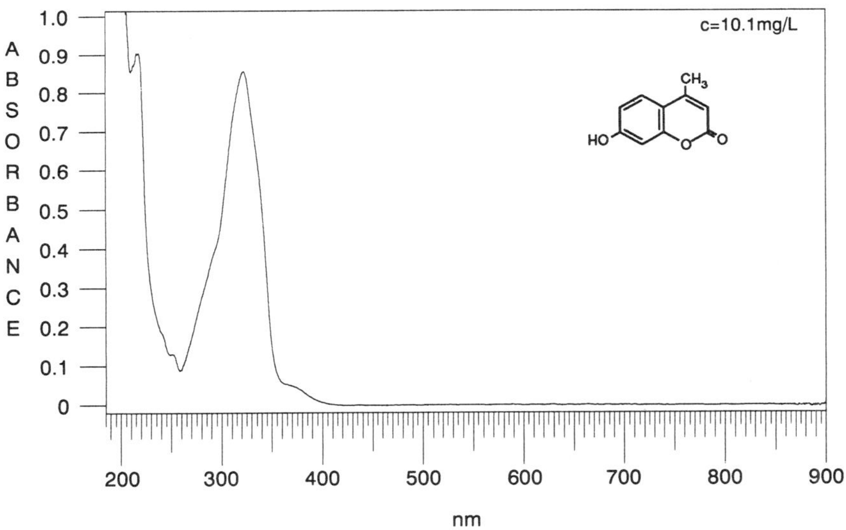

This off-white, fluorescent coumarin derivative (HMC) is practically insoluble in water and slightly soluble in ethanol. A fluorescent indicator with two ranges, HMC has a fluorescent-transition interval of pH 0.0 (green) to pH 2.0 (weak blue) and a fluorescent-transition interval of pH 6.5 (weak blue) to pH 8.0 (blue).

It is used as a standard for the fluorometric determination of enzymes active in biological systems. Thus, it is important in studies involving various metabolic diseases which are manifested by the absence of specific glycosidases responsible for the degradation of glycoproteins or glycolipids. Since test specimens are often very limited, a sensitive assay system is required.

This requirement is met by high-sensitivity, fluorescent-measurement techniques which treat the specimen with the appropriate glycosidase releasing HMC, which is quantified by comparing with a reference standard.[1-3]

High-purity HMC (essentially free of impurities as determined by HPLC) is used as a laser dye for pulsed and CW operation. Absorption λ_{max} (in ethanol) is 372nm, while the optimum lasing wavelength is 456nm. The tunable range is 430-490nm. Fluorescence λ_{max} is 445nm in basic ethanol.[4, 5]

References:

1) Brewster, M.A.; Whaley, S.A.; Kane, A.C. *Clin. Chem.* **1974,** *20*, 383-386.
2) Johnson, D.L.; Del Monte, M.A.; Cotlier, E.; Desnick, R.J. *Clin. Chim. Acta* **1975,** *63*, 81-90.
3) Beyhyl, F.E.; Sinsharay, A. In *Microsomes, Drug Oxid., Chem. Carcinogens;* Int. Symp. Microsomes Drug Oxid., 4th ed.; Coon, M.J.; Conney, A.H.; Estabrook, R.W., Eds.; Academic: New York, 1980; Vol. 1, p 111; Chem. Abstr. **1980,** *93*, 127810q.
4) Tuccio, S.A.; Drexhage, K.H.; Reynolds, G.A. *Opt. Commun.* **1973,** *7* (3), 248.
5) Konjevic, R.; Konjevic, N. *ibid* . **1977,** *23*, 187.

General references:

Aldrich FT-IR Library	**1**(2), 325C
Sigma FT-IR Library	**1**(2), 167A
IR Library	**3**, 1045C
Safety	**2**(1), 1937C
NMR Library	**2**(2), 309D
Merck Index	**11,** 4792

HYDROXY NAPHTHOL BLUE, disodium salt

3-Hydroxy-4-(2-hydroxy-4-sulfo-1-naphthylazo)-2,7-naphthalene-disulfonic acid, disodium salt

Aldrich Cat. No.:	21,991-6
Sigma Prod. No.:	H2516
Grade:	A.C.S. reagent

Mol. Form.:	$C_{20}H_{12}N_2Na_2O_{11}S_3$
F.W.:	598.50
Appearance:	Dark-blue crystals
λ_{max}:	650nm in methanol +10ml water
Solubility:	H_2O 100mg/ml
	EGME 7mg/ml
	EtOH 0.7mg/ml

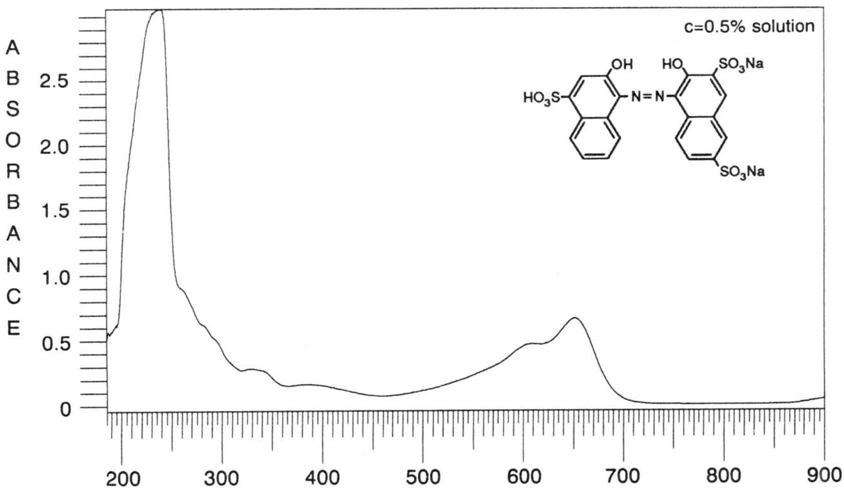

Hydroxy Naphthol Blue is a very water-soluble, anionic monoazo dye that is prepared by coupling diazotized 4-amino-3-hydroxy-1-naphthalenesulfonic acid to R acid (2-naphthol-3,6-disulfonic acid, disodium salt).

Deposited on sodium chloride crystals, it is used as an analytical reagent for determining calcium.[1]

There are no reported industrial applications.

Reference:

1) *Reagent Chemicals,* 7th ed.; American Chemical Society: Washington, DC, 1986; p 334.

General references:

Sigma FT-IR Library	**1**(2), 151C
Safety	**2**(1), 1943B

8-HYDROXYQUINOLINE-5-SULFONIC ACID monohydrate

Aldrich Cat. No.:	H5,875-7
Sigma Prod. No.:	H1129
Purity:	98%

CAS No.:	[*84-88-8* (anhydrous)]
Mol. Form.:	$C_9H_7NO_4S \cdot H_2O$
F.W.:	243.24
m.p.:	>300°C
Appearance:	Yellow powder
λ_{max}:	315nm in methanol
Solubility:	H_2O 0.6mg/ml
	EGME 0.3mg/ml
	EtOH 0.1mg/ml

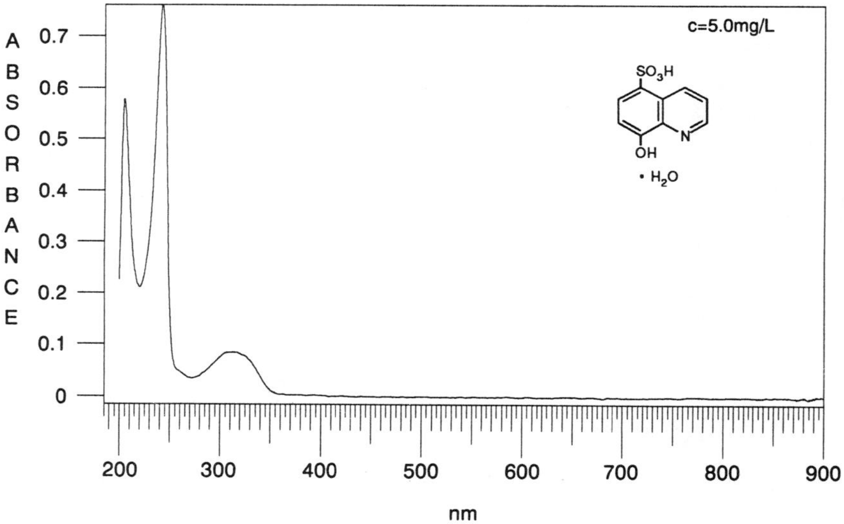

This chromogenic reagent is used for the spectrophotometric determination of manganese.[1] It is soluble in boiling hot water and virtually insoluble in ethanol.

Reference:

1) Pal, B.K.; Ryan, D.E. *Anal. Chim. Acta* **1969,** *47*, 35.

General references:

Aldrich FT-IR Library	**1**(2), 871D
Sigma FT-IR Library	**1**(2), 152C
IR Library	**3**, 1398A
NMR Library	**2**(2), 748A
Safety	**2**(1), 1961D
Merck Index	**11**, 4780

INDIGO

Indigo Blue • Vat Blue 1

Aldrich Cat. No.: 22,929-6
Sigma Prod. No.: I-1881
Grade: Synthetic

CAS No.: [*482-89-3*]
C.I. No.: 73000
Mol. Form.: $C_{16}H_{10}N_2O_2$
F.W.: 262.27
Appearance: Violet powder
λ_{max}: 602nm in chloroform
Solubility: H_2O 0.2mg/ml
EGME 0.4mg/ml
EtOH 0.3mg/ml

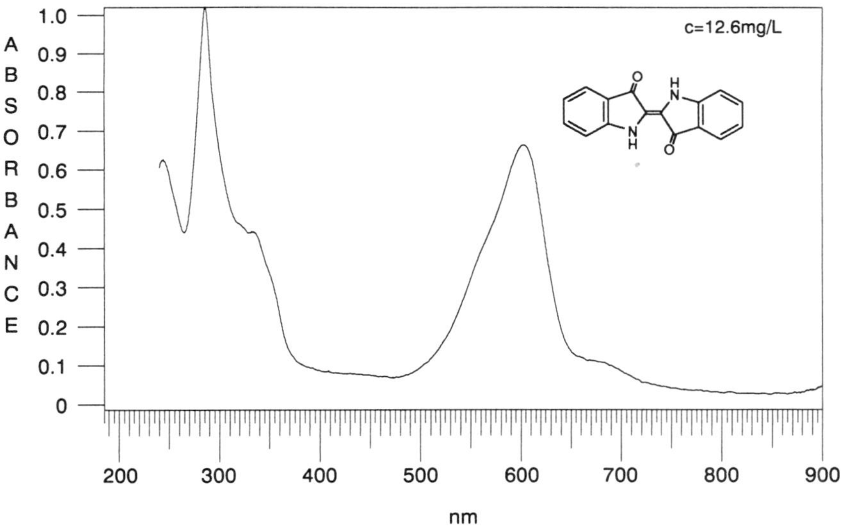

This indigoid dye is, most likely, one of the first dyes known to man. It was formerly isolated from various plants belonging to the genus *Indigofera*, where it is found as a colorless glucoside, indican.

The current product of commerce is synthetic. Lubs[1] reported that the most widely used process utilized *N*-phenylglycine treated with sodium amide to yield indoxyl, which then was oxidized to yield Indigo. This process can be carried out at relatively low temperatures with good yields.

Presently, practically all of the Indigo produced is used in the textile industry to dye blue jeans and other blue denim products.

There is no evidence in the literature that it has been used as a biological stain.

Reference:

1) *The Chemistry of Synthetic Dyes and Pigments;* Lubs, H.A., Ed.; A.C.S. Monograph Series; Krieger: Malabar, 1955; p 555.

General references:

Aldrich FT-IR Library	**1**(2), 394B
Sigma FT-IR Library	**1**(2), 153A
Safety	**2**(1), 1979B
Merck Index	**11,** 4855
Colour Index	3rd ed., 1971

INDIGO CARMINE

Acid Blue 74

Aldrich Cat. No.:	13,116-4
Grade:	Certified
CAS No.:	*[860-22-0]*
C.I. No.:	73015
Mol. Form.:	$C_{16}H_8N_2Na_2O_8S_2$
F.W.:	466.36
Appearance:	Purple powder
λ_{max}:	608nm in water
Solubility:	H_2O 3mg/ml
	EGME 9mg/ml
	EtOH 3mg/ml

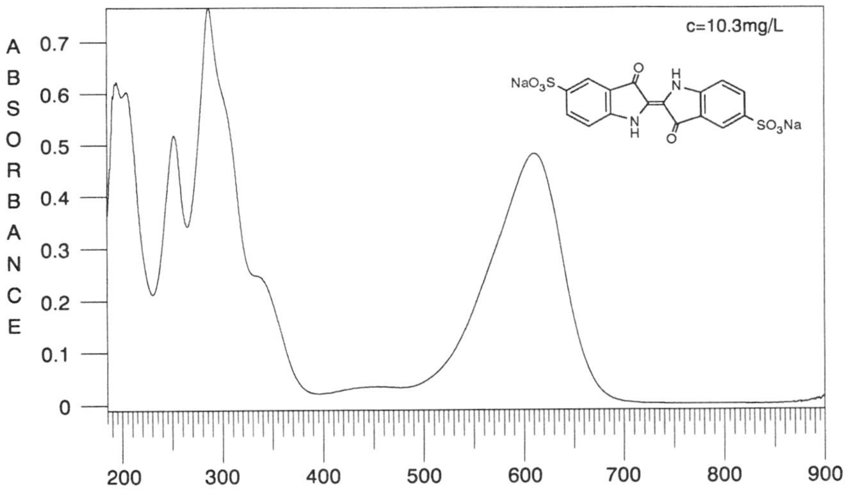

Indigo Carmine is a blue, anionic indigoid dye that is slightly soluble in water and the lower alcohols. The *Colour Index* distinguishes between the product herein described obtained by the sulfonation of synthetic Indigo (which is listed as Acid Blue 74, C.I. 73015) and the one obtained from natural Indigo by sulfonation. The latter, with the same structure as C.I. 73015, is listed as C.I. Natural Blue 2, C.I. 75781. It has indicator properties with a visual-transition interval of pH 11.5 (blue) to pH 14 (yellow).

Indigo Carmine is of little value as a textile dye because it is too fugitive. However, the *Colour Index* indicates that it has been used to dye wool directly and silk from either a formic acid or sulfuric acid bath, and that wool and nylon are printed directly.

Indigo Carmine has been used as a contrast stain for plasma when red nuclear stains were used and also has been used on occasion with Hematoxylin. Lillie[1] reported good results when it was used in a Van Gieson-type procedure; he also observed its inability to stain collagen in the Masson-trichrome method. *Staining Procedures*[2] describes a trichrome method with Basic Fuchsin and picric acid for demonstrating "nuclei and basic forms in whole mounts, protozoans and normal and altered tissues in sections of infected hosts."

References:

1) Lillie, R.D. *J. Technol. Methods* **1945,** *25*, 1-47.

2) Clark, G. *Staining Procedures,* 4th ed.; Williams and Wilkins: Baltimore, 1981; p 300.

General references:

Aldrich FT-IR Library	**1**(2), 1040C
IR Library	**3**, 1487H
Merck Index	**11**, 4856
Colour Index	3rd ed., 1971

INDOCYANINE GREEN

Cardio-Green • Fox Green

Aldrich Cat. No.: 22,886-9
Sigma Prod. No.: I-2633

CAS No.: [*3599-32-4*]
Mol. Form.: $C_{43}H_{47}N_2NaO_6S_2$
F.W.: 774.98
Appearance: Green powder
λ_{max}: 775nm in water

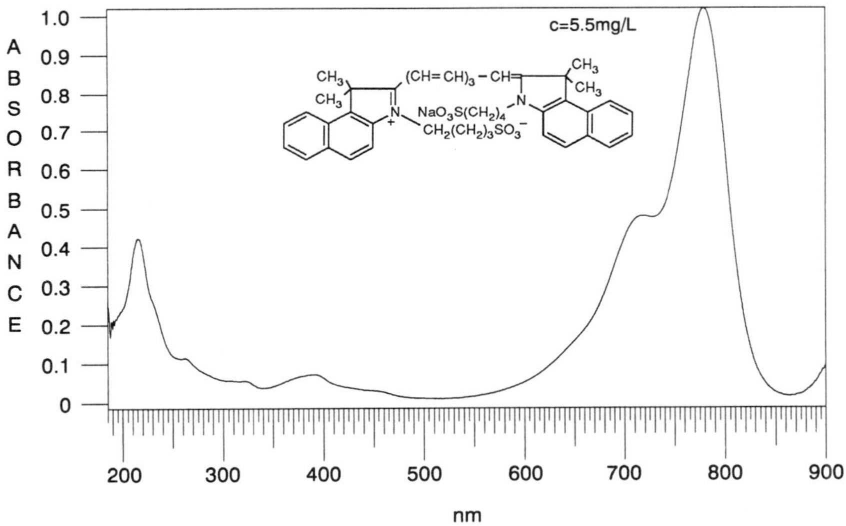

Indocyanine Green is a tricarbocyanine dye that absorbs very little below 700nm and peaks in the near infrared region at ~800nm. Heseltine and Brooker[1] describe a preparative method.

The dye is soluble in water and less soluble in ethanol. Solutions, however, are stable only for a few hours and cannot be kept overnight, greatly limiting the dye's utility.

Highly purified material meeting *USP XX*[2] standards is used as a diagnostic aid for blood-volume determinations, and cardiac- and liver-function tests.

Cost, lack of stability and limited absorbance in the visible region practically preclude the use of Indocyanine Green as a biological stain.

References:

1) Heseltine, D.W.; Brooker, L.G.S. U.S. Patent 2 895 955, 1959.

2) *The United States Pharmacopeia,* twentieth version; United States Pharmacopeial Convention; Mack: Easton, 1979; p 399.

General references:

Aldrich FT-IR Library	**1**(2), 1043A
Sigma FT-IR Library	**1**(2), 153D
Safety	**2**(1), 1981A
Merck Index	**11**, 4868
Colour Index	3rd ed., 1971

INDOINE BLUE

Basic Blue 16

Aldrich Cat. No.: 20,081-6
Sigma Prod. No.: I-1756

CAS No.: [*4569-88-4*]
C.I. No.: 12210
Mol. Form.: $C_{30}H_{24}ClN_5O$
F.W.: 506.01
Appearance: Brown crystalline powder
λ_{max}: 598nm in methanol
Solubility: H_2O 20mg/ml
EGME 40mg/ml
EtOH 30mg/ml

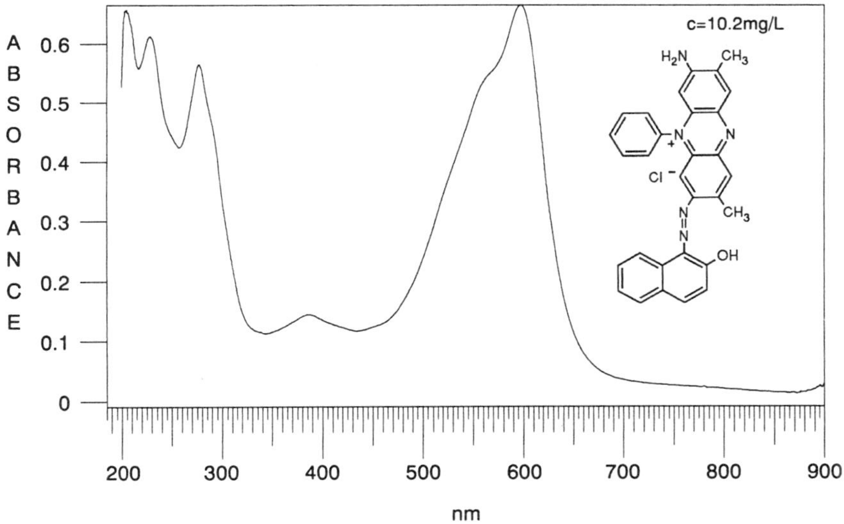

Indoine Blue is a water- and ethanol-soluble, basic monoazo dye that is made by diazotizing Safranine O and coupling the resulting diazonium salt to 2-naphthol. It dyes silk, paper and leather directly, while cotton must first be mordanted with tannin. It is one of a small number of dyes that display thermotropic properties. They can be incorporated into inks for use as a sterilization indicator. Substances marked with such an ink undergo a permanent color change (in this case, the change is from blue to red) in the presence of steam at elevated temperatures.[1]

Reference:

1) Robowetz, S.E.; Specht, J.D.; Murray, L.J. U. S. Patent 4 155 895, 1979.

General references:

IR Library	**3**, 1471A
Safety	**2**(1), 1981B
Colour Index	3rd ed., 1971

2-(4-IODOPHENYL)-3-(4-NITROPHENYL)-5-PHENYLTETRAZOLIUM CHLORIDE hydrate

Aldrich Cat. No.:	I-1040-6
Sigma Prod. No.:	I-8377

CAS No.:	[*146-68-9*]
Mol. Form.:	$C_{19}H_{13}ClIN_5O_2 \cdot xH_2O$
F.W.:	505.70
m.p.:	231°C (dec.)
Appearance:	Off-white to pale-yellow powder
Solubility:	H_2O 4mg/ml
	EGME 30mg/ml
	EtOH 8mg/ml

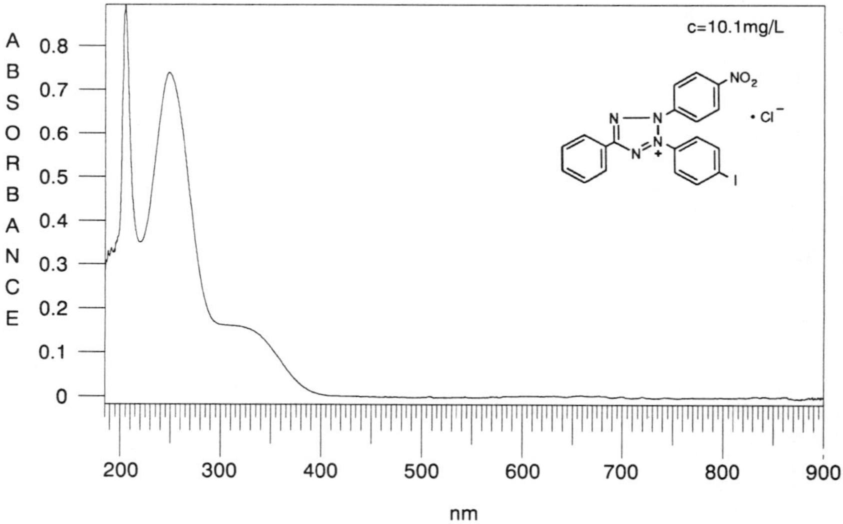

This tetrazolium dye precursor forms a purple formazan dye on reduction. It is used as a histochemical reagent for the determination of dehydrogenases[1] and in an improved method for succinate dehydrogenase activity.[2] It is slightly soluble in water and alcohol.

References:

1) Babson, A.L.; Babson, S.R. *Clin. Chem.* **1973,** *19*, 766.
2) Green, J.D.; Nasahara, H.T. *J. Histochem. Cytochem.* **1980,** *28*, 408.

General references:

Aldrich FT-IR Library	**1**(2), 634A
Sigma FT-IR Library	**1**(2), 661C
IR Library	**3**, 1228B
NMR Library	**2**(2), 502C

JANUS GREEN B

Aldrich Cat. No.:	20,167-7
Sigma Cat. No.:	J2125
Grade:	Certified
CAS No.:	*[2869-83-2]*
C.I. No.:	11050
Mol. Form.:	$C_{30}H_{31}ClN_6$
F.W.:	511.07
Appearance:	Black powder
λ_{max}:	660(395)nm in 50% ethanol
Solubility:	H_2O 30mg/ml
	EGME 10mg/ml
	EtOH 5mg/ml

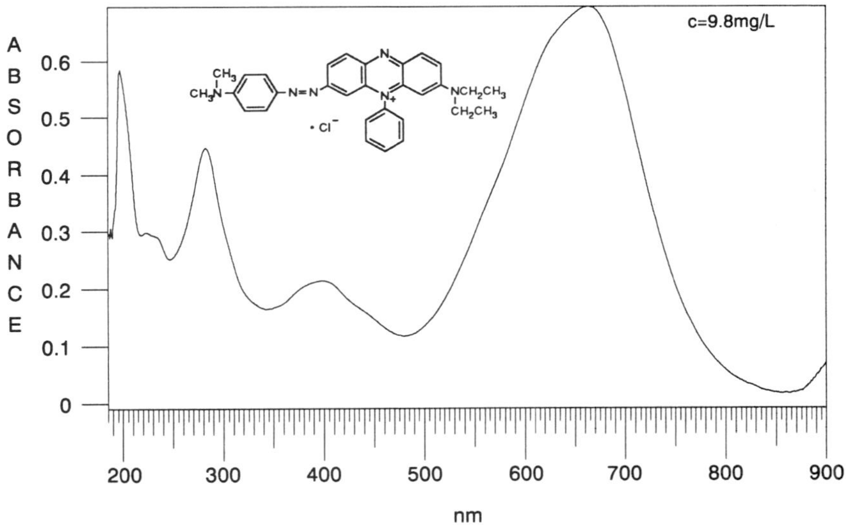

This complex, bluish-green, cationic azine monoazo dye is soluble in water and slightly soluble in alcohol. It is used for dyeing wool-cotton unions from an acid bath without first mordanting the cotton. Probably its most important application, however, is as a catalyst for the electrodeposition of copper in the electronic and automotive industries.[1]

The most important biological application is as a stain for demonstrating mitochondria. To this end, it is used supravitally with Neutral Red to stain living blood cells, and it is certified by the BSC for use in this application. *Staining Procedures*[2] provides methods and literature references for demonstrating mitochondria and for the dye's use as a counterstain in a modified Ziehl-Neelsen procedure.

References:

1) Brown, H.; Woods, H.; Fellows, R.A. U.S. Patent 2 882 209, 1959.

2) Clark, G. *Staining Procedures,* 4th ed.; Williams and Wilkins: Baltimore, 1981; pp 281, 446.

General references:

Aldrich FT-IR Library	**1**(2), 1006A
Sigma FT-IR Library	**1**(2), 156D
IR Library	**3**, 1471B
Merck Index	**11**, 5138
Colour Index	3rd ed., 1971

JENNER STAIN

Aldrich Cat. No.: 86,119-7
Sigma Prod. No.: J1875
Grade: Certified

CAS No.: [*62851-42-7*]
Appearance: Dark-green powder
λ_{max}: 651(522)nm in methanol
Solubility: H_2O 2mg/ml
EGME 6mg/ml
EtOH 3mg/ml

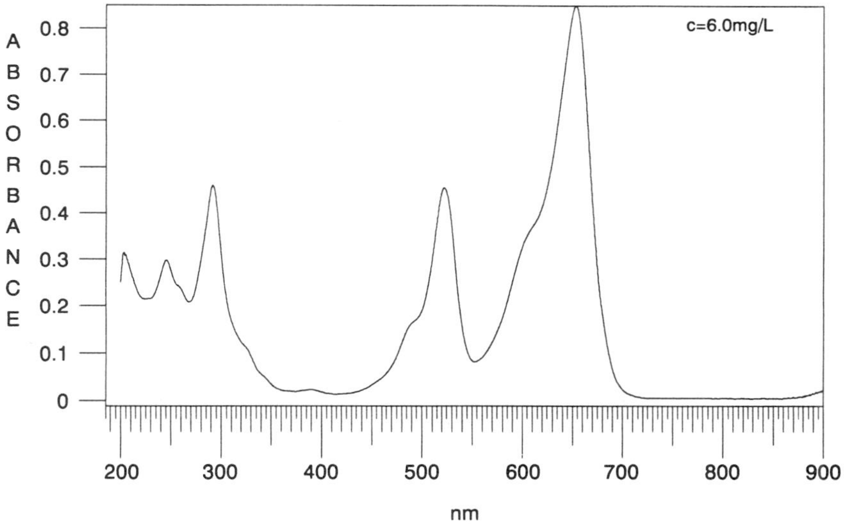

Jenner Stain is a neutral stain discovered by Jenner, an Englishman, about 1899.[1] A German equivalent was reported by May and Grünwald.[2]

Very slightly soluble in water and slightly soluble in methanol and ethanol, it is made by adding an aqueous solution of Eosin Y to an aqueous solution of unoxidized Methylene Blue until a filter-paper runout shows a very slight excess of Methylene Blue. The resulting precipitate is then washed with water to remove free dyes and inorganic salts formed in the combination.

Jenner Stain has no industrial application and limited application as a biological stain, since it does not contain Polychrome Methylene Blue. However, it does polychrome on aging in solution ("ripening"). Therefore, it has been included in some Romanowsky-type blood formulations; the theory is that it increases the shelf life of a "ripening" stock solution by replenishing the azure components that have spontaneously over-oxidized with their necessary less-oxidized precursors.

Currently, the BSC certifies Jenner Stain for use in staining fresh, peripheral blood by the Nocht[3] method.

References:

1) Jenner, L. *Lancet* **1899,** *1*, 370.

2) May, R.; Grünwald, L. *Zentralbl. Inner. Med.* **1902,** *11*, 265-270.

3) Nocht, H. *Zentralbl. Bakteriol.* **1898,** *24*, 839-843.

LACMOID

Aldrich Cat. No.: 27,472-0
Sigma Prod. No.: L7512

Appearance: Black powder
λ_{max}: 611nm in methanol
Solubility: H_2O 30mg/ml
EGME 10mg/ml
EtOH 20mg/ml

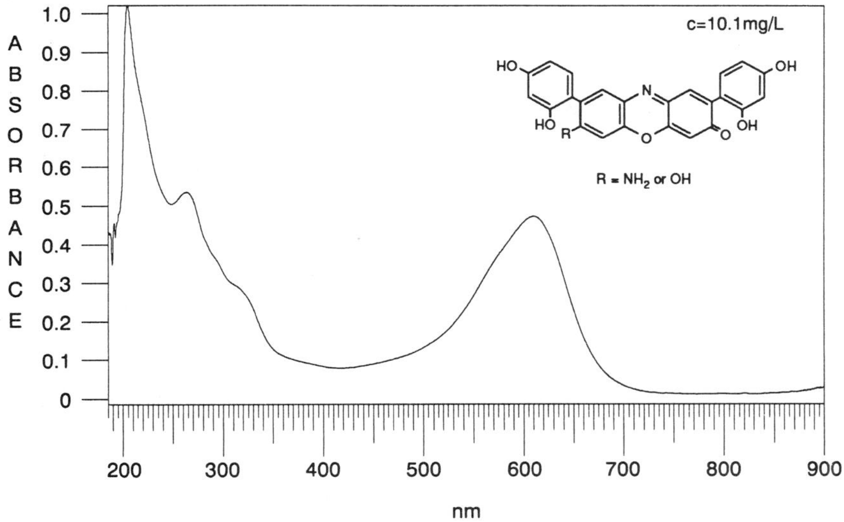

Lacmoid is an oxazine dye that is made by heating resorcinol, sodium nitrite and water (100:5:5) in an oil bath until all of the ammonia emanating from the reaction has been depleted. The resulting melt is digested in hot water and a crude product is precipitated from the filtrate by adjusting the pH to ~5.8 with hydrochloric acid. Further purification is accomplished by an ethanol recrystallization.

Soluble in methanol, ethanol, acetic acid and acetone and slightly soluble in ether, Lacmoid has pH indicator properties. The visual-transition interval is pH 4.4 (red) to pH 6.4 (blue)[1] and it is suitable as an indicator for mineral acids, strong bases and alkaloids, but not suitable for carbonates, weak bases and organic acids.

Staining Procedures[2] provides recipes and methods that utilize Lacmoid. These are for plant material containing callose, as a counterstain for Hematoxylin suitable for general plant tissue, as a neutral counterstain used with Hematoxylin for phloem, and the use of Lacmoid and Martius Yellow, after Nebel in 1931, for demonstrating callose in pollen tubes.

References:

1) Kolthoff, I.M. *Acid-Base Indicators*; Macmillan: New York, 1937; p 160.

2) Clark, G. *Staining Procedures,* 4th ed.; Williams and Wilkins: Baltimore, 1981; pp 45, 317-319, 330, 363.

General references:

Sigma FT-IR Library **1**(2), 157C
Safety **2**(1), 2060B
Merck Index **11**, 5211

LEISHMAN STAIN

Aldrich Cat. No.:	20,020-4
Sigma Prod. No.:	L6254
CAS No.:	[*12627-53-1*]
Appearance:	Dark-green with metallic crystals
λ_{max}:	642(522)nm in methanol
Solubility:	H_2O 4mg/ml
	EGME 20mg/ml
	EtOH 4mg/ml

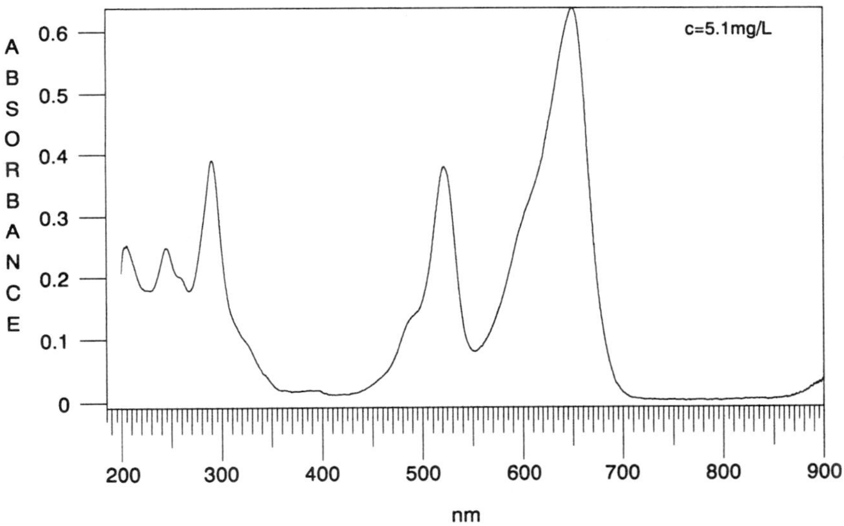

Leishman Stain[1] and Wright Stain,[2] both neutral stains, are probably the two most widely used Romanowsky-type dyes, with Wright Stain the dye of choice in the United States and Leishman Stain preferred in England and Europe. Initially, they differed only slightly, with the variation limited to the method in which the Methylene Blue was polychromed. This detail is even less significant today since oxidative methods have improved. Thus, it is doubtful that either commercial product is made using precisely the same method of oxidation described by its inventor. If they were, they would probably be inferior to products made by the more recent methods.

Leishman Stain is used generally in hematology and tropical medicine to differentiate and identify leucocytes, malarial parasites and trypanosomes. It has no industrial application.

References:

1) Leishman, W.B. *Br. Med. J.* **1901,** *2*, 757-758.
2) Wright, J.H. *J. Med. Res.* **1902,** *7*, 138-144.

General references:

Aldrich FT-IR Library	1(2), 1044D
Sigma FT-IR Library	1(2), 157D
IR Library	3, 1489G

LEUCO CRYSTAL VIOLET

Aldrich Cat. No.:	21,921-5
Sigma Prod. No.:	L5760
CAS No.:	[*603-48-5*]
Mol. Form.:	$C_{25}H_{31}N_3$
F.W.:	373.55
m.p.:	175-177°C
Appearance:	White to very pale lavender crystalline powder
λ_{max}:	261nm in 0.1*N* hydrochloric acid in methanol
Solubility:	H_2O 0.3mg/ml
	EGME 5mg/ml
	EtOH 0.6mg/ml

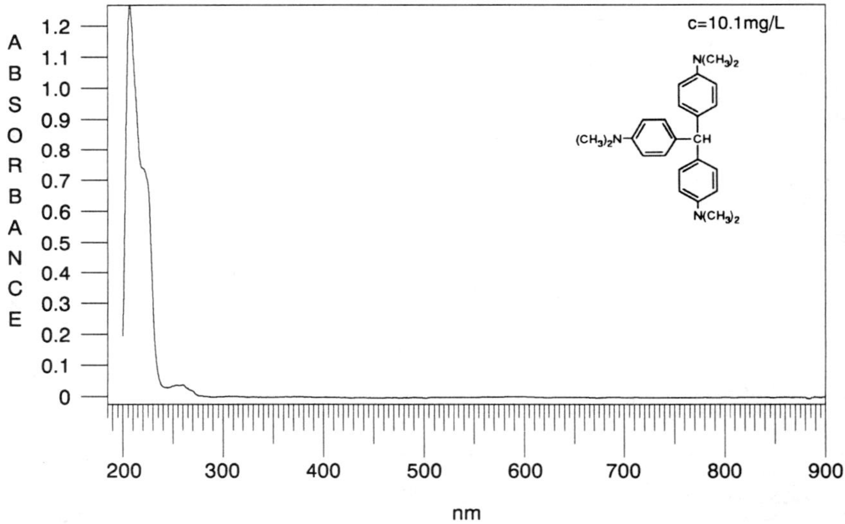

See description on pages 239 and 240 under Crystal Violet.

General references:

Aldrich FT-IR Library	**1**(2), 1030C
Sigma FT-IR Library	**1**(2), 158A
IR Library	**3**, 1484A
Safety	**2**(1), 2121D

LEUCOMALACHITE GREEN

4,4'-Benzylidenebis(*N,N*-dimethylaniline)

Aldrich Cat. No.:	12,566-0
Sigma Prod. No.:	L7257
CAS No.:	[*129-73-7*]
Mol. Form.:	$C_{23}H_{26}N_2$
F.W.:	330.48
Appearance:	Off-white to light-brown powder
λ_{max}**:**	262nm in chloroform
Solubility:	H_2O <0.1mg/ml
	EGME 30mg/ml
	EtOH 4mg/ml

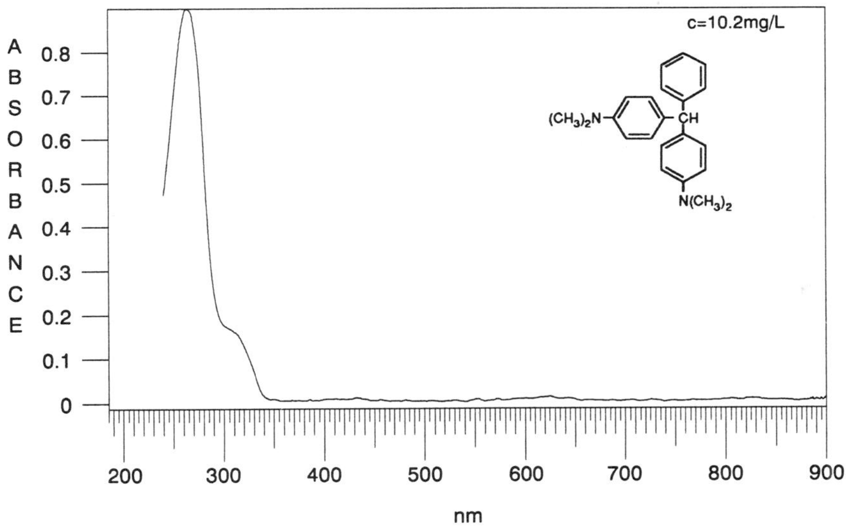

Leucomalachite Green, as the name implies, is the colorless precursor of Malachite Green (Aldrich Cat. No. 86,121-9, 21,302-0 and 22,910-5 described on pp 436-439). It is somewhat soluble in the lower aliphatic alcohols and insoluble in water.

Synthesis is accomplished by condensing two moles of dimethylaniline with one of benzaldehyde in the presence of hydrochloric acid. The neutralized and water-washed condensation product is dried and recrystallized from alcohol. Malachite Green Carbinol base can be prepared by treating the foregoing neutralized and washed Leucomalachite Green before recrystallization with an oxidizing agent such as lead peroxide, and the various salts are prepared from the carbinol by treating it with the appropriate acid.

The alcohol-recrystallized Leucomalachite Green is used as a reagent for the quantitative colorimetric microdetermination of hemoglobin and other heme compounds.[1] This product must be protected from light and the atmosphere to prevent oxidation to the Malachite Green Carbinol base.

Reference:

1) Ahlquist, D.A.; Schwartz, S. *Clin. Chem.* **1975,** *21*, 362.

General references:

Aldrich FT-IR Library	**1**(2), 1247A
Sigma FT-IR Library	**1**(2), 158B
Safety	**2**(1), 2121C

LIGHT GREEN SF YELLOWISH
Acid Green 5

Aldrich Cat. No.: 86,120-0
Sigma Prod. No.: L1886
Grade: Certified

CAS No.: [*5141-20-8*]
C.I. No.: 42095
Mol. Form.: $C_{37}H_{34}N_2Na_2O_9S_3$
F.W.: 792.86
m.p.: 288°C (dec.)
Appearance: Deep-purple powder
λ_{max}: 630(422)nm in water
Solubility: H_2O 100mg/ml
EGME 100mg/ml
EtOH 2mg/ml

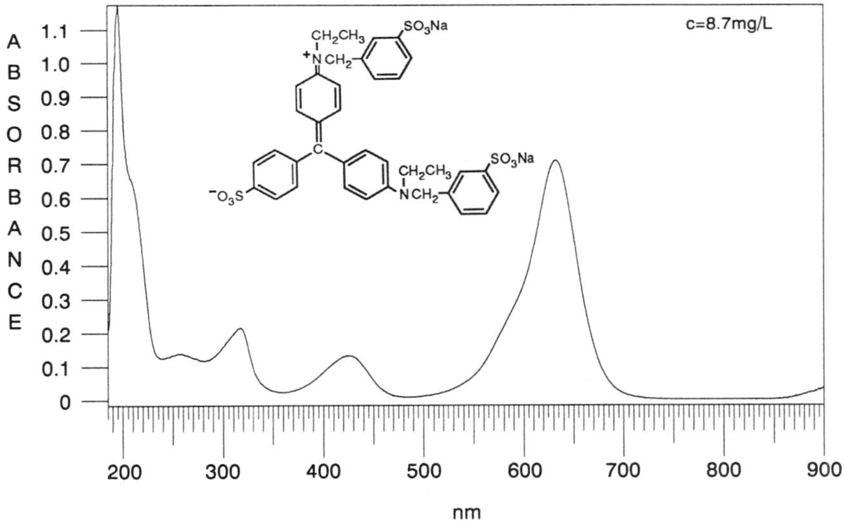

Light Green SF is a bluish-green, anionic triphenylmethane dye that is very soluble in water and slightly soluble in ethanol.

Industrial applications include dyeing wool directly, nylon from a formic acid bath, silk from an acetic acid, formic acid or sulfuric acid bath and chromed leather. The barium lake is used in paper coatings and printing inks.

Twort[1] combined Light Green SF with Neutral Red to form the neutral dye named for the inventor. Twort Stain was first used to stain animal parasites, microorganisms and tissues. It has since been used to stain bacteria, algae and yeast in various methods.

Light Green SF is an important contrast stain for plasma. In this application, it is used with a contrasting nuclear stain such as Hematoxylin or Safranin. Lillie[2] reported that Light Green SF stains collagen fibers very satisfactorily when substituted for Aniline Blue in Masson's trichrome modification of Mallory's method. Using Masson's trichrome staining method, he reported satisfactory staining quality, while less encouraging performance was observed when it was used in the Van Gieson and Masson-Van Gieson methods. It is a component of Papanicolaou EA stain solution. The formulation also contains Eosin Y, Bismarck Brown Y and phosphotungstic acid.

Light Green SF has been used interchangeably with Fast Green FCF in a number of applications, although the latter has proven to be more permanent and less susceptible to light deterioration.

Staining Procedures[3] provides methods and a bibliography for the following applications: Papanicolaou stain, Rae's modified Masson trichrome stain for insect tissue, with Aldehyde Fuchsin for neurosecretory products in insects, as a substitute for Fast Green FCF to show cellulose walls and cytoplasm in plant tissue, chromosomes in root tips or other meristemic tissue and as a counterstain in a modified methenamine-silver nitrate method for demonstrating endamoeba in sections.

References:

1) Twort, F.W. *J. State Med.* (London) **1924,** *32*, 351-355.
2) Lillie, R.D. *J. Technol. Methods* **1945,** *25*, 1-47.
3) Clark, G. *Staining Procedures,* 4th ed.; Williams and Wilkins: Baltimore, 981; pp 209, 260, 268, 330, 350, 472.

General references:

IR Library	**3,** 1485E
Safety	**2**(1), 2126D
Merck Index	**11**, 5361
Colour Index	3rd ed., 1971

LISSAMINE GREEN B

Acid Green 50 • Wool Green S

Aldrich Cat. No.:	19,958-3
Sigma Prod. No.:	L6382
CAS No.:	[*3087-16-9*]
C.I. No.:	44090
Mol. Form.:	$C_{27}H_{25}N_2NaO_7S_2$
F.W.:	576.63
Appearance:	Black powder
λ_{max}:	633nm in water
Solubility:	H_2O 40mg/ml EGME 20mg/ml EtOH 3mg/ml

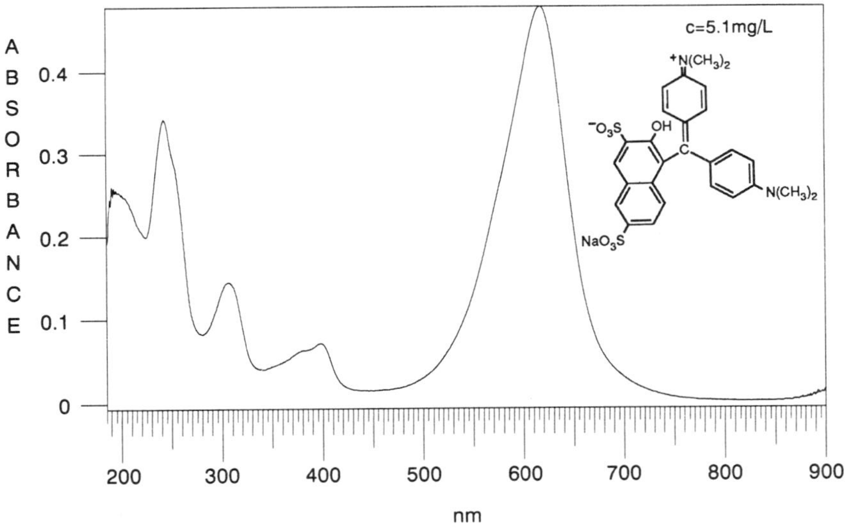

Lissamine Green B is a bluish-green, acid diphenylnaphthylmethane dye with an affinity for proteinaceous substrates and a lesser attraction to cellulosic material. It is soluble in water and slightly soluble in alcohol. Industrially, it is used to dye wool, silk (from a sulfuric acid bath), to print wool and silk directly and to coat paper, dye leather and surface-stain wood.

It has been substituted for Fast Green FCF without any real advantage. Lillie[1] found it performed well as a collagen stain in Masson's trichrome and as a counterstain in the Van Gieson method. However, it was inferior when used as a component in Van Gieson's trichrome stain. See *Staining Procedures*[2] for the Lillie modification of Masson's trichrome method.

References:

1) Lillie, R.D. *J. Technol. Methods* **1945,** *25*, 1-47.

2) Clark, G. *Staining Procedures,* 4th ed.; Williams and Wilkins: Baltimore, 1981; p 120.

General references:

Aldrich FT-IR Library	**1**(2), 1032A
Sigma FT-IR Library	**1**(2), 158D
IR Library	**3**, 1484G
Colour Index	3rd ed., 1971

LITMUS

Aldrich Cat. No.: 19,972-9
Sigma Prod. No.: L7382
Grade: Indicator

CAS No.: [*1393-92-6*]
Appearance: Bluish-black powder
λ_{max}: 575nm in water
Solubility: H_2O 50mg/ml
EGME 10mg/ml
EtOH 3mg/ml

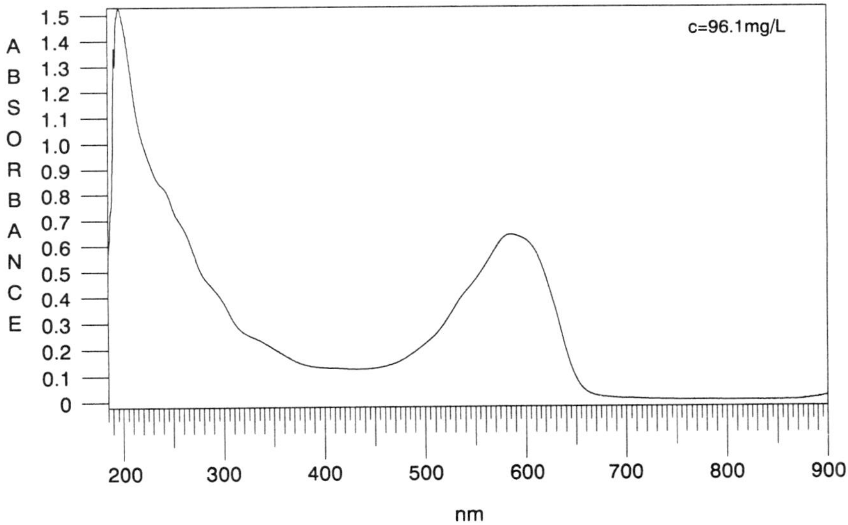

Litmus is a weak natural dye that is developed from various orcinol-containing lichen; *Lecanora tartarea*, *Rocella fuciformis* and *Variolaria dealbata* are among the more important ones. These are found on trees and cliffs on the coasts of the Mediterranean Sea, Spain, France, Sweden, California, Corsica and Madagascar.[1] When these lichen are allowed to ferment in the presence of air and ammonia, orcein is formed. When either potassium carbonate or potassium hydroxide is added to the preceding recipe, Litmus is the end product. Musso and Kramer[2] reported that the principal coloring matter in Litmus is Azolitmus, a substituted 7-hydroxy-2-phenoxazone derivative. This material is responsible to a large degree for the indicator properties displayed by Litmus. However, HPLC reveals that there are several other products closely related to Azolitmus present in lesser amounts.

At one time, Litmus, with a broad visual-transition interval of pH 4.8 (red) to pH 8.3 (blue), was an important indicator. It has been used in the form of a test paper, as an indicator (for both oxidation-reduction and pH), in culture media and to color beverages. Today, however, it has been largely replaced by more precise and stable synthetic indicators such as Nitrazine Yellow, Bromothymol Blue and other sulfonephthalein indicators, depending on the application.

Litmus has not been used as a histological stain because of its sensitivity to oxidation-reduction.

References:

1) *The Merck Index,* 11th ed.; Budavari, S. *et al.*, Eds.; Merck: Rahway, 1989; Monograph 5426.

2) Musso, H.; Kramer, H. *Chem. Ber.* **1959,** *92*, 751-753.

General references:

Aldrich FT-IR Library	**1**(2), 1045A
Sigma FT-IR Library	**1**(2), 159A
IR Library	**3**, 1489H

LUCIFER YELLOW CH

Aldrich Cat. No.: 86,150-2
Sigma Prod. No.: L0251

CAS No.: [*67769-47-5*]
Mol. Form.: $C_{13}H_9Li_2N_5O_9S_2$
F.W.: 457.25
Appearance: Orange powder
λ_{max}: 428(348)nm in water

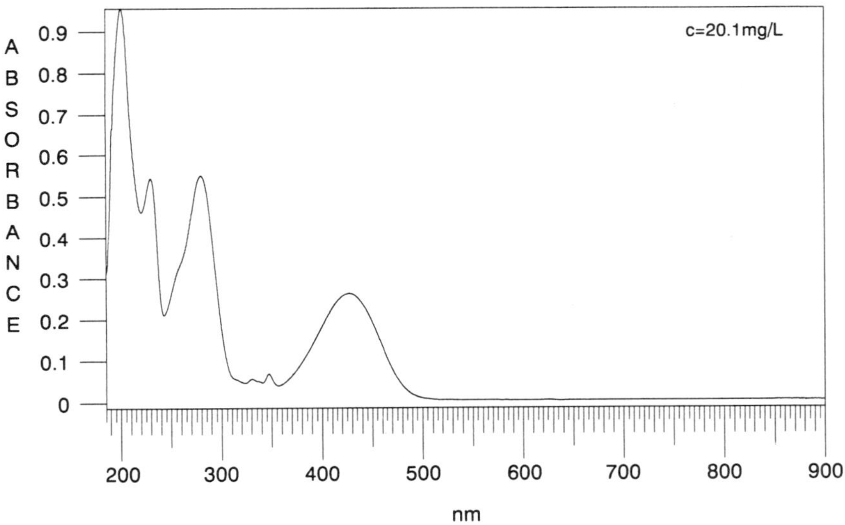

Lucifer Yellow CH and Lucifer Yellow VS, highly fluorescent, reactive, naphthalimide dyes, were first prepared by Stewart[1,2] from FIAT Brilliant Sulfoflavin FF (Acid Yellow 7, C.I. 56205). While both dyes combine covalently with tissue to become an integral part of the substrate, they differ in mode of attachment and the type of site required for the linkage. *Conn's*[3] states that the CH with a hydrazine function binds with substrates bearing carbonyl groups and the VS (vinyl sulfone) function reacts with amines, hydroxyl groups and sulfhydryl groups. Stewart[4] indicates that Lucifer Yellow CH, which possesses a free hydrazido group, reacts with aliphatic aldehydes at room temperature to bind the dye to the tissue. He presumes the attachment is through the hydrazido group. Both dyes are very water soluble (Stewart[1] reported the CH to be 50% soluble, w/w) and insoluble in ethanol.

Stewart[1] synthesized the Lucifer dyes for use as dye-coupling tracers. They are reported to have a fluorescence intensity of up to 100 times as great as Procion Yellow M-4R, a reactive dye not fully identified (a dichlorotriazinylmonoazopyrazolone) and widely used prior to the discovery of the

Lucifers in the same application. Bailey *et al.* used the VS for the fluoroimmunoassay of albumin in serum[5] and for the homogeneous fluoroimmunoassay of albumin plasma.[6] Racowsky and Satterlie[7] used the CH for electrical coupling between hamster oocytes and cumulous cells during meiotic maturation *in vivo* and *in vitro*.

References:

1) Stewart, W. *Cell* **1978,** *14*, 741-759.

2) *Idem J. Am. Chem. Soc.* **1981,** 103, 7615-7620.

3) *Conn's Biological Stains,* 9th ed.; Lillie, R.D., Ed.; Williams and Wilkins: Baltimore, 1977; p 242.

4) Stewart, W. *Nature* **1981,** *292,* 17-21.

5) Bailey, M.P.; Bernard, F.R.; Riley, C. *Ann. Clin. Biochem.* **1983,** *20*, 213-216.

6) *Idem ibid.* **1984,** *21*, 59-63.

7) Racowsky, C.; Satterlie, R.A. *Develop. Biol.* **1985,** *108*, 191-202.

General references:

Aldrich FT-IR Library	**1**(2), 955B
Sigma FT-IR Library	**1**(2), 159B
IR Library	**3**, 1445B
Safety	**2**(1), 2163D
Merck Index	**11**, 5466
Colour Index	3rd ed., 1971

LUCIFER YELLOW VS

Dilithium 4-amino-*N*-[3-(vinylsulfonyl)phenyl]naphthalimide-3,6-disulfonate

Aldrich Cat. No.: 86,205-3
Sigma Prod. No.: L3510

CAS No.: *[71231-14-6]*
Mol. Form.: $C_{20}H_{12}Li_2N_2O_{10}S_3$
F.W.: 550.40
Appearance: Dark-yellow, fibrous powder
λ_{max}: 427(277)nm in water

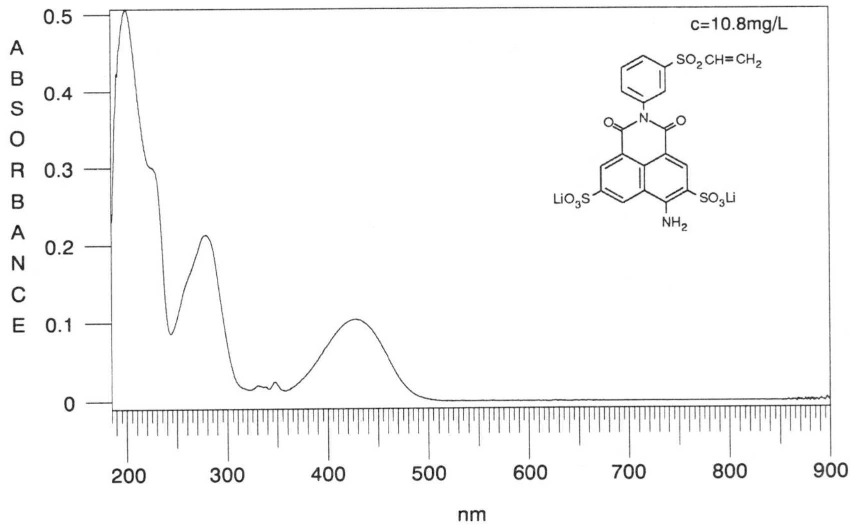

See Lucifer Yellow CH on pages 427 and 428 for description.

General references:

Sigma FT-IR Library **1**(2), 159C
Safety **2**(1), 2164A

LUMICHROME

7,8-Dimethylalloxazine

Aldrich Cat. No.: 10,321-7
Sigma Prod. No.: L7507

CAS No.: [*1086-80-2*]
Mol. Form.: $C_{12}H_{10}N_4O_2$
F.W.: 242.24
m.p.: >300°C
Appearance: Green powder with chunks
λ_{max}: 392(350)nm in methanol
Solubility:

H_2O	0.6mg/ml
EGME	0.8mg/ml
EtOH	0.5mg/ml

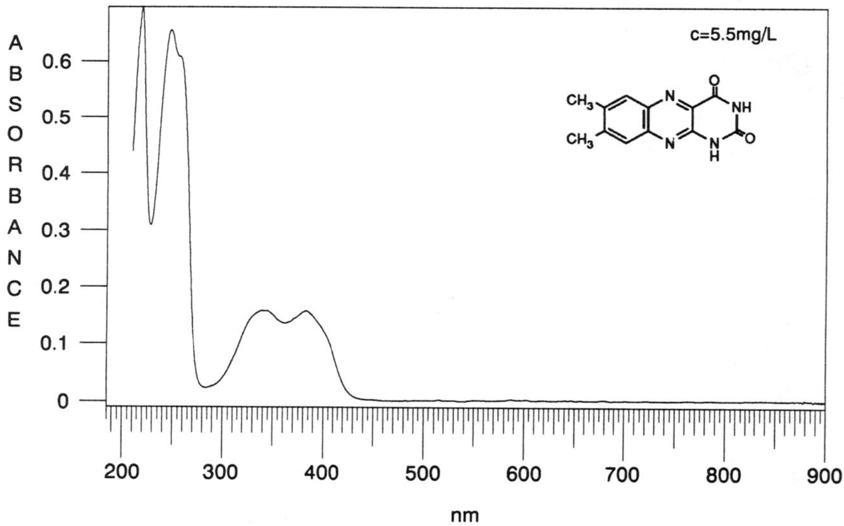

When riboflavine is irradiated, its sugar moiety is cleaved, freeing the residual Lumichrome molecule. According to *The Merck Index,*[1] a number of researchers have reported synthetic routes.

Lumichrome is a pale-yellow organic compound with an azine chromophore. It is sparingly soluble in methanol, hot 90% ethanol and chloroform and very slightly soluble in water. Aqueous, alcoholic and chloroform solutions display a blue fluorescence. Although Lumichrome absorbs in the visible region starting at approximately 430nm and increasing down scale, its absorption maxima are in the ultraviolet region with a primary peak at 392nm and a secondary peak at 350nm.

Reference:

1) *The Merck Index,* 11th ed.; Budavari, S. *et al.*, Eds.; Rahway, 1989; Monograph 5468.

General references:

Aldrich FT-IR Library	**1**(2), 894D
Sigma FT-IR Library	**1**(2), 160A
IR Library	**3**, 1414C
NMR Library	**2**(2), 768B

LUXOL® BRILLIANT GREEN BL

Solvent Green 11

Aldrich Cat. No.:	27,726-6
CAS No.:	[*71799-14-9*]
Appearance:	Purple crystalline powder
λ_{max}:	624nm in methanol
Solubility:	H_2O 4mg/ml
	EGME 30mg/ml
	EtOH 9mg/ml

® Reg. trademark of E. I. du Pont de Nemours & Co., Inc.

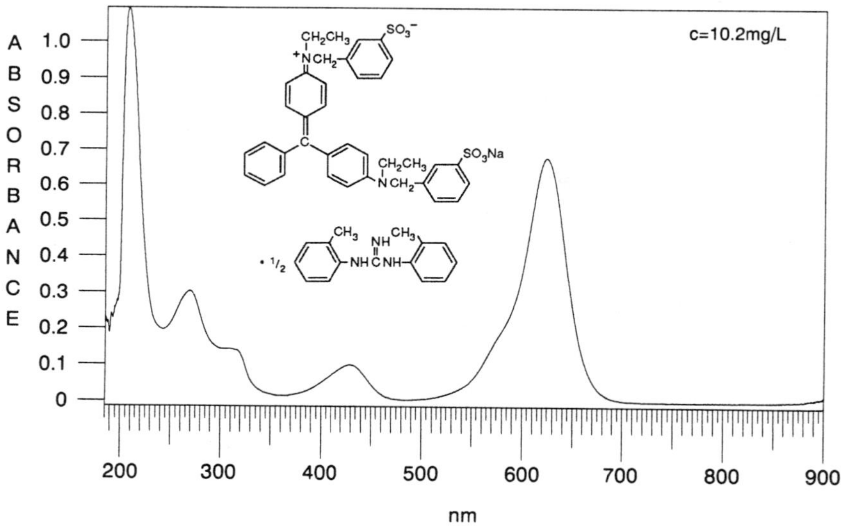

Luxol Brilliant Green BL is a solvent-soluble dye that is made by adding an aqueous solution of 1,3-di-*o*-tolylguanidine to an aqueous solution of Acid Green 3 (2/1). The resulting precipitate is filtered and washed free of inorganic salts with deionized water.

Slightly soluble in water, methanol and ethanol, it has been used industrially to color lacquers, metal-foil lacquers, plastics, shellac products and non-grain-raising wood stains, despite its poor lightfastness (1-2).

Christie[1] describes a method that utilizes Luxol Brilliant Green BL in combination with Calcomine Orange 2RS (Aldrich Cat. No. 27,752-5) as a stain for plant-pathology studies.

Reference:

1) Christie, R.G. *Virology* **1967**, *31*, 268.

General reference:

Colour Index 3rd ed., 1971

LUXOL® FAST BLUE ARN
Solvent Blue 37

Aldrich Cat. No.: 22,933-4
Sigma Prod. No.: S3259

CAS No.: [*12226-74-3*]
Appearance: Dark-purple powder
λ_{max}: 576nm in methanol
Solubility: H_2O 1mg/ml
EGME 20mg/ml
EtOH 70mg/ml

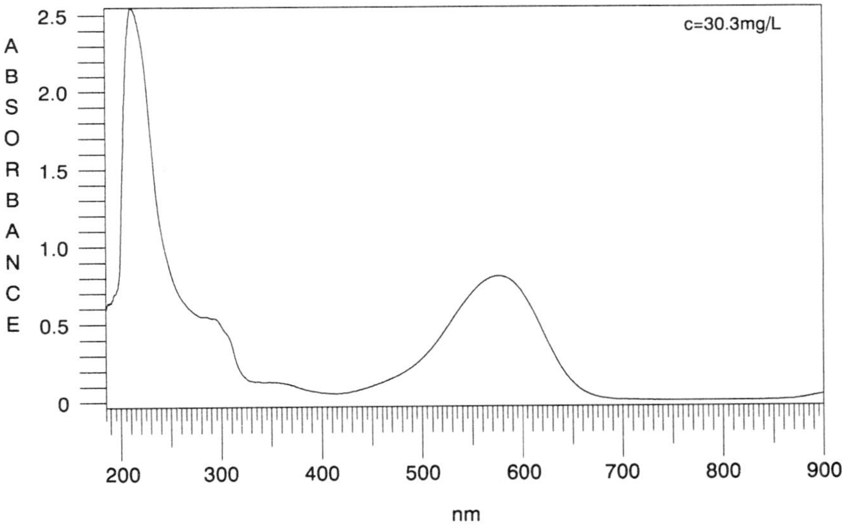

This blue solvent dye was marketed by du Pont until they discontinued manufacturing dyes in the early 1970s.

Although the exact composition has been closely guarded, it is believed to be a diarylguanidine salt of a sulfonated azo dye.

Soluble in methyl Cellosolve, ethanol and methanol and slightly soluble in acetone, it is used to color lacquers and ball-point-pen inks.

Salthouse[1] used it to stain myelin and phospholipids with "improved" results.

Reference:

1) Salthouse, T.N. *Stain Technol.* **1962,** *37*, 313-316.

General reference:

Sigma FT-IR Library **1**(2), 213A

LUXOL® FAST BLUE MBSN

C.I. Solvent Blue 38

Aldrich Cat. No.:	22,934-2
Sigma Prod. No.:	S3382
CAS No.:	[*1328-51-4*]
C.I. No.:	74180 (amine salt)
Appearance:	Blue powder
λ_{max}:	666nm in methanol
Solubility:	H_2O 2mg/ml
	EGME 20mg/ml
	EtOH 30mg/ml

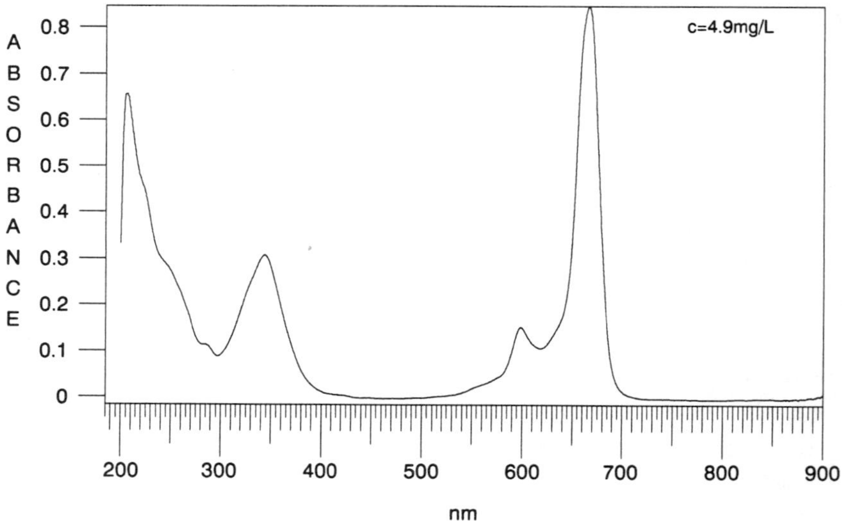

Like Luxol Fast Blue ARN, this blue solvent dye was manufactured and marketed by du Pont until the early 1970s.

Salthouse,[1] apparently from private correspondence with du Pont, considered this product to be a diarylguanidine salt of a sulfonated copper phthalocyanine pigment.

The *Colour Index* lists it as very soluble in methyl Cellosolve, ethanol and methanol (our laboratory found it to be only moderately soluble in these solvents) and slightly soluble in acetone. Salthouse reported it to be soluble in methanol, ethanol and the propanols and very soluble in glycols, ethanolamine, choline and phospholipids.

Industrially, it is used to dye leather and to color lacquers and ball-point-pen inks.

Staining Procedures[2] provides a recipe and method, taken from Lockard and Reers,[3] for using Luxol Fast Blue MBSN in combination with Neutral Red as a myelin-sheath stain. The same source also provides a recipe and method, after Kluver and Barrera,[4] for staining myelinated nerve fibers of both the central and peripheral nervous systems. An option is to use Darrow Red, Aldrich Cat. No. 21,188-5, as a counterstain.

References:

1) Salthouse, T.N. *Stain Technol.* **1962,** *37*, 313-316.

2) Clark, G. *Staining Procedures,* 4th ed.; Williams and Wilkins: Baltimore, 1981; pp 35-36, 147-148.

3) Lockard, I.; Reers, B.L. *Stain Technol.* **1962,** *37*, 13-16.

4) Kluver, H.; Barrera, E. *J. Neuropathol. Exp. Neurol.* **1953,** *12*, 400-403.

General references:

Sigma FT-IR Library 1(2), 213B

Colour Index 3rd ed., 1971

MALACHITE GREEN CARBINOL base
Solvent Green 1

Aldrich Cat. No.:	22,910-5
Sigma Prod. No.:	M8635
CAS No.:	[*510-13-4*]
C.I. No.:	42000:1
Mol. Form.:	$C_{23}H_{26}N_2O$
F.W.:	346.48
m.p.:	112-114°C
Appearance:	Grey powder
λ_{max}**:**	446(259)nm in 1*N* hydrochloric acid
Solubility:	H_2O 0.3mg/ml
	EGME 20 mg/ml
	EtOH 10 mg/ml

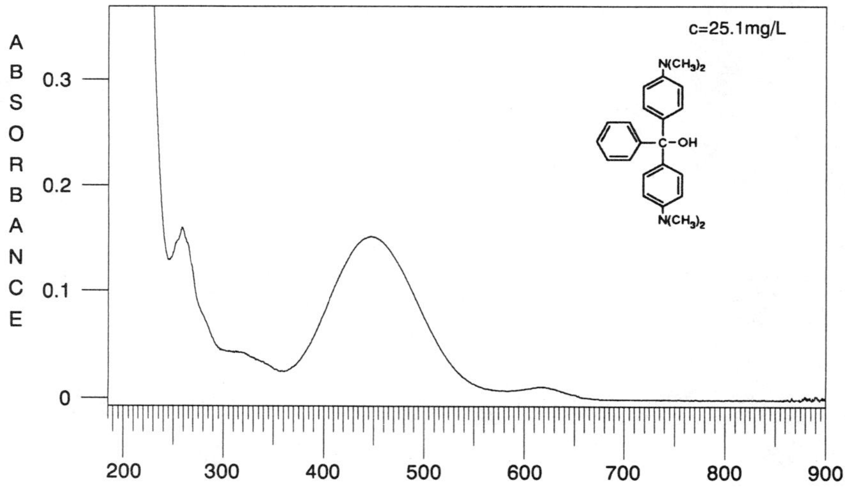

See description on pages 438 and 439 under Malachite Green oxalate.

General references:

Aldrich FT-IR Library	**1**(1), 1248C
Sigma FT-IR Library	**1**(2), 161B
IR Library	**3**, 1464H
NMR Library	**2**(1), 1058A
Safety	**2**(2), 2182A

MALACHITE GREEN CARBINOL hydrochloride

Aldrich Cat. No.: 21,302-0
Sigma Prod. No.: M9636

Mol. Form.: $C_{23}H_{26}N_2O \cdot HCl$
F.W.: 382.94
Appearance: Blue-green powder
λ_{max}: 615(425)nm in water
Solubility: H_2O 60mg/ml
EGME 30mg/ml
EtOH 20mg/ml

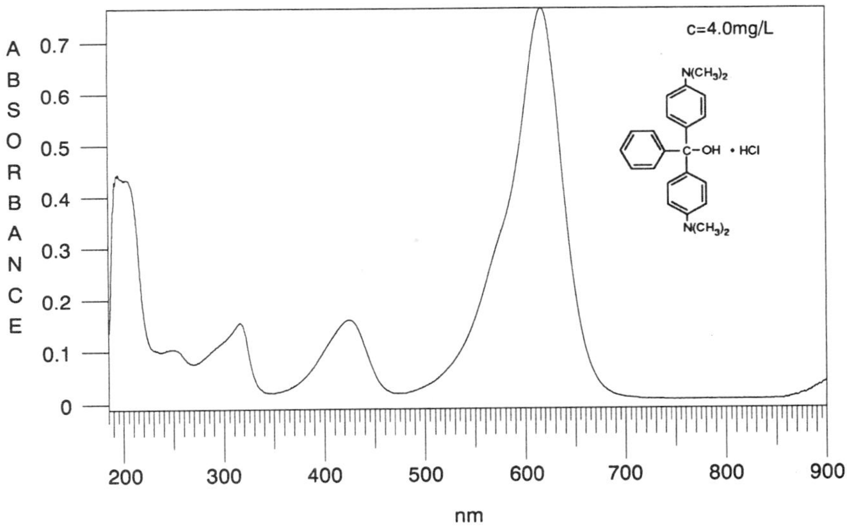

See description on pages 438 and 439 under Malachite Green oxalate.

General references:

Aldrich FT-IR Library	1(2), 1029C
Sigma FT-IR Library	1(2), 160C
IR Library	**3**, 1483D
Safety	**2**(2), 2182B

MALACHITE GREEN oxalate
Basic Green 4

Aldrich Cat. No.:	86,121-9
Sigma Prod. No.:	M9015
Grade:	Certified
CAS No.:	[*2437-29-8*]
C.I. No.:	42000
Mol. Form.:	$C_{46}H_{50}N_4 \cdot 2HC_2O_4^- \cdot C_2H_2O_4$
F.W.:	927.03
m.p.:	164°C (dec.)
Appearance:	Green crystalline powder
λ_{max}**:**	614(425)nm in water
Solubility:	H_2O 40mg/ml
	EGME 40mg/ml
	EtOH 50mg/ml

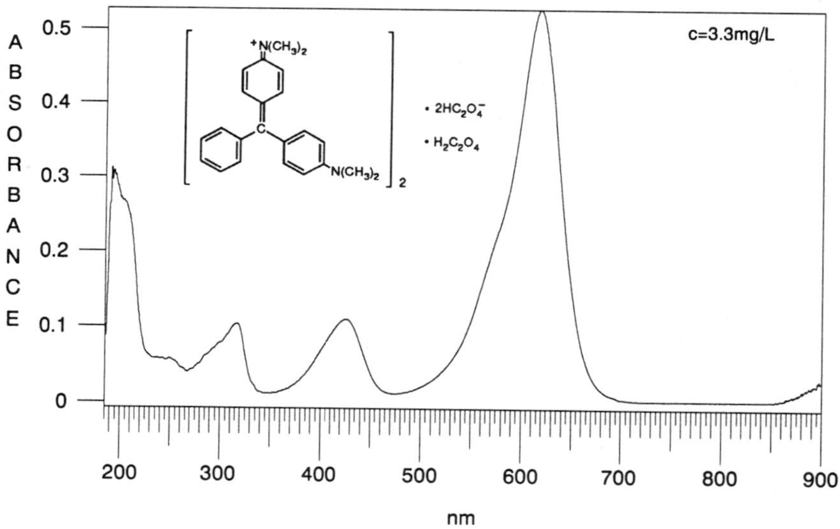

Malachite Green (MG) is a cationic triphenylmethane dye. A preparative method is described under Leucomalachite Green, (Aldrich Cat. No. 12,566-0 on pages 419 and 420).

Being asymmetrical, it has two axes of polarizability. These axes differ in degree and lie at right angles to each other. This configuration is responsible for the secondary absorption peak displayed at λ_{max} 425nm. The primary peak occurs at 614nm.

Malachite Green is available commercially as the oxalate salt and the carbinol hydrochloride salt. The oxalate salt is the form usually certified by the

BSC. However, the chloride is the salt most frequently described in the literature; although, for all practical purposes, the chloride form exists only in solution and as an impurity in the carbinol hydrochloride form. It should be noted that material labeled as the chloride probably is the zinc chloride double salt.

If a non-toxic (free of oxalate and zinc) form is desired for fish tanks or similar uses demanding low toxicity, the carbinol hydrochloride is the form of choice. This product is a colorless dye precursor that rapidly loses a mole of water intramolecularly when dissolved in water or one of the short-chain alcohols to form the chloride salt of the dye.

The *Colour Index*[1] lists the following industrial applications for the dye: dyeing wool, leather and jute; printing silk and cotton. A lake is formed with phosphotungstomolybdic acid which, after processing, is used as a pigment (C.I. Pigment 4, C.I. 42,000:2).

Malachite Green displays acid-base-indicator properties with visual-transition intervals extending from pH 0.0 (yellow) to pH 2.0 (green) and pH 11.6 (blue-green) to pH 14.0 (colorless).

Staining Procedures[2] lists methods for the following biological-stain applications: Pianese IIIB stain for fungi in plant tissue; Schaeffer and Fulton method of spore staining; the Laybourn modification of Albert stain for diphtheria and the Gimenez *Rickettsia* stain. Alexander[3] used Malachite Green, in combination with Acid Fuchsin and Orange G, as a trichrome stain to differentially stain aborted pollen green, and non-aborted red.

The carbinol base (Aldrich Cat. No. 22,910-5) is used to color fats, oils and waxes.

References:

1) *Colour Index,* 1st ed., Society of Dyers and Colourists: Bradford, Yorkshire, 1924.

2) Clark, G. *Staining Procedures,* 4th ed.; Williams and Wilkins: Baltimore, 1981; pp 372, 390, 405, 414.

3) Alexander, M.P. *Stain Technol.* **1969,** *44*, 117-122.

General references:

Aldrich FT-IR Library	1(2), 1029A
Sigma FT-IR Library	1(2), 161A
IR Library	**3**, 1483B
Safety	2(2), 2182C
Colour Index	3rd ed., 1971

MARTIUS YELLOW monohydrate

Acid Yellow 24 • 2,4-Dinitro-1-naphthol, sodium salt

Aldrich Cat. No.:	28,781-4
Sigma Prod. No.:	M6897
Purity:	98%

CAS No.:	*[605-69-6]*
C.I. No.:	10315
Mol. Form.:	$C_{10}H_5N_2NaO_5 \cdot H_2O$
F.W.:	274.16
Appearance:	Orange powder
λ_{max}:	432nm in methanol
Solubility:	H_2O 4mg/ml
	EGME 20mg/ml
	EtOH 4mg/ml

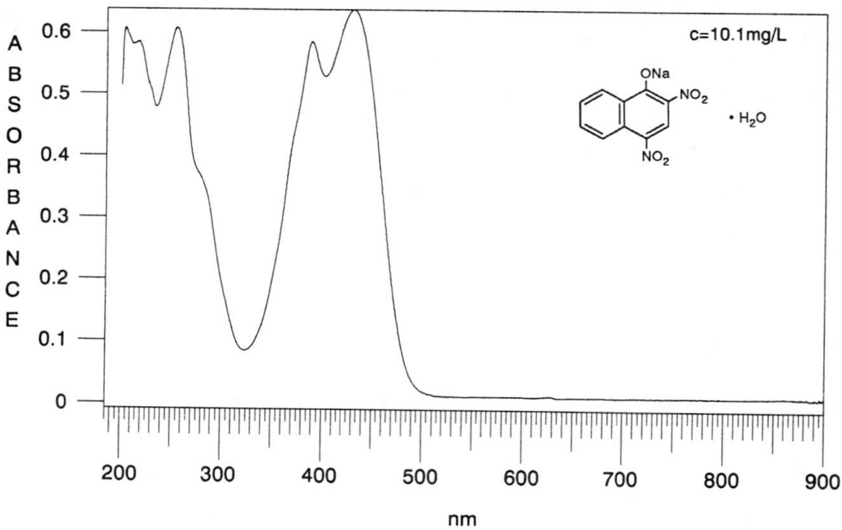

Martius Yellow was discovered by Karl Alexander Martius in 1868 according to Fieser.[1] *Conn's*[2] indicates that Martius was the second discoverer and that Ganahl actually discovered it in 1856.

Industrially, it has been used for mothproofing and coloring varnish. Current industrial application is confined to the preparation of light filters for photomicrography.

This yellow dinitro dye is prepared by nitrating 1-naphthol and converting the reaction product to the sodium salt. It is slightly soluble in water and ethanol. The *Colour Index* lists the ammonium salt as being soluble in ethanol; however, no appreciable differences in solubility between the sodium and ammonium salts have been noted.

Martius Yellow is a pH indicator of dubious value. Long[3] listed a visual-transition interval of pH 2.0 (colorless) to pH 3.2 (yellow). Over the years, our laboratories have observed the following visual-transition interval: pH 2.0 (yellow) to pH 4.0 (darker yellow).

Conn's indicates that currently it is used extensively by plant pathologists in studying sections of tissue infected by fungi. *Staining Procedures*[4] describes a method using Martius Yellow and Lacmoid to demonstrate callose in pollen tubes. *Staining Procedures*[5] provides a recipe and method using premixed Martius Yellow, Malachite Green and Acid Fuchsin for use in fungus-infected plant tissue to differentiate hyphae from host tissue. This is taken from a Pianese (1896) cancer stain adapted by Vaughn in 1914, the details of which were from a personal communication from L.W. Sharp to the editor, G. Clark. From the third edition to the fourth edition of *Staining Procedures*, the editor had concluded that Martius Yellow was of little if any value in the stain mixture.

References:

1) Fieser, L.F. *Experiments in Organic Chemistry,* 3rd ed. revised; Heath: Boston, 1957; p 234
2) *Conn's Biological Stains,* 9th ed.; Lillie, R.D., Ed.; Williams and Wilkins: Baltimore, 1977; p 75.
3) Long , C. *Biochemist's Handbook;* Van Nostrand: Princeton, 1961.
4) Clark, G. *Staining Procedures,* 4th ed.; Williams and Wilkins: Baltimore, 1981; p 363.
5) Clark, G. *Staining Procedures,* 3rd ed.; Williams and Wilkins: Baltimore, 1973; p 243.

General reference:

Colour Index 3rd ed., 1971

MAY-GRÜNWALD STAIN

Aldrich Cat. No.:	20,543-5
Sigma Prod. No.:	M6901

Appearance:	Dark-green powder
λ_{max}:	645(522)nm in methanol
Solubility:	H_2O 2mg/ml
	EGME 20mg/ml
	EtOH 6mg/ml

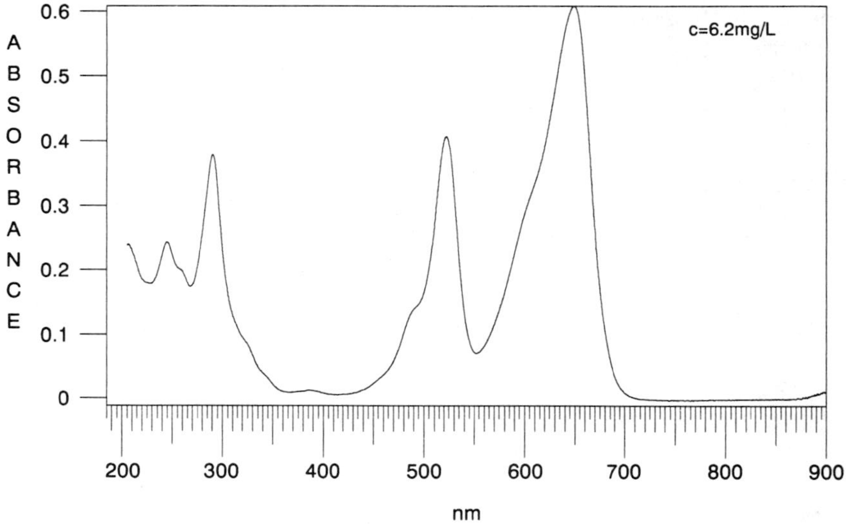

May-Grünwald Stain[1] is a neutral stain that is prepared by adding an aqueous solution of Eosin Y to an aqueous solution of Methylene Blue. Thus, it is the German equivalent of Jenner Stain.[2]

Freshly prepared methanolic solutions of Methylene Blue eosinate yield an inferior stain unsuitable for Romanowsky-type-stain applications. However, on aging, these solutions are prone to slow, spontaneous oxidation. A "ripened" (aged) solution of Methylene Blue eosinate, depending on the degree of polychroming that has occurred, may function satisfactorily as a Romanowsky-type blood stain. Early stains made exclusively from either Jenner or May-Grünwald stains required as much as six months "ripening" before they yielded a satisfactory blood stain. Aging is not necessary with modern-day chemically polychromed Romanowsky stains.

It appears that the only function Jenner Stain or May-Grünwald Stain can serve in a Romanowsky-type dye mixture is to replenish the azure components that have been spontaneously over-oxidized with their necessary less-oxidized precursors. Theoretically, this should enhance the shelf life of the stock solution.

It has been observed that contemporary Jenner Stain is made from Methylene Blue and Eosin Y, while many lots of present-day commercial May-Grünwald Stain are apparently made from Polychromed Methylene Blue and Eosin Y.

The BSC does not certify May-Grünwald Stain; however, it does certify Jenner Stain for use in staining fresh peripheral blood by the Nocht method.[3]

References:

1) May, R.; Grünwald, L. *Zentralbl. Inner Med.* **1902,** *11*, 265-270.

2) Jenner, L. *Lancet* **1899,** *1*, 370.

3) Nocht, H. *Zentralbl. Bakteriol.* **1898,** *24*, 839-843.

General reference:

Sigma FT-IR Library **1**(2), 161D

MERBROMIN

2',7'-Dibromo-4'-(hydroxymercuri)fluorescein, disodium salt • Mercurochrome

Aldrich Cat. No.:	19,959-1
Sigma Prod. No.:	M7011
CAS No.:	[*129-16-8*]
Mol. Form.:	$C_{20}H_8Br_2HgNa_2O_6$
F.W.:	750.67
Appearance:	Green crystals
λ_{max}**:**	507nm in water
Solubility:	H_2O 20mg/ml
	EGME 9mg/ml
	EtOH 10mg/ml

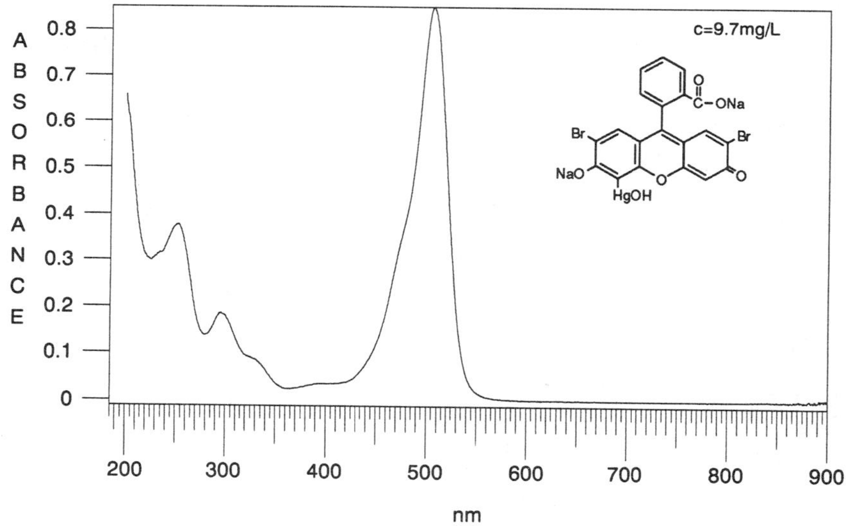

Merbromin is a red xanthene dye that is soluble in water. *Merck*[1] states that a 0.5% aqueous solution has a pH of 8.8 and that 1g is soluble in 50g of ethanol, while 8.4g is soluble in 50g of methanol.

At one time, Mercurochrome was widely advertised as an antiseptic. An abraded- or punctured-skin area that did not display the carmine-red color typical of the product was considered by many, particularly children, not to be properly treated. In fact, it had such an appeal that competitors felt compelled to color their colorless, disinfectant or germicide solutions with dyes such as Eosin Y or Orange G to mimic this quality. Since its true effectiveness as an antiseptic has been established, it finds little, if any, use in this application today.

Conn's[2] cites a number of histological- and hematological-stain applications. Gurr[3] indicates that it is an excellent substitute for Eosin as a counterstain to Hematoxylin (H&E stain). He points out that it is much more stable to light than Eosin and does not fade even when exposed to strong sunlight.

References:

1) *The Merck Index,* 11th ed.; Budavari, S. *et al.*, Eds.; Merck: Rahway, 1989; Monograph 5757.

2) *Conn's Biological Stains,* 9th ed.; Lillie, R.D., Ed.; Williams and Wilkins: Baltimore, 1977; p 340.

3) Gurr, E. *Synthetic Dyes in Biology, Medicine and Chemistry*; Academic: London, 1971; pp 193-194.

General references:

Aldrich FT-IR Library	**1**(2), 1015B
Sigma FT-IR Library	**1**(2), 162A
IR Library	**3**, 1475H
Safety	**2**(2), 2206D

METANIL YELLOW

Acid Yellow 36

Aldrich Cat. No.:	20,202-9
Sigma Prod. No.:	M7276
CAS No.:	[*587-98-4*]
C.I. No.:	13065
Mol. Form.:	$C_{18}H_{14}N_3NaO_3S$
F.W.:	375.38
Appearance:	Gold powder
λ_{max}:	414nm in methanol
Solubility:	H_2O 8mg/ml
	EGME 20mg/ml
	EtOH 7mg/ml

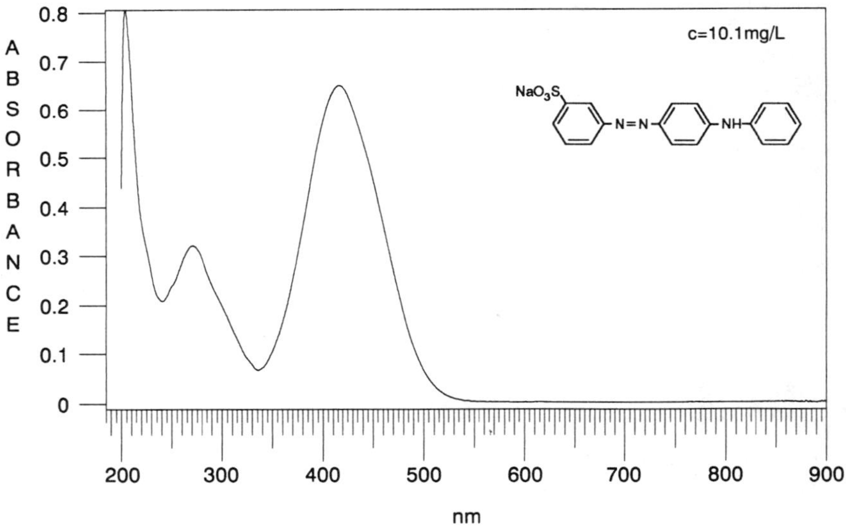

Metanil Yellow is an anionic monoazo dye that is moderately soluble in benzene, methyl Cellosolve and ether and slightly soluble in water, ethanol and acetone. It has indicator properties with a visual-transition interval of pH 1.2 (red) to 2.3 (yellow). An Indicator Chart and information on the preparation of solutions for use as an indicator can be found in the back of this book.

Industrial uses include dyeing wool, silk and jute-like fibers directly, coloring paper, lacquers and polishes, staining wood and preparing heavy-metal salts for pigments.

Staining Procedures[1] lists methods and a bibliography for the use of Mucicarmine with Hematoxylin and Metanil Yellow for demonstrating mucin, and for use in Gridley's method for demonstrating fungi in sections. Lillie[2] reported further development of Masson's Trichrome method for staining connective tissue.

References:

1) Clark, G. *Staining Procedures,* 4th ed.; Williams and Wilkins: Baltimore, 1981; pp 213, 469.

2) Lillie, R.D. *J. Technol. Methods* **1945,** *25*, 1-47.

General references:

Aldrich FT-IR Library	**1**(2), 978C
Sigma FT-IR Library	**1**(2), 162D
IR Library	**3**, 1457G
Safety	**2**(2), 2224B
Merck Index	**11**, 5833
Colour Index	3rd ed., 1971

METHYLENE BLUE

Basic Blue 9 • 3,7-Bis(dimethylamino)phenothiazin-5-ium chloride

Aldrich Cat. No.:	86,124-3	M4,490-7
Sigma Prod. No.:	M9140	M8126
Grade:	Certified	——

CAS No.: [*61-73-4*]
C.I. No.: 52015
Mol. Form.: $C_{16}H_{18}ClN_3S \cdot 3H_2O$
F.W.: 373.90
Appearance: Green powder
λ_{max}: 661nm in water
Solubility: H_2O 50mg/ml
EGME 60mg/ml
EtOH 70mg/ml

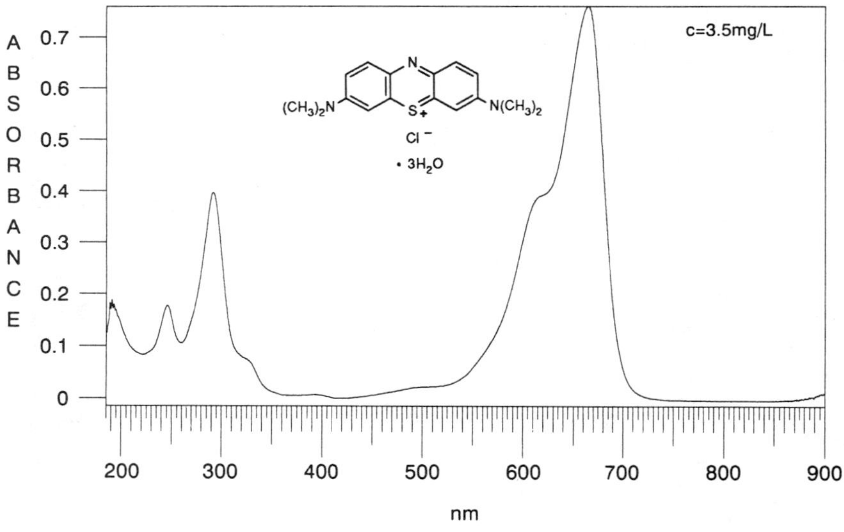

Methylene Blue is a cationic thiazine dye. The chloride salt form is soluble in water, ethanol, methyl Cellosolve and ethylene glycol. It is difficult to isolate from solution, requiring prolonged crystallizing time which necessitates tying up valuable equipment and space. On the other hand, the zinc chloride double salt (Aldrich Cat. No. 23,409-5) is readily isolated from solution by the addition of zinc chloride. This economy constitutes the bulk of its advantages, since it is less soluble in all ordinary solvents and is more toxic than the chloride.

Methylene Blue thiocyanate also is isolated readily and in a pure, anhydrous state. This is the form of choice for use as an oxidation-reduction indicator in milk analysis. It too is less soluble than the chloride and is not normally used as a biological stain.

For most biological-stain applications and all medicinal uses, the chloride is the salt form of choice. A BSC-certified grade is recommended for the former and a USP grade is required in the United States for the latter.

Medicinally, Methylene Blue serves as an antiseptic in veterinary medicine and, being antimethemoglobinemic, as an antidote for cyanide poisoning in humans and animals.

Probably its most important application is as an *in vitro* diagnostic in bacteriology, cytology, hematology and histology. Methylene Blue serves as a precursor to the Methylene Azures and a component of the various Romanowsky-type blood stains (Wright, Leishman, Giemsa and others). These have been important diagnostic aids in hematology and parasitology since the turn of the century. While pure Methylene Blue is not a good nuclear stain, it readily undergoes oxidative demethylation to form the various Methylene Azures which *are* good nuclear stains. Wittekind and Gehring[1] consider Azure B to be the best nuclear stain in this group. Spontaneous oxidation or "ripening" occurs in Methylene Blue solutions resulting in the formation of Polychrome Methylene Blue. This process can be accelerated by the addition of hydrogen peroxide, potassium dichromate, silver oxide or some other suitable oxidizing agent. The solvent dye, Methylene Violet Bernthsen (7-dimethylamino-3-oxophenothiazine), is formed when aqueous Methylene Blue solutions are boiled with alkalis or alkali carbonates. (See under Methylene Violet Bernthsen, pages 453 and 454.)

Since space precludes elaboration on all of the staining methods involving Methylene Blue, it is suggested that *Staining Procedures*[2] be consulted. A rather complete spectrum of these uses along with references, recipes and methods is provided.

References:

1) Wittekind, D.H.; Gehring, T. *Histochem. J.* **1985,** *17*, 263-289.

2) Clark, G. *Staining Procedures,* 4th ed.; Williams and Wilkins: Baltimore, 1981.

General references:

Aldrich FT-IR Library	**1**(2), 887D
Sigma FT-IR Library	**1**(2), 163D
IR Library	**3**, 1409H
Safety	**2**(2), 2344C
Merck Index	**11**, 5979
Colour Index	3rd ed., 1971

METHYLENE BLUE, zinc chloride double salt monohydrate

Aldrich Cat. No.: 23,409-5

CAS No.: [*26283-09-0*] (anhydrous)
C.I. No.: 52015
Mol. Form.: $C_{16}H_{18}ClN_3S \cdot {}^1/_2\, ZnCl_2 \cdot H_2O$
F.W.: 406.01
Appearance: Shiny, red-brown, crystalline powder
λ_{max}: 664nm in water
Solubility: H_2O 10mg/ml
EGME 1mg/ml
EtOH 0.5mg/ml

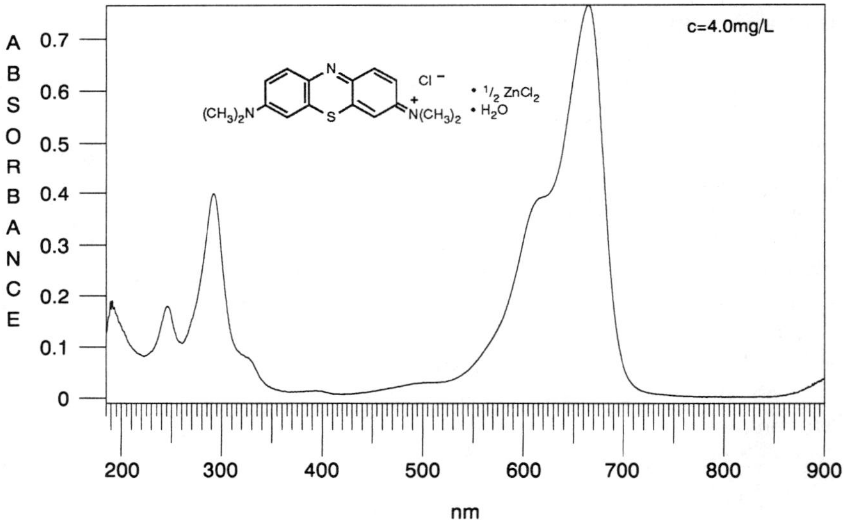

See description on pages 448 and 449 under Methylene Blue.

General references:

Aldrich FT-IR Library **1**(2), 1037D
Colour Index 3rd ed., 1971

METHYLENE GREEN, zinc chloride double salt

Basic Green 5

Aldrich Cat. No.: 21,110-9
Sigma Prod. No.: M7766

CAS No.: *[6722-15-2]*
C.I. No.: 52020
Mol. Form.: $C_{16}H_{17}ClN_4O_2S \cdot {}^1/_2\ ZnCl_2$
F.W.: 433.00
Appearance: Dark-brown powder
λ_{max}: 657(618)nm in pH 5.8 buffer
Solubility: H_2O 30mg/ml
EGME 3mg/ml
EtOH 1mg/ml

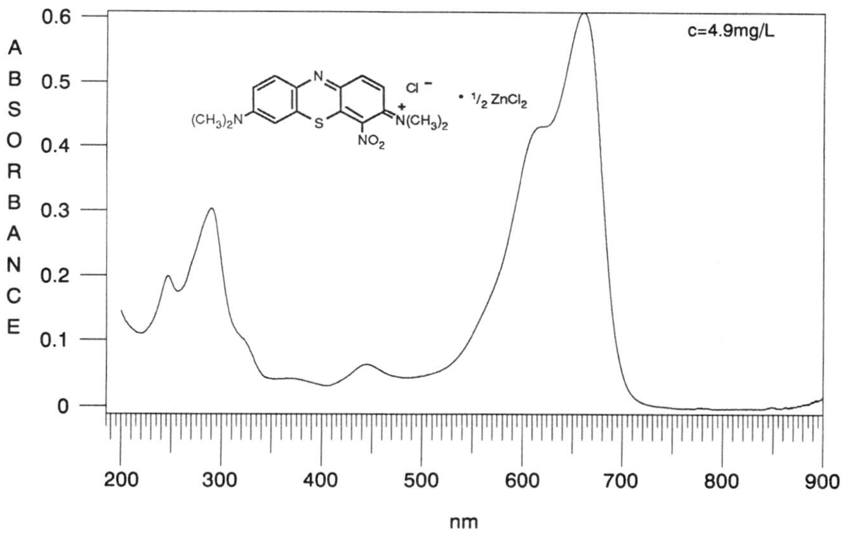

Methylene Green is a metachromatic, cationic thiazine dye that is made by the mononitration of Methylene Blue. This is accomplished by nitrosating Methylene Blue with nitrous acid followed by treatment with nitric acid to oxidize the nitroso group to a nitro function. The dye is then isolated as the zinc chloride double salt with a dye content of approximately 65%.

The *Colour Index*[1] reported Methylene Green to be the most fast basic-green dye known. At that time, it was used to top logwood-loaded silk to attain a good black shade. On a limited scale, it is currently used commercially to dye paper and leather.

Conn's[2] reported its occasional use by botanists as a substitute for

Methyl Green in combination with Eosin for staining fixed chromatin in wood. Lison[3] used it as a vital stain for insects. Lillie[4] observed that mast cells of the mouse and rat which failed to color with the periodic acid-Schiff reaction responded to Methylene Green.

References:

1) *The Colour Index,* 1st ed., Society of Dyers and Colourists: Bradford, Yorkshire, 1924.

2) *Conn's Biological Stains,* 9th ed.; Lillie, R.D., Ed.; Williams and Wilkins: Baltimore, 1977; p 426.

3) Lison, L. *Z. Zellorsch.* **1938,** *28*, 179-209.

4) Lillie, R.D. *Anat. Rec.* **1950,** *108*, 239-253.

General references:

Aldrich FT-IR Library **1**(2), 1038D
Sigma FT-IR Library **1**(2), 164A
Colour Index 3rd ed., 1971

METHYLENE VIOLET Bernthsen

Aldrich Cat. No.:	21,631-3
Sigma Prod. No.:	M1138
Grade:	Certified
CAS No.:	*[2516-05-4]*
C.I. No.:	52041
Mol. Form.:	$C_{14}H_{12}N_2OS$
F.W.:	256.33
m.p.:	216°C (dec.)
Appearance:	Dark-green powder
λ_{max}:	580nm in 50% ethanol +10ml 0.1*N* hydrochloric acid
Solubility:	H_2O 0.6mg/ml
	EGME 7mg/ml
	EtOH 4mg/ml

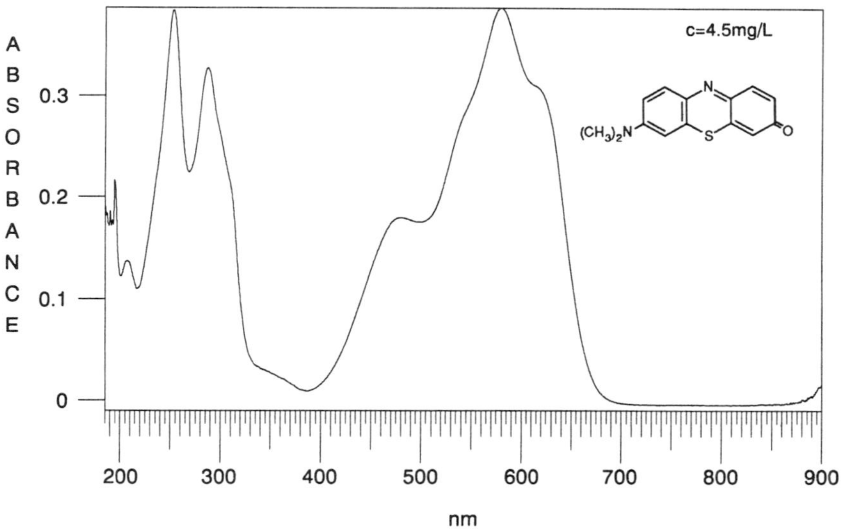

When Methylene Blue is treated with acid dichromate to prepare Polychrome Methylene Blue (a variable mixture), a small percent of Methylene Violet Bernthsen (MVB) is formed, along with larger amounts of the various azures, Thionin, thiolines and thionols.

MVB, a basic, metachromatic, thiazine dye, is produced in up-to-40% yield by treating a dilute ammoniacal solution of Methylene Blue with potassium dichromate. After the ammonia is driven off by heating with an alkali carbonate, the resulting product is recrystallized from 1,2-dichloroethane. This product, the free base, is soluble to some degree in most common organic solvents and virtually insoluble in water.

The free base is the form shown in the *Colour Index* and *Conn's*.[1] It is also the form certified by the BSC.

Gurr[2,3] shows two slightly different structures as the chloride salt. Both are highly unlikely. It is possible he was working with the hydrochloride-salt form which is soluble in water and readily soluble in ethanol and methanol.

Conn's stated "MVB plays an important part in the nuclear and granule staining of Polychrome Methylene Blue stain. A definite quantity of this dye is employed in the tetrachrome stain of McNeal. It is thought to be responsible for the sky blue assumed by the cytoplasm of lymphocytes in such blood stains." Lillie[4] dissented from this position. He attributed this staining quality to Azure B. Marshall *et al.*[5] demonstrated that it is possible to get a balanced bloodstain and the Romanowsky-Giemsa effect with just three pure components: Azure B, Methylene Blue and Eosin Y. Wittekind *et al.*[6] eliminated Methylene Blue from this mixture, leaving only Azure B and Eosin Y. The staining quality was essentially unaltered. As a result, Lillie's position was reinforced since the "sky blue" lymphocytic cytoplasm could be demonstrated in the absence of MVB.

From these recent studies, it can be concluded that MVB is of no value as a component in Romanowsky-type blood stain mixtures. In fact, its presence in such mixtures is probably detrimental, due to its inferior solubility. This lesser solubility prompts the precipitation of MVB on standing, which appears to facilitate the precipitation of other beneficial components as well and results in a reduced shelf life of the solution.

From its solubility characteristics, it appears MVB might be of some value as a lipid stain in applications similar to those for which Nile Red is used.

Gurr[2] stated that MVB is used as an oxidation-reduction indicator.

MVB has been confused with the azine dye, C.I. 50205 (Basic Violet 5) and Methylene Violet RRA or 3RA, both formerly used as textile dyes. They do not display the same staining quality and cannot be used interchangeably.

References:

1) *Conn's Biological Stains,* 9th ed.; Lillie, R.D., Ed.; Williams and Wilkins: Baltimore, 1977; pp 429-430.

2) Gurr, E. *Encyclopedia of Microscopic Stains* ; Neill: Edinburgh, 1960; p 271.

3) Gurr, E. *Synthetic Dyes*; Academic: London, 1971; p 60.

4) Lillie, R.D. *Stain Technol.* **1944,** *19,* 55-58.

5) Marshall, P.N.; Bentley, S.A.; Lewis, S.M. *J. Clin. Pathol.* **1975,** *28,* 920-923.

6) Wittekind, D.; Kretschmer, V.; Lohr, W. *Blut* **1976,** *32*, 71-78.

General references:

Aldrich FT-IR Library **1**(2), 1036D
Sigma FT-IR Library **1**(2), 164B
Colour Index 3rd ed., 1971

METHYLENE VIOLET 3RAX

3-Amino-7-(diethylamino)-5-phenylphenazinium chloride
• *N,N*-Diethylphenosafranine

Aldrich Cat. No.:	30,750-5
Sigma Prod. No.:	M8274

CAS No.:	*[4569-86-2]*
C.I. No.:	50206
Mol. Form.:	$C_{22}H_{23}ClN_4$
F.W.:	378.91
m.p.:	285°C (dec.)
Appearance:	Green crystalline powder
λ_{max}:	557nm in water
Solubility:	H_2O 10mg/ml
	EGME 20mg/ml
	EtOH 5mg/ml

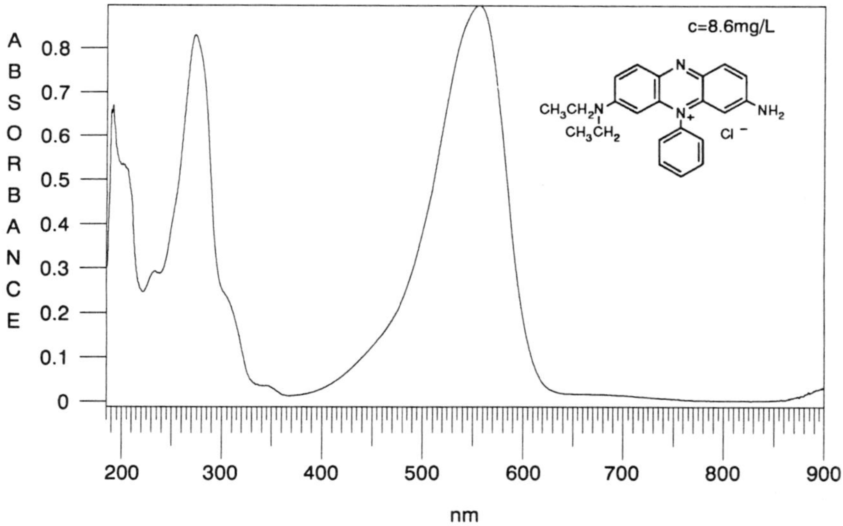

This reddish-violet, cationic azine dye is made by oxidizing a mixture of *N,N*-diethyl-1,4-phenylenediamine and aniline.

Slightly soluble in water, ethanol and methyl Cellosolve, it is used as an intermediate to make cationic monoazo dyes such as Janus Green B (C.I. 11050) and Copying Black (C.I. 11090).

Conn's[1] states that it is occasionally used as a fresh diazonium salt in histochemistry.

Reference:

1) *Conn's Biological Stains,* 9th ed.; Lillie, R.D., Ed.; Williams and Wilkins: Baltimore, 1977; p 383.

METHYL EOSIN
Solvent Red 44

Aldrich Cat. No.:	19,955-9
Sigma Prod. No.:	M4886
CAS No.:	*[23391-49-3]*
C.I. No.:	45385
Mol. Form.:	$C_{21}H_9Br_4NaO_5$
F.W.:	683.93
Appearance:	Red powder
λ_{max}:	520nm in water
Solubility:	H_2O 30mg/ml
	EGME 10mg/ml
	EtOH 10mg/ml

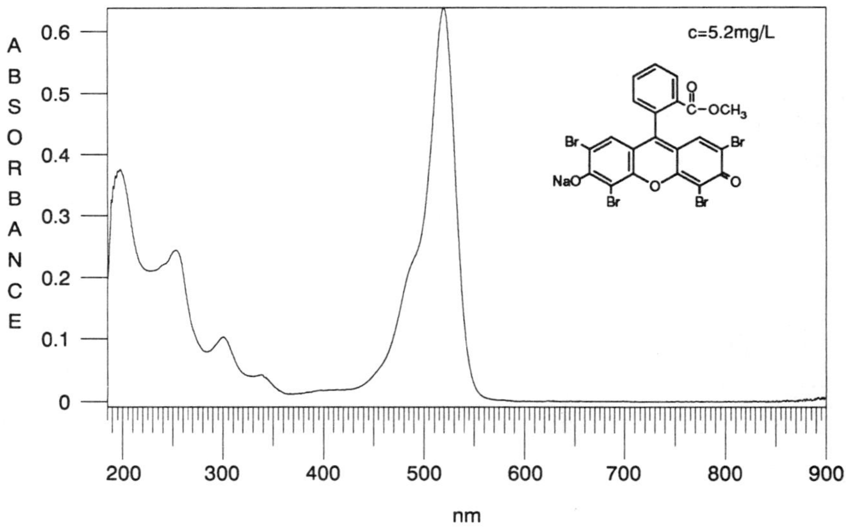

Methyl Eosin is a red xanthene dye that is made from Eosin Y by esterifying its carboxyl function with methanol. It is practically insoluble in cold water, soluble in hot water remaining in solution on cooling and somewhat soluble in cold alcohol. Its primary industrial application is coloring spirit varnishes used to coat metal foils.

Apparently, it has many of the same properties as Ethyl Eosin and probably can be substituted for Ethyl Eosin in biological-stain applications.

General references:

Aldrich FT-IR Library	**1**(2), 1012A
Sigma FT-IR Library	**1**(2), 164C
IR Library	**3**, 1474C
Colour Index	3rd ed., 1971

4-METHYLESCULETIN

6,7-Dihydroxy-4-methylcoumarin • 6-Hydroxy-4-methylumbelliferone

Aldrich Cat. No.: 11,990-3
Sigma Prod. No.: D6275
Purity: 98%

CAS No.: *[529-84-0]*
Mol. Form.: $C_{10}H_8O_4$
F.W.: 192.17
m.p.: 274-276°C
Appearance: Light-brown powder
λ_{max}: 348(292)nm in methanol
Solubility: H_2O 0.1mg/ml
EGME 20mg/ml
EtOH 6mg/ml

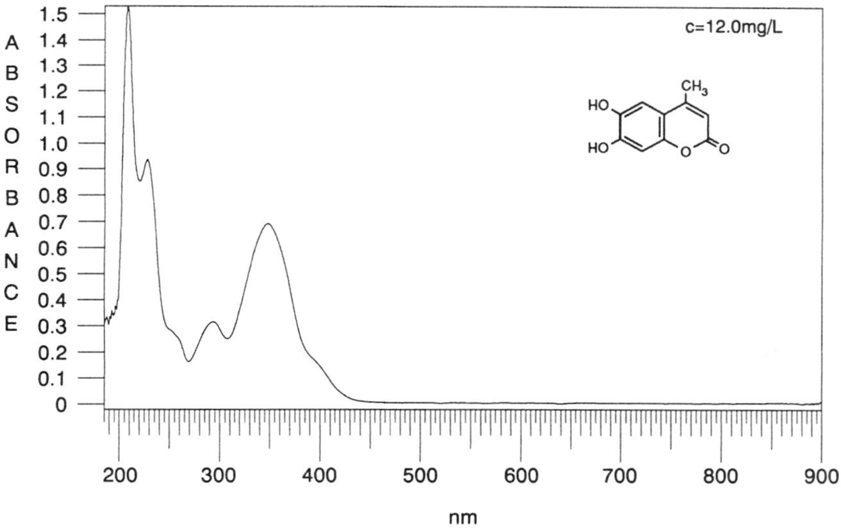

This coumarin derivative is very slightly soluble in water, slightly soluble in ethanol and soluble in ethylene glycol, chloroform, ether and alkali hydroxide solutions. A fluorescent indicator, its fluorescent-transition interval extends from pH 4.0 (colorless) to pH 6.2 (blue).

There is no evidence in the literature that 4-methylesculetin has been used as a biological stain.

General references:

Aldrich FT-IR Library	1(2), 326A
Sigma FT-IR Library	1(2), 476B
IR Library	**3**, 1045D
NMR Library	1(2), 476B

METHYL GREEN, zinc chloride salt

Ethyl Green

Aldrich Cat. No.:	19,808-0	32,382-9
Sigma Prod. No.:	M5015	M6776
Grade:	Certified	Crystal Violet free
CAS No.:	*[7114-03-6]*	
C.I. No.:	42590	
Mol. Form.:	$C_{27}H_{35}Cl_2N_3 \cdot XZnCl_2$	
F.W.:	472.51	
m.p.:	233°C (dec.)	>300°C
Appearance:	Dark, brownish-red powder	Green powder
λ_{max}:	629(423)nm in water	635(420)nm in water
Solubility:	H_2O 70mg/ml	
	EGME 4mg/ml	
	EtOH 1mg/ml	

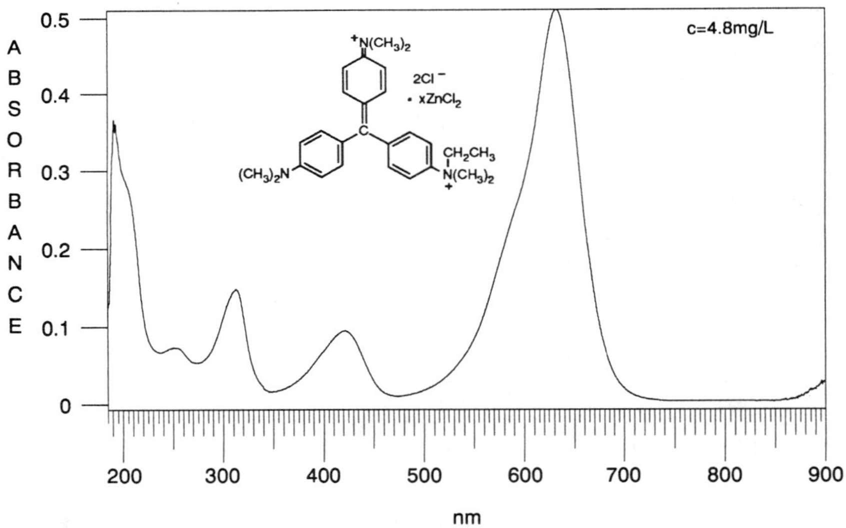

Traditional Methyl Green (Basic Green 20, C.I. 42585) apparently has not been produced commercially for many years. However, its deficiencies, unwittingly, are associated with contemporary Methyl Green (C.I. 45290). The latter is readily available.

Both of the Sigma/Aldrich products listed above under C.I. 42590, are basic triphenylmethane-type dyes. They are also rare examples of dicationic dyes. Among the common dyes, this property is shared only with Bismarck Brown Y and Bismarck Brown R.

Both the traditional and the contemporary Methyl Greens are equally effective stains when they are pure, and structurally they differ only slightly. In fact, the difference is so subtle that it is practically impossible to detect spectrophotometrically in pure samples. However, it is difficult to make and keep traditional Methyl Green pure. This stems from the reluctance of the reaction to go to completion, due to the gaseous nature of the methylating reagent, methyl chloride. Conversely, decomposition occurs much more readily, since methyl chloride is the product eliminated. Both deficiencies result in the traditional product being contaminated with Crystal Violet, the precursor of the Methyl Greens. Although Crystal Violet can be removed from traditional Methyl Green by extraction with chloroform, it is gradually regenerated, necessitating subsequent purification. Contemporary Methyl Green, on the other hand, is made with a liquid ethyl bromide, and the ethylation goes to completion quite smoothly. This fact can be verified when the progress of the reaction is followed by thin-layer chromatography. This product is also more stable than its traditional counterpart. Material stored for prolonged periods has not displayed evidence of decomposition.

It has been reported[1-3] that, unless the dye is pure, cell and tissue components that should be stained a brilliant green will stain a rather greyish green. Solutions made with pure Methyl Green are stable for many months.

Although nuclei are stained with Methyl Green, most nucleoli are not. This gives some insight into the reason that Methyl Green exhibits a superior staining synergism with Pyronin Y. Pyronin Y has an affinity for many cell and tissue components that are not stained by Methyl Green. The nucleoli, the cytoplasm of plasma cells, plasmablasts and large basophilic stem cells are all pyraninophilic.[4] RNA probably is responsible for the pyraninophilia exhibited by most bacteria.

Nonspecific staining of both Methyl Green and Pyronin Y is suppressed by increasing the salt concentration in the staining solutions. However, this has much less effect on DNA and RNA than other cell components. Scott[5] has suggested that the staining of DNA by this method is partly due to forces other than electrostatic bonds (*e.g.*, the high salt concentration reduces the repulsion of the ionic charge and allows the weaker hydrogen bonding to prevail).

Ahlquist[6] suggested a simple, fast technique using Methyl Green-Pyronin for routine staining on biopsies of lymphoid glands and to enable the pathologist to identify different pyraninophilic cells.

McCann[7] used Methyl Green to demonstrate the sites of developing cartilage in human embryos. In order to avoid the opacity of pigments indigenous to the tissue, the specimen was first bleached in a hydrogen peroxide-ethanol mixture and then stained with Methyl Green. The developing cartilage sites were stained a deep green and the surrounding tissue, which did not retain the dye, was almost transparent.

The addition of DNA to dilute solutions of Methyl Green results in a shift upscale of the absorption maximum from 635nm to 645nm. Kurnick[8] did not observe this shift with RNA or depolymerized DNA.

References:

1) Pearse, A.G.E. *Histochemistry Theoretical and Applied;* Churchill: London, 1968.
2) Augulis, V.; Sepinwall, J. *Stain Technol.* **1971**, *46*, 137-143.
3) Kurnick, N.B. *ibid.* **1952**, *27*, 233.
4) Mori, Y.; Lennert, K. *Electron Microscopic Atlas of Lymph Node Cytology;* Springer: Berlin, 1969.
5) Scott, J.E. *Histochemie* **1967**, *9*, 30.
6) Ahlquist, J.; Anderson, L. *Stain Technol.* **1971**, *46*, 17.
7) McCann, J.A. *ibid.* **1972**, *47*, 263.
8) Kurnick, N.B. *J. Gen. Phys.* **1950**, *23*, 243.

General references:

IR Library **3**, 1483H
Safety **2**(2), 2356D

METHYL ORANGE
Acid Orange 52

Aldrich Cat. No.:	23,410-9	11,451-0
Sigma Prod. No.:	——	M1400
Grade:	Certified	A.C.S. reagent
Dye content:	~95%	~85%

CAS No.:	[*547-58-0*]
C.I. No.:	13025
Mol. Form.:	$C_{14}H_{14}N_3NaO_3S$
F.W.:	327.34
m.p.:	>300°C
Appearance:	Orange powder
λ_{max}:	507nm in water + 0.5ml 1*N* hydrochloric acid
Solubility:	H_2O 4mg/ml
	EGME 20mg/ml
	EtOH 0.3mg/ml

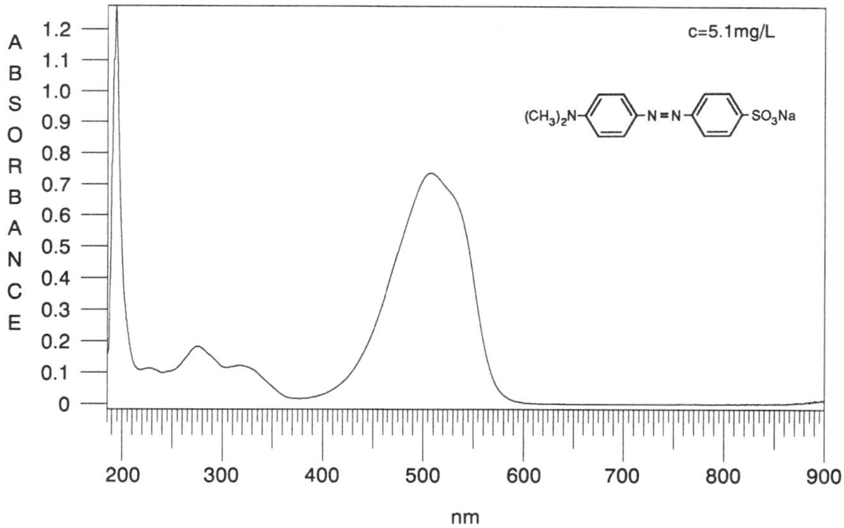

Methyl Orange is an amphoteric, monoazo dye that is made by coupling dimethylaniline to diazotized sulfanilic acid and converting the coupled product to the sodium salt.

Slightly soluble in water, it displays useful indicator properties and is sensitive to mineral acids while not affected by carbonates. *Reagent Chemicals*[1] lists it with a visual-transition interval of pH 3.2 (pink or red) to pH 4.4 (yellow).

Methyl Orange is very seldom used in textile applications because it is fugitive and sensitive to acids; however, on occasion, it has been used to dye wool and silk from an acid bath.

The BSC certifies it for use as a stain for pituitary acidophils with the Wilson-Ezrin method, in the Kreyberg stain and in Flemming's triple stain on plant material.

Reference:

1) *Reagent Chemicals,* 7th ed.; American Chemical Society: Washington, DC, 1986; pp 434-435.

General references:

Aldrich FT-IR Library	**1**(2), 972C
Sigma FT-IR Library	**1**(2), 165D
IR Library	**3**, 1455B
Safety	**2**(2), 2397C
Merck Index	**11**, 6019
Colour Index	3rd ed., 1971

METHYL PURPLE INDICATOR, solution in water

Aldrich Cat. No.: 27,733-9
Sigma Prod. No.: M8391

CAS No.: [*1340-02-9*]
Appearance: Green liquid
λ_{max}: 410(620)nm in methanol

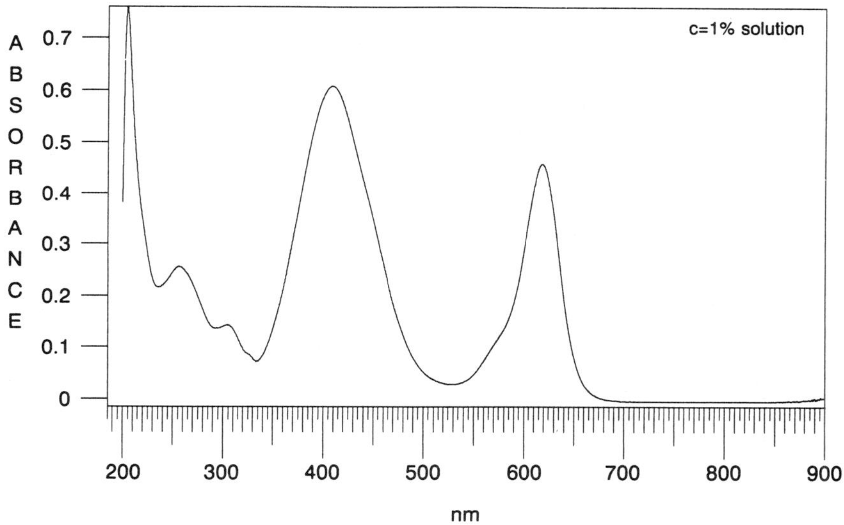

This mixed-indicator solution has a visual-transition interval (VTI) that extends from pH 4.8 (purple) to 5.4 (green) with a dirty, blue-grey mid range that is readily distinguished from the end point. While Methyl Purple was designed as an indicator for Kjeldahl-nitrogen determinations, it has been used in other applications that demand an indicator with a sharp end point in the described VTI.

General reference:

Safety **2**(2), 2427C

METHYL RED

Acid Red 2 • 2-[4-(Dimethylamino)phenylazo]benzoic acid

Aldrich Cat. No.:	25,019-8
Sigma Prod. No.:	M1525
Grade:	A.C.S. reagent Crystals

CAS No.:	[*493-52-7*]
C.I. No.:	13020
Mol. Form.:	$C_{15}H_{15}N_3O_2$
F.W.:	269.31
m.p.:	179°C
Appearance:	Maroon crystalline powder
λ_{max}:	410nm in methanol
Solubility:	H_2O 0.1mg/ml EGME 6mg/ml EtOH 2mg/ml

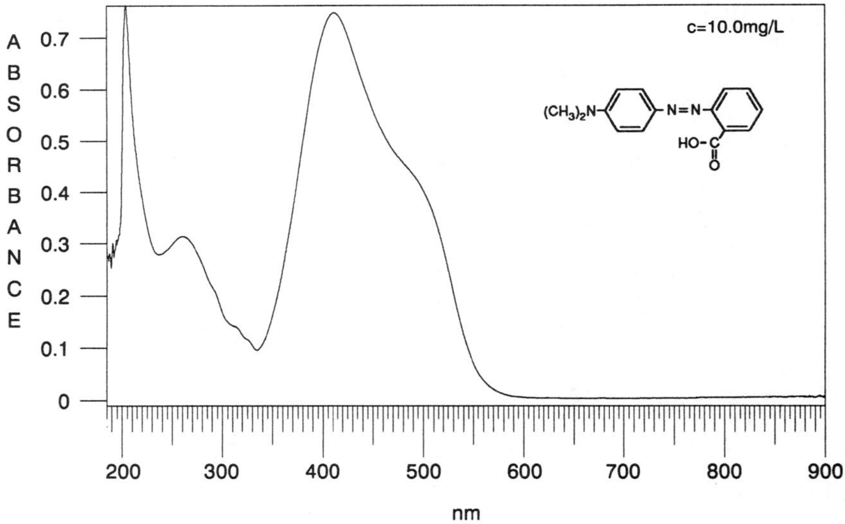

Methyl Red is an amphoteric monoazo dye that is made by coupling dimethylaniline to diazotized anthranilic acid.

There are three forms of this dye and all are used as pH indicators. The visual-transition interval for the three forms is pH 4.2 (pink or red) to pH 6.2 (yellow). *Reagent Chemicals*[1] recommends the crystalline (free form), which has a melting point of 179-182°C and is soluble in benzene and slightly soluble in alcohol, for nonaqueous titrations, particularly when an aprotic solvent is used. The hydrochloride-salt form is slightly soluble in alcohol and water. It is the choice for aqueous titrations and suitable for nonaqueous titrations when the medium is an amphiprotic solvent. The sodium salt is soluble in water and alcohol. It can be used as an indicator in aqueous and amphiprotic solvents. Methyl Red is very susceptible to reduction, which necessitates a quick interpretation of the rapidly fading end-point.

Methyl Red is of little practical value industrially because of its indicator properties and the ease with which it is reduced.

Carter[2] used it as a vital stain for protozoa.

References:

1) *Reagent Chemicals,* 7th ed.; American Chemical Society: Washington, DC, 1986; pp 437-438.

2) Carter, J.S. *J. Exp. Zool.* **1933,** *65*, 159-179.

General references:

Aldrich FT-IR Library **1**(2), 971C
Sigma FT-IR Library **1**(2), 166A
Merck Index **11**, 6037

METHYL RED hydrochloride

Aldrich Cat. No.:	86,123-5
Sigma Prod. No.:	M4400
Grade:	A.C.S. reagent

CAS No.:	*[63451-28-5]*
Mol. Form.:	$C_{15}H_{15}N_3O_2 \cdot HCl$
F.W.:	305.77
m.p.:	175°C (dec.)
Appearance:	Blue crystalline powder
λ_{max}:	493nm in methanol + 1ml 1*N* hydrochloric acid
Solubility:	H_2O 2mg/ml
	EGME 7mg/ml
	EtOH 5mg/ml

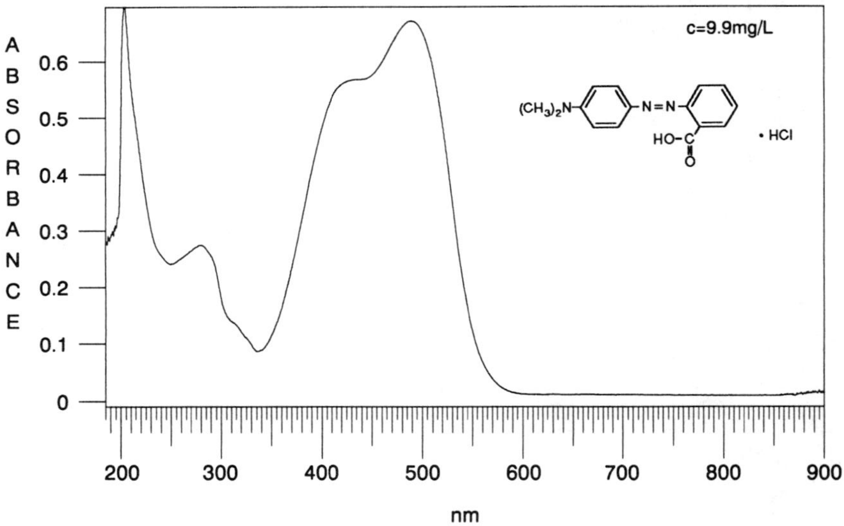

See description on pages 464 and 465 under Methyl Red.

General references:

Aldrich FT-IR Library	**1**(2), 968C
Sigma FT-IR Library	**1**(2), 166B
IR Library	**3**, 1451G

METHYL RED, sodium salt

Aldrich Cat. No.:	11,450-2
Sigma Prod. No.:	M1650
Grade:	A.C.S. reagent
CAS No.:	[*845-10-3*]
Mol. Form.:	$C_{15}H_{14}N_3NaO_2$
F.W.:	291.29
Appearance:	Orange powder
λ_{max}:	437nm in methanol
Solubility:	H_2O 70mg/ml
	EGME 60mg/ml
	EtOH 40mg/ml

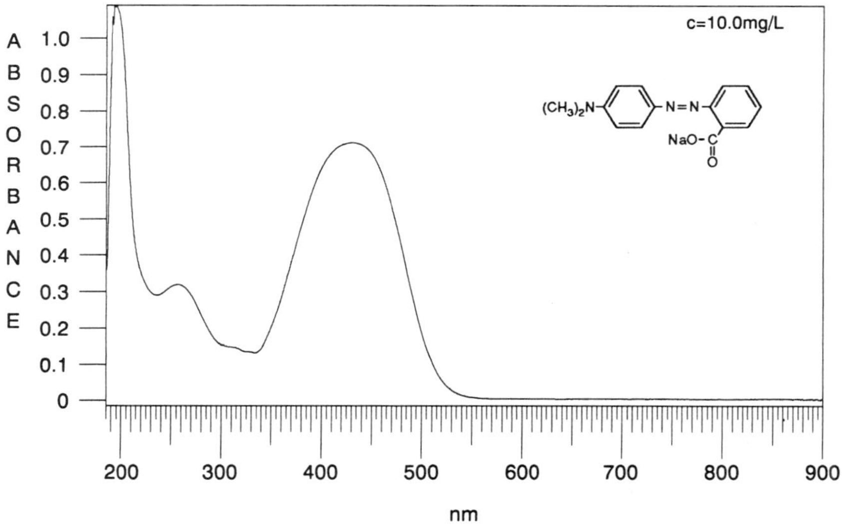

See description on pages 464 and 465 under Methyl Red.

General references:

Aldrich FT-IR Library	**1**(2), 971D
Sigma FT-IR Library	**1**(2), 166C
IR Library	**3**, 1454F
NMR Library	**2**(2), 980B

METHYLTHYMOL BLUE, water soluble

3,3'-Bis[*N,N*-di(carboxymethyl)aminomethyl]thymolsulfonephthalein, sodium salt

Aldrich Cat. No.: B4,200-4
Sigma Prod. No.: M2255

CAS No.: [*32469-43-5*]
Mol. Form.: $C_{37}H_{39}N_2Na_5O_{13}S$
F.W.: 866.62
Appearance: Dark-green powder
λ_{max}: 438nm in water
Solubility: H_2O 70mg/ml
EGME 40mg/ml
EtOH 1mg/ml

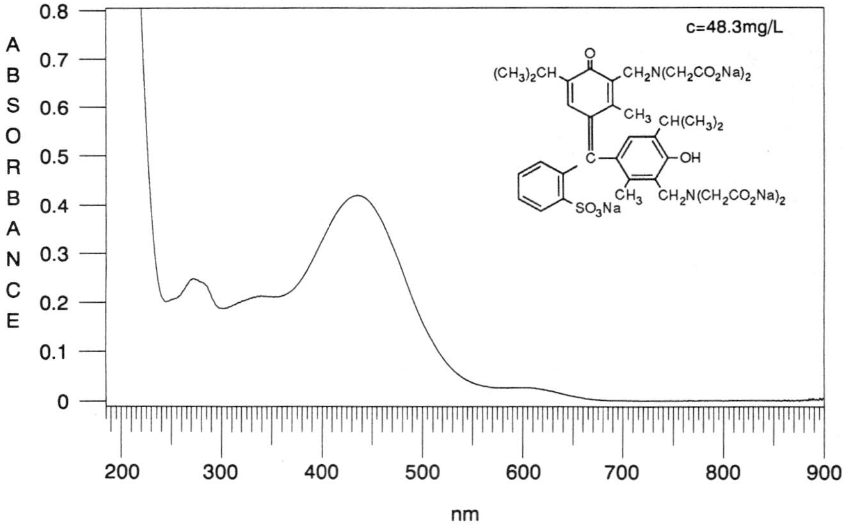

Methylthymol Blue is a water-soluble, sulfonephthalein dye that is used as a chelatometric indicator for the determination of zinc. There is no evidence in the literature that it has been used industrially.

General references:

Aldrich FT-IR Library **1**(2), 1028B
Sigma FT-IR Library **1**(2), 166D
IR Library **3**, 1482F

METHYL VIOLET B base
Solvent Violet 8

Aldrich Cat. No.: 24,221-7

CAS No.: [*52080-58-7*]
C.I. No.: 42535B
Appearance: Black powder
λ_{max}: 580nm in methanol
Solubility: H_2O 0.4mg/ml
EGME 70mg/ml
EtOH 30mg/ml

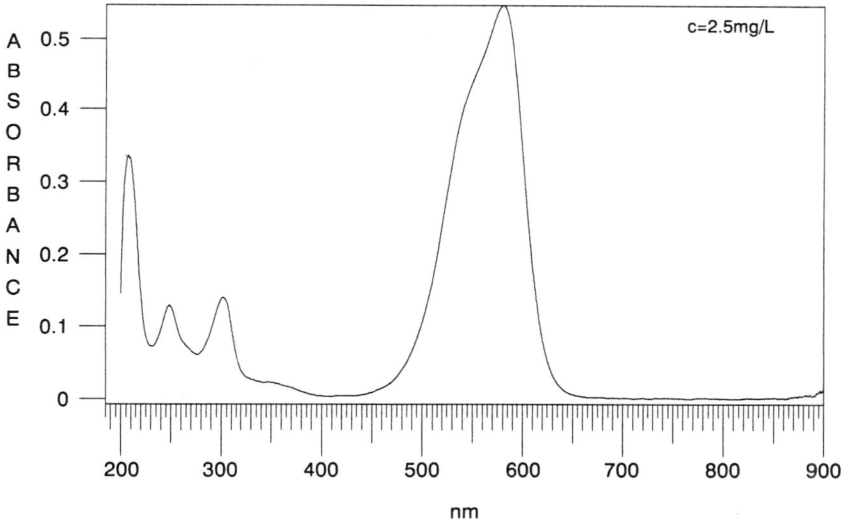

See description on pages 470 and 471 under Methyl Violet 2B.

General Reference:
Aldrich FT-IR Library **1**(2), 964A

METHYL VIOLET 2B
Basic Violet 1

Aldrich Cat. No.:	19,809-9	85,840-4
Sigma Prod. No.:	——	M4511
Grade:	Certified	Indicator
Dye content:	~85%	~75%

CAS No.:	[*8004-87-3*]	
C.I. No.:	42535	
m.p.:	137°C (dec.)	
Appearance:	Green crystalline powder	
λ_{max}:	584nm in 50% ethanol	
Solubility:	H_2O 50mg/ml	20mg/ml
	EGME 50mg/ml	40mg/ml
	EtOH 60mg/ml	9mg/ml

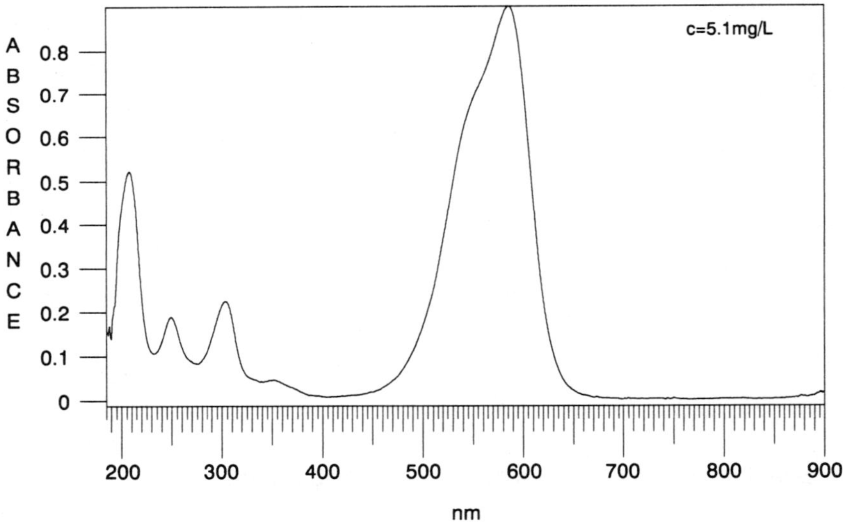

Methyl Violet is a purple, basic triphenylmethane dye. The certified and indicator grades are composed of a mixture of the hydrochloride salts of the more highly methylated forms of pararosaniline (tetra, penta and hexa), predominantly the penta. It is soluble in water and the lower alcohols. It dyes cotton (tannin-mordanted), wool, bast fibers and silk directly. However, it is more widely used in non-textile applications such as leather, photographic filters, spirit inks, lacquers and paper, the latter being probably the largest use. Methyl Violet base is the free-base form of the same homologous mixture of dyes. It is very slightly soluble in water and soluble in the lower alcohols. It is used to dye oleic, stearic and other fatty acids, typewriter ribbons, copying papers, printing and ball-point inks and lacquers.

For biological applications, the same homologous dye mixture is sold in Europe under the name of Gentian Violet. This is an archaic designation in the United States where Crystal Violet (hexamethylpararosaniline) is supplied when Gentian Violet is requested. *Conn's*[1] states that certified Methyl Violet 2B is checked for all the same applications as Crystal Violet except as a Gram stain. The results are described as being a "redder hue."

Reference:

1) *Conn's Biological Stains,* 9th ed.; Lillie, R.D., Ed.; Williams and Wilkins: Baltimore, 1977; pp 585-586.

General references:

Aldrich FT-IR Library	**1**(2), 963C
Sigma FT-IR Library	**1**(2), 167C
IR Library	**3**, 1448G
Safety	**2**(2), 2464B
Colour Index	3rd ed., 1971

METHYL YELLOW

N,N-Dimethyl-4-phenylazoaniline • Solvent Yellow 2

Aldrich Cat. No.:	11,449-9
Sigma Prod. No.:	D6760
Grade:	Indicator
CAS No.:	[*60-11-7*]
C.I. No.:	11020
Mol. Form.:	$C_{14}H_{15}N_3$
F.W.:	225.30
m.p.:	111°C (dec.)
Appearance:	Orange-yellow powder
λ_{max}:	408(256) in methanol

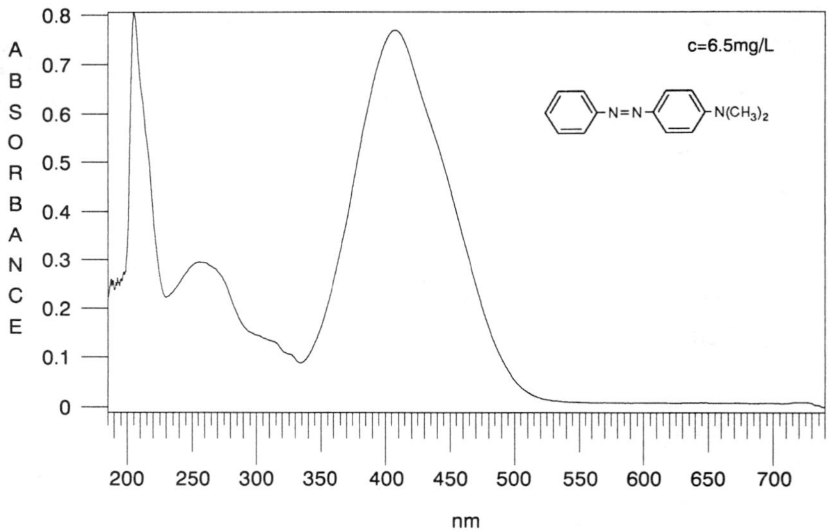

Methyl Yellow is a monoazo solvent dye that is made by coupling dimethylaniline to diazotized aniline. It is insoluble in water and soluble in alcohol, benzene, chloroform, ether, petroleum ether, oils and mineral acids. It was formerly used to color polishes, waxes, polystyrene, petroleum and soaps. Methyl Yellow displays indicator properties with a visual-transition interval of pH 2.9 (red) to pH 4.0 (yellow). *Merck*[1] states that it is used to determine free hydrochloric acid in gastric juices and to spot-test peroxidized fats.

Methyl Yellow is an OSHA-controlled carcinogen.

Reference:

1) *The Merck Index,* 11th ed.; Budavari, S. *et al.*, Eds.; Merck: Rahway, 1989; Monograph 3218

General references:

IR Library	**3,** 1452G
Safety	**2**(2), 2464D

MORDANT BLUE 9

Acid Fast Chromium Blue RR

Aldrich Cat. No.: 19,521-9
Sigma Prod. No.: M5011

CAS No.: *[3624-68-8]*
C.I. No.: 14855
Mol. Form.: $C_{16}H_9ClN_2Na_2O_8S_2$
F.W.: 502.82
Appearance: Black-brown powder
λ_{max}: 516nm in water + 1ml 1*N* hydrochloric acid
Solubility: H_2O 50mg/ml
EGME 5mg/ml
EtOH 0.6mg/ml

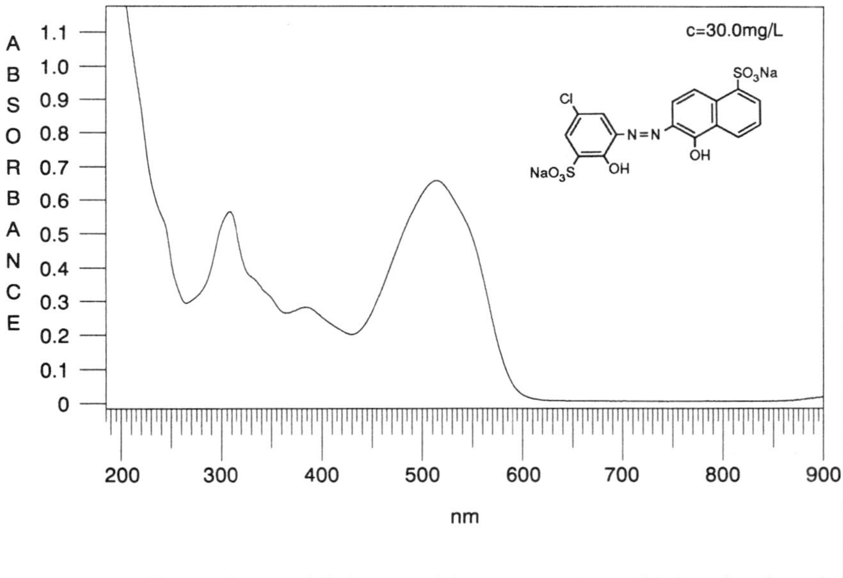

Mordant Blue 9 is a reddish-navy-blue, monoazo acid dye that is soluble in water and very slightly soluble in alcohol. It dyes wool, mordanted silk and nylon with afterchroming. Wool is vigoureux printed with a chromium mordant.

There is no evidence that it has been used as a biological stain.

General references:

Aldrich FT-IR Library	**1**(2), 982B
Sigma FT-IR Library	**1**(2), 168B
IR Library	**3**, 1459H
Colour Index	3rd ed., 1971

MORDANT BROWN 1

5-[2,4-Diamino-5-(2-hydroxy-5-nitrophenylazo)phenylazo]-
1-naphthalenesulfonic acid, sodium salt • Metachrome Brown BC, EB

Aldrich Cat. No.: 19,517-0
Sigma Prod. No.: M5136

CAS No.: [*3564-15-6*]
C.I. No.: 20110
Mol. Form.: $C_{22}H_{16}N_7NaO_6S$
F.W.: 529.47
Appearance: Dark-brown powder
λ_{max}: 373(487)nm in water
Solubility: H_2O 60mg/ml
EGME 30mg/ml
EtOH 5mg/ml

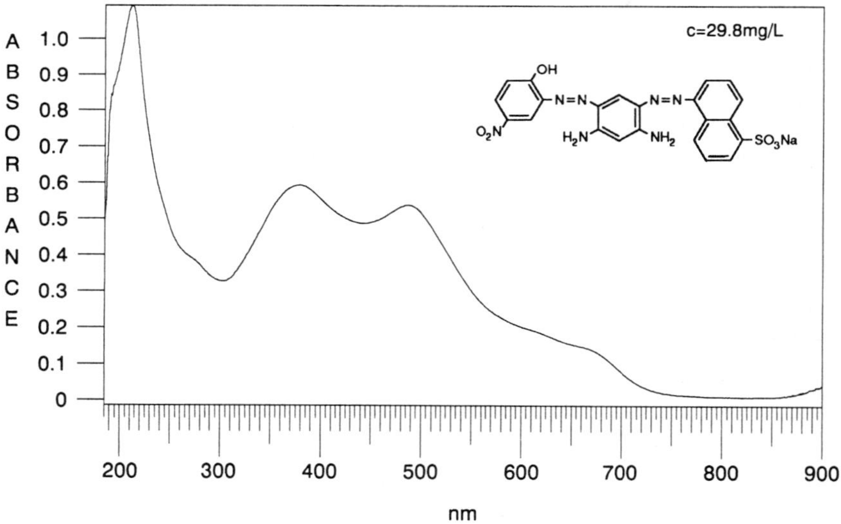

This deep-brown disazo dye is prepared by first coupling diazotized 2-amino-4-nitrophenol to 1,3-phenylenediamine (in the 4-position) followed by coupling diazotized 1-aminonaphthalene-5-sulfonic acid to the same nucleus (in the 6-position). It is soluble in water and methyl Cellosolve and slightly soluble in acetone and ethanol. Industrially, it is used to dye chromed leather, loose wool and hat bodies. Textile applications include dyeing wool by mordanting, metachroming or afterchroming, silk by afterchroming or metachroming and nylon by afterchroming. Printing wool requires a mordant, and printing cotton requires chromium acetate.

There is no evidence in the literature to indicate that this dye has been used as a biological stain.

General references:

Aldrich FT-IR Library **1**(2), 990B
Sigma FT-IR Library **1**(2), 168C
IR Library **3**, 1464A
Colour Index 3rd ed., 1971

MORDANT BROWN 4

2-(4,6-Diamino-*m*-tolylazo)-4,6-dinitrophenol

Aldrich Cat. No.: 19,515-4
Sigma Prod. No.: M5261

CAS No.: [*6247-27-4*]
C.I. No.: 11335
Mol. Form.: $C_{13}H_{12}N_6O_5$
F.W.: 332.28
Appearance: Black powder
λ_{max}: 500(374)nm in ethanol
Solubility: H_2O 60mg/ml
EGME 0.3mg/ml
EtOH 1mg/ml

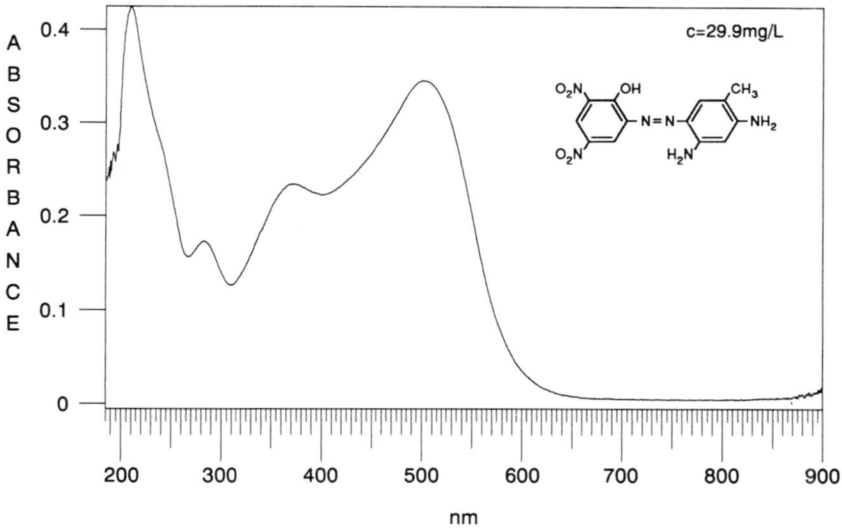

This reddish-brown, basic monoazo dye is prepared by coupling diazotized picramic acid to 2,4-tolylenediamine. It is soluble in hot water and slightly soluble in ethanol, methyl Cellosolve and acetone. Industrially, it is used to dye chromed leather, wool (by mordanting, metachroming or afterchroming), silk by afterchroming and nylon by either mordanting or afterchroming.

There is no evidence in the literature that it has been used as a biological stain.

General references:

Aldrich FT-IR Library **1**(2), 967D
Sigma FT-IR Library **1**(2), 168D
IR Library **3**, 1451D
Colour Index 3rd ed., 1971

MORDANT BROWN 6

4,6-Dinitro-4'-methyl-2,2'-azodiphenol

Aldrich Cat. No.:	19,520-0
Sigma Prod. No.:	M5386
CAS No.:	*[6247-28-5]*
C.I. No.:	11875
Mol. Form.:	$C_{13}H_{10}N_4O_6$
F.W.:	318.25
Appearance:	Brown-black powder
λ_{max}:	400nm in water
Solubility:	H_2O 50mg/ml
	EGME 20mg/ml
	EtOH 2mg/ml

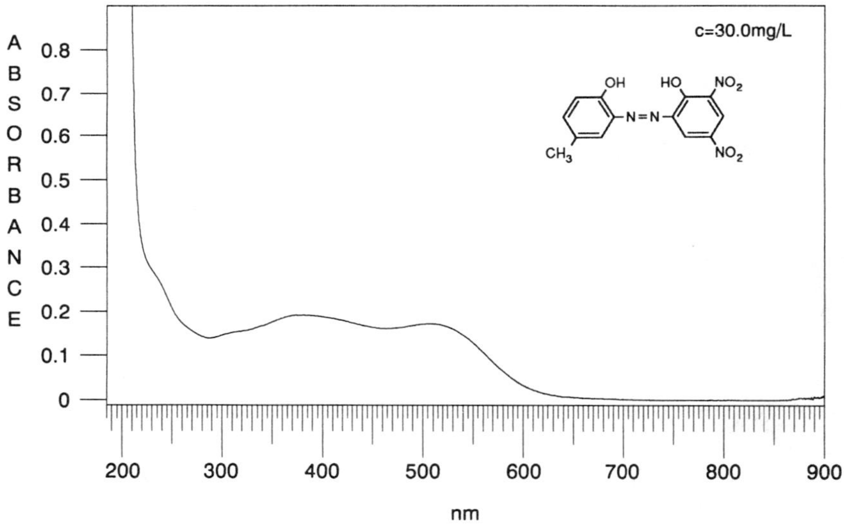

This deep-brown monoazo dye is made by coupling diazotized picramic acid to *p*-cresol. It is soluble in hot water and slightly soluble in ethanol and acetone. Industrially, it is used to dye wool (mordanted, metachromed or afterchromed), silk (mordanted or afterchromed), nylon (afterchromed), chromed leather, loose fur and hat bodies.

There is no evidence in the literature that this dye has been used as a biological stain.

General references:

Aldrich FT-IR Library	**1**(2), 967C
Sigma FT-IR Library	**1**(2), 169A
IR Library	**3**, 1451B
Colour Index	3rd ed., 1971

MORDANT BROWN 24

5'-(3,5-Dinitro-2-hydroxyphenylazo)-6'-hydroxy-*m*-acetotoluidide • Metachrome Brown 6G

Aldrich Cat. No.: 19,519-7
Sigma Prod. No.: M5511

CAS No.: [*6370-46-3*]
C.I. No.: 11880
Mol. Form.: $C_{15}H_{13}N_5O_7$
F.W.: 375.30
Appearance: Black-brown powder
λ_{max}: 400nm in water
Solubility: H_2O 60mg/ml
EGME 10mg/ml
EtOH 5mg/ml

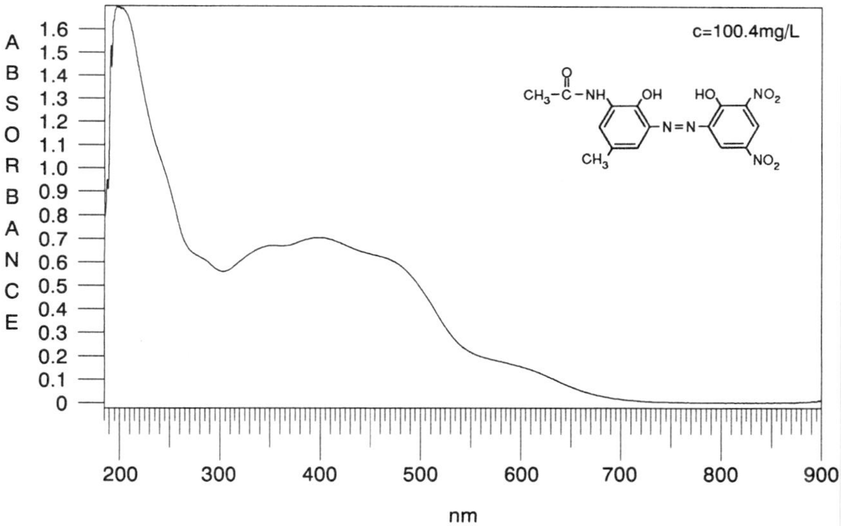

This yellowish-brown monoazo dye is prepared by coupling diazotized picramic acid with 6'-hydroxy-*m*-acetotoluidide. It is soluble in water and slightly soluble in ethanol. Wool can be dyed by mordanting, metachroming or afterchroming, silk by mordanting and afterchroming and nylon by afterchroming. Other industrial applications include dyeing loose wool, fur for hats, anodized aluminum and chromed leather.

There is no evidence in the literature that this dye has been used as a biological stain.

General references:

Aldrich FT-IR Library 1(2), 978A
Sigma FT-IR Library 1(2), 169B
Colour Index 3rd ed., 1971

MORDANT BROWN 33

Acid Anthracene Brown RH • 2,4-Diamino-5-(2-hydroxy-5-nitrophenylazo)benzenesulfonic acid, sodium salt

Aldrich Cat. No.:	19,518-9
Sigma Prod. No.:	M5636
CAS No.:	*[3618-62-0]*
C.I. No.:	13250
Mol. Form.:	$C_{12}H_{10}N_5NaO_6S$
F.W.:	375.27
Appearance:	Red-brown powder
λ_{max}:	442nm in water
Solubility:	H_2O 20mg/ml
	EGME 30mg/ml
	EtOH 1mg/ml

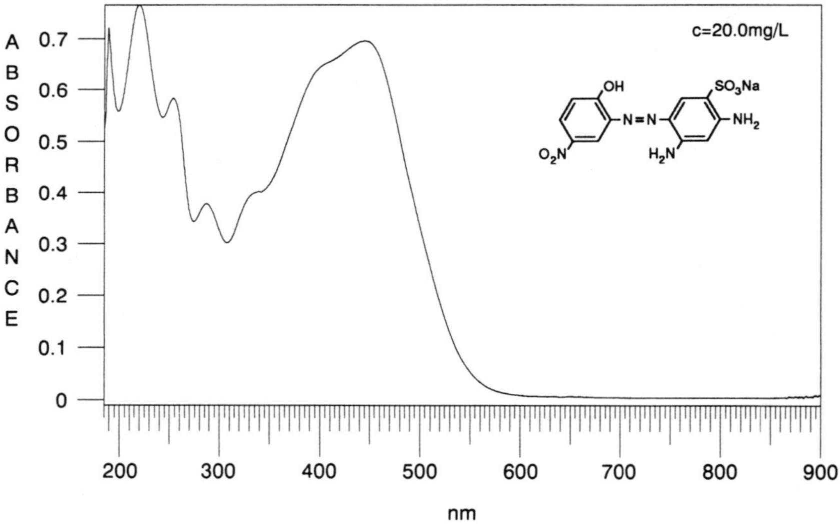

This yellowish-brown monoazo dye is prepared by coupling diazotized 2-amino-4-nitrophenol to 2,4-diaminobenzenesulfonic acid followed by conversion to the sodium salt. It is soluble in water and methyl Cellosolve, slightly soluble in acetone and very slightly soluble in ethanol. Wool is dyed by mordanting, metachroming, or afterchroming, silk by afterchroming or mordanting and nylon by afterchroming. Printing wool and silk requires mordanting. Fully chromed leather is dyed a red-brown.

There is no evidence in the literature that this dye has been used as a biological stain.

General references:

Aldrich FT-IR Library	**1**(2), 977D
Sigma FT-IR Library	**1**(2), 169C
IR Library	**3**, 1457E
Colour Index	3rd ed., 1971

MORDANT BROWN 48

2-(5-Chloro-2,4-diaminophenylazo)-4,6-dinitrophenol • Metachrome Brown BC

Aldrich Cat. No.:	19,514-6
Sigma Prod. No.:	M5761

CAS No.:	*[6232-53-7]*
C.I. No.:	11300
Mol. Form.:	$C_{12}H_9ClN_6O_5$
F.W.:	352.70
Appearance:	Dark-green crystals
λ_{max}**:**	492nm in water
Solubility:	H_2O 40mg/ml
	EGME 10mg/ml
	EtOH 3mg/ml

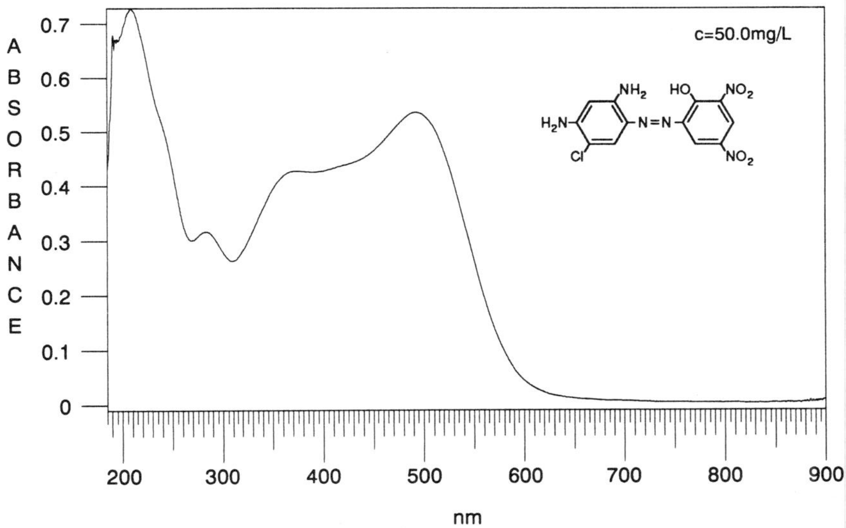

This reddish-brown, basic monoazo dye is prepared by coupling diazotized picramic acid to 4-chloro-1,3-phenylenediamine. It is soluble in hot water and slightly soluble in ethanol. Industrially, it is used to dye wool (by mordanting, metachroming or afterchroming) and chromed leather and to print basic, chromium-chloride-mordanted silk and nylon afterchromed.

There is no evidence that this dye has been used as a biological stain.

General references:

Aldrich FT-IR Library	**1**(2), 968A
Sigma FT-IR Library	**1**(2), 169D
IR Library	**3**, 1451E
Colour Index	3rd ed., 1971

MORDANT ORANGE 1

Alizarin Yellow R

Aldrich Cat. No.: 19,507-3
Sigma Prod. No.: A4146

CAS No.: [*2243-76-7*]
C.I. No.: 14030
Mol. Form.: $C_{13}H_9N_3O_5$
F.W.: 287.23
m.p.: >300°C
Appearance: Brown powder
λ_{max}: 385nm in methanol
Solubility: H_2O 10mg/ml
EGME 30mg/ml
EtOH 3mg/ml

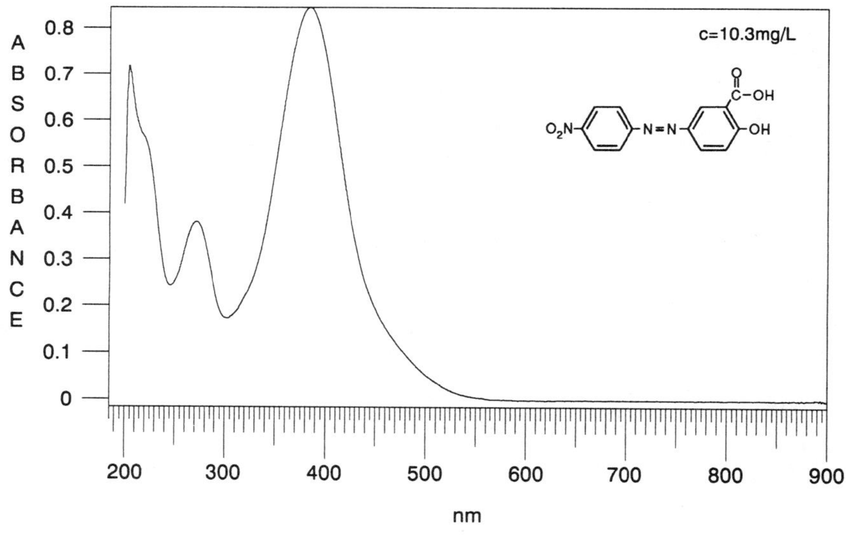

This yellowish-orange, monoazo dye is prepared by coupling diazotized 4-nitroaniline to salicylic acid. It is soluble in methyl Cellosolve, somewhat soluble in water and slightly soluble in ethanol and acetone. Industrially, it is used to dye wool, nylon (afterchromed), silk (afterchromed) and chromed leather, and to print chromium-mordanted wool, silk and cotton. In the laboratory, it is used primarily as a pH indicator, both as a component of the mixed, broad-spectrum Universal indicator and separately as a close-range indicator. The visual-transition interval for the latter is pH 10.1 (yellow) to pH

12.1 (orange-red). The indicator test solution normally used is prepared by dissolving 0.1% (w/v) dye in deionized water. See the pH Indicator Chart in the back of this book for more detailed information.

There is no evidence that it has been used as a biological stain.

General references:

Aldrich FT-IR Library	**1**(2), 972B
Sigma FT-IR Library	**1**(2), 96B
IR Library	**3**, 1265E
Colour Index	3rd ed., 1971

MORDANT ORANGE 6

Chrome Orange GR

Aldrich Cat. No.: 21,460-4

CAS No.: [*3564-27-0*]
C.I. No.: 26520
Mol. Form.: $C_{19}H_{12}N_4Na_2O_6S$
F.W.: 470.38
Appearance: Orange powder
λ_{max}: 381nm in water
Solubility: H_2O 40mg/ml
EGME 6mg/ml
EtOH 0.7mg/ml

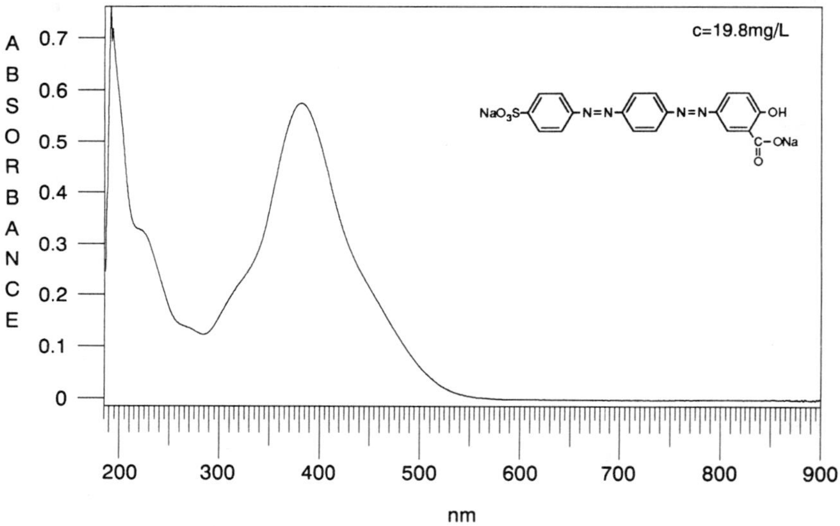

This anionic disazo dye is prepared by coupling diazotized 4-(4-aminophenylazo)benzenesulfonic acid to salicylic acid and converting the coupling product to the disodium salt.

Soluble in water (yellow to orange) and very slightly soluble in ethanol and acetone, it is used to dye wool, silk and nylon (all afterchromed) and to print chromium-mordanted wool, silk and cotton. Chromed leather is dyed a bright orange.

Long[1] lists this dye as an acid-base indicator with a visual-transition interval of pH 10.5 (yellow) to pH 12.0 (red).

There is no evidence that it has been used as a biological stain.

Reference:

1) Long, C. *Biochemist's Handbook;* Van Nostrand: Princeton, 1961.

General references:

Aldrich FT-IR Library **1**(2), 979B
IR Library **3**, 1458B
Colour Index 3rd ed., 1971

MORDANT ORANGE 10

Aldrich Cat. No.:	19,508-1
Sigma Prod. No.:	M5886
CAS No.:	*[6406-37-7]*
C.I. No.:	26560
Mol. Form.:	$C_{20}H_{14}N_4Na_2O_6S$
F.W.:	484.40
Appearance:	Rust-orange powder
λ_{max}:	386nm in water
Solubility:	H_2O 40mg/ml
	EGME 40mg/ml
	EtOH 1mg/ml

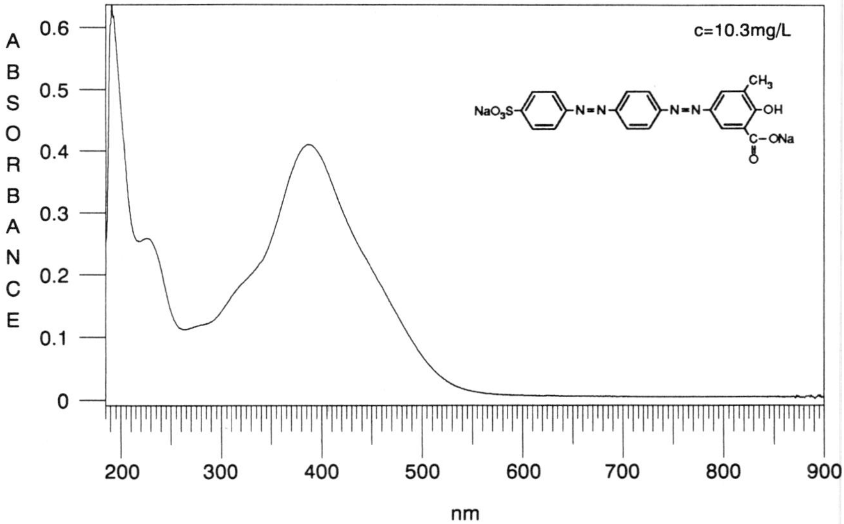

This reddish-orange, acid disazo dye is prepared by coupling diazotized 4-(4-aminophenylazo)benzenesulfonic acid to 2,3-cresotic acid and converting the coupling product to the disodium salt. It is soluble in water and very slightly soluble in the lower alcohols. Industrially, it is used to dye mordanted wool and chromed leather and to print chromium-mordanted wool and silk.

There is no evidence in the literature that it has been used as a biological stain.

General references:

Aldrich FT-IR Library	**1**(2), 978D
Sigma FT-IR Library	**1**(2), 170A
IR Library	**3**, 1457H
Colour Index	3rd ed., 1971

MORDANT RED 19

Metachrome Red 5G

Aldrich Cat. No.: 19,506-5
Sigma Prod. No.: M6011

CAS No.: [*1934-24-3*]
C.I. No.: 18735
Mol. Form.: $C_{16}H_{12}ClN_4NaO_5S$
F.W.: 430.81
Appearance: Red powder
λ_{max}: 413nm in water
Solubility: H_2O 20mg/ml
EGME 30mg/ml
EtOH 20mg/ml

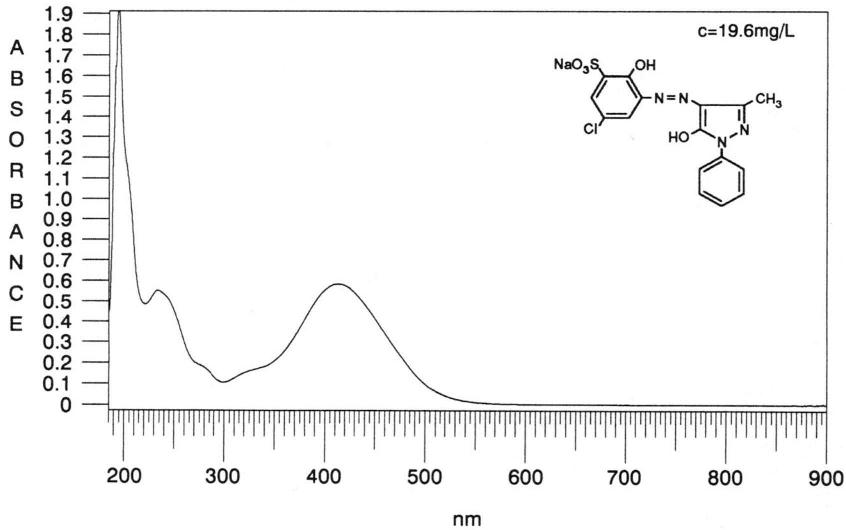

This red, acid monoazo pyrazolone dye is soluble in water, ethanol, acetone and methyl Cellosolve. It is prepared by coupling diazotized 6-amino-4-chloro-1-phenol-2-sulfonic acid to 3-methyl-1-phenyl-5-pyrazolone and converting the coupling product to the sodium salt. Industrially, it is used to dye chromium-mordanted wool and silk, nylon (afterchromed) and chromed leather and to print chromium-mordanted wool.

There is no evidence in the literature that it has been used as a biological stain.

General references:

Aldrich FT-IR Library	**1**(2), 1000C
Sigma FT-IR Library	**1**(2), 170B
IR Library	**3**, 1468E
Colour Index	3rd ed., 1971

MORDANT YELLOW 7

3-Methyl-5-(4-sulfophenylazo)salicylic acid, disodium salt

Aldrich Cat. No.:	19,513-8
Sigma Prod. No.:	M7386
CAS No.:	*[6408-91-9]*
Mol. Form.:	$C_{14}H_{10}N_2Na_2O_6S$
F.W.:	380.29
m.p.:	285°C (dec.)
Appearance:	Gold-brown powder
λ_{max}:	365nm in methanol
Solubility:	H_2O 20mg/ml
	EGME 50mg/ml
	EtOH 4mg/ml

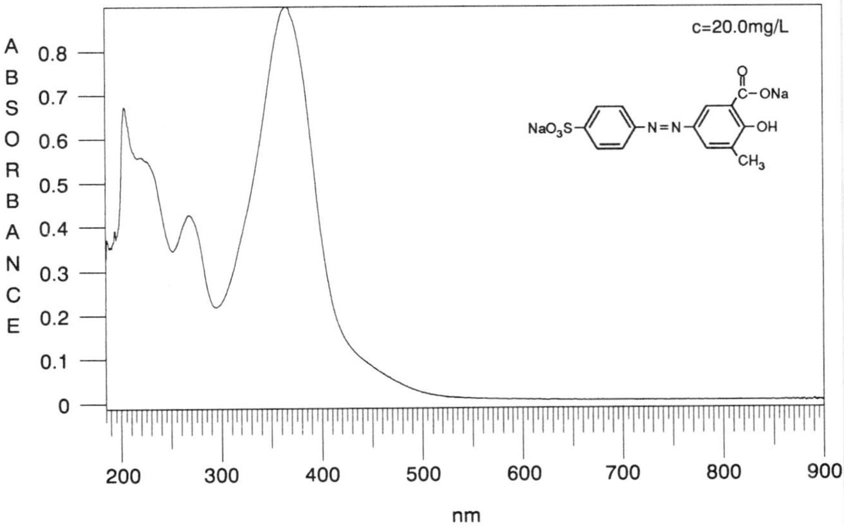

This dull-yellow, anionic monoazo dye is prepared by coupling diazotized sulfanilic acid to 3-methylsalicylic acid and converting the coupling product to the disodium salt. It is soluble in water and methyl Cellosolve and slightly soluble in ethanol.

Industrially, it is used to dye wool (by mordanting, metachroming or afterchroming) and fully chromed leather. Silk and cotton are printed by chromium mordanting.

There is no evidence in the literature that it has been used as a biological stain.

General references:

Aldrich FT-IR Library	**1**(2), 977C
Sigma FT-IR Library	**1**(2), 170C
IR Library	**3**, 1457D
Colour Index	3rd ed., 1971

MORDANT YELLOW 10

5-(4-Sulfophenylazo)salicylic acid, disodium salt

Aldrich Cat. No.:	19,510-3
Sigma Prod. No.:	M6136
CAS No.:	[*6054-99-5*]
C.I. No.:	14010
Mol. Form.:	$C_{13}H_8N_2Na_2O_6S$
F.W.:	366.26
Appearance:	Yellow powder
λ_{max}:	354nm in methanol
Solubility:	H_2O 10mg/ml
	EGME 30mg/ml
	EtOH 3mg/ml

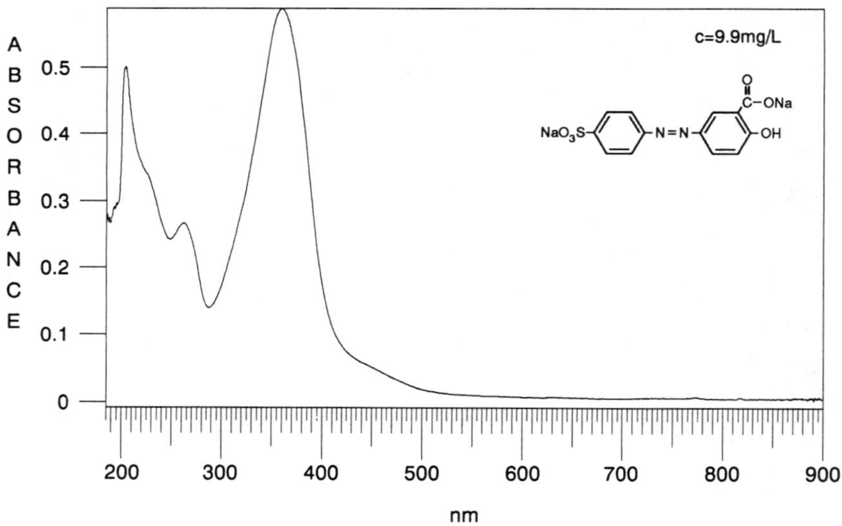

This greenish-yellow, anionic monoazo dye is made by coupling diazotized sulfanilic acid to salicylic acid and converting to the disodium salt. It is soluble in methyl Cellosolve, slightly soluble in water and chloroform and very slightly soluble in ethanol. Industrially, it is used to dye wool, acetate, cellulose (by mordanting, metachroming or afterchroming) and chromed leather, and to print chromium-mordanted wool, silk and cotton.

While there is no reference to its use as a biological stain in the literature, its structure and solubility suggest that it may have some potential as a contrast stain.

General references:

Aldrich FT-IR Library	**1**(2), 977A
Sigma FT-IR Library	**1**(2), 170D
IR Library	**3**,1457B
Colour Index	3rd ed., 1971

MORDANT YELLOW 12

5-(4-Aminophenylazo)salicylic acid, sodium salt

Aldrich Cat. No.: 19,511-1
Sigma Prod. No.: M6261

CAS No.: [*6470-98-0*]
C.I. No.: 14045
Mol. Form.: $C_{13}H_{10}N_3NaO_3$
F.W.: 279.23
Appearance: Gold-brown powder
λ_{max}: 380nm in water
Solubility: H_2O 40mg/ml
EGME 20mg/ml
EtOH 10mg/ml

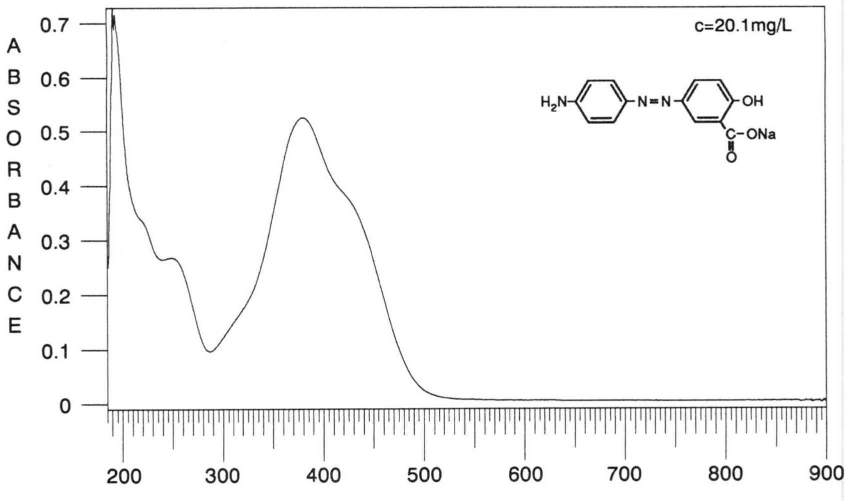

This bright-yellow, acid monoazo dye is made from C.I. Mordant Orange I. This is readily accomplished by sulfide reduction of the nitro group. Mordant Yellow 12 is soluble in water and slightly soluble in ethanol and acetone. Industrially, it is used to dye wool, cellulose and acetate (by mordanting, metachroming or afterchroming), mordanted silk, afterchromed nylon and fully chromed leather, and to print chromium-mordanted wool and cotton.

There is no evidence that it has been used as a biological stain.

General references:

Aldrich FT-IR Library **1**(2), 968D
Sigma FT-IR Library **1**(2), 171A
IR Library **3**, 1451H
Colour Index 3rd ed., 1971

MTT – METHYLTHIAZOLYLDIPHENYL TETRAZOLIUM

3-(4,5-Dimethylthiazol-2-yl)-2,5-diphenyl-2*H*-tetrazolium bromide

Aldrich Cat. No.:	13,503-8
Sigma Prod. No.:	M2128
Purity:	98%

CAS No.:	[*298-93-1*]
Mol. Form.:	$C_{18}H_{16}BrN_5S$
F.W.:	414.33
m.p.:	195°C (dec.)
Appearance:	Yellow powder
λ_{max}**:**	378nm in methanol
Solubility:	H_2O 10mg/ml
	EGME 20mg/ml
	EtOH 20mg/ml

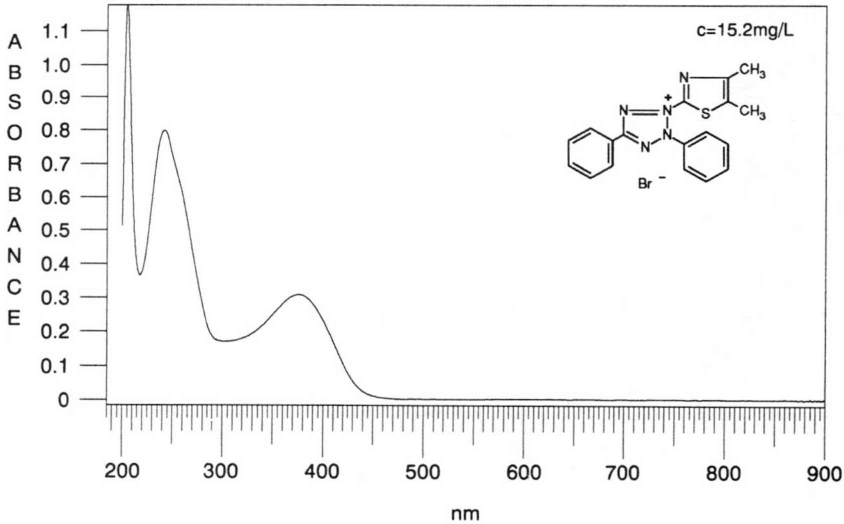

MTT is a colorless monotetrazole developed by Pearse.[1] While it is more readily reduced in an oxidoreductase system than the neotetrazolium (NT) analog, the formazan produced is lipid soluble, necessitating the use of a metal chelate. Cobalt forms a fine, purple-black, insoluble chelate (pigment) with MTT in such a system. Other colored pigments are formed with copper and nickel.[2,3] Troyer [4] reported that, like other older tetrazolium salts, MTT has been replaced largely by Nitro Blue Tetrazolium Chloride (Nitro BT).

References:

1) Pearse, A.G.E. *J. Histochem. Cytochem.* **1957,** *5*, 515-527.

2) Léric, H.; Kaplan, J.-C.; Broun, G. *Clin. Chim. Acta.* **1970,** *29,* 523.

3) Schauenstein, E.; Höfler-Bergthler, E. *Monatsh. Chem.* **1972,** *103*, 1271.

4) Troyer, H. *Principles and Techniques of Histochemistry;* Little, Brown: Boston, 1980; p 378.

General references:

Aldrich FT-IR Library	**1**(2), 633B
Sigma FT-IR Library	**1**(2), 171C
IR Library	**3**,1227G
NMR Library	**2**(2), 501D
Safety	**2**(2), 2479B

NAPHTHOCHROME GREEN
Mordant Green 31

Aldrich Cat. No.:	20,682-2
CAS No.:	*[5715-76-4]*
C.I. No.:	44530
Mol. Form.:	$C_{29}H_{16}Na_2O_6$
F.W.:	506.43
Appearance:	Brown-red powder
λ_{max}:	362nm in water
Solubility:	H_2O 60mg/ml
	EGME 20mg/ml
	EtOH 8mg/ml

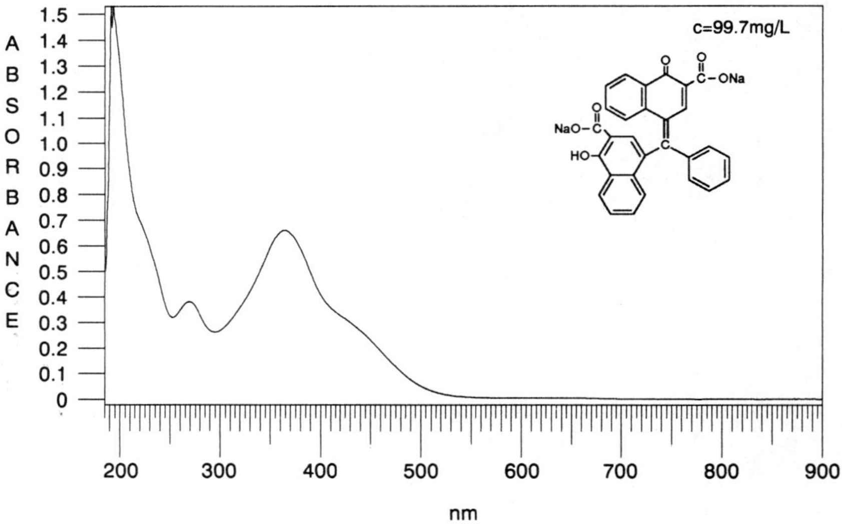

Naphthochrome Green is an anionic dinaphthylphenylmethane dye that is soluble in water and slightly soluble in alcohol with a yellow-brown color. It dyes wool, silk and nylon green using chromic acetate as a mordant, and prints green on chromium-mordanted wool.

Denz[1] used Naphthochrome Green for the histochemical detection of beryllium.

Reference:

1) Denz, F.A. *Q. J. Microsc. Sci.* **1949,** *90*, 317-321.

General references:

Aldrich FT-IR Library	**1**(2), 1031D
IR Library	**3**, 1484F
Colour Index	3rd ed., 1971

NAPHTHOL AS

Azoic Coupling Component 2 • 3-Hydroxy-2-naphthanilide

Aldrich Cat. No.:	22,935-0
Sigma Prod. No.:	N1375
Purity:	99%

CAS No.:	*[92-77-3]*
C.I. No.:	37505
Mol. Form.:	$C_{17}H_{13}NO_2$
F.W.:	263.30
m.p.:	247-250°C
Appearance:	Beige powder
λ_{max}:	394nm in 1*N* sodium hydroxide

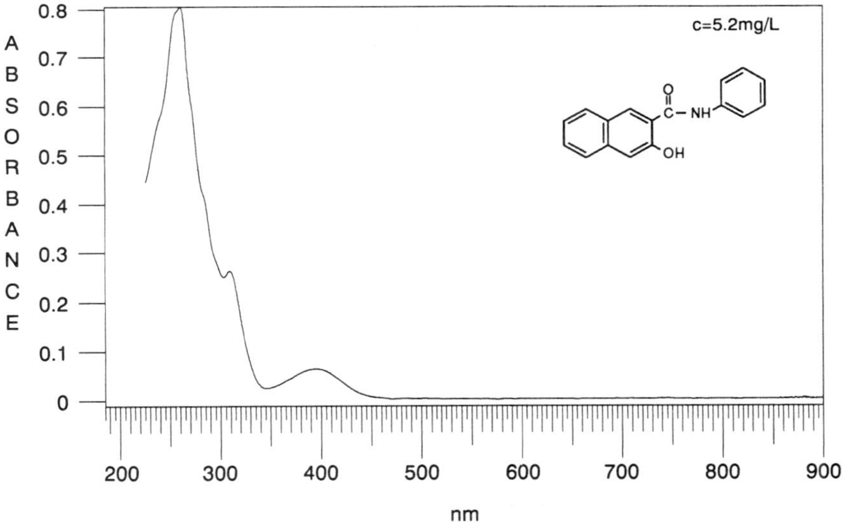

Naphthol AS is a dye intermediate that is used to pad textiles for the subsequent *in situ* development (coupling) of a dye into the fabric when treated with a suitable Azoic Diazo Component (stable diazonium salt) or other diazonium salt. It is soluble in hot nitrobenzene and alkali solutions, sparingly soluble in alcohol and insoluble in water.

Naphthol AS is used to prepare the phosphate and acetate esters, as well as the other derivatives, that are used as reagents for the demonstration of phosphatases and esterases present in histochemical substrates.

General references:

Aldrich FT-IR Library	**1**(2), 384A
Sigma FT-IR Library	**1**(2), 172B
Safety	**2**(2), 2494B
Colour Index	3rd ed., 1971

NAPHTHOL AS acetate

3-Acetoxy-2-naphthanilide

Aldrich Cat. No.:	22,936-9
Sigma Prod. No.:	N1500
Purity:	99%
CAS No.:	[*1163-67-3*]
Mol. Form.:	$C_{19}H_{15}NO_3$
F.W.:	305.34
m.p.:	160-162°C
Appearance:	White powder with chunks

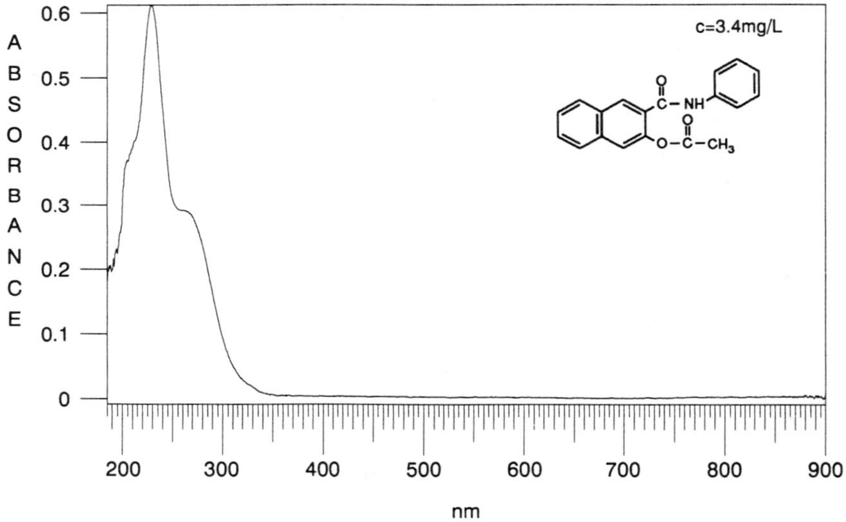

While this dye intermediate is not listed in the *Colour Index* as a coupling component, it can be used in a similar manner.

As would be expected, the solubility characteristics and chemical reactivity of this acetic acid ester of Naphthol AS (Aldrich Cat. No. 22,935-0) have been altered in comparison to its precursor.

Naphthol AS acetate is moderately soluble in water and serves as a hydrolase substrate in histochemistry. Gomori[1] reported using it for the localization of esterase.

Reference:

1) Gomori, G. *Microscopic Histochemistry;* University of Chicago: Chicago, 1952.

General references:

Aldrich FT-IR Library	**1**(2), 379D
Sigma FT-IR Library	**1**(2), 172C
NMR Library	**2**(2), 357B
Safety	**2**(2), 2494D

NAPHTHOL AS BI

Azoic Coupling Component 45 • 7-Bromo-3-hydroxy-*N*-(2-methoxyphenyl)-2-naphthalenecarboxamide

Aldrich Cat. No.:	86,003-4
Sigma Prod. No.:	N4130
CAS No.:	[*1237-75-8*]
C.I. No.:	37566
Mol. Form.:	$C_{18}H_{14}BrNO_3$
F.W.:	372.23
m.p.:	180-183°C
Appearance:	Off-white powder
λ_{max}:	305nm in *N,N*-dimethylformamide

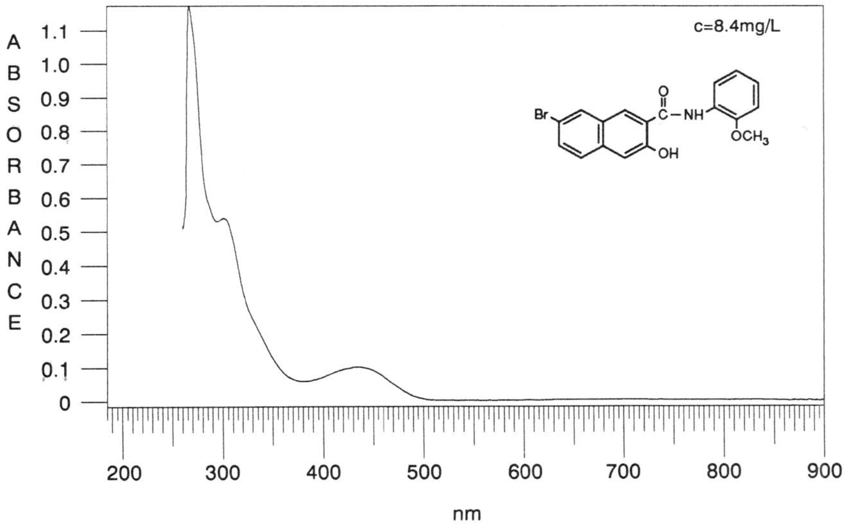

This colorless dye intermediate (Azoic Coupling Component) is used to pad fabric which then is treated with a stable diazonium salt (Azoic Diazo Component) for the *in situ* dye formation on the fabric (usually cellulosic). Since there are a number of Azoic Diazo Components to choose from, there is some latitude for choosing the desired color.

Naphthol AS BI is used to prepare esters with phosphoric, acetic and sulfuric acids. The esters act as hydrolase substrates in histochemical reactions. Burstone[1,2] used the phosphoric acid ester with good results for the demonstration of acid phosphatase.

References:

1) Burstone, M.S. *J. Histochem. Cytochem.* **1959,** *4*, 130-139.

2) *Idem ibid*. **1961,** *9*, 146-153.

General references:

Aldrich FT-IR Library	**1**(2), 380A
Sigma FT-IR Library	**1**(2), 173A
Safety	**2**(2), 2495B
Colour Index	3rd ed., 1971

NAPHTHOL AS BI phosphate

Aldrich Cat. No.: 22,937-7
Sigma Prod. No.: N2125
Purity: 98%

CAS No.: [*1919-91-1*]
Mol. Form.: $C_{18}H_{15}BrNO_6P$
F.W.: 452.21
Appearance: Off-white powder

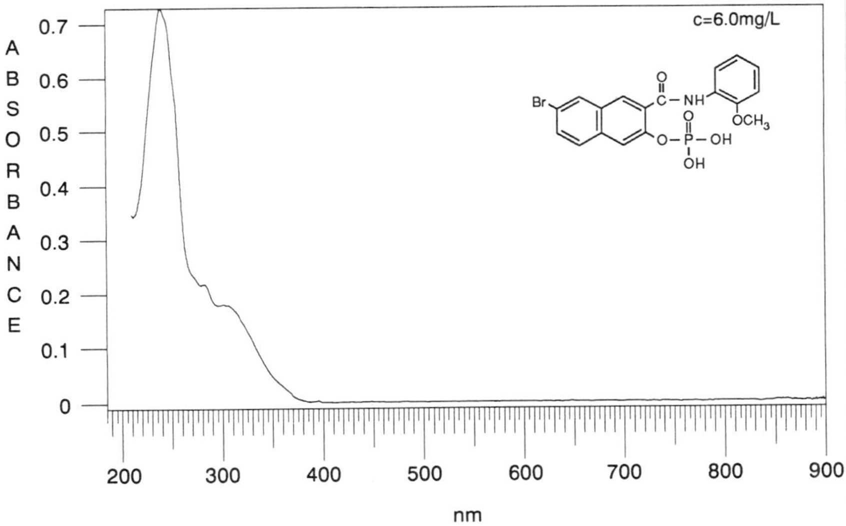

While this dye intermediate is not listed in the *Colour Index* as a coupling component, it can be used in a similar manner.

As would be expected, the solubility characteristics and chemical reactivity of this phosphoric acid ester of Naphthol AS BI (Aldrich Cat. No. 86,003-4) differ from its precursor.

Moderately soluble in water, it serves as a hydrolase substrate in histochemistry. Burstone[1,2] considered it an excellent substrate for acid phosphatases.

References:

1) Burstone, M.S. *J. Histochem. Cytochem.* **1959,** *4*, 130-139.

2) *Idem ibid.* **1961,** *9*, 146-153.

General references:

Aldrich FT-IR Library	**1**(2), 563A
Sigma FT-IR Library	**1**(2), 174A
Safety	**2**(2), 2496B

NAPHTHOL BLUE BLACK

Acid Black 1 • Amido Black 10B

Aldrich Cat. No.:	19,524-3
Sigma Prod. No.:	N3005
CAS No.:	[*1064-48-8*]
C.I. No.:	20470
Mol. Form.:	$C_{22}H_{14}N_6Na_2O_9S_2$
F.W.:	616.50
Appearance:	Dark-brown powder
λ_{max}:	618nm in water
Solubility:	H_2O <0.1mg/ml
	EGME 50mg/ml
	EtOH 3mg/ml

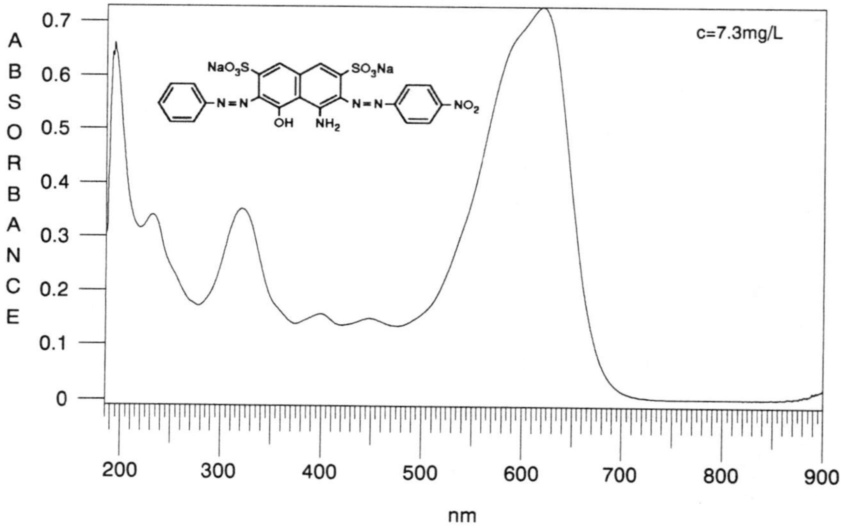

Naphthol Blue Black is an acid disazo dye that is made by coupling one mole each of *p*-nitroaniline and aniline to one mole of tetraazotized 8-amino-1-naphthol-3,6-disulfonic acid (H acid). The *p*-nitroaniline coupling is accomplished at a pH of 5-6 while the aniline is coupled at a pH of 8-10. The product is soluble in methyl Cellosolve, slightly soluble in ethanol and acetone and soluble in water with difficulty. It is used industrially to dye wool, silk (acetic acid or sulfuric acid bath) and nylon (formic acid bath) and to print directly on wool, silk and nylon. Non-textile uses include dyeing paper, leather, anodized aluminum, casein buttons and urea-melamine molding powder, and in the preparation of wood stains and writing inks.

In the laboratory, it is used as a stain to aid in protein separation utilizing chromatography and electrophoresis. *Staining Procedures*[1] describes a method for staining plant material which is designed to show total carbohydrates and proteins.

Reference:

1) Clark, G. *Staining Procedures,* 4th ed.; Williams and Wilkins: Baltimore, 1981; pp 323-324.

General references:

Aldrich FT-IR Library	**1**(2), 990C
Sigma FT-IR Library	**1**(2), 181C
IR Library	**3**,1464B
Colour Index	3rd ed., 1971

NAPHTHOL GREEN B

Acid Green 1

Aldrich Cat. No.:	11,991-1
Sigma Prod. No.:	N7257

CAS No.:	[*19381-50-1*]
C.I. No.:	10020
Mol. Form.:	$(C_{10}H_5NNaO_5S)_3Fe$
F.W.:	878.47
Appearance:	Dark-green powder
λ_{max}**:**	714nm in water
Solubility:	H_2O 30mg/ml
	EGME 9mg/ml
	EtOH 0.9mg/ml

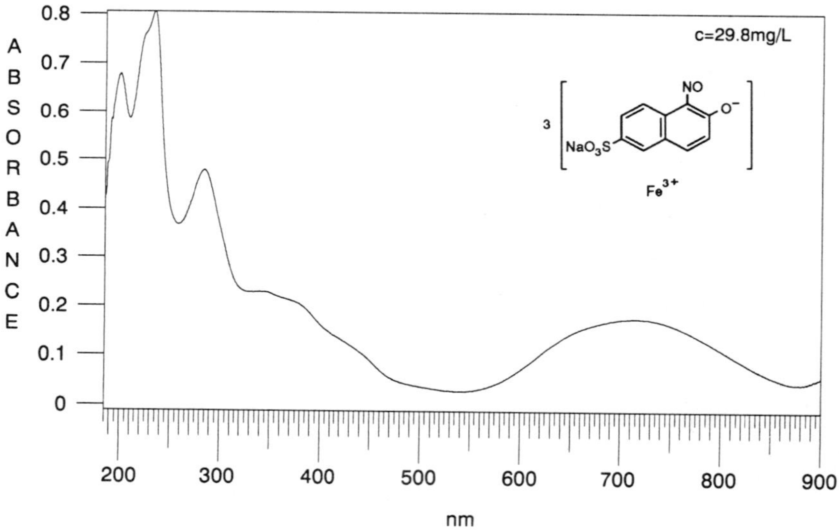

Naphthol Green B is a yellowish-green, acid nitroso dye that is soluble in water, with decomposition occurring on prolonged storage of the solution. It is made by nitrosating 2-naphthol-6-sulfonic acid in the 1-position, and converting to the sodium salt. Three moles of the salt are complexed to an iron(III) nucleus. Industrially, it has been used to dye wool, nylon, paper and chromed leather and color anodized aluminum and soap. The iron(III)-complexed barium salt is a pigment (Pigment Green 12, C.I. 10020:1) that is used to color printing inks and lacquers.

Conn's[1] suggests that it may be the Naphthol Green dye used by Volkmann and Strauss in 1934 and Mollier in 1938 as a combined muscle, collagen and elastin-fiber stain.

Reference:

1) *Conn's Biological Stains,* 9th ed.; Lillie, R.D., Ed.; Williams and Wilkins: Baltimore, 1977; p 71.

General references:

Aldrich FT-IR Library	**1**(2), 954B
Sigma FT-IR Library	**1**(2), 182A
Colour Index	3rd ed., 1971

NAPHTHOL YELLOW S

Acid Yellow 1 • 8-Hydroxy-5,7-dinitro-2-naphthalenesulfonic acid, disodium salt

Aldrich Cat. No.: 19,962-1
Sigma Prod. No.: F7754

CAS No.: [*846-70-8*]
C.I. No.: 10316
Mol. Form.: $C_{10}H_4N_2Na_2O_8S$
F.W.: 358.20
Appearance: Yellow powder
λ_{max}: 428(392)nm in water
Solubility: H_2O 70mg/ml
EGME 10mg/ml
EtOH 0.3mg/ml

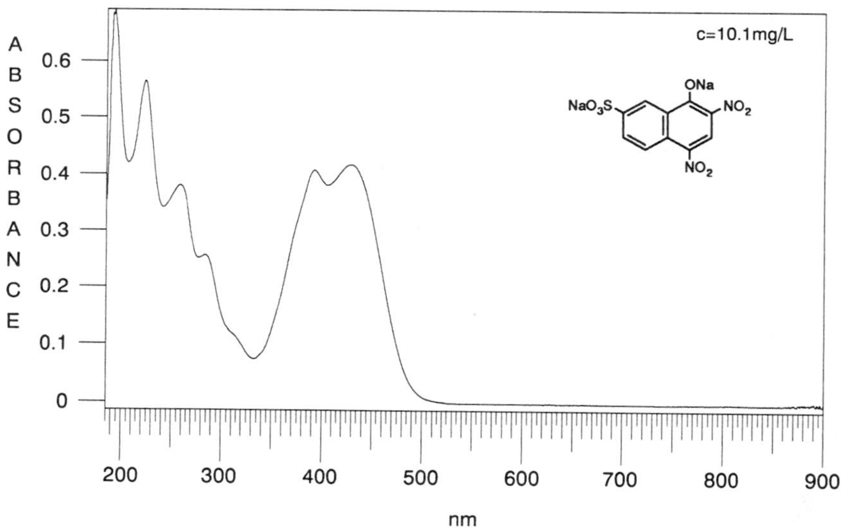

Naphthol Yellow S is an anionic nitro dye. It is the disodium salt of flavianic acid, and it is considerably more water soluble than the acid. Conversely, flavianic acid is more soluble in alcohol than Naphthol Yellow S.

Industrial applications for Naphthol Yellow S include the dyeing of wool, leather, furs and paper, coloring soap, the formation of heavy-metal lakes for use as pigments in printing inks and for coloring rubber.

Staining Procedures[1] describes a method for staining connective tissue by Clark after Van Gieson who used Naphthol Yellow S along with Trypan Blue and Acid Fuchsin in 1971. It is designed to show collagen (including very fine fibrils), muscle and cornified epithelium. *Conn's*[2] cites Bergmann[3] who used flavianic acid to precipitate arginine in protein hydrolysates, and Lillie *et al.*[4] who substituted it for picric acid in combination with Acid Fuchsin in a collagen stain.

References:

1) Clark, G. *Staining Procedures,* 4th ed.; Williams and Wilkins: Baltimore, 1981; p 115.
2) *Conn's Biological Stains,* 9th ed.; Lillie, R.D., Ed.; Williams and Wilkins: Baltimore, 1977; pp 75-76.
3) Bergmann, M. *J. Biol. Chem.* **1935,** *119,* 471-479.
4) Lillie, R.D.; Gutierrez, A; Palmer, R.W. *Anat. Rec.* **1967,** *159,* 165-370.

General references:

Aldrich FT-IR Library	**1**(2), 954A
Sigma FT-IR Library	**1**(2), 561C
IR Library	**3**, 1444F
Safety	**2**(2), 1685A
Merck Index	**11**, 6313
Colour Index	3rd ed., 1971

α-NAPHTHOLPHTHALEIN

Aldrich Cat. No.: 19,960-5
Sigma Prod. No.: N8257

CAS No.: [*596-01-0*]
Mol. Form.: $C_{28}H_{18}O_4$
F.W.: 418.45
m.p.: 238-240°C
Appearance: Brown powder
λ_{max}: 648nm in 0.1*N* sodium hydroxide

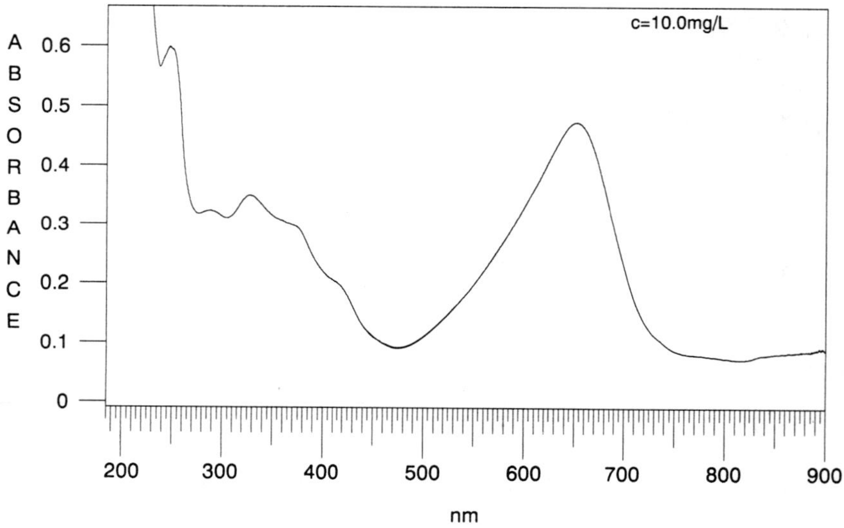

α-Naphtholphthalein is made by condensing one mole of naphthalic anhydride with two moles of α-naphthol. It is normally isolated as the lactone, which is colorless when pure. Typical of the phthaleins, it possesses valuable indicator properties with a visual-transition interval of pH 7.3 (pinkish yellow) to pH 8.7 (greenish blue). It is soluble in alcohol and practically insoluble in water.

Due to its sensitivity to pH and its fugitive nature, α-naphtholphthalein is totally unsatisfactory for use as an industrial dye and a biological stain.

General references:

Aldrich FT-IR Library	**1**(2), 1006C
Sigma FT-IR Library	**1**(2), 767A
IR Library	**3**, 1471D
Safety	**2**(2), 2206D
Merck Index	**11**, 6309

α-NAPHTHYL RED

Nubian Yellow TB • Solvent Yellow 4

Aldrich Cat. No.: 26,157-2
Sigma Prod. No.: N3761

CAS No.: [*131-22-6*]
C.I. No.: 11350
Mol. Form.: $C_{16}H_{13}N_3$
F.W.: 247.30
Appearance: Dark blue-green powder
λ_{max}: 439nm in methanol
Solubility: H_2O 0.1mg/ml
EGME 8mg/ml
EtOH 6mg/ml

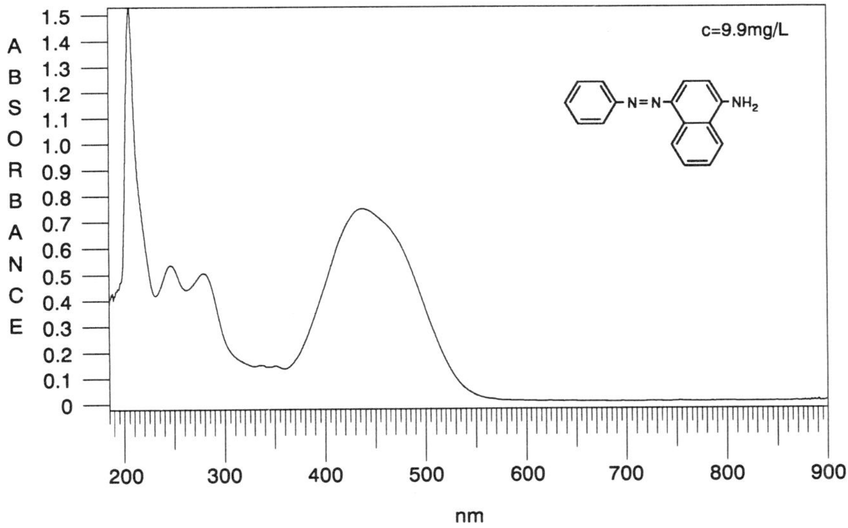

This yellow, solvent monoazo dye is prepared by coupling diazotized aniline to 1-naphthylamine. It is virtually insoluble in water, soluble in acetone, benzene, ethyl acetate, toluene, xylene, mineral oil, ligroin, paraffin, linseed oil, oleic and stearic acids and slightly soluble in ethanol.

Kolthoff[1] and Long[2] listed it as an indicator with a visual-transition interval of pH 3.7 (red) to pH 5.0 (yellow); a 0.1% solution in ethanol is the suggested concentration for indicator use.

There is no evidence that it has been used as a biological stain.

References:

1) Kolthoff, I.M.; Rosenblum, C. *Acid-Base Indicators;* Macmillan: New York, 1937; p 146.
2) Long, C. *Biochemist's Handbook* ; Van Nostrand: Princeton, 1960.

General references:

Aldrich FT-IR Library **1**(2), 182B
Safety **2**(2), 2521A

NEUTRAL RED

Basic Red 5 • Toluylene Red

Aldrich Cat. No.:	86,125-1
Sigma Prod. No.:	N6634
Grade:	Certified
CAS No.:	[*553-24-2*]
C.I. No.:	50040
Mol. Form.:	$C_{15}H_{17}ClN_4$
F.W.:	288.78
m.p.:	290°C (dec.)
Appearance:	Very dark-green powder
λ_{max}:	540nm in 50% ethanol + 0.5ml acetic acid

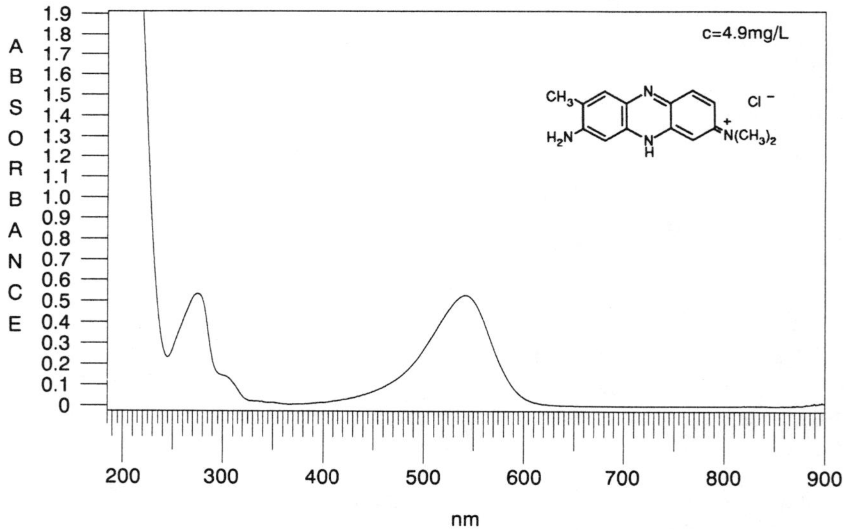

This cationic azine dye, which is slightly soluble in water and ethanol , is of little commercial importance. The chloride-salt form is available with a dye concentration of 60-70%, although the BSC will certify material with a minimum dye content of 50% if it meets other criteria including staining quality and low toxicity.

Neutral Red finds limited use as a pH indicator with a visual-transition interval extending from pH 6.8 (red) to pH 8.0 (yellow). *Conn's*[1] indicates that it may be useful as an indicator for the adjustment of alkaline-cyanide solutions to approximate neutrality by the cautious addition of 1.0*N* acetic acid to the first blush of pink.

Neutral Red is a weak nuclear stain that is used extensively in biological-stain applications, including vital staining. The supravital staining of fresh peripheral blood is probably one of its more important and demanding applications. This application is used by the BSC to qualify samples of Neutral Red. Although "ripened" solutions of Neutral Red have been used as a Nissl stain, it is not as permanent in this application as Cresyl Violet acetate.

By allowing plankton to be exposed to a solution of one part of Neutral Red in 100,000 parts of sea-water, Crippen and Perrier[2] demonstrated that dead cells were stained red and live cells remained unchanged.

An aqueous solution of Neutral Red, pH 6.5-7.0, was used by Kirk[3] to demonstrate lipids fluorochromatically. Some specimens displayed blue-green fluorochromicity; others were bright yellow. Twort[4] combined it with Light Green to form a neutral dye complex, Twort stain. This product is used to stain parasites in tissue.

References:

1) *Conn's Biological Stains,* 9th ed.; Lillie, R.D., Ed.; Williams and Wilkins: Baltimore, 1977; p 378.

2) Crippen, R.W.; Perrier, J.L. *Stain Technol.* **1974,** *49*, 97-104.

3) Kirk, P.W., Jr. *ibid.* **1970,** *45*, 1-4.

4) Twort, F.W. *J. State Med.* (London) **1924,** *32*, 351-355.

General references:

Aldrich FT-IR Library	**1**(2), 887C
Sigma FT-IR Library	**1**(2), 183C
Safety	**2**(2), 2533D
Merck Index	**11**, 6395

NEW COCCINE

Acid Red 18 • Victoria Scarlet 3R

Aldrich Cat. No.:	19,973-7
Sigma Prod. No.:	N6882
CAS No.:	*[2611-82-7]*
C.I. No.:	16255
Mol. Form.:	$C_{20}H_{11}N_2Na_3O_{10}S_3$
F.W.:	604.48
Appearance:	Maroon powder
λ_{max}**:**	506(335)nm in water
Solubility:	H_2O 80mg/ml EGME 80mg/ml EtOH 0.9mg/ml

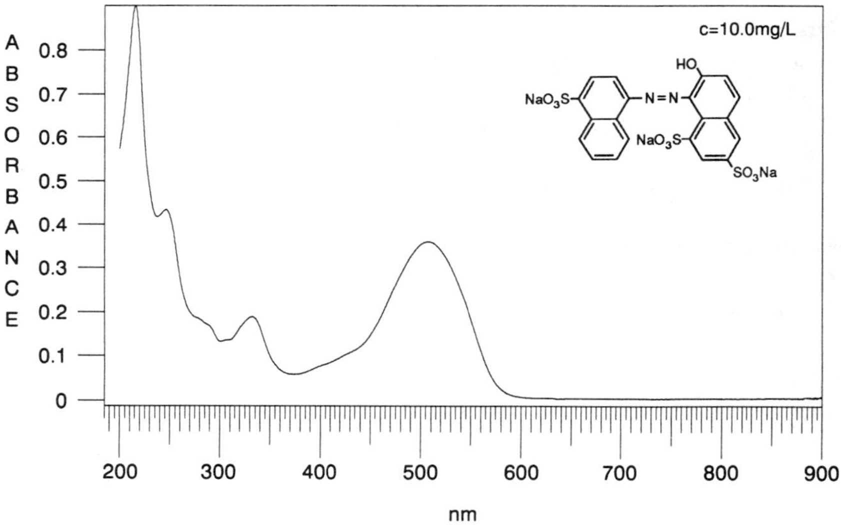

New Coccine is a red, anionic azo dye that is made by coupling diazotized naphthionic acid to G acid (2-naphthol-6,8-disulfonic acid) and converting the coupling product to the trisodium salt. It is soluble in water and methyl Cellosolve and very slightly soluble in ethanol. It is an important textile dye that is used to dye wool a bright red directly, nylon from an acetic acid or formic acid bath and silk from an acetic acid or sulfuric acid bath. Wool, silk and nylon are printed directly. Other industrial uses are as a dye for inks, for dyeing paper and leather, coloring plastics and staining wood and, formerly, as a colorant for drugs and cosmetics.

New Coccine is included in Lendrum's[1] study on dyes used to stain fibrin.

Reference:

1) Lendrum, A.D.; Fraser, D.S.; Slidders, W.; Henderson, R. *J. Clin. Pathol.* **1962,** *15*, 401-413.

General references:

Aldrich FT-IR Library	**1**(2), 988D
Sigma FT-IR Library	**1**(2), 184A
IR Library	**3**, 1463C
Safety	**2**(2), 2534A
Colour Index	3rd ed., 1971

NEW FUCHSIN

Basic Violet 2 • Magenta III

Aldrich Cat. No.:	22,931-8
Sigma Prod. No.:	N0638

CAS No.:	*[3248-91-7]*
C.I. No.:	42520
Mol. Form.:	$C_{22}H_{23}N_3 \cdot HCl$
F.W.:	365.91
Appearance:	Green crystalline powder
λ_{max}:	553nm in 50% ethanol
Solubility:	H_2O 20mg/ml
	EGME 20mg/ml
	EtOH 20mg/ml

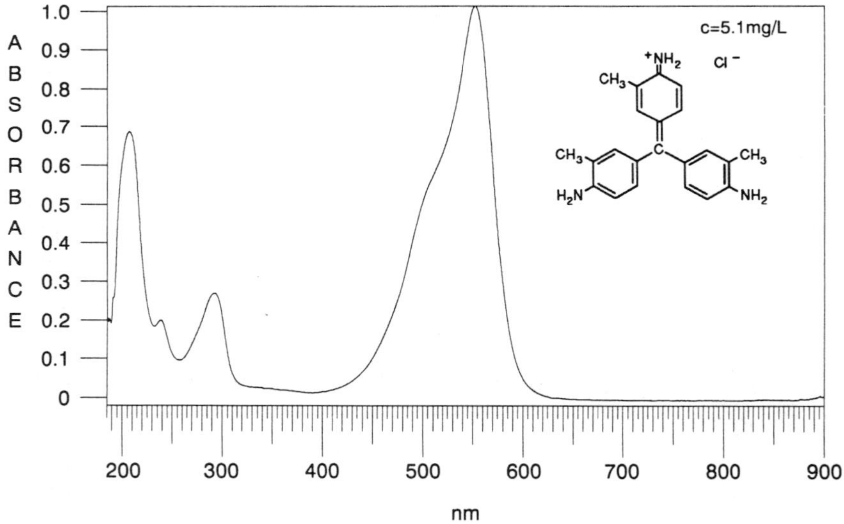

This reddish-violet cationic triphenylmethane dye is prepared by condensing 4,4'-methylenebis(*o*-toluidine) with *o*-toluidine and its hydrochloride in the presence of vanadium pentoxide.

New Fuchsin, like Pararosaniline, is symmetrical. However, Pararosaniline, which is at the lower extreme of the homologous series, is devoid of methyl substituents and New Fuchsin, at the higher end, has a methyl substituent on each of its three phenyl groups (3,3',3"). This symmetry makes it possible to produce both dyes in a pure state directly, while the asymmetrical members that complete the homologous series, Rosaniline and Magenta II, require chromatographic separation to attain equivalent purity.

New Fuchsin is soluble in both water and ethanol. Industrial uses include dyeing acrylics, cotton and chromed leather. However, the textile applications are limited because the dye is light fugitive (1-2 fastness).

Chance[1] used New Fuchsin in combination with Congo Red to stain the cell walls of bacteria.

Reference:

1) Chance, H.L. *Stain Technol.* **1953,** *27*, 253-258.

General references:

Aldrich FT-IR Library	**1**(2), 963D
Safety	**2**(2), 2534B
Colour Index	3rd ed., 1971

NEW METHYLENE BLUE N, zinc chloride double salt

Basic Blue 24

Aldrich Cat. No.:	20,209-6
Sigma Prod. No.:	B4631
CAS No.:	[*6586-05-6*]
C.I. No.:	52030
Mol. Form.:	$C_{18}H_{22}ClN_3S \cdot {}^1/_2\, ZnCl_2$
F.W.:	416.05
Appearance:	Dark-brown powder
λ_{max}:	630(591)nm in water
Solubility:	H_2O 40mg/ml
	EGME 10mg/ml
	EtOH 10mg/ml

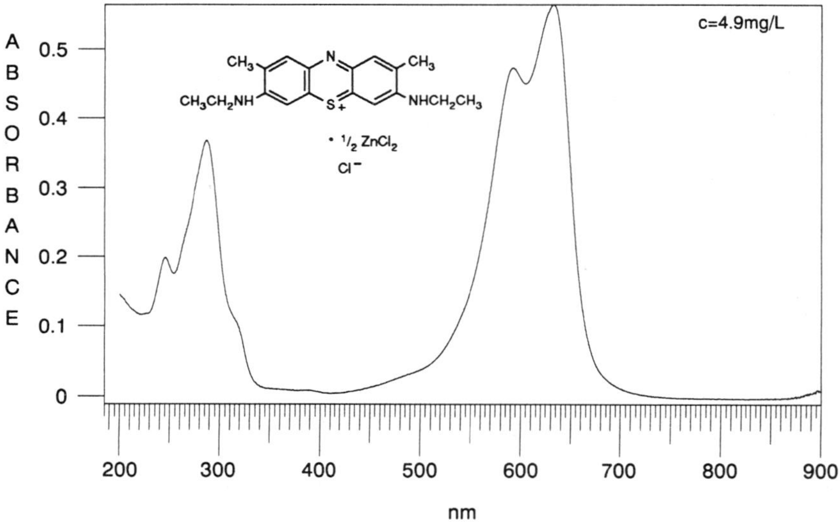

This blue, cationic thiazine dye is soluble in water and slightly soluble in ethanol. Currently, there is little demand for this dye in industrial applications. The *Colour Index* lists New Methylene Blue as an illuminating color used in discharges since it withstands reduction in this process. The same source reports the direct- or tannin-mordant dyeing of silk, wool, bast fibers, paper and leather.

Semi-commercial lots of the zinc chloride double salt (New Methylene Blue N) are currently manufactured for use as a biological stain. This material has a dye content of approximately 90%.

Staining Procedures[1] provides a method adapted from that reported by Brecher in 1949 for using New Methylene Blue as a reticulocyte stain for the identification and enumeration of reticulocytes in peripheral blood.

Reference:

1) Clark, G. *Staining Procedures,* 4th ed.; Williams and Wilkins: Baltimore, 1981; p 177.

General references:

Aldrich FT-IR Library	**1**(2), 1039C
Sigma FT-IR Library	**1**(2), 103B
IR Library	**3**, 1487D
Safety	**2**(2), 2534C
Colour Index	3rd ed., 1971

NICKEL PHTHALOCYANINETETRASULFONIC ACID, tetrasodium salt

Aldrich Cat. No.: 27,490-9

CAS No.: [*27835-99-0*]
Mol. Form.: $C_{32}H_{12}N_8Na_4NiO_{12}S_4$
F.W.: 979.43
Appearance: Blue powder
λ_{max}: 657nm in water
Solubility: H_2O 2mg/ml
EGME 0.5mg/ml
EtOH 2mg/ml

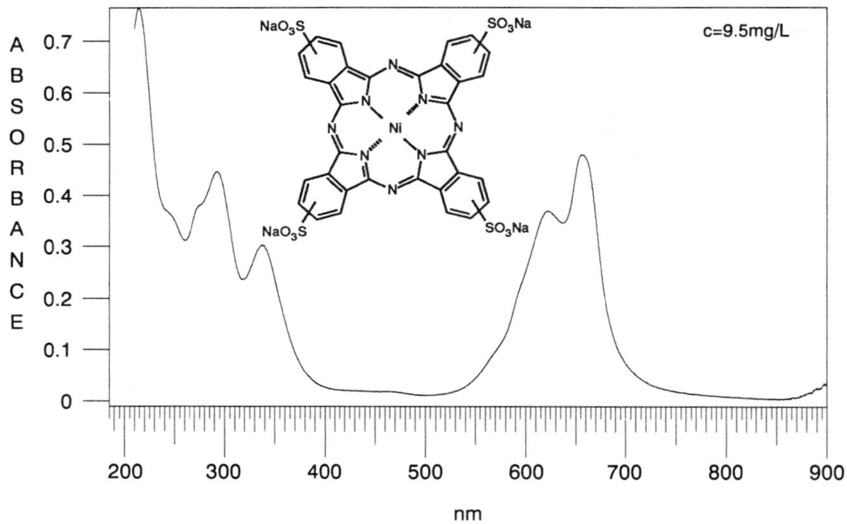

This bright-blue anionic phthalocyanine dye is prepared by condensing 4-sulfophthalic acid with cyanoguanidine in the presence of nickel chloride at 180-200°C.

While the utility of this experimental dye has not been thoroughly investigated, testing it in the same applications that its copper counterpart has satisfied is suggested.

General reference:

Safety **2**(2), 2539D

NIGROSIN, alcohol soluble
Solvent Black 5

Aldrich Cat. No.: 21,168-0

CAS No.: [*11099-03-9*]
C.I. No.: 50415
Appearance: Black powder
λ_{max}: 565nm in ethanol
Solubility: H_2O 1mg/ml
EGME 20mg/ml
EtOH 5mg/ml

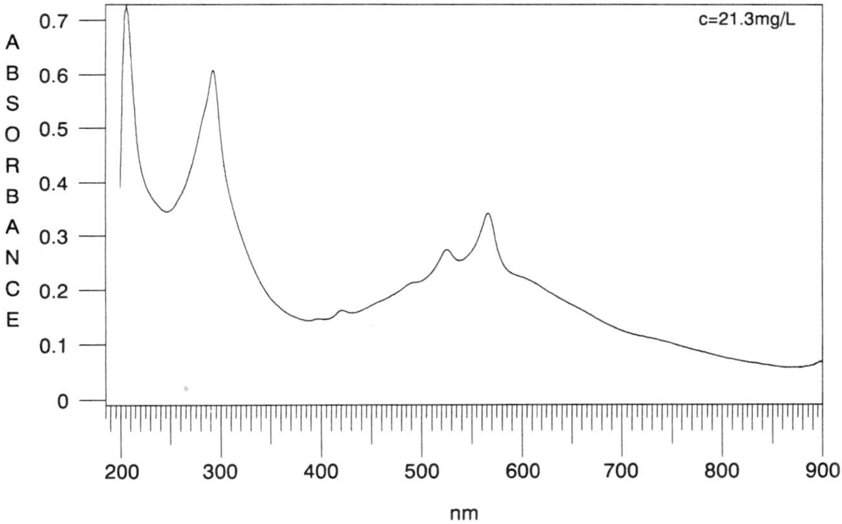

Nigrosin, alcohol soluble (AS) (Aldrich Cat. No. 21,168-0), is an indulin-like dye mixture that is prepared by heating (180-200°C) nitrobenzene, aniline and aniline hydrochloride in the presence of an iron or copper catalyst. The resulting product is virtually insoluble in water, slightly soluble in ethanol (bluish-black), benzene and toluene and very soluble in stearic and oleic acids. Industrially, this blue-black solvent dye is used to color alcoholic solvents, varnishes, lacquers (especially shellac), wood stains, marker-pencil inks, printing inks and spirited finishes for leather.

Nigrosin, water soluble (WS), is prepared by sulfonating Nigrosin AS. However, this blue-black product must be blended with 10-20% of a yellow or orange dye, such as C.I. Acid Orange 10, in order to attain a black suitable for use as a biological stain. The blended product is soluble in water, somewhat soluble in methyl Cellosolve and ethylene glycol and slightly soluble in ethanol.

It has been used primarily as a negative stain in place of India ink, for spirochetes, bacteria, protozoa and fungi. Dorner[1] used it to contrast the nuclear stain Fuchsin to demonstrate bacterial spores.

Staining Procedures[2] describes a method, taken from Dienst and Sanderson in 1936, for the negative staining of spirochetes.

References:

1) Dorner, W. *Lait* **1926,** *6*, 8-12.

2) Clark, G. *Staining Procedures,* 4th ed.; Williams and Wilkins: Baltimore, 1981; p 412.

General reference:

Colour Index 3rd ed., 1971

NIGROSIN, water soluble
Acid Black 2

Aldrich Cat. No.: 19,828-5
Sigma Prod. No.: N4754
Grade: Certified

CAS No.: [*8005-03-6*]
C.I. No.: 50420
Appearance: Black crystalline powder
λ_{max}: 570nm in 50% ethanol
Solubility: H_2O 50mg/ml
EGME 20mg/ml
EtOH 4mg/ml

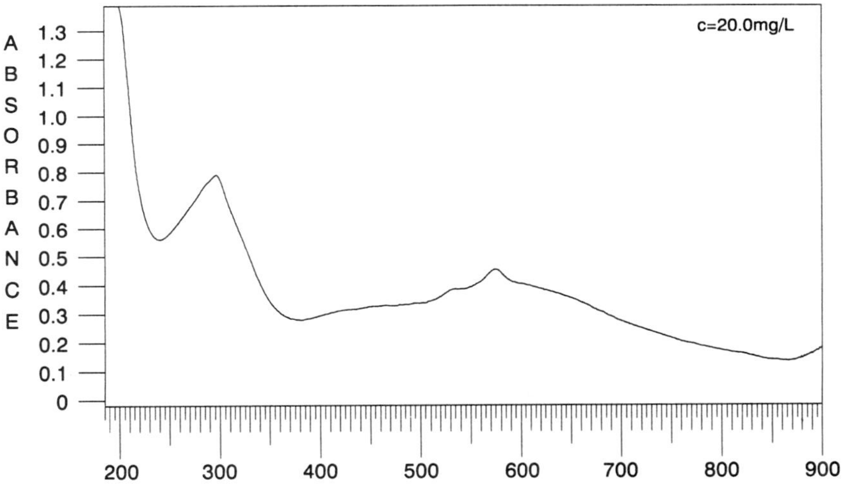

See description on pages 513 and 514 under Nigrosin, alcohol soluble.

General references:

Aldrich FT-IR Library	**1**(2), 1042B
Sigma FT-IR Library	**1**(2), 184C
IR Library	**3**, 1488E
Safety	**2**(2), 2543A
Colour Index	3rd ed., 1971

NILE BLUE A

Nile Blue sulfate

Aldrich Cat. No.: 12,147-9
Sigma Prod. No.: N0766
Grade: Certified

CAS No.: [*3625-57-8*]
C.I. No.: 51180
Mol. Form.: $C_{40}H_{40}N_6O_6S$
F.W.: 732.84
m.p.: >300°C (dec.)
Appearance: Green powder
λ_{max}: 633nm in 50% ethanol
Solubility: H_2O 20mg/ml
EGME 8mg/ml
EtOH 3mg/ml

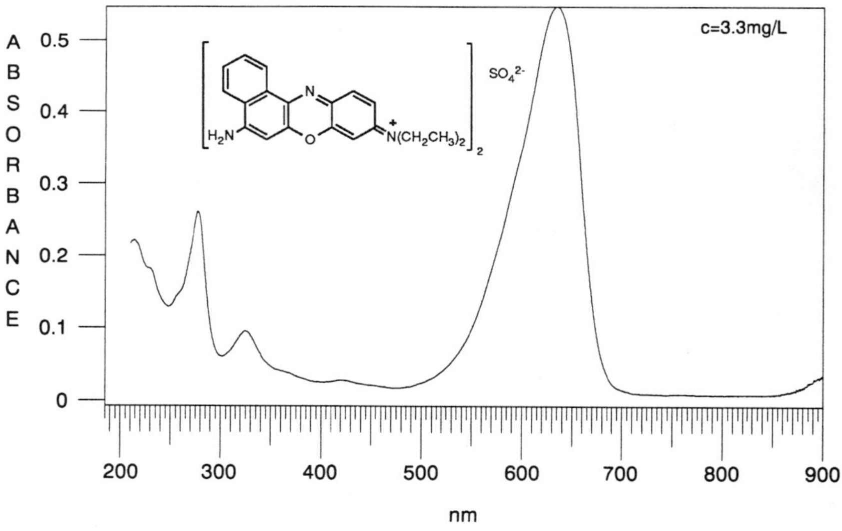

Nile Blue, a cationic oxazine dye, is available as both the sulfate and chloride salts. If greater water solubility is desired, the sulfate salt is the form of choice; and this is the form usually certified by the BSC. The λ_{max} normally displayed at 635nm is shifted to 470nm when the solution is made highly acidic (pH 1-2). This is the result of both amino groups becoming protonated.[1] In concentrated aqueous solutions, aggregation of the dye is indicated by the appearance of a shoulder at approximately 590nm.

Nile Blue is a highly fluorescent dye which has been found to lase in the 670-740nm region. As is typical of all oxazines, Nile Blue is exceptionally resistant to photochemical decay.

Nile Blue has been used by a number of workers as a lipid stain. However, in order for the Nile Blue to function properly as a lipid dye, it cannot be too pure. For this purpose, it must contain a small amount of the hydrolytic product Nile Red, a closely related oxazone dye. Nile Blue deficient in the red dye can be altered optimally by subjecting it to hydrolysis with dilute sulfuric acid. This results in the formation of small percentages of the oxazine free base and the oxazone derivative (both are red). Neutral lipids will dissolve only the oxazone and free base out of aqueous solutions of Nile Blue. Acidic lipids will dissolve the oxazone and combine with the free base to form blue lipid-soluble compounds. Triglycerides, free or dissolved in hydrocarbons, are colored red by the oxazine salts and much less by the free base. Solid lecithin is stained deep blue with the oxazine base or with concentrated or dilute aqueous Nile Blue solutions; it is not affected by the oxazone.[2]

Menschik[3] introduced a Nile Blue staining technique that is unique to phospholipids.

Staining Procedures[4] describes a method adapted from Lillie[5] for the differentiation of melanins and lipofuscins.

References:

1) Dunnington, M.J. *Stain Technol.* **1968**, *43*, 243.

2) Pearse, A.G.E. *Histochemistry: Theoretical and Applied*; J.A. Churchill: London, 1968.

3) Menschik, Z. *Stain Technol.* **1953**, *28*, 13.

4) Clark, G. *Staining Procedures,* 4th ed.; Williams and Wilkins: Baltimore, 1981; p 190.

5) Lillie, R.D. *J. Histochem. Cytochem.* **1956**, *4*, 377-381.

General references:

Aldrich FT-IR Library	**1**(2), 1035C
Sigma FT-IR Library	**1**(2), 185A
Safety	**2**(2), 2543B

NILE BLUE chloride

Aldrich Cat. No.: 22,255-0
Sigma Prod. No.: N5383

CAS No.: [*2381-85-3*]
C.I. No.: 51180
Mol. Form.: $C_{20}H_{20}ClN_3O$
F.W.: 353.85
Appearance: Green powder
λ_{max}: 638nm in 50% ethanol
Solubility: H_2O 2mg/ml
EGME 4mg/ml
EtOH 2mg/ml

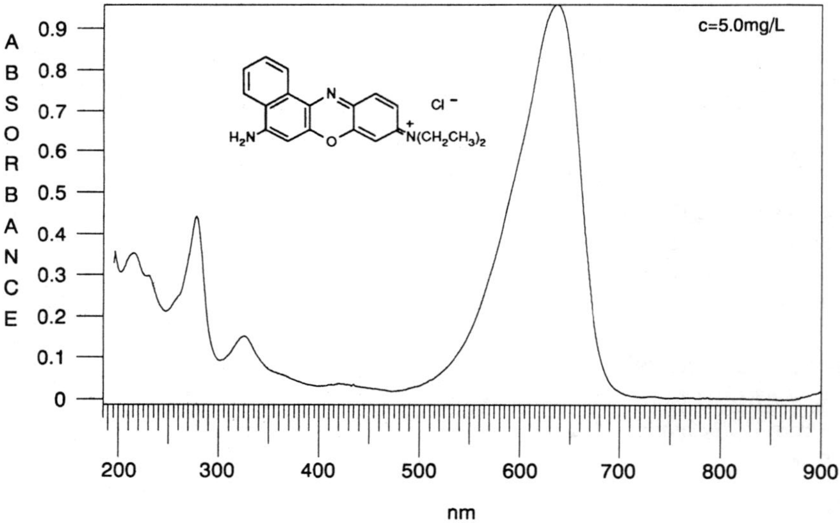

See description on pages 516 and 517 under Nile Blue A.

General references:

Aldrich FT-IR Library	**1**(2), 1035B
Sigma FT-IR Library	**1**(2), 184D
IR Library	**3**, 1486C
Safety	**2**(2), 2543C

NILE RED
Nile Blue A Oxazone

Aldrich Cat. No.:	29,839-5
Sigma Prod. No.:	N3013
CAS No.:	[*7385-67-3*]
Mol. Form.:	$C_{20}H_{18}N_2O_2$
F.W.:	318.38
m.p.:	203-205°C
Appearance:	Green crystalline powder
λ_{max}:	553(310)nm in methanol
Solubility:	H_2O 0.2mg/ml
	EGME 3mg/ml
	EtOH 1mg/ml

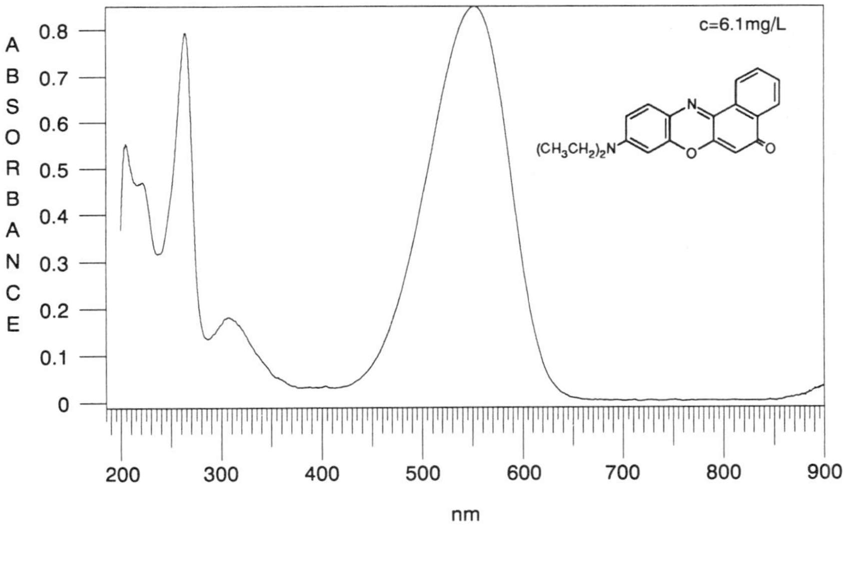

Nile Red has been little more than a laboratory curiosity for many years although the structure of this solvent oxazone dye has been known almost as long as its oxazine precursor (Nile Blue A, C.I. 51180). Indeed, some of the staining qualities attributed to the latter are at least enhanced by the presence of Nile Red as an impurity, if not due entirely to its existence. Nile Blue must contain several percent of Nile Red to perform satisfactorily in the Lillie[1] lipofuscins staining method. The BSC has found that lots essentially free of Nile Red do not perform satisfactorily in their testing protocol.

Thorpe[2] prepared Nile Red by boiling a Nile Blue solution with dilute sulfuric acid overlayed with toluene. The fat- and solvent-soluble Nile Red, generated as the primary amino function, is gradually replaced hydrolytically with an oxygen atom, then is taken up in the toluene layer and separated. Smith and Mair[3] used this method to prepare a fat stain that subsequently was identified with the senior investigator, Lorrain Smith Fat Stain.

Interest in the obscure dye was stimulated in the middle 1980s when Greenspan, Fowler *et al.*[4-6] utilized it as a vital stain in both fluoresence microscopy and flow cytometry, for the visualization of lipid droplets in the cytoplasm of living cells. Fowler, Brown and Greenspan[7] reported its use as a fluorochrome for the rapid determination of lipids by thin-layer chromatography. Fowler and Jamison[8] suggested staining whole tissue such as atherosclerotic blood vessels.

References:

1) Lillie, R.D. *J. Histochem. Cytochem.* **1956,** *4*, 377-381.

2) Thorpe, J.F. *J. Chem. Soc. Trans.* **1907,** *91*, 324-336.

3) Smith, J.L.; Mair, W. *Scand. Arch. Physiol.* **1911,** *25*, 245-255.

4) Greenspan, P.; Fowler, S.D. *J. Lipid Res.* **1985,** *26*, 781.

5) Fowler, S.D.; Greenspan, P. *J. Histochem. Cytochem.* **1985,** *33*, 833.

6) Greenspan, P.; Mayer, E.P.; Fowler, S.D. *J. Cell Biol.* **1985,** *100*, 965.

7) Fowler, S.D.; Brown, W.J.; Greenspan, P. *Fed. Proc.* **1986,** *45*(6), 1669.

8) Fowler, S.D.; Jamison, J. *Kodak Lab. Chem. Bull.* **1985,** *56*(3), 1.

NITRAZINE YELLOW

Delta Yellow Indicator • Nitrazine

Aldrich Cat. No.:	20,018-2
Sigma Prod. No.:	N7007
Grade:	Indicator
CAS No.:	[*5423-07-4*]
C.I. No.:	14890
Mol. Form.:	$C_{16}H_8N_4Na_2O_{11}S_2$
F.W.:	542.37
Appearance:	Brown powder
λ_{max}:	586nm in 0.1*N* sodium hydroxide
Solubility:	H_2O 20mg/ml
	EGME 20mg/ml
	EtOH 1mg/ml

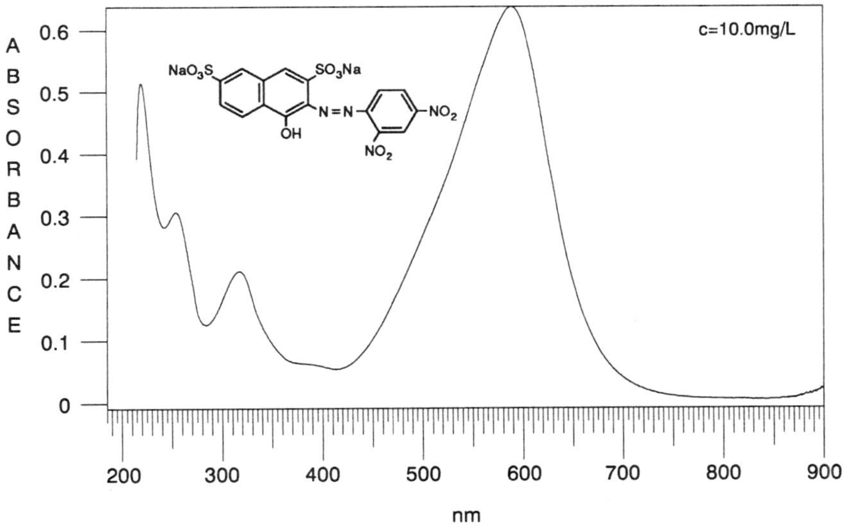

Nitrazine Yellow is an anionic monoazo dye that is made by coupling diazotized 2,4-dinitroaniline to R salt (1-naphthol-3,6-disulfonic acid, disodium salt). It is soluble in water and acetic acid and very slightly soluble in ethanol. An important pH indicator with a range useful for physiological systems, Nitrazine Yellow has a visual-transition interval from pH 6.0 (bright yellow) to pH 7.2 (bright blue) with a grey-green midpoint. Thus, it covers the same range as Litmus with a greater sensitivity and a sharper color contrast.

Lillie[1] substituted Nitrazine for Xylidine Ponceau in the Masson trichrome stain.

Reference:

1) Lillie, R.D. *Stain Technol.* **1940,** *15*, 17-22.

General references:

Aldrich FT-IR Library	**1**(2), 985A
Sigma FT-IR Library	**1**(2), 185C
IR Library	**3,** 1461A
Safety	**2**(2), 2546A

NITRO BLUE TETRAZOLIUM CHLORIDE monohydrate

NBT • Nitro BT

Aldrich Cat. No.:	N1,540-5
Sigma Prod. No.:	N6876
CAS No.:	*[298-83-9]*
Mol. Form.:	$C_{40}H_{30}Cl_2N_{10}O_6 \cdot H_2O \cdot CH_3OH$
F.W.:	867.80
m.p.:	205°C (dec.)
Appearance:	Yellow powder
λ_{max}:	260nm in water
Solubility:	H_2O 10mg/ml
	EGME 20mg/ml
	EtOH 5mg/ml

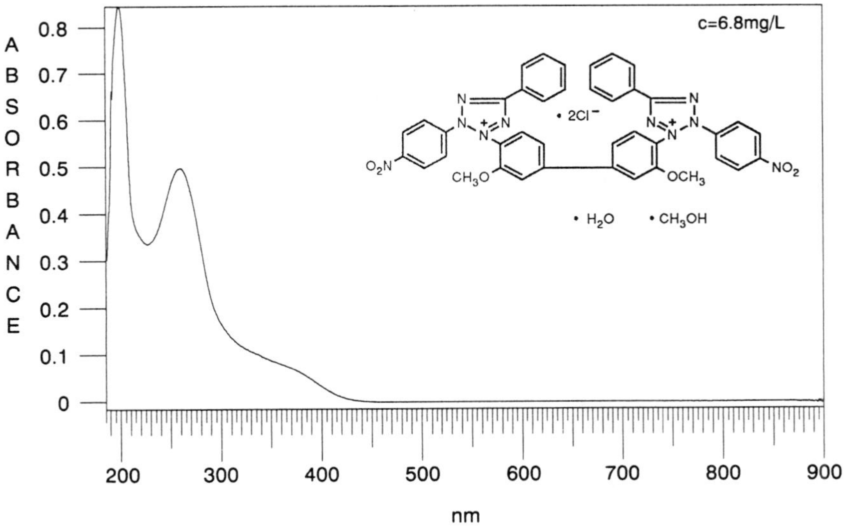

Nitro BT is a dye precursor that forms a formazan dye on reduction. It is somewhat soluble in water and less soluble in ethanol. This system is used as a diagnostic tool with biological substrates which possess enzyme systems that elicit the reduction of the precursor to form the blue dye. Thus, it is used for the determination of dehydrogenase,[1-3] alkaline phosphatase,[4] threonine deaminase[5] and for the detection of other oxidases.[6]

References:

1) Cotton, F.A. *Chem. Rev.* **1955,** *55*, 555.

2) Tsou, K-C.; Cheng, C-S.; Nachlas, M.M.; Seligman, A.M. *J. Am. Chem. Soc.* **1956,** *78*, 6139.

3) Fine, I.H.; Costello, L.A. *Methods Enzymol.* **1963,** *6*, 958.

4) *The Lancet*, Oct. 23, 1971, p. 909 and references 1-11 cited therein.

5) Feldberg, R.S.; Datta, P. *Science* **1970,** *170*, 1414.

6) Feinstein, R.N.; Lindahl, R. *Anal. Biochem.* **1973,** *56*, 353.

General references:

Aldrich FT-IR Library	**1**(2), 634B
Sigma FT-IR Library	**1**(2), 186A
IR Library	**3**, 1228C
Safety	**2**(2), 502D

4-(4-NITROPHENYLAZO)RESORCINOL
Magneson

Aldrich Cat. No.: 11,466-9
Sigma Prod. No.: N8007

CAS No.: [*74-39-5*]
Mol. Form.: $C_{12}H_9N_3O_4$
F.W.: 259.22
m.p.: 185°C (dec.)
Appearance: Red-orange powder
λ_{max}: 432nm in methanol
Solubility: H_2O 1mg/ml
EGME 60mg/ml
EtOH 4mg/ml

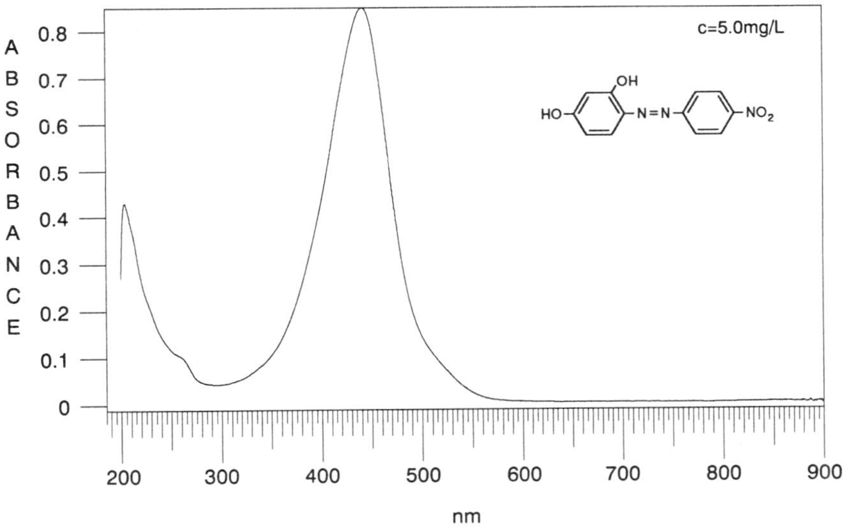

Magneson is a monoazo dye that is made by coupling diazotized 4-nitroaniline to resorcinol. It is very slightly soluble in water, slightly soluble in the lower alcohols and soluble in dilute sodium hydroxide solutions. It has no disclosed use industrially. However, it is ideally structured to form chelates. As the name implies, it is used as a reagent for the detection of magnesium, with which it yields a bright-blue color in alkaline solutions. In addition, it is used to determine molybdenum, which is indicated by the formation of a red-violet complex.[1]

Reference:

1) *The Merck Index,* 11th ed.; Budavari, S. *et al.*, Eds.; Merck: Rahway, 1989; Monograph 5578.

General references:

Aldrich FT-IR Library	**1**(2), 970B
Sigma FT-IR Library	**1**(2), 186D
IR Library	**3**, 1454A

NITRO RED

Aldrich Cat. No.: 20,890-6

CAS No.: [*56431-61-9*]
Mol. Form.: $C_{16}H_{10}N_4Na_2O_9S_2$
F.W.: 512.39
Appearance: Black powder
λ_{max}: 572nm in water
Solubility: H_2O 40mg/ml
EGME 9mg/ml
EtOH 5mg/ml

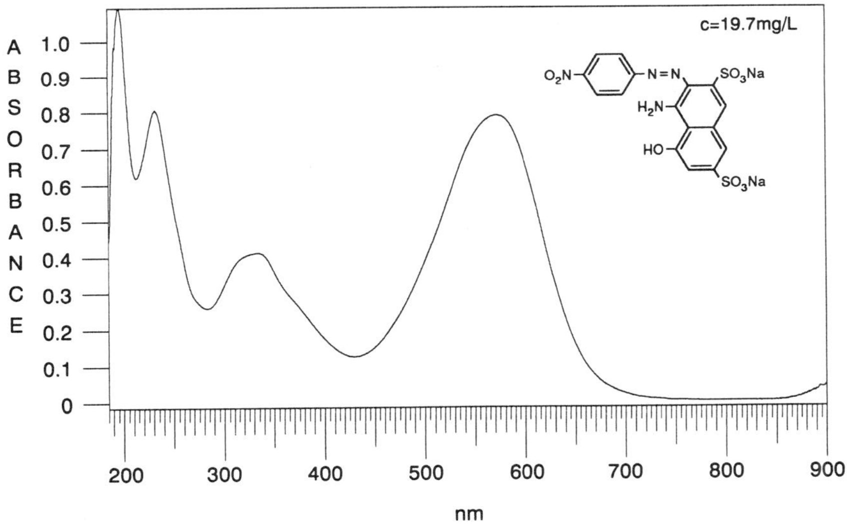

Nitro Red is a monoazo dye that is made by coupling diazotized 4-nitroaniline to H acid (1-amino-8-hydroxynaphthalene-3,6-disulfonic acid). The coupling product is then converted to the disodium salt. It is soluble in water and slightly soluble in the lower alcohols.

Sippel[1,2] found it quite useful as a histochemical stain for the detection of thiol and disulfide groups.

References:

1) Sippel, T.O. *Histochem. J.* **1973,** *5*, 413.
2) *Idem ibid.* **1978,** *10*, 585-609.

General references:

Aldrich FT-IR Library **1**(2), 984A
IR Library **3**, 1460E

2-NITROSO-1-NAPHTHOL

Aldrich Cat. No.:	11,470-7
Sigma Prod. No.:	N1255
Purity:	95%
CAS No.:	[*132-53-6*]
Mol. Form.:	$C_{10}H_7NO_2$
F.W.:	173.17
m.p.:	156-158°C (dec.)
Appearance:	Greenish-gold powder
λ_{max}:	262nm in chloroform
Solubility:	H_2O 0.5mg/ml
	EGME 20mg/ml
	EtOH 6mg/ml

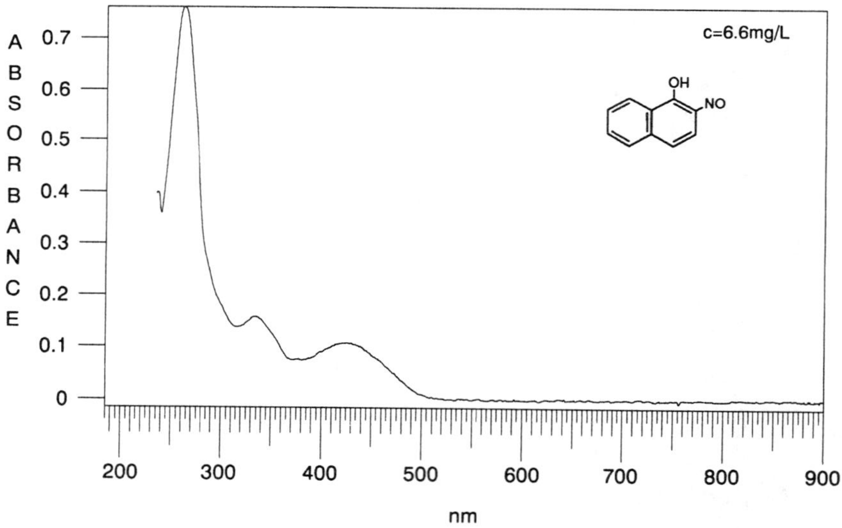

This solvent nitroso dye is made by nitrosating 1-naphthol. Unlike the 1,2-isomer, it is of no industrial value. It is practically insoluble in water, slightly soluble in alcohol and soluble in ether, carbon disulfide, glacial acetic acid and alkali solutions.

2-Nitroso-1-naphthol has been used for the histochemical demonstration of tyrosine residues utilizing fluorescence microscopy.[1,2]

References:

1) Sundler, F.; Larsson, L.-I.; Hakanson, R. *Histochemistry* **1976,** *50*, 39.

2) Ekman, R.; Hakanson, R.; Sundler, F.; Thorell, J. *J. Histochem. Cytochem.* **1980,** *28*, 401.

General references:

Aldrich FT-IR Library	**1**(1), 1327C
Sigma FT-IR Library	**1**(2), 187A
IR Library	**3**, 802G
NMR Library	**2**(1), 1129B
Safety	**2**(2), 2603A

1-NITROSO-2-NAPHTHOL-3,6-DISULFONIC ACID, disodium salt hydrate

Nitroso-R-salt

Aldrich Cat. No.:	N2,538-9
Sigma Prod. No.:	N5507

CAS No.:	*[525-05-3]*
Mol. Form.:	$C_{10}H_5NNa_2O_8S_2 \cdot xH_2O$
F.W.:	377.26
m.p.:	>300°C
Appearance:	Yellow powder
λ_{max}:	424nm in 0.1*N* sodium hydroxide
Solubility:	H_2O 20mg/ml
	EGME 0.8mg/ml
	EtOH 0.6mg/ml

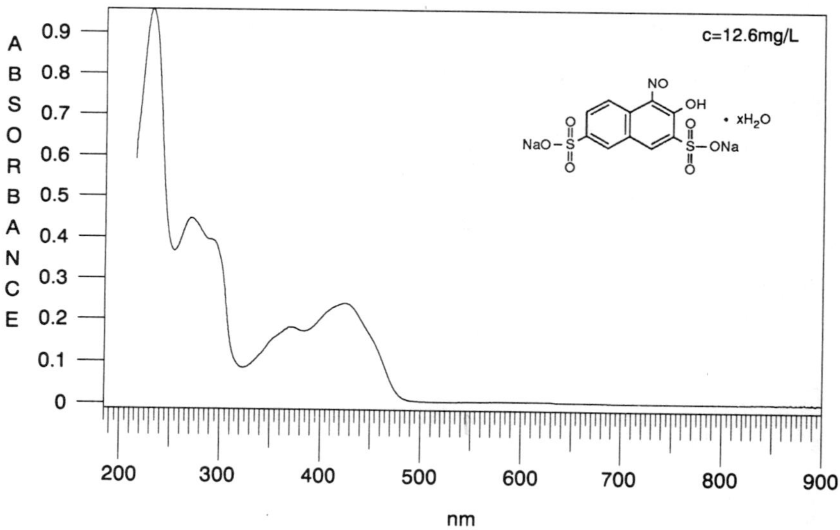

This nitroso dye is made by first nitrosating R acid (2-naphthol-3,6-disulfonic acid) and then converting the resulting product to the disodium salt.

The dye is slightly soluble in cold water, more soluble in hot water and very slightly soluble in methanol and ethanol. Nitroso-R-salt is used as a reagent for the detection of cobalt and potassium.[1]

Reference:

1) *The Merck Index,* 11th ed.; Budavari, S. *et al.*, Eds.; Merck: Rahway, 1989; Monograph 6565.

General references:

Aldrich FT-IR Library	**1**(2), 501A
Sigma FT-IR Library	**1**(2), 801D
IR Library	**3**, 1159F
NMR Library	**2**(2), 822C
Safety	**2**(2), 2603B

NUCLEAR FAST RED

Calcium Red • Helio Fast Rubin BBL

Aldrich Cat. No.:	22,911-3
Sigma Prod. No.:	N8002
CAS No.:	*[6409-77-4]*
C.I. No.:	60760
Mol. Form.:	$C_{14}H_8NNaO_7S$
F.W.:	357.28
m.p.:	>300°C
Appearance:	Dark-red powder
λ_{max}:	535(505)nm in water
Solubility:	H_2O 6mg/ml
	EGME 3mg/ml
	EtOH 0.4mg/ml

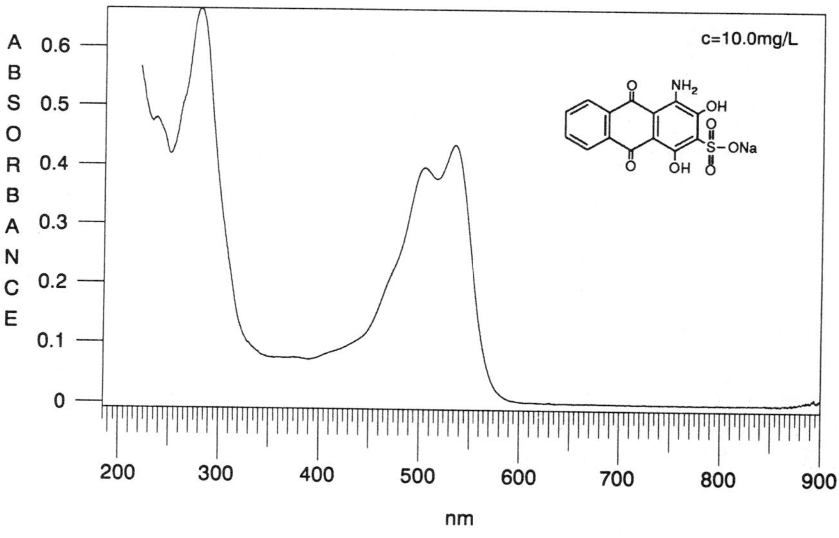

This anionic anthraquinone dye is prepared by treating Purpurin with sodium sulfite, under pressure, to form Pigment Red 84 (C.I. 58210), which then is heated with ammonia to yield the desired product.

Slightly soluble in water, it forms heavy-metal lakes (pigments) which have been used to prepare printing inks and to manufacture wallpaper.

McGee-Russell[1] used it, under the name of Calcium Red, to determine calcium in tissue. The names Calcium Red and Nuclear Fast Red (structure undisclosed) were not linked for nine years. Sams and Davies,[2] surveying commercial samples of Nuclear Fast Red for their suitability as a nuclear counterstain in radioautographs of developing cartilage and bone, concluded that a George Gurr, Ltd. sample, bearing a label with a description that included C.I. 60760, was identical to the Calcium Red used in the McGee-Russell study.

References:

1) McGee-Russell, S.M. *J. Histochem. Cytochem.* **1958,** *6*, 22-42.

2) Sams, A.; Davies, F.M.R. *Stain Technol.* **1967,** *42*, 269-272.

General references:

Aldrich FT-IR Library 1(2), 511A

Sigma FT-IR Library 1(2), 187D

Colour Index 3rd ed., 1971

OIL BLUE N

Solvent Blue 14

Aldrich Cat. No.:	20,525-7
Sigma Prod. No.:	O8376
CAS No.:	*[2646-15-3]*
C.I. No.:	61555
Mol. Form.:	$C_{24}H_{30}N_2O_2$
F.W.:	378.52
m.p.:	210°C (dec.)
Appearance:	Dark-violet-maroon powder
λ_{max}:	637(591)nm in ethanol
Solubility:	H_2O 2mg/ml
	EGME 1mg/ml
	EtOH 0.4mg/ml

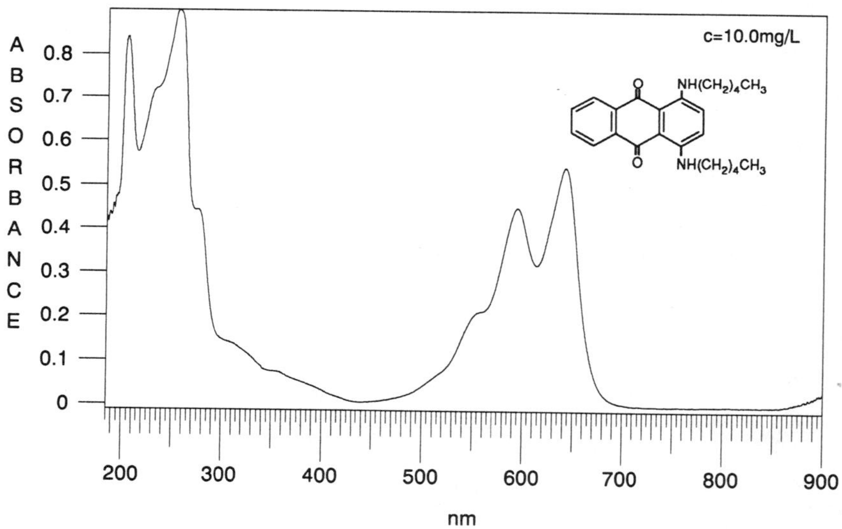

This blue, aminoanthraquinone, solvent dye is made by reacting leucoquinizarin with amylamine. Oil Blue N is soluble in acetone, butyl acetate, and toluene and very slightly soluble in methyl Cellosolve, ethanol and water.

Despite poor lightfastness, it has been used to color petroleum, hydrocarbons, oils, fats and waxes.

Oil Blue N has been used very little as a biological stain. *Conn's*[1] reports the following uses: as a rubber stain in plant histology by both Wittenberger and Addicott in 1944 and as a stain in animal histology by Lillie in 1945 and by Quay in 1954.

Reference:

1) *Conn's Biological Stains,* 9th ed.; Lillie, R.D., Ed.; Williams and Wilkins: Baltimore, 1977; p 443.

General references:

Aldrich FT-IR Library	**1**(2), 1017B
Sigma FT-IR Library	**1**(2), 188A
IR Library	**3**, 1476G
Colour Index	3rd ed., 1971

OIL RED EGN
Solvent Red 26

Aldrich Cat. No.:	23,411-7
Sigma Prod. No.:	O2003
CAS No.:	[*4477-79-6*]
C.I. No.:	26120
Mol. Form.:	$C_{25}H_{22}N_4O$
F.W.:	394.48
m.p.:	430°C
Appearance:	Dark-green powder
λ_{max}:	521nm in chloroform
Solubility:	H_2O 2mg/ml
	EGME 30mg/ml
	EtOH 5mg/ml

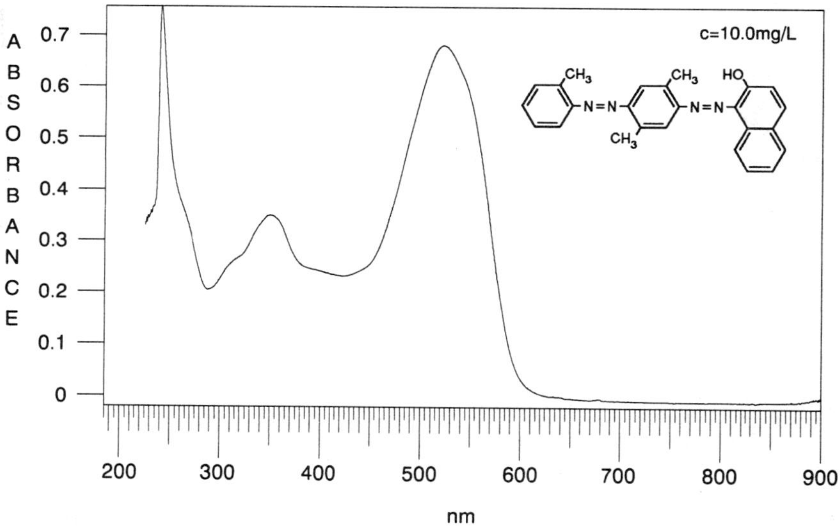

Oil Red EGN is a neutral disazo dye that is prepared by coupling diazotized *o*-toluidine with 2,5-xylidine, diazotizing the product of the coupling and coupling it in turn to 2-naphthol.

Very soluble in benzene, toluene and acetone, soluble in oleic acid, stearic acid, linseed oil and ethyl acetate and slightly soluble in ethanol, mineral oil and paraffin wax, it is used to color fats, oils, waxes, petroleum products and acrylic resins.

The structure and solubilities indicate that it may have some use as a fat stain in histology; however, there is no direct literature citations to this effect.

General references:

Aldrich FT-IR Library	**1**(2), 975C
Sigma FT-IR Library	**1**(2), 188B
Safety	**2**(2), 2651C
Colour Index	3rd ed., 1971

OIL RED O

Solvent Red 27

Aldrich Cat. No.:	19,819-6
Sigma Prod. No.:	O0625
Grade:	Certified
CAS No.:	[*1320-06-5*]
C.I. No.:	26125
Mol. Form.:	$C_{26}H_{24}N_4O$
F.W.:	408.51
m.p.:	120°C (dec.)
Appearance:	Dark greenish-brown powder
λ_{max}:	518(359)nm in toluene
Solubility:	H_2O 4mg/ml
	EGME 30mg/ml
	EtOH 7mg/ml

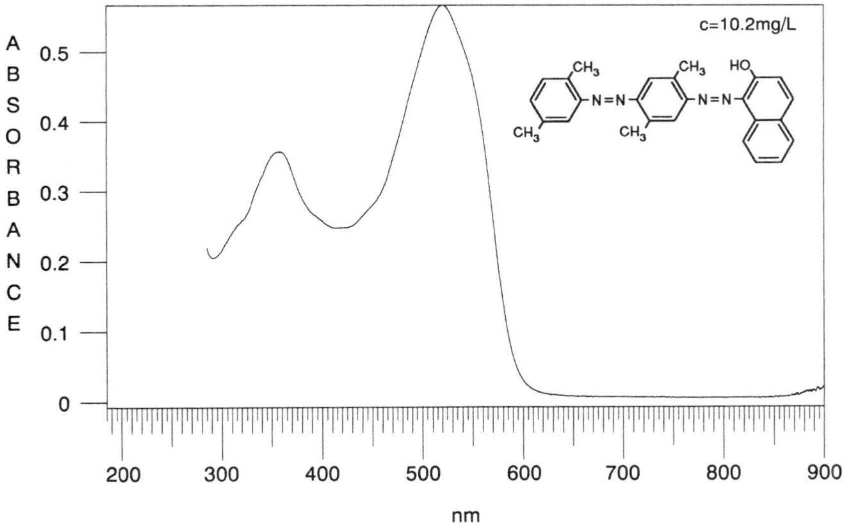

Oil Red O is a disazo solvent dye with a +8.99 Hansch *pi* value[1] and good lightfastness. Used commercially to color oils, fats, waxes, oil varnishes and cosmetics, it is very soluble in benzene and toluene, soluble in acetone, methyl Cellosolve, ethyl acetate, linseed oil, stearic acid and xylene, and less soluble in mineral oil and paraffin wax.

Oil Red O has been certified by the BSC since August of 1984 and is checked by the Commission according to the the Lillie-Ashburn[2] method for staining lipids in frozen sections. *Staining Procedures*[3] provides methods and a bibliography for the following: Proescher's Oil Red with pyridine for demonstrating lipid degeneration in the central nervous system; fat stain in supersaturated isopropanol for demonstrating fat in animal tissue; the Oil Red O method adapted by Lillie and Fullmer[4] from Lillie and Ashburn.[2]

References:

1) Horobin, R.W. *Acta Histochemica* **1981,** *5 (Suppl.-Band XXIV),* 237-246.
2) Lillie, R.D.; Ashburn, L.L. *Arch. Pathol.* **1943,** *36*, 432-435.
3) Clark, G. *Staining Procedures,* 4th ed.; Williams and Wilkins: Baltimore, 1981; pp 193, 194, 228.
4) Lillie, R.D.; Fullmer, H.M. *Histopathologic Technic and Practical Histochemistry,* 4th ed.; McGraw-Hill: New York, 1976.

General references:

Aldrich FT-IR Library	**1**(2), 976B
Sigma FT-IR Library	**1**(2), 188C
IR Library	**3**, 1456G
Safety	**2**(2), 2651D
Colour Index	3rd ed., 1971

ORANGE G

Acid Orange 10

Aldrich Cat. No.:	86,128-6
Sigma Prod. No.:	O7252
Grade:	Certified
CAS No.:	[*1936-15-8*]
C.I. No.:	16230
Mol. Form.:	$C_{16}H_{10}N_2Na_2O_7S_2$
F.W.:	452.38
Appearance:	Red-orange powder
λ_{max}:	475nm in water
Solubility:	H_2O 80mg/ml
	EGME 40mg/ml
	EtOH 3mg/ml

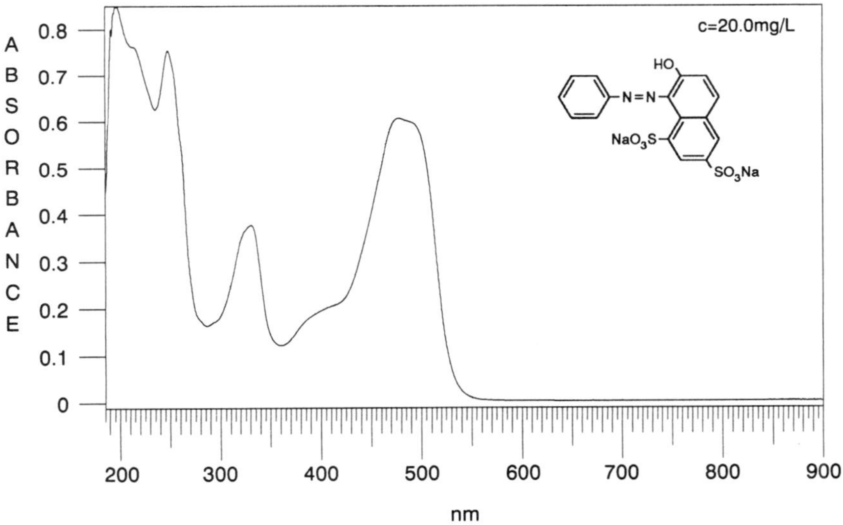

Orange G is an anionic monoazo dye that is made by coupling diazotized aniline to G acid (2-naphthol-6,8-disulfonic acid) and converting the coupling product to the disodium salt.

Soluble in water and methyl Cellosolve and slightly soluble in ethanol, it dyes wool directly and silk from either a formic acid or sulfuric acid bath. Wool and silk can be printed directly. Non-textile industrial uses include dyeing paper, staining wood, and coloring inks and copying pencils. Fully chromed leather is dyed a dull orange.

Orange G is an important biological stain that is used widely as a background or contrast stain for a number of important nuclear stains such as

Hematoxylin, Safranin O, Crystal Violet, Methyl Green and Basic Fuchsin. It is an excellent plasma stain and an important component of one of the stain solutions used in the Papanicolaou method for detecting vaginal, cervical and uterine cancer.

Staining Procedures[1] provides methods and a bibliography for the following applications using Orange G: as a Mallory's collagen stain with Aniline Blue; in Heidenhain's azan modification; in Kornhauser's "Quad" Stain; as a component in the Orcein and Feulgen techniques to demonstrate somatic mitosis; in Papanicolaou stain; in Mallory-Heidenhain azan; in Gomori's modification for simultaneously demonstrating α, β and D cells in the islets of Langerhans; in the performic acid-Alcian Blue, periodic acid Schiff's reagent, Orange G method for the human pituitary gland; to demonstrate collagen fibers in oyster tissue; in the modified Barbeito-Lopez trichrome stain for oyster pathology; in Aldehyde Fuchsin for neurosecretory products in insects; in the oxidative Schiff Aniline Blue-Orange G to demonstrate coccidians in tissue; in Mallory's triple stain; in the Safranin-Orange G stain to demonstrate plant cell walls and protoplasts; as a counterstain with various nuclear stains for examining general plant tissue and in Thionin and Orange G Stain for demonstrating fungi and bacteria in tissue.

Reference:

1) Clark, G. *Staining Procedures,* 4th ed.; Williams and Wilkins: Baltimore, 1981; pp 117, 118, 124, 125, 197, 209, 218, 219, 250, 256-258, 262, 268, 269, 292, 301, 321, 324, 330, 355, 373.

General references:

Aldrich FT-IR Library	**1**(2), 982C
Sigma FT-IR Library	**1**(2), 189B
IR Library	**3**, 1460A
Safety	**2**(2), 2655D
Colour Index	3rd ed., 1971

ORANGE II

Acid Orange 7

Aldrich Cat. No.: 19,523-5
Sigma Prod. No.: O8126
Grade: Certified

CAS No.: [*633-96-5*]
C.I. No.: 15510
Mol. Form.: $C_{16}H_{11}N_2NaO_4S$
F.W.: 350.33
Appearance: Orange-brown powder
λ_{max}: 483nm in water
Solubility: H_2O 130mg/ml
EGME 120mg/ml
EtOH 4mg/ml

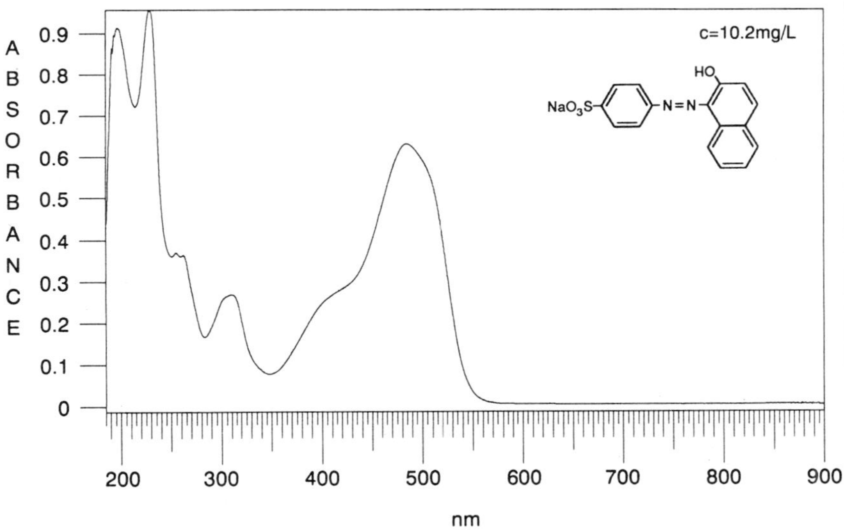

Orange II is an anionic monoazo dye that is made by coupling diazotized sulfanilic acid to 2-naphthol and converting the coupling product to the sodium salt. Very soluble in water and very slightly soluble in ethanol, it has limited value as an indicator since the color changes are subtle and hard to interpret (the upper range is more usable than the lower). The visual-transition intervals are pH 7.4 (amber) to pH 8.6 (orange) and pH 10.2 (orange) to pH 11.8 (red).

Commercially, it is used to dye wool directly, nylon from a formic acid bath and silk from a broken degumming liquor. Wool, silk and nylon can be printed directly. Fully chromed leather is dyed a bright orange. The heavy-metal salts of Orange II are used for paper coating.

Orange II is on the BSC list of certified dyes; however, very little, if any, is certified. It has been substituted for both Methyl Orange and Orange G in certain histological applications. *Staining Procedures*[1] provides a method, after Newcomer,[2] utilizing Orange II in combination with Crystal Violet for staining pollen tube cultures to demonstrate chromosomes.

References:

1) Clark, G. *Staining Procedures,* 4th ed.; Williams and Wilkins: Baltimore, 1981; p 364.

2) Newcomer, E.H. *Stain Technol.* **1938,** *13,* 89-91.

General references:

Safety	**2**(2), 2656A
Colour Index	3rd ed., 1971

ORCEIN

Aldrich Cat. No.:	11,417-0	21,350-0
Sigma Prod. No.:	O3626	——
Grade:	Natural	Synthetic
	Natural Red 28	Certified
CAS No.:	[*1400-62-0*]	
Appearance:	Brown-black powder	Brown crystalline powder
λ_{max}:	579nm in 0.1*N* sodium hydroxide	
Solubility:	H_2O 10mg/ml	
	EGME 4mg/ml	
	EtOH 1mg/ml	

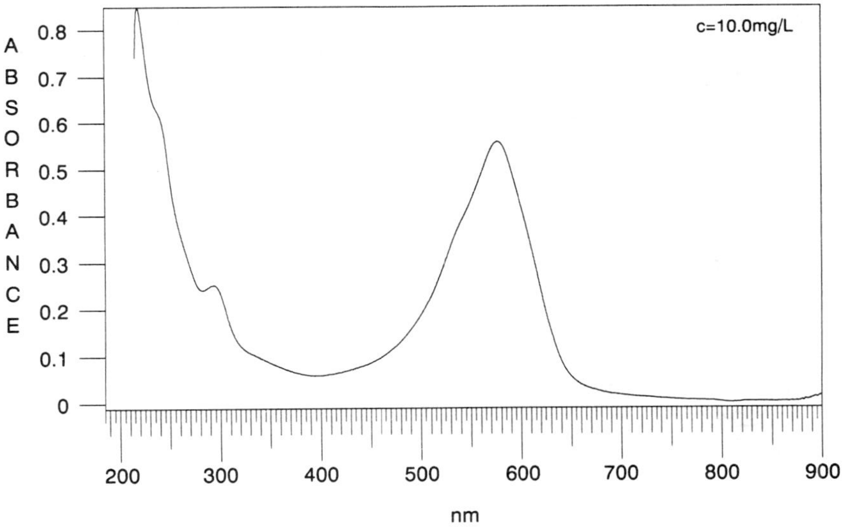

Orcein was first derived from natural sources. Lichen, such as *Lecanora tinctoria* and *Rocella tinctoria*, abundant in Madagascar, the Azores, the Canary Islands and the Cape of Good Hope, contain lecanoric and other closely related, colorless acids that serve as Orcein precursors. The stain is developed by a series of reactions: Lecanoric acid is hydrolyzed to orcelinic acid (2,4-dihydroxy-6-methylbenzoic acid) which in turn is hydrolyzed to orcinol. Finally, the color is developed when the orcinol solution is subjected to air oxidation in the presence of ammonia. Litmus, which is described separately in this book (pages 425 and 426), has the same natural origin as orcein. However, the orcinol air-oxidation must be carried out in the presence of sodium or potassium carbonate.

More recently, Orcein has been produced synthetically by treating ammoniacal orcinol with hydrogen peroxide. Synthetic Orcein, like natural Orcein, is not a pure substance. Indeed, Musso,[1] studying synthetic Orcein, reported separating 14 different closely related components by chromatography. These included: α-, β- and γ-aminophenoxazines, α-, β- and γ-hydroxyphenoxazines and as much as 10% γ-aminoorceimine (7-aminophenoxazine containing two orcinol moieties, one each in the 3- and 6-positions).

Synthetic Orcein is soluble in water, slightly soluble in methyl Cellosolve and very slightly soluble in ethanol. While Orcein has indicator properties, it has no apparent industrial application.

Staining Procedures[2] provides a bibliography and methods for the following *in vitro* biological applications: Darrow's elastic fiber stain modification using synthetic Orcein; Fraenkel's Orcein method for elastic tissue; Kornhauser's "Quad" Stain; Orcein and Feulgen techniques using corneal epithelium to demonstrate mammalian somatic mitosis; synthetic Orcein demonstrating chromosomes using the salivary-gland cells of dipteran and the meristemic cells of root tips; Crozier method for insect chromosomes; LaCour method for acetic-Orcein demonstration of insect chromosomes; Dyer's lactopropionic-Orcein method for root tips and pollen mother cells in a wide range of plants; and Smith's propionic acid-Orcein root-tip smear technique designed to show chromosomes.

References:

1) Musso, H. *Planta Med.* **1960,** *8*, 432-446.

2) Clark, G. *Staining Procedures,* 4th ed.; Williams and Wilkins: Baltimore, 1981; pp 123-124, 197-198, 258, 266, 348-349.

General references:

Aldrich FT-IR Library	**1**(2), 1036B	1(2), 1036C
Sigma FT-IR Library	**1**(2), 189C	1(2), 189D
IR Library	**3**, 1486E	—
Merck Index	**11**, 6818	

PALATINE CHROME BLACK 6BN

Mordant Black 17

Aldrich Cat. No.:	21,673-9
Sigma Prod. No.:	H1878
CAS No.:	[*2538-85-4*]
C.I. No.:	15705
Mol. Form.:	$C_{20}H_{13}N_2NaO_5S$
F.W.:	416.39
Appearance:	Brown-purple powder
λ_{max}**:**	569nm in water
Solubility:	H_2O 30mg/ml
	EGME 70mg/ml
	EtOH 10mg/ml

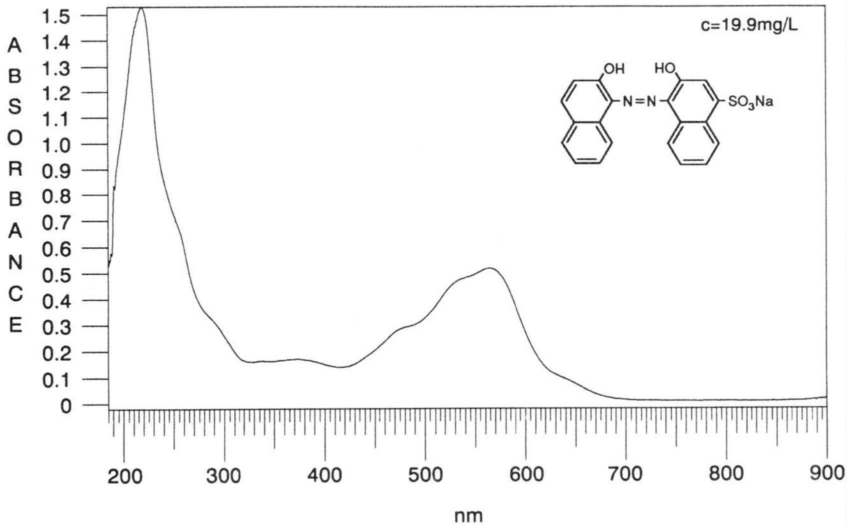

Palatine Chrome Black 6BN is an anionic, monoazo mordant dye that is made by coupling diazotized 1,2,4-acid (1-amino-2-naphthol-4-sulfonic acid) to 2-naphthol and converting the coupling product to the sodium salt. The resulting product is ideally structured (with a hydroxy group *ortho* to each of the two azo nitrogens) for chelate formation with chromium and certain other metals.

Soluble in water (red) and methyl Cellosolve, slightly soluble in ethanol and insoluble in other common organic solvents, it dyes afterchromed wool, silk and nylon a dull reddish-navy blue, prints on wool with a chromium mordant and dyes loose wool, fur, and hat bodies when chromium is used as a mordant.

Conn's[1] states that, like Hematoxylin and other metal-binding dyes, Palatine Chrome Black 6BN stains eleidin, keratohyalin, rodent elastin and eosinophil-leucocyte granules dark red, selectively, on extended exposure. Erythrocytes are stained less intensely.

Reference:

1) *Conn's Biological Stains,* 9th ed.; Lillie, R.D., Ed.; Williams and Wilkins: Baltimore, 1977; p 115.

General references:

Aldrich FT-IR Library **1**(2), 987A
Sigma FT-IR Library **1**(2), 151D
IR Library **3**, 1462F
Colour Index 3rd ed., 1971

PALATINE FAST BLACK WAN

Acid Black 52

Aldrich Cat. No.: 20,184-7
Sigma Prod. No.: P5019

CAS No.: [*5610-64-0*]
C.I. No.: 15711
Mol. Form.: $C_{60}H_{36}Cr_2N_9Na_3O_{21}S_3$
F.W.: 1488.01
Appearance: Black powder
λ_{max}: 588nm in water
Solubility: H_2O 30mg/ml
EGME 20mg/ml
EtOH 20mg/ml

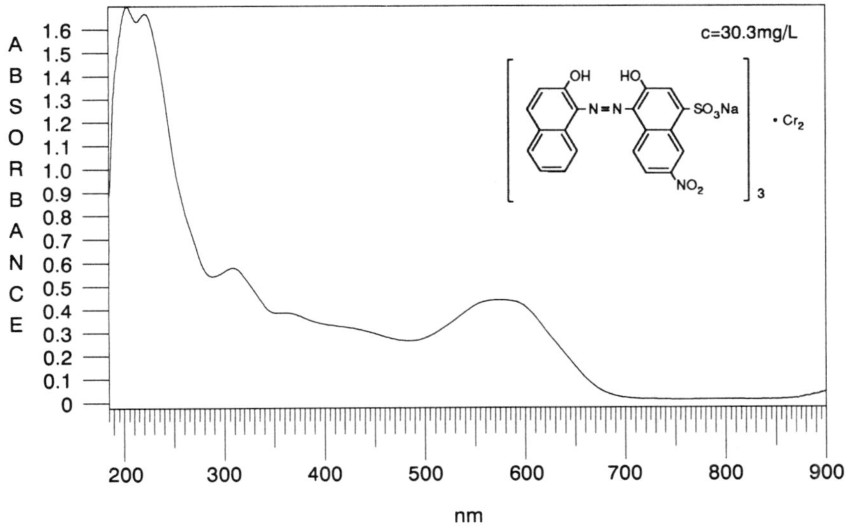

Palatine Fast Black WAN is a metallized, anionic monoazo dye. It is prepared by treating Salicin Black EAG (Mordant Black 1, C.I. 15710) with chromium formate and combining this product with untreated dye in a ratio of approximately 2:1 w/w. The dye is soluble in water, ethanol and methyl Cellosolve, slightly soluble in acetone and insoluble in all other organic solvents. Industrially, it is used to dye wool, silk, nylon and chromed leather and to color anodized aluminum.

There is no evidence that it has been used as a biological stain.

General references:

Aldrich FT-IR Library	**1**(2), 989A
Sigma FT-IR Library	**1**(2), 190A
IR Library	**3**, 1463D
Colour Index	3rd ed., 1971

PALATINE FAST YELLOW BLN

Acid Yellow 54

Aldrich Cat. No.:	20,179-0
Sigma Prod. No.:	P5144
CAS No.:	[*10127-05-6*]
C.I. No.:	19010
Mol. Form.:	$C_{18}H_{13}N_4Na_3O_9S_2$ • xCr
F.W.:	562.42
Appearance:	Rust powder
λ_{max}:	440nm in water
Solubility:	H_2O 50mg/ml
	EGME 20mg/ml
	EtOH 8mg/ml

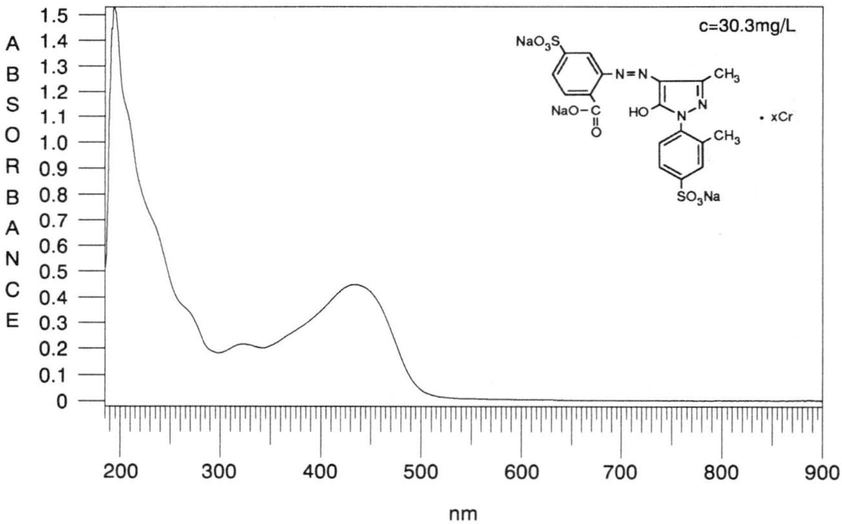

Palatine Fast Yellow BLN is an anionic monoazo dye that forms complexes with chromium and other metal ions. The commercial product, usually metallized with chromium, is soluble in water and slightly soluble in ethanol and acetone. It dyes wool, silk and nylon directly and prints wool and silk directly.

Bedrick[1] used it to polychromatically stain both benign and malignant tumors.

Reference:

1) Bedrick, A.E. *Stain Technol.* **1970,** *45*, 273-276.

General references:

Aldrich FT-IR Library	**1**(2), 1001D
Sigma FT-IR Library	**1**(2), 190B
IR Library	**3,** 1469B
Colour Index	3rd ed., 1971

PARA RED

1-(4-Nitrophenylazo)-2-naphthol • Pigment Red 1

Aldrich Cat. No.:	10,099-4
Sigma Prod. No.:	P7019
Grade:	Biological stain

CAS No.:	[*6410-10-2*]
C.I. No.:	12070
Mol. Form.:	$C_{16}H_{11}N_3O_3$
F.W.:	293.28
m.p.:	248-252°C
Appearance:	Red powder
λ_{max}:	488nm in toluene
Solubility:	H_2O 1mg/ml
	EGME 2mg/ml
	EtOH 1mg/ml

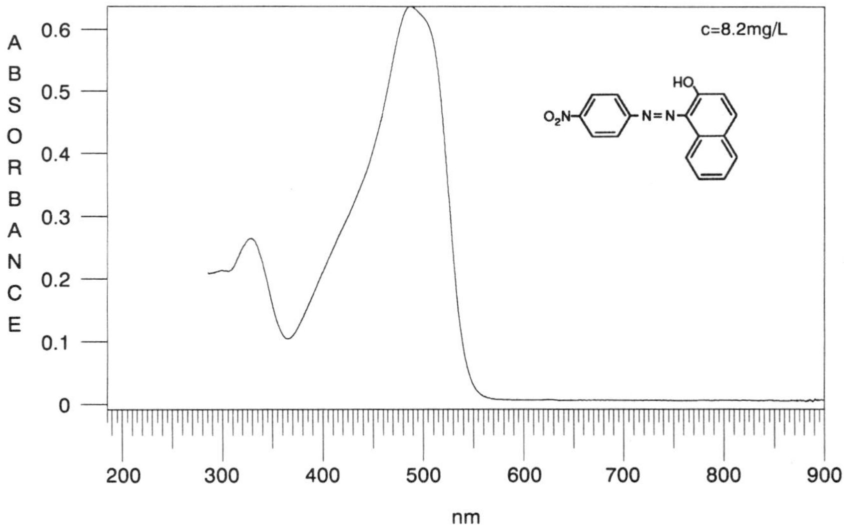

Para Red is a dull-red pigment with a high tinctorial strength due to the combined effect of two chromophores, a nitro group and a monoazo linkage, working in concert with the fundamental aromatic structure of the molecule. It is made by coupling diazotized 4-nitroaniline with 2-naphthol. It is soluble in hot 10% NaOH, slightly soluble in toluene (crystals that melt at 256°C are formed on cooling the toluene solution) and very slightly soluble in boiling ethanol.

Para Red is used in paper coatings and printing inks, alkyd-resin enamels and lacquers, with limitations due to its solubility. In addition, it is used to color bookcloth, leather finishes, tinplate printing inks, rubber, plastics and vinyl products.

There is no evidence in the literature that it has been used as a biological stain.

General references:

Aldrich FT-IR Library	**1**(2), 971A
Sigma FT-IR Library	**1**(2), 198B
IR Library	**3**, 1454C
Colour Index	3rd ed., 1971

PARAROSANILINE acetate

Aldrich Cat. No.: 21,594-5
Sigma Prod. No.: P4896
Dye content: 90%

CAS No.: [*6035-94-5*]
Mol. Form.: $C_{21}H_{21}N_3O_2$
F.W.: 347.42
m.p.: 203°C (dec.)
Appearance: Green powder
λ_{max}: 545nm in 50% ethanol
Solubility: H_2O 30mg/ml
EGME 10mg/ml
EtOH 20mg/ml

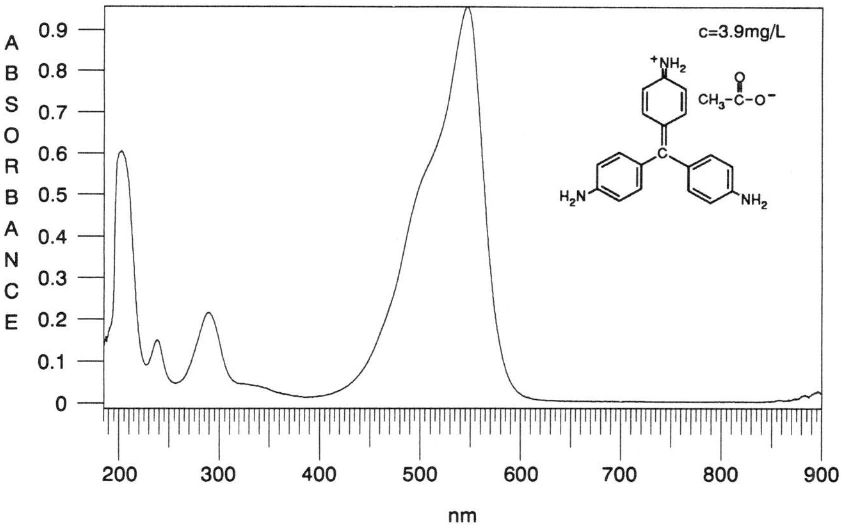

See description on pages 126-128 under Basic Fuchsin.

General references:

Aldrich FT-IR Library	**1**(2), 1030A
Sigma FT-IR Library	**1**(2), 190D
IR Library	**3**, 1483F
Safety	**2**(2), 2682B

PARAROSANILINE base

Aldrich Cat. No.: 21,560-0
Sigma Prod. No.: P7632
Dye content: 95%

CAS No.: [*25620-78-4*]
Mol. Form.: $C_{19}H_{19}N_3O$
F.W.: 305.38
m.p.: 205°C (dec.)
Appearance: Purple crystalline powder
λ_{max}: 544nm in ethanol
Solubility: H_2O 0.8mg/ml
EGME 10mg/ml
EtOH 3mg/ml

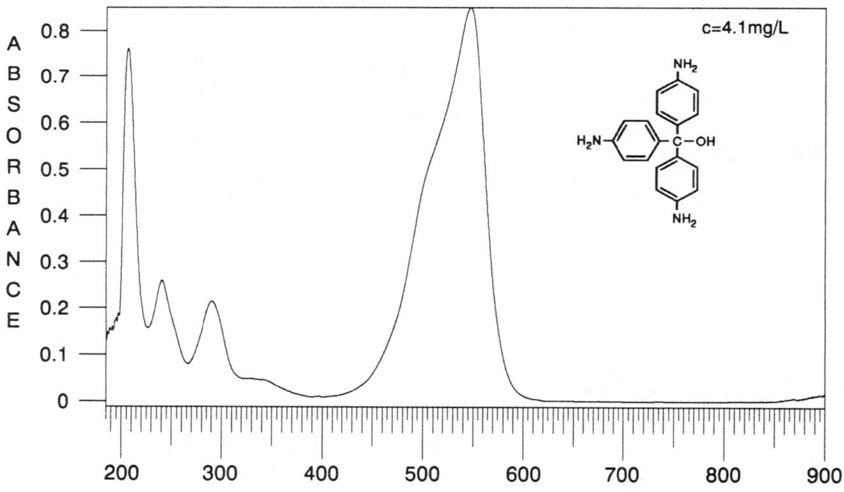

See description on pages 126-128 under Basic Fuchsin.

General references:

Aldrich FT-IR Library	**1**(1), 1247D
Sigma FT-IR Library	**1**(2), 190C
IR Library	**3**, 745G
NMR Library	**2**(1), 1015D
Safety	**2**(2), 2682A

PATENT BLUE VF
Acid Blue 1

Aldrich Cat. No.:	19,821-8
Sigma Prod. No.:	P6396
Grade:	Indicator Biological stain
CAS No.:	[*129-17-9*]
C.I. No.:	42045
Mol. Form.:	$C_{27}H_{31}N_2NaO_6S_2$
F.W.:	566.68
Appearance:	Dark bluish-green powder
λ_{max}:	635(410)nm in water
Solubility:	H_2O 30mg/ml EGME 30mg/ml EtOH 7mg/ml

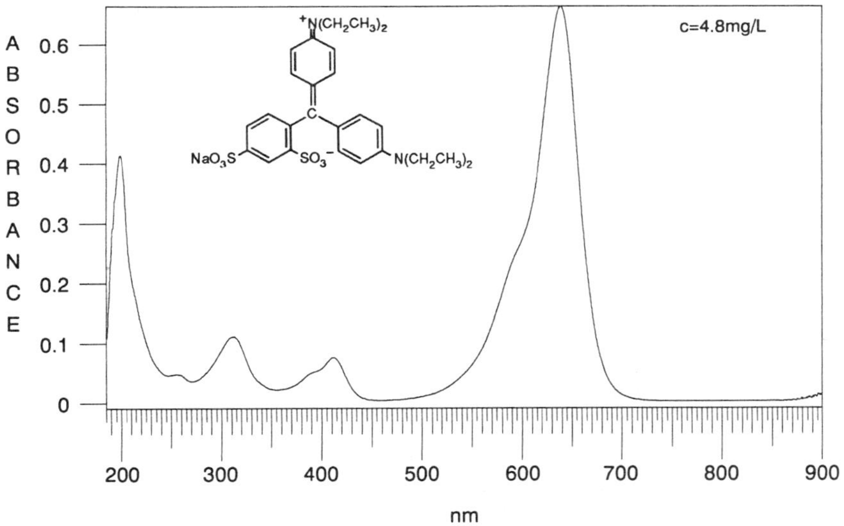

This anionic triphenylmethane dye is used to dye wool directly and silk from either a sulfuric acid or acetic acid bath. It dyes fully chromed leather a bright blue, prints wool and silk directly and its heavy-metal salts are used as pigments.

Patent Blue displays acid-base-indicator properties, turning orange in the presence of concentrated hydrochloric acid and deep blue when sodium hydroxide is added. *Conn's*[1] believes it is suitable for use in the zinc leuco-hemoglobin method.

Reference:

1) *Conn's Biological Stains,* 9th ed.; Lillie, R.D., Ed.; Williams and Wilkins: Baltimore, 1977; pp 252-253.

General references:

Aldrich FT-IR Library	**1**(2), 1028C
Sigma FT-IR Library	**1**(2), 192B
IR Library	**3**, 1482H
Safety	**2**(2), 2687D
Merck Index	**11**, 8968
Colour Index	3rd ed., 1971

PHENANTHRENEQUINONE

Aldrich Cat. No.:	15,650-7
Sigma Prod. No.:	P2896
Purity:	99+%
CAS No.:	[*84-11-7*]
Mol. Form.:	$C_{14}H_8O_2$
F.W.:	208.22
m.p.:	209-211°C
Appearance:	Burnt-orange powder
λ_{max}:	270(325)nm in chloroform
Solubility:	H_2O 0.4mg/ml
	EGME 7mg/ml
	EtOH 1mg/ml

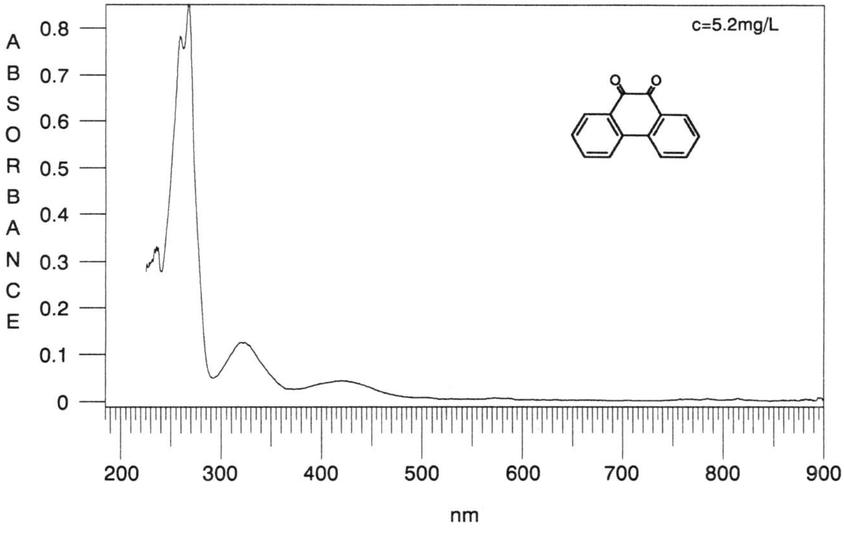

This orange-red crystalline chromogen is virtually insoluble in water and soluble in benzene, ether, glacial acetic acid and hot ethanol (only very slightly soluble in cold ethanol). It has been used as a histochemical reagent for the demonstration of arginine residues by fluorescence.[1,2]

References:

1) Sundler, F.; Hakanson, R. *Histochemistry* **1978,** *56*, 221.

2) Ekman, R.; Hakanson, R.; Sundler, F.; Thorell, J. *J. Histochem. Cytochem.* **1980,** *28*, 401.

General references:

Aldrich FT-IR Library	**1**(2), 82B
Sigma FT-IR Library	**1**(2), 824C
IR Library	**3**, 897H
Safety	**2**(2), 2737D
Merck Index	**11,** 7169

1,10-PHENANTHROLINE

Aldrich Cat. No.:	13,137-7
Purity:	99+%
CAS No.:	*[66-71-7]*
Mol. Form.:	$C_{12}H_8N_2$
F.W.:	180.21
m.p.:	114-117°C
Appearance:	White powder
λ_{max}:	264nm in methanol
Solubility:	H_2O 2mg/ml
	EGME 30mg/ml
	EtOH 20mg/ml

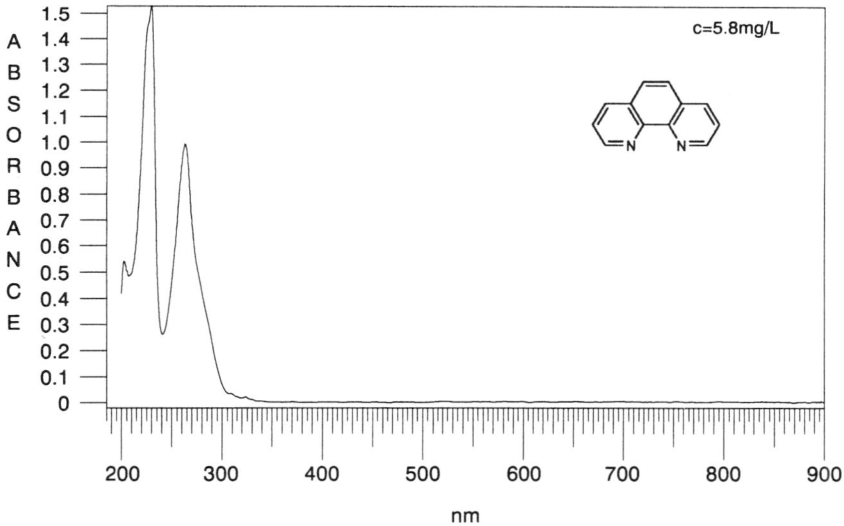

1,10-Phenanthroline is a heterocyclic compound that is prepared by a double-Skraup reaction on 1,2-phenylenediamine. *Vogel's*[1] shows a modified-Skraup synthesis for preparing 1,10-phenanthroline that takes precautions to avoid excessive oxidation of the highly sensitive intermediate 1,2-phenylenediamine.

1,10-Phenanthroline is an important oxidation-reduction indicator. It is used as an indicator in the titration of Grignard reagents, and it forms chelates with iron, nickel, ruthenium, silver and other metals. The ability of 1,10-phenanthroline to form complexes with these metals makes it possible to use it as a reagent for the determination of these metals.[2-4] *Reagent Chemicals*[5] provides specifications and test methods for utilizing 1,10-phenanthroline monohydrate as a redox indicator and as a reagent for determining iron.

References:

1) *Vogel's Textbook of Practical Organic Chemistry*, 4th ed.; Furniss, B.S.; Hannaford, A.J.; Rogers, V.; Smith, P.W.G.; Tatchell, A.R., Eds.; Longman: New York, 1978; p 912.
2) Krishna Murtie, G.S.R.; Moharir, A.V.; Sarma, V.A.K. *Microchem. J.* **1970,** *15*, 585.
3) O'Laughlin, J.W. *Anal. Chem.* **1982,** *54*, 178.
4) Watson, S.C.; Eastham, J.F. *J. Organomet. Chem.* **1967,** *9,* 165.
5) *Reagent Chemicals,* 7th ed.; American Chemical Society: Washington, DC, 1986; pp 466-467.

General references:

Aldrich FT-IR Library	**1**(2), 889D
IR Library	**3**, 1410H
Safety	**2**(2), 2738A
Merck Index	**11**, 7169

1,10-PHENANTHROLINE monohydrate

Aldrich Cat. No.:	P1,280-4	32,005-6
Sigma Prod. No.:	P9375	——
Grade:	——	A.C.S. reagent
Purity:	99+%	99%

CAS No.:	[*5144-89-8*]
Mol. Form.:	$C_{12}H_8N_2 \cdot H_2O$
F.W.:	198.23
m.p.:	100-103°C
Appearance:	White crystalline powder
λ_{max}:	264nm in ethanol
Solubility:	H_2O 2mg/ml
	EGME 30mg/ml
	EtOH 20mg/ml

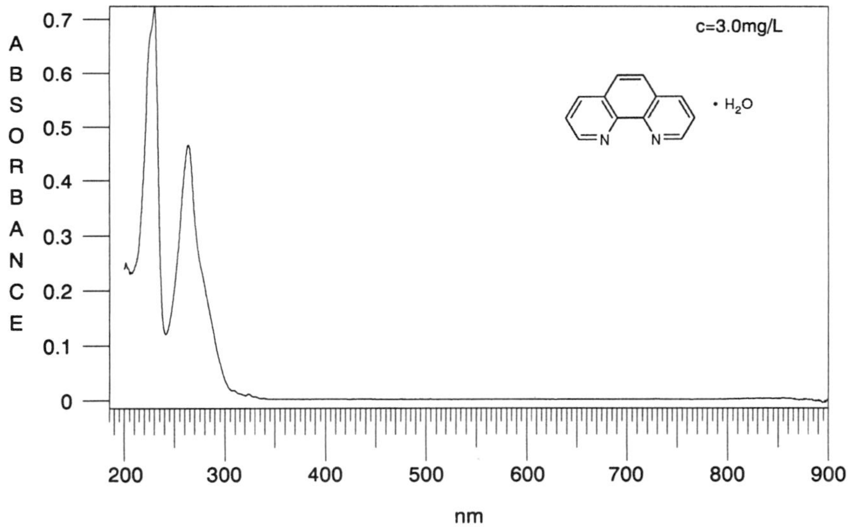

See description on pages 557 and 558 under 1,10-phenanthroline.

General references:

Aldrich FT-IR Library	**1**(2), 890A
Sigma FT-IR Library	**1**(2), 192C
IR Library	**3**, 1411A
Safety	**2**(2), 2738B
Merck Index	**11**, 7169

1,10-PHENANTHROLINE monohydrochloride monohydrate

Aldrich Cat. No.: P1,300-2
Purity: 97%

CAS No.: [*3829-86-5*]
Mol. Form.: $C_{12}H_8N_2 \cdot HCl \cdot H_2O$
F.W.: 234.69
m.p.: 224-225°C
Appearance: White crystalline powder with chunks
λ_{max}: 264nm in methanol
Solubility: H_2O 600mg/ml
EGME 80mg/ml
EtOH 40mg/ml

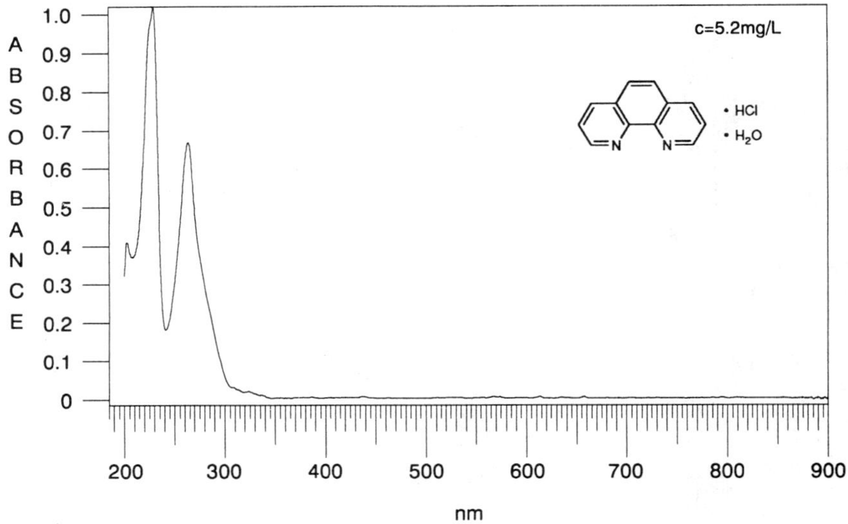

See description on pages 557 and 558 under 1,10-phenanthroline.

General references:

Aldrich FT-IR Library	**1**(2), 890B
IR Library	**3**, 1411B
NMR Library	**2**(2), 764A
Safety	**2**(2), 2738C

PHENAZINE METHOSULFATE

N-Methylphenazonium methosulfate • PMS

Aldrich Cat. No.:	P1,340-1
Sigma Prod. No.:	P9625
Purity:	99%
CAS No.:	[*299-11-6*]
Mol. Form.:	$C_{14}H_{14}N_2O_4S$
F.W.:	306.34
m.p.:	158-160°C (dec.)
Appearance:	Green crystalline powder
λ_{max}**:**	386nm in water
Solubility:	H_2O 50mg/ml
	EGME 30mg/ml
	EtOH 10mg/ml

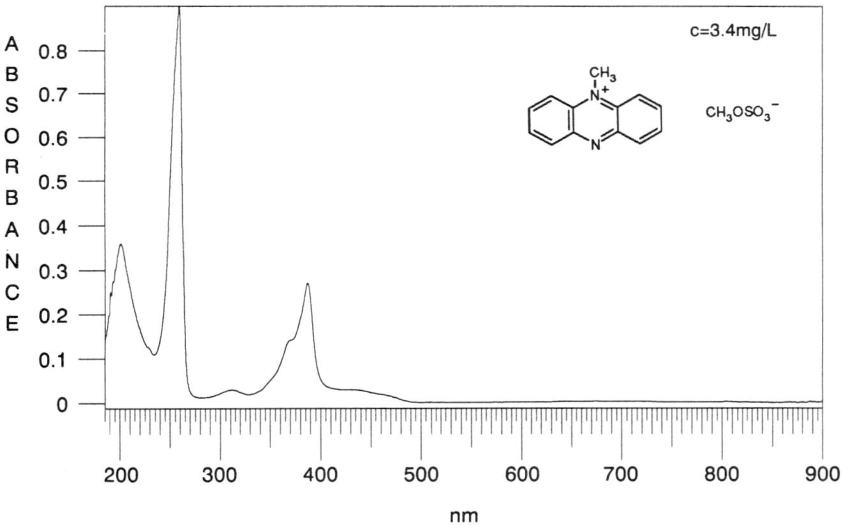

Kehrman and Havas[1] treated the chromogen phenazine with dimethyl sulfate, using nitrobenzene as a solvent, to obtain phenazine methosulfate.

This water-soluble compound, which is also slightly soluble in alcohol, is not a dye as it lacks an auxochrome. However, it has been used widely as an electron-transfer agent in histochemical enzyme reactions since Farber and Bueding[2] first used it to reactivate succinate dehydrogenase. Studies on the detection of specific dehydrogenases were then carried out.[3-5] Its roles as an acceptor[6] and accelerator[7] in dehydrogenase-tetrazolium reductions were also studied.

References:

1) Kehrman, F.; Havas, E. *Ber. Dtsch. Chem. Ges.* **1913,** *46*, 341-352.

2) Farber, E.; Bueding, E. *J. Histochem. Cytochem.* **1956,** *4*, 357-362.

3) Babson, A.L.; Babson, S.R. *Clin. Chem.* **1973,** *19*, 766.

4) Vaughan, H.; Newsholme, E.A. *Biochem. J.* **1970,** *116*, 31P.

5) Nachlas, M.M.; Margulies, S.I.; Seligman, A.M. *J. Biol. Chem.* **1960,** *235*, 499.

6) Zaugg, W.S. *ibid.* **1964,** *239*, 3964.

7) Kolb, E.; Fleisher, G.A.; Larner, J. *Biochemistry* **1970,** *9*, 4372.

General references:

Aldrich FT-IR Library	**1**(2), 885A
Sigma FT-IR Library	**1**(2), 193A
IR Library	**3**,1407F
Safety	**2**(2), 2739B
Merck Index	**11**, 6024

PHENOLPHTHALEIN

3,3-Bis(4-hydroxyphenyl)-1(3*H*)-isobenzofuranone • 3,3-Bis(4-hydroxyphenyl)phthalide

Aldrich Cat. No.:	10,594-5
Sigma Prod. No.:	P2292
Grade:	A.C.S. reagent

CAS No.:	[*77-09-8*]
Mol. Form.:	$C_{20}H_{14}O_4$
F.W.:	318.33
m.p.:	261-263°C
Appearance:	White powder
λ_{max}**:**	552(374)nm in water + 2ml 1*N* sodium hydroxide
Solubility:	H_2O 1mg/ml
	EGME 100mg/ml
	EtOH 30mg/ml

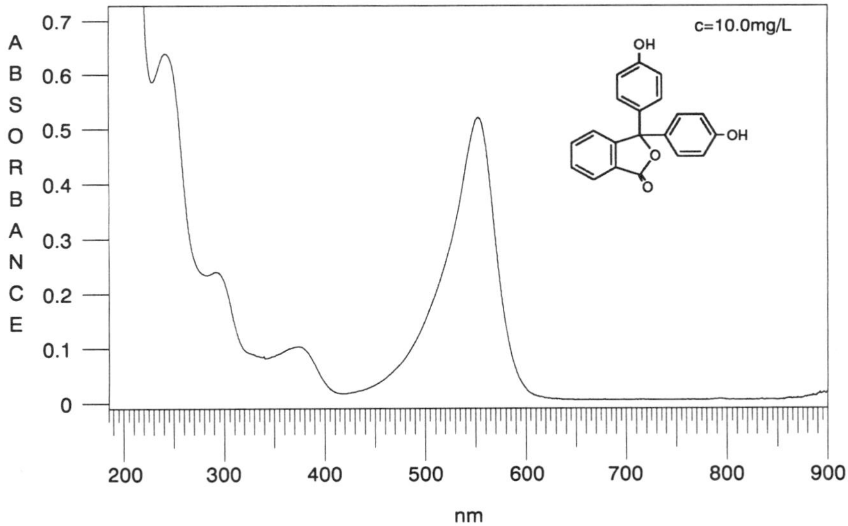

Phenolphthalein, a valuable acid-base indicator, is made by condensing one mole of phthalic anhydride with two moles of phenol. The condensation product is usually isolated as the colorless lactone form. The disodium salt form is red due to the opening of the lactone ring and the concomitant paraquinoid-chromophore formation. However, with the addition of enough sodium hydroxide to form the trisodium carbinol salt, it becomes colorless again due to the elimination of the paraquinoidal structure. *Reagent Chemicals*[1] lists a visual-transition interval of pH 8.0 (colorless) to pH 10 (red).

There is no evidence in the literature that it has been used as a biological stain.

Reference:

1) *Reagent Chemicals,* 7th ed.; American Chemical Society: Washington, DC, 1986; pp 468-469.

General references:

Aldrich FT-IR Library	**1**(2), 1006B
Sigma FT-IR Library	**1**(2), 193B
IR Library	**3**, 1471C
Merck Index	**11**, 7208

PHENOL RED

Phenolsulfonephthalein

Aldrich Cat. No.: 11,452-9
Sigma Prod. No.: P2167
Grade: A.C.S. reagent

CAS No.: [*143-74-8*]
Mol. Form.: $C_{19}H_{14}O_5S$
F.W.: 354.38
Appearance: Dark-red powder
λ_{max}: 557(360)nm in water + 20ml boric buffer, pH 9
Solubility: H_2O 3mg/ml
EGME 10mg/ml
EtOH 4mg/ml

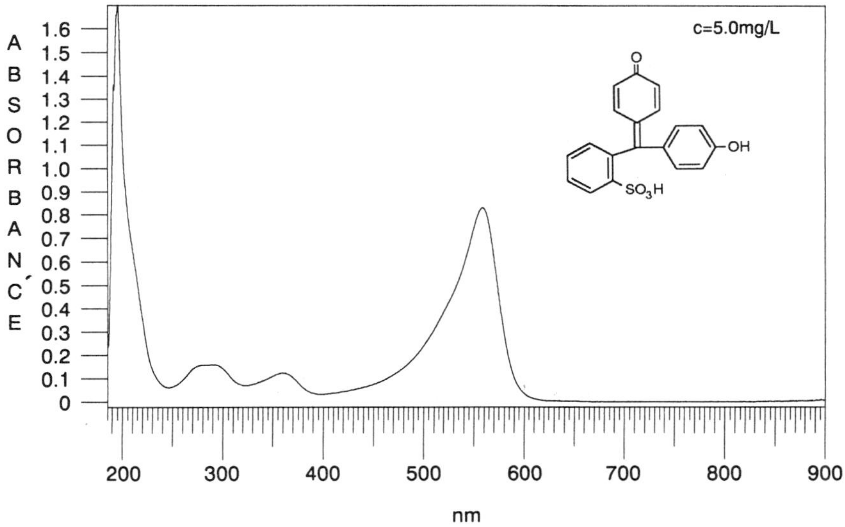

Phenol Red is an important member of the sulfonephthalein list of acid-base indicators. It is made by condensing two moles of phenol with one mole of *o*-sulfobenzoic acid anhydride. The free-acid form is slightly soluble in water, somewhat more soluble in methanol, ethanol and acetic acid and readily soluble in alkaline solutions. The monosodium salt, or water-soluble form, is freely soluble in water and the lower alcohols. *Reagent Chemicals*[1] lists test methods and specifications for Phenol Red, including a visual-transition interval from pH 6.8 (yellow) to pH 8.2 (red). The pK is 7.9. The Indicator Chart in the back of this book has more detailed information on the preparation of test solutions for using Phenol Red as an indicator.

Phenol Red is the precursor for Bromophenol Blue, a sulfonephthalein that has been used extensively in the photocopying industry. The *USP*-grade of free-acid Phenol Red is used as a diagnostic aid for renal determinations,[2] and the sodium salt is used widely in culture media for identifying acid-producing bacteria.

References:

1) *Reagent Chemicals,* 7th ed.; American Chemical Society: Washington, DC, 1986; pp 469-470.

2) *The United States Pharmacopeia,* twentieth version; United States Pharmacopeial Convention; Mack: Easton, 1979; pp 612-613.

General references:

Aldrich FT-IR Library **1**(2), 1015D

Sigma FT-IR Library **1**(2), 194C

IR Library **3**, 1476A

Merck Index **11**, 7213

PHENOL RED, water soluble

Aldrich Cat. No.: 11,453-7
Sigma Prod. No.: P2417
Grade: A.C.S. reagent

CAS No.: [*34487-61-1*]
Mol. Form.: $C_{19}H_{13}NaO_5S$
F.W.: 376.36
m.p.: 285°C
Appearance: Shiny dark-red powder
λ_{max}: 423nm in methanol
Solubility: H_2O 100mg/ml
EGME 200mg/ml
EtOH 200mg/ml

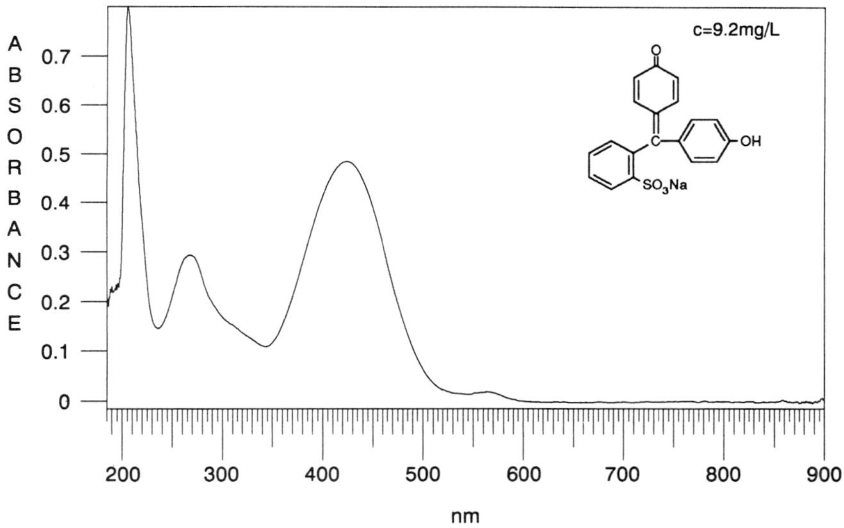

See description on pages 565 and 566 under Phenol Red.

General references:

Aldrich FT-IR Library	**1**(2), 1016A
Sigma FT-IR Library	**1**(2), 194D
IR Library	**3**, 1476B
Merck Index	**11**, 7213

4-PHENYLAZOANILINE

Aniline Yellow • Solvent Yellow 1

Aldrich Cat. No.:	18,636-8
Sigma Prod. No.:	P7144
Purity:	98%
CAS No.:	[*60-09-3*]
C.I. No.:	11000
Mol. Form.:	$C_{12}H_{11}N_3$
F.W.:	197.24
m.p.:	123-126°C
b.p.:	>360°C
Appearance:	Medium-brown powder
λ_{max}:	386nm in methanol
Solubility:	H_2O 10mg/ml
	EGME 20mg/ml
	EtOH 50mg/ml

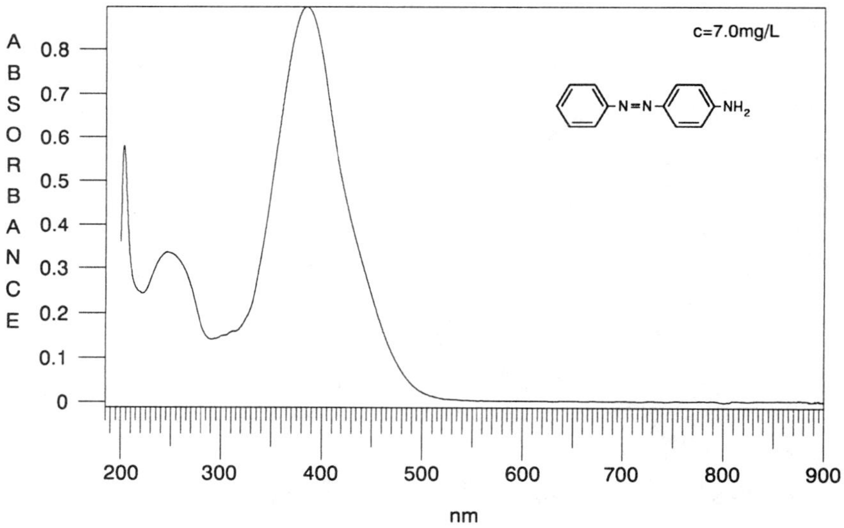

4-Phenylazoaniline is a yellow solvent dye with a monoazo chromophore. It is slightly soluble in water and soluble in methyl Cellosolve, ethyl acetate, acetone, benzene, carbon tetrachloride, ethanol, oleic acid, toluene, stearic acid and xylene. The hydrochloride-salt form is very slightly soluble in water, methanol and ethanol.

The free base (Solvent Yellow 1) is used in spirit lacquers, varnishes, waxes, oil stains and styrene resins. Aniline Yellow is also used as an intermediate to prepare other dyes, including certain Indulins, Nigrosins and diazos. *Conn's*[1] indicates that it may be used as a fat dye.

Reference:

1) *Conn's Biological Stains,* 9th ed.; Lillie, R.D., Ed.; Williams and Wilkins: Baltimore, 1977; p 86.

General references:

Aldrich FT-IR Library	**1**(2), 969D
Sigma FT-IR Library	**1**(2), 195B
IR Library	**3**, 1452E
NMR Library	**2**(2), 979D
Safety	**2**(2), 2762C
Merck Index	**11**, 430
Colour Index	3rd ed., 1971

4-PHENYLAZOANILINE hydrochloride

Aldrich Cat. No.:	15,006-1
Sigma Prod. No.:	P7269
Purity:	97%

CAS No.:	[*3457-98-5*]
Mol. Form.:	$C_{12}H_{12}ClN_3$
F.W.:	233.70
m.p.:	227-228°C
Appearance:	Dark-blue crystalline powder
λ_{max}:	385nm in methanol
Solubility:	H_2O 1mg/ml
	EGME 2mg/ml
	EtOH 0.5mg/ml

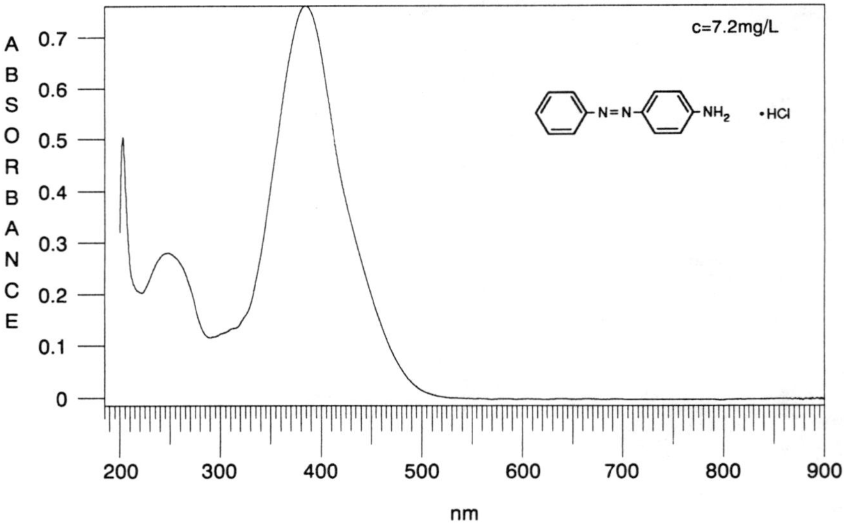

See description on pages 568 and 569 under 4-phenylazoaniline.

General references:

Aldrich FT-IR Library	**1**(2), 970A
Sigma FT-IR Library	**1**(2), 195C
IR Library	**3**, 1452F
Safety	**2**(2), 2762D
Merck Index	**11**, 430

4-(PHENYLAZO)DIPHENYLAMINE

Aldrich Cat. No.:	22,916-4
Purity:	97%
CAS No.:	[*101-75-7*]
Mol. Form.:	$C_{18}H_{15}N_3$
F.W.:	273.34
m.p.:	89-91°C
Appearance:	Orange powder
λ_{max}**:**	411(272)nm in ethanol
Solubility:	H_2O <0.1mg/ml
	EGME 50mg/ml
	EtOH 30mg/ml

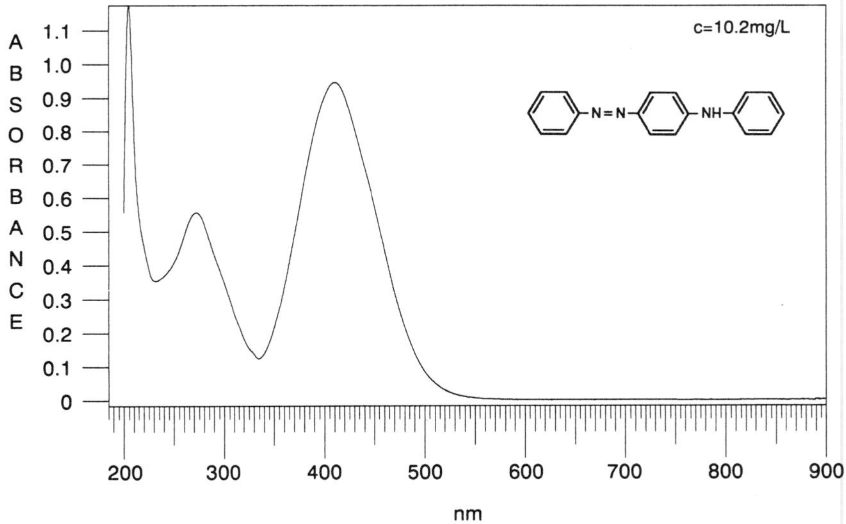

This orange-yellow, monoazo solvent dye is insoluble in water, but soluble in ethanol and methanol.

Made by coupling benzenediazonium chloride to diphenylamine, it is used industrially as a colorant in lithography and, in the laboratory, primarily as a pH indicator. The visual-transition interval is pH 1.2 (red) to pH 2.5 (yellow).

There is no evidence in the literature that it has been used as a biological stain.

General reference:

Safety **2**(2), 2763A

4-PHENYLAZOMALEINANIL

Aldrich Cat. No.:	P1,860-8
Purity:	97%
CAS No.:	[*16201-96-0*]
Mol. Form.:	$C_{16}H_{11}N_3O_2$
F.W.:	277.28
m.p.:	162-165°C
Appearance:	Orange powder
λ_{max}:	329nm in chloroform
Solubility:	H_2O 0.1mg/ml
	EGME 20mg/ml
	EtOH 7mg/ml

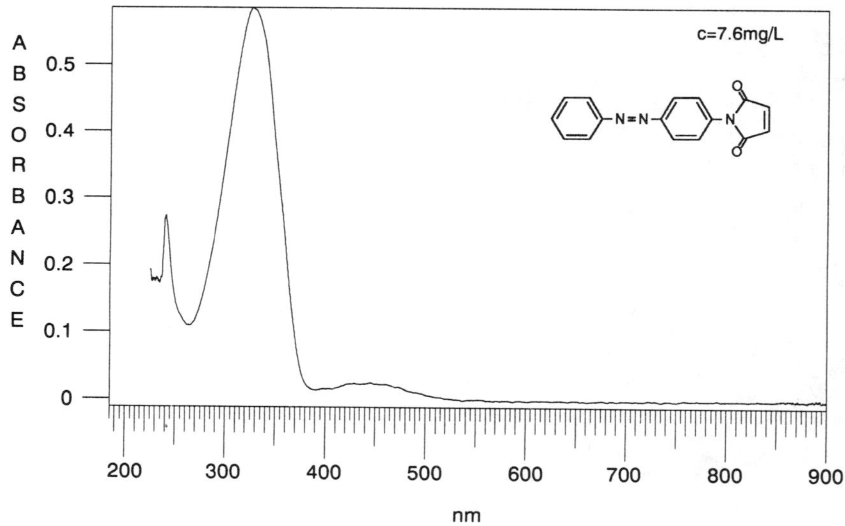

4-Phenylazomaleinanil is an orange, crystalline (needles) dienophile that undergoes the Diels-Alder reaction with conjugated dienes to give colored, crystalline adducts.[1] Fieser and Fieser[2] stated that "adducts from butadiene, cyclopentadiene, 1,3-hexadiene and cosmene, are obtainable under mild conditions (20-50°C), and that they are high-melting compounds which crystallize well."

It appears from its structure that 4-phenylazomaleinanil could be considered a solvent dye. However, there is not enough solubility information in the literature to make this a valid judgment. It is virtually insoluble in water and slightly soluble in methanol and ethanol.

References:

1) Naylor, P.; Whiting, M.C. *J. Chem. Soc.* **1955**, 2970.

2) Fieser, L.; Fieser, M. *Reagents for Organic Synthesis*, Vol. 1, Wiley: New York, 1967; p 833.

General references:

Aldrich FT-IR Library	**1**(2), 403C
IR Library	**3**,1099E
NMR Library	**2**(2), 378B

4-PHENYLAZOPHENOL

Solvent Yellow 7

Aldrich Cat. No.:	13,108-3
Sigma Prod. No.:	S4016
Purity:	98%
CAS No.:	[*1689-82-3*]
C.I. No.:	11800
Mol. Form.:	$C_{12}H_{10}N_2O$
F.W.:	198.23
m.p.:	150-152°C
Appearance:	Gold powder
λ_{max}:	347nm in methanol
Solubility:	H_2O 1mg/ml
	EGME 300mg/ml
	EtOH 50mg/ml

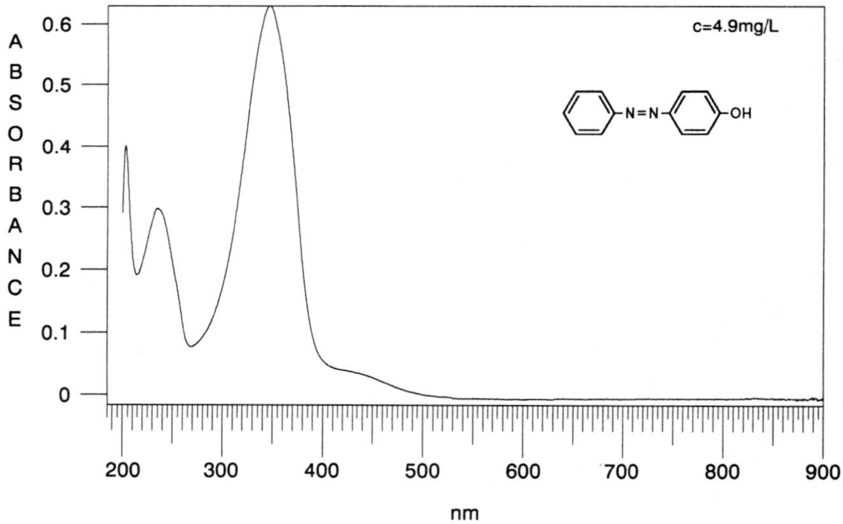

4-Phenylazophenol is a monoazo solvent dye that is made by coupling benzenediazonium chloride to phenol. It imparts a bright-yellow hue. It is soluble in waxes, linseed oil, ethanol, ethyl acetate, carbon tetrachloride, toluene, mineral oil and acetone and is very soluble in both oleic and stearic acids. Industrially, it is used to color varnishes, fats, waxes, resins and soaps.

There is no evidence that it has been used as a biological stain; however, it appears to have the requisite qualities to function as a good fat and lipid stain in biological applications.

General references:

Aldrich FT-IR Library	**1**(2), 965B
Sigma FT-IR Library	**1**(2), 213D
IR Library	**3**, 1449G
NMR Library	**2**(2), 979C
Safety	**2**(2), 2763C
Colour Index	3rd ed., 1971

9-PHENYL-2,3,7-TRIHYDROXY-6-FLUORONE

Fluorone Black • 2,6,7-Trihydroxy-9-phenylisoxanthen-3-one

Aldrich Cat. No.: P3,680-0
Grade: Technical

CAS No.: [*975-17-7*]
Mol. Form.: $C_{19}H_{12}O_5$
F.W.: 320.30
m.p.: >300°C
Appearance: Orange powder
λ_{max}: 552(303)nm in 0.1*N* sodium hydroxide
Solubility: H_2O 2mg/ml
EGME 4mg/ml
EtOH 3mg/ml

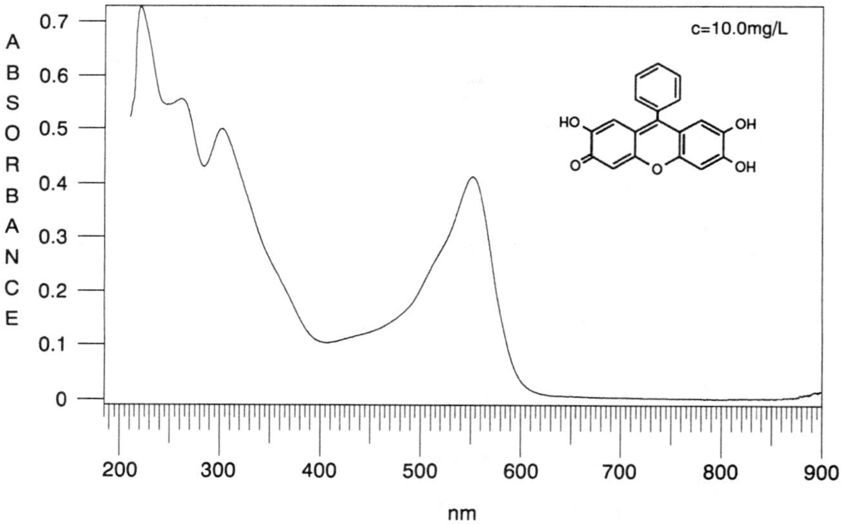

This mordant xanthene dye is soluble in hot water, hot alcohol (only slightly soluble when water and alcohol are cold), ethyl acetate, acetone and alkali solutions. Lillie *et al.*[1] used the dye mordanted with iron as a substitute for Iron Hematoxylin as a nuclear stain.

There are no reported industrial applications.

Reference:

1) Lillie, R.D.; Pizzolato, P.; Donaldson, P.T. *Stain Technol.* **1975,** *50*, 127-131.

General references:

Aldrich FT-IR Library **1**(2), 98D
Sigma FT-IR Library **1**(2), 842D

PHLOXINE B
Acid Red 92

Aldrich Cat. No.:	19,827-7
Sigma Prod. No.:	P4030
Grade:	Certified
CAS No.:	[*18472-87-2*]
C.I. No.:	45410
Mol. Form.:	$C_{20}H_2Br_4Cl_4Na_2O_5$
F.W.:	829.66
Appearance:	Brick-red powder
λ_{max}:	548(510)nm in 50% ethanol
Solubility:	H_2O 90mg/ml
	EGME 60mg/ml
	EtOH 10mg/ml

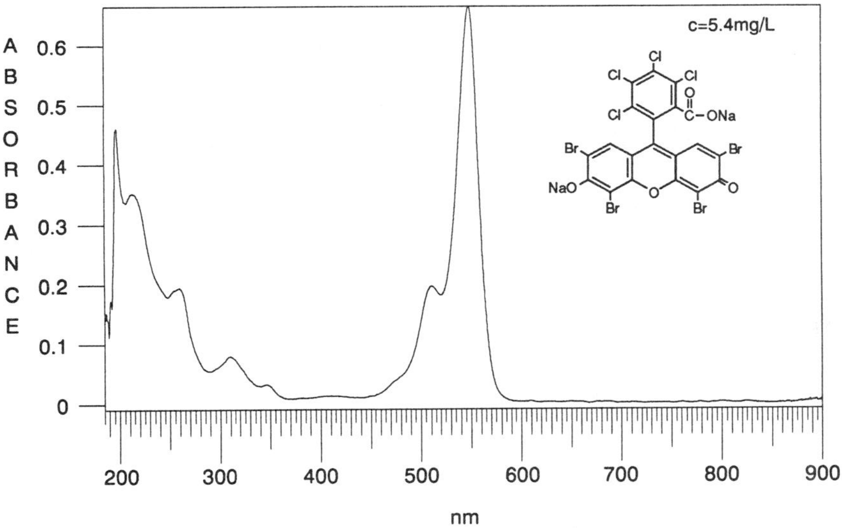

Phloxine B is a red, anionic, xanthene dye that is made by condensing two moles of resorcinol with one mole of tetrachlorophthalic anhydride to form 4,5,6,7-tetrachlorofluorescein. The tetrachlorofluorescein is then tetrabrominated in the 2',4',5',7'-positions and that product is converted to the disodium salt.

Phloxine B is soluble in water and ethylene glycol and slightly soluble in ethanol and methanol. Industrially, it is used to dye wool from a weak-acid bath, nylon from a formic acid bath and silk from a neutral or acetic acid bath, to print wool and silk directly and to color stamping inks, lacquers and paper. Previously, an especially purified grade (D & C Red 28) was used in drugs and cosmetics in the United States.

Phloxine B displays indicator properties with a visual-transition interval from pH 1.1 (colorless) to pH 3.3 (purple). It also has been used as a fluorescent indicator with a fluorescent-transition interval from pH 3.4 (colorless) to pH 5.0 (yellow). An Indicator Chart including Phloxine B and information regarding the preparation of an indicator test solution using this dye can be found in the back of this book.

Conn's[1] cites a number of instances where Phloxine B has been used mistakenly for other closely related xanthene dyes or vice versa. Apparently, this was due to mislabeling.

Phloxine B has been substituted for Eosin Y in Romanowsky-type stain formulations. However, Wittekind[2] states that the Romanowsky-Giemsa effect (RGE) is not displayed when Phloxine is the xanthene component of such a formulation.

Staining Procedures[3] provides methods and a bibliography for the following applications: Chrome Hematoxylin-Phloxine for demonstrating *beta* cells; counterstain to Hematoxylin in Bargmann's Chrome Hematoxylin for neurosecretory substances; Night Blue-Phloxine relief stain designed to show cilia, flagella and cortical patterns and Lendrum Stain for inclusion bodies in sections.

References:

1) *Conn's Biological Stains,* 9th ed.; Lillie, R.D., Ed.; Williams and Wilkins: Baltimore, 1977; pp 347-349.

2) Wittekind, D. *Clin. Lab. Haematol.* **1979,** *1*, 247-262.

3) Clark, G. *Staining Procedures,* 4th ed.; Williams and Wilkins: Baltimore, 1981; pp 224, 253, 287, 463.

General references:

Aldrich FT-IR Library **1**(2), 1013A

Sigma FT-IR Library **1**(2), 197D

IR Library **3**, 1474G

PINACYANOL chloride

Pinacyanol • Quinaldine Blue • Sensitol Red

Aldrich Cat. No.: 20,171-5
Sigma Prod. No.: P0392

CAS No.: [*2768-90-3*]
Mol. Form.: $C_{25}H_{25}ClN_2$
F.W.: 388.94
m.p.: 270°C (dec.)
Appearance: Shiny green powder
λ_{max}: 604(560)nm in ethanol
Solubility: H_2O 7mg/ml
EGME 40mg/ml
EtOH 20mg/ml

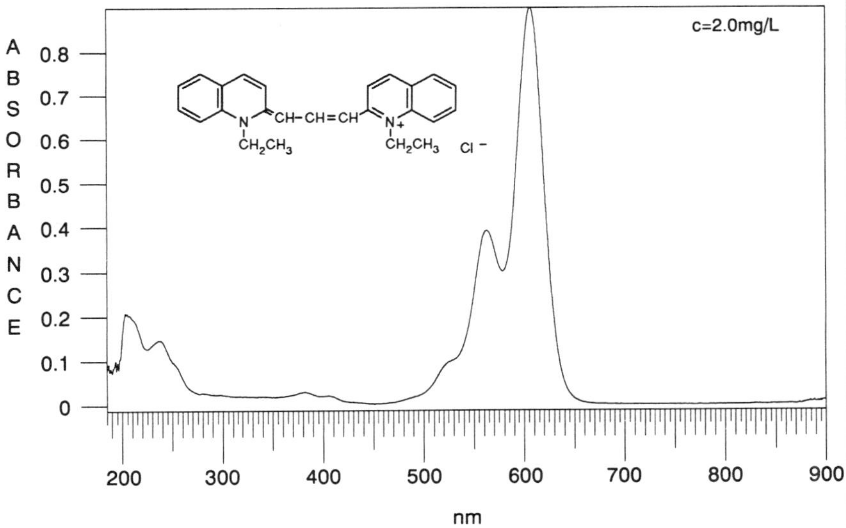

Pinacyanol chloride is a cationic, polymethine quinoline dye that is used primarily as a sensitizer in color photography. It is a particularly good sensitizer in the yellow, orange and red regions. The first edition of *Colour Index*[1] listed the iodide-salt form; however, it has not been listed in subsequent editions.

Pinacyanol is made by condensing one mole each of ethylquinaldinium iodide, ethylquinolinium iodide and formaldehyde, in the presence of potassium hydroxide. The resulting product is converted to the the chloride salt with hydrochloric acid. This salt form is slightly soluble in water and soluble in ethanol and methyl Cellosolve.

Merck[2] indicates that Pinacyanol is used as a diagnostic aid in obstetrics.

Conn's[3] cites several biological-stain applications: staining frozen sections; as a supravital stain for mitochondria in blood cells and staining mast cells with a stain solution prepared from Pinacyanol erythrosinate, a neutral stain, which is made by neutralizing an aqueous solution of Pinacyanol with an aqueous solution of the anionic dye Erythrosin.

References:

1) *Colour Index,* 1st ed., Society of Dyers and Colourists: Bradford, Yorkshire, 1924.

2) *The Merck Index,* 11th ed.; Budavari, S. *et al.*, Eds.; Merck: Rahway, 1989; Monograph 8056.

3) *Conn's Biological Stains,* 9th ed.; Lillie, R.D., Ed.; Williams and Wilkins: Baltimore, 1977; p 365.

General references:

Aldrich FT-IR Library	**1**(2), 1040D
Sigma FT-IR Library	**1**(2), 851C
IR Library	**3**, 1488A
Safety	**2**(2), 2840A

PLASMOCORINTH B

Mordant Blue 13

Aldrich Cat. No.: 20,172-3
Sigma Prod. No.: P8253

CAS No.: [*1058-92-0*]
C.I. No.: 16680
Mol. Form.: $C_{16}H_9ClN_2Na_2O_9S_2$
F.W.: 518.82
Appearance: Brownish-black powder
λ_{max}: 527nm in water
Solubility: H_2O 50mg/ml
EGME 50mg/ml
EtOH 5mg/ml

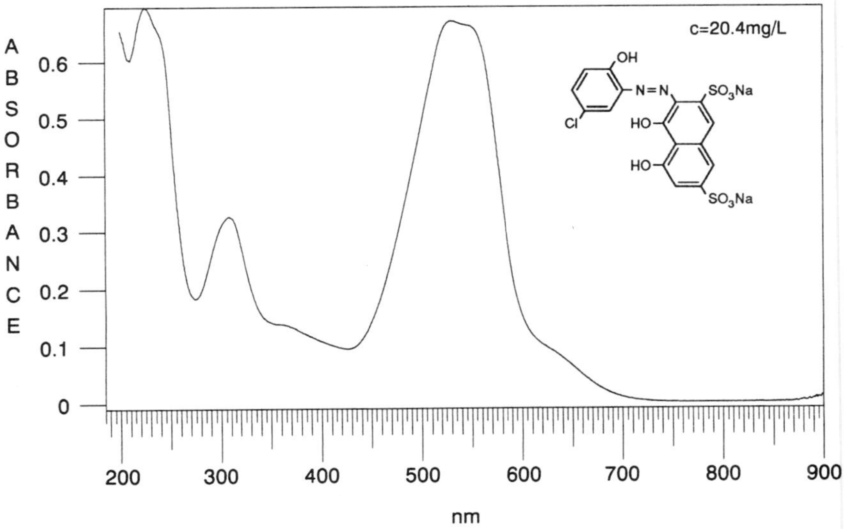

Plasmocorinth B is an anionic monoazo dye that is soluble in water and methyl Cellosolve and slightly soluble in ethanol. With adjacent hydroxyl groups on both extremities of the azo bond, it readily chelates with chromium and other metal ions. It dyes afterchromed wool, silk and nylon (low substantivity) a navy to reddish-navy blue. Textile printing is accomplished by the vigoureux system. Industrially, it is used to dye anodized aluminum.

The literature does not report Plasmocorinth B as a biological stain; however, its structure indicates it might be of some value as a fluorochrome counterstain in the Hall and Hansen[1] Fluorescein-tagged antibody techniques.

Reference:

1) Hall, C.T.; Hansen, P.A. *Bakteriol.* **1962,** *184*, 548-554.

General references:

Aldrich FT-IR Library **1**(2), 983A
Sigma FT-IR Library **1**(2), 198D
IR Library **3**, 1460C

PONCEAU S
Acid Red 112

Aldrich Cat. No.: 14,119-4
Sigma Prod. No.: P3504

CAS No.: [*6226-79-5*]
C.I. No.: 27195
Mol. Form.: $C_{22}H_{12}N_4Na_4O_{13}S_4$
F.W.: 760.58
Appearance: Brown powder
λ_{max}: 520(352)nm in water
Solubility: H_2O 40mg/ml
EGME 10mg/ml
EtOH 1mg/ml

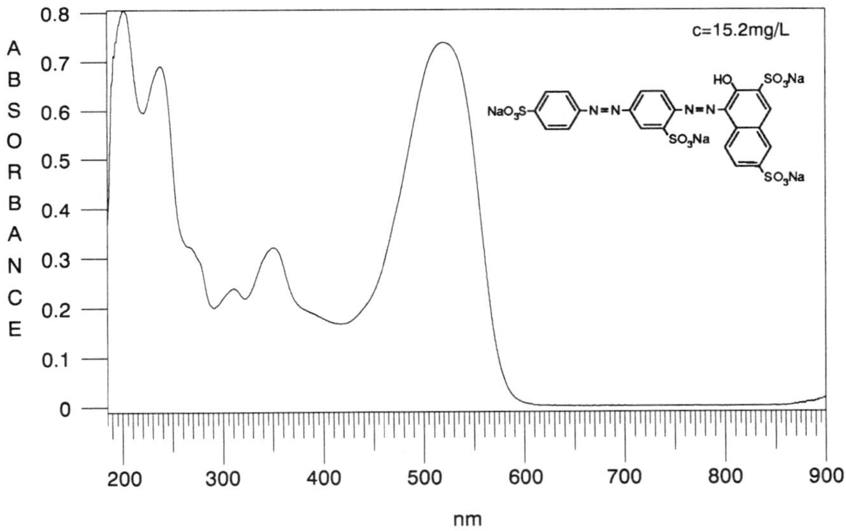

Ponceau S is a red, water-soluble, anionic diazo dye that is made by coupling diazotized 4-amino-1,1'-azobenzene-3,4'-disulfonic acid (Acid Yellow 9, C.I. 13015) with 2-naphthol-3,6-disulfonic acid (R acid) and converting the coupling product to the tetrasodium salt. While the *Colour Index* lists it as similar to Acid Red 112, (C.I. 27195) with no known manufacturer, material identical to C.I. 27195, with an early *Colour Index* number of C.I. 282,[1] currently is manufactured and sold as Ponceau S. This material, produced primarily for laboratory use, has been available continuously for over thirty years.

Ponceau S is used widely in gel electrophoresis. Curtis[2] reported it to be superior to Acid Fuchsin in the Van Gieson Stain. Ruth[3] and Leach[4] concurred with Curtis, citing greater stability and resistance to fading.

References:

1) *Colour Index,* 1st ed.; Society of Dyers and Colourists: Bradford, Yorkshire, 1924.

2) Curtis, F. *C.R. Soc. Biol. (Paris)* **1905,** *58*, 1038-1040.

3) Ruth, E.B. *Stain Technol.* **1946,** *21*, 27-30.

4) Leach, E.H. *ibid.* **1946,** *21*, 107-109.

General references:

Aldrich FT-IR Library	**1**(2), 989B
Sigma FT-IR Library	**1**(2), 200A
IR Library	**3**, 1463E

PONCEAU SS

Acid Red 150

Aldrich Cat. No.:	19,974-5
Sigma Prod. No.:	A6520
CAS No.:	[*6226-78-4*]
C.I. No.:	27190
Mol. Form.:	$C_{22}H_{14}N_4Na_2O_7S_2$
F.W.:	556.49
Appearance:	Green crystalline powder
λ_{max}:	514(351)nm in water
Solubility:	H_2O 10mg/ml
	EGME 4mg/ml
	EtOH 0.8mg/ml

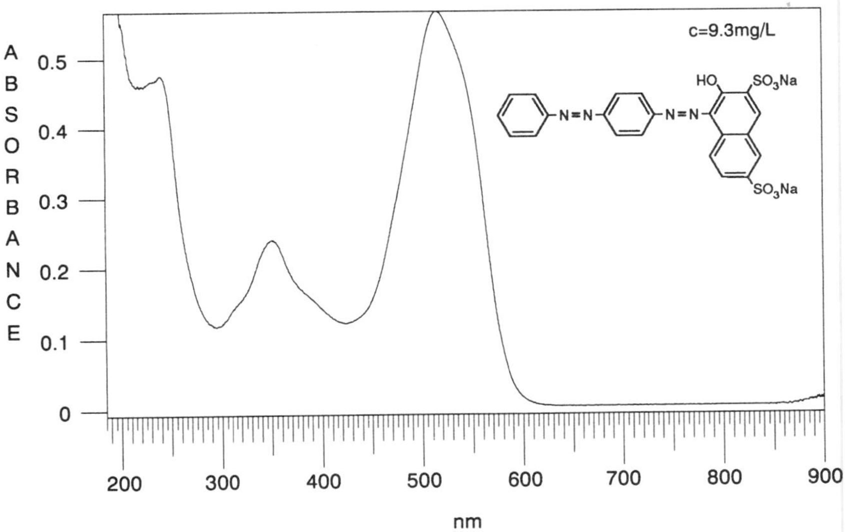

Ponceau SS is a bright-red, acid diazo dye that is made by coupling diazotized 4-phenylazoaniline with R acid (2-naphthol-3,6-disulfonic acid) and converting the coupling product to the disodium salt. It is slightly soluble in water and methyl Cellosolve and virtually insoluble in ethanol and other organic solvents. Wool is dyed from a weak-acid bath, and nylon and silk are dyed from a formic or acetic acid bath.

There is no evidence that Ponceau SS has been used as a biological stain.

General references:

Aldrich FT-IR Library	**1**(2), 990A
Sigma FT-IR Library	**1**(2), 88D
IR Library	**3**, 1463H
Colour Index	3rd ed., 1971

POTASSIUM INDIGOTRISULFONATE

5,5',7-Indigotrisulfonic acid, tripotassium salt

Aldrich Cat. No.: 23,408-7
Sigma Prod. No.: I-3007

CAS No.: [*67627-18-3*]
Mol. Form.: $C_{16}H_7N_2K_3O_{11}S_3$
F.W.: 616.74
Appearance: Dark-blue crystalline powder with chunks
λ_{max}: 600(311)nm in water
Solubility: H_2O 7mg/ml
EGME 2mg/ml
EtOH 0.2mg/ml

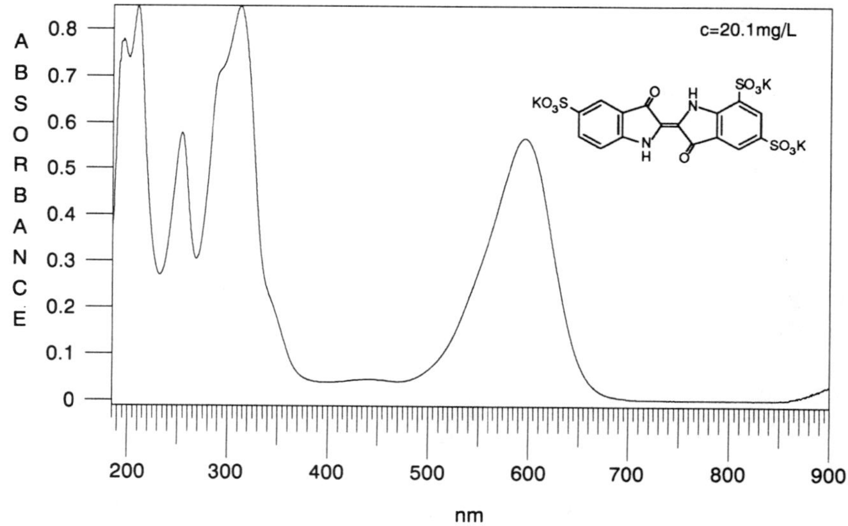

This anionic indigoid dye is made by heating (160-180°C) Indigo Carmine with oleum, purifying the product and converting it to the tripotassium salt.

Slightly soluble in water and very slightly soluble in ethanol, potassium indigotrisulfonate is used primarily as a reagent for the determination of ozone in water. Bader and Hoigne[1] describe a method for determining ozone in water which they consider to be superior to other methods. They describe it as "very sensitive, precise, fast, specific, and easy to handle" at a pH below 4.0.

This dye has no *Colour Index* number, and there is no evidence in the literature that it has been used industrially or as a biological stain.

Reference:

1) Bader, H.; Hoigne, J. *Water Res.* **1981,** *15*, 449.

General references:

Aldrich FT-IR Library	**1**(2), 507A
Sigma FT-IR Library	**1**(2), 153C
Safety	**2**(2), 2905B

PRIMULIN

Direct Yellow 59

Aldrich Cat. No.:	20,686-5
Sigma Prod. No.:	P7522
CAS No.:	*[30113-37-2]*
C.I. No.:	49000
Mol. Form.:	$C_{21}H_{14}N_3NaO_3S_3$
F.W.:	475.55
Appearance:	Yellow-gold powder
λ_{max}**:**	356nm in water
Solubility:	H_2O 20mg/ml
	EGME 30mg/ml
	EtOH 10mg/ml

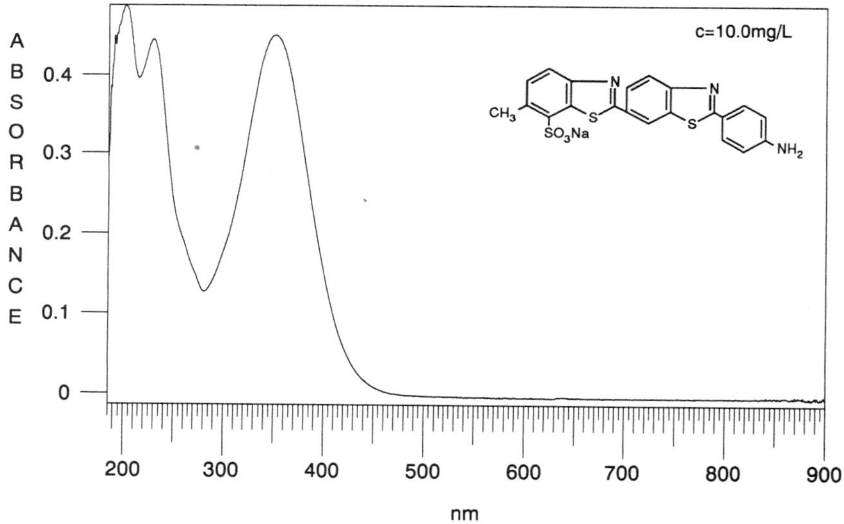

Primulin is an anionic thiazole dye that is moderately soluble in water. Its aqueous solutions display a pale-yellow color when observed in visible light. While under ultraviolet light, a blue fluorescence is apparent.

The blue-fluorescent staining quality of Primulin was described by Pick[1] when he used the dye as a vital stain. Hahn von Dorsche and Opitz[2] used the same characteristic for observing the selective fluorochroming of mast cells at pH 8.0.

Van der Kooy and Kuypers[3] reported that Primulin is useful in a simple, retrograde double-labeling procedure for studying axonal branching in combination with 4',6-diamidino-2-phenylindole dihydrochloride (DAPI, Aldrich Cat. No. 21,708-5) and Evans Blue (Aldrich Cat. No. 20,633-4).

References:

1) Pick, J. *Microsc. Acta* **1935,** *51*, 338-351.

2) Hahn von Dorsche, H.; Opitz, M. *Acta Histochem.* **1970,** *36*, 74-85.

3) Van der Kooy, D.; Kuypers, H.G.J.M. *Science* **1979,** *204*, 873.

General references:

Aldrich FT-IR Library	**1**(2), 1039B
Sigma FT-IR Library	**1**(2), 200B
IR Library	**3**, 1487C

PRUSSIAN BLUE

Ferric ferrocyanide • Pigment Blue 27

Aldrich Cat. No.:	23,412-5
CAS No.:	[*14038-43-8*]
C.I. No.:	77510
Mol. Form.:	$C_{18}Fe_7N_{18}$
F.W.:	859.25
Appearance:	Purple powder
λ_{max}:	694(260)nm in water
Solubility:	H_2O 6mg/ml
	EGME 10mg/ml
	EtOH 20mg/ml

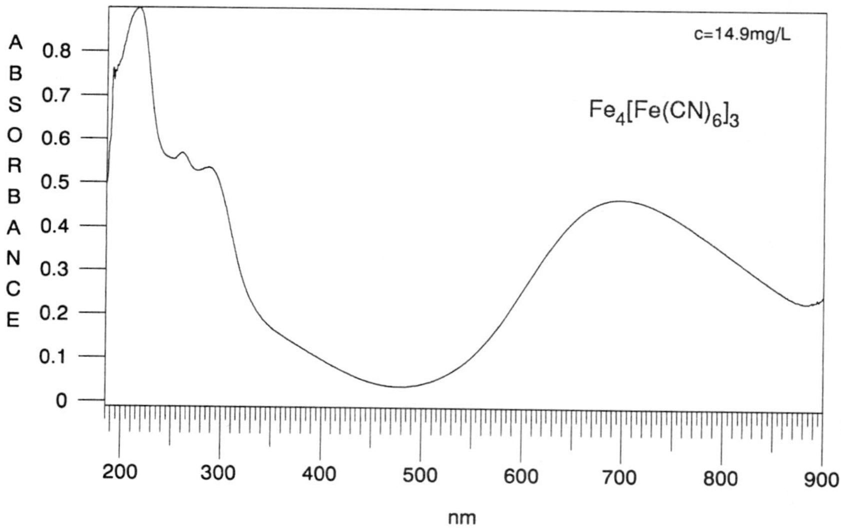

Prussian Blue is not a dye but an inorganic pigment that is made by adding a ferric chloride solution to a solution of potassium ferrocyanide and washing the precipitate free of the water-soluble inorganic salts formed in the reaction. It is slightly soluble in water and organic solvents, and soluble in oxalic acid solutions only to be reprecipitated suddenly on exposure to sunlight. Industrial applications include coloring paints, alkyd-resin enamels, linoleum, leather, cloth, printing inks, carbon paper and typewriter ribbon. Prussian Blue possesses a high tinctorial value with excellent light stability, but poor resistance to alkali.

Prussian Blue is injected into the vascular and lymphatic systems of anatomical specimens to facilitate differentiation from surrounding tissue. *Staining Procedures* [1] provides recipes and methods for the demonstration of nonhemoglobin, cytoplasmic iron, according to Sundberg and Broman (1955) and for the demonstration of hemosiderin according to Highman and from Lillie and Fullmer (1976), after the work originally done by Perle in 1867. Prussian Blue is formed *in situ* in both of these methods.

Reference:

1) Clark, G. *Staining Procedures,* 4th ed.; Williams and Wilkins: Baltimore, 1981; pp 178, 202.

General references:

Merck Index	**11**, 3964
Colour Index	3rd ed., 1971

PURPURIN

Natural Red 16, 8

Aldrich Cat. No.:	22,914-8
Sigma Prod. No.:	P0274
CAS No.:	[*81-54-9*]
C.I. No.:	58205 (synthetic)
	75410 (natural)
Mol. Form.:	$C_{14}H_8O_5$
F.W.:	256.22
m.p.:	253-256°C
Appearance:	Red-brown powder
λ_{max}:	485(521)nm in methanol
Solubility:	H_2O 0.5mg/ml
	EGME 7mg/ml
	EtOH 2mg/ml

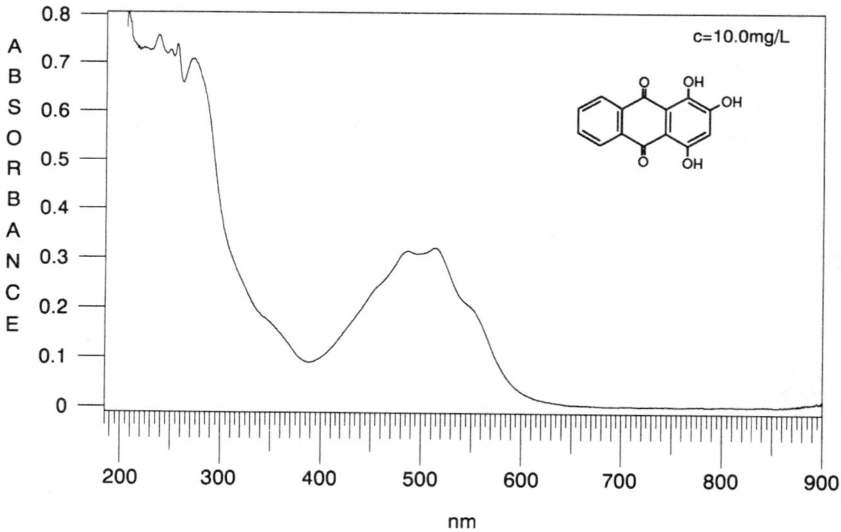

Purpurin is an anthraquinone dye that is synthesized by oxidizing Alizarin with manganese dioxide in the presence of sulfuric acid. It is virtually insoluble in water, soluble in ammonium and alkali hydroxides and very slightly soluble in ethanol.

The naturally occurring variety, Natural Red 16, is the principal coloring matter in Munjeet, and Natural Red 8 is present as a glucoside in the madder root (*Rubia tinctorum*).

Previously, Purpurin was used as a dye for cotton mordanted with aluminum (scarlet) and chromium (dull red with lightfastness inferior to

Alizarin). Currently, it is used as an intermediate to manufacture certain acid and mordant dyes.

Merck[1] lists it as a reagent for the detection of boron. In histological applications, the same source lists it as a nuclear stain and a reagent for detecting insoluble calcium salts inside the cell.

Reference:

1) *The Merck Index,* 11th ed.; Budavari, S. *et al.*, Eds.; Merck: Rahway, 1989; Monograph 7962

General references:

Aldrich FT-IR Library	**1**(2), 89A
Sigma FT-IR Library	**1**(2), 201A
Safety	**2**(2), 2985D
Colour Index	3rd ed., 1971

4-(2-PYRIDYLAZO)RESORCINOL, monosodium salt hydrate
PAR

Aldrich Cat. No.: 17,826-8
Sigma Prod. No.: P8019

Mol. Form.: $C_{11}H_8N_3NaO_2 \cdot xH_2O$
F.W.: 237.20
Appearance: Brown-orange powder
λ_{max}: 411nm in methanol
Solubility: H_2O 20mg/ml
EGME 40mg/ml
EtOH 10mg/ml

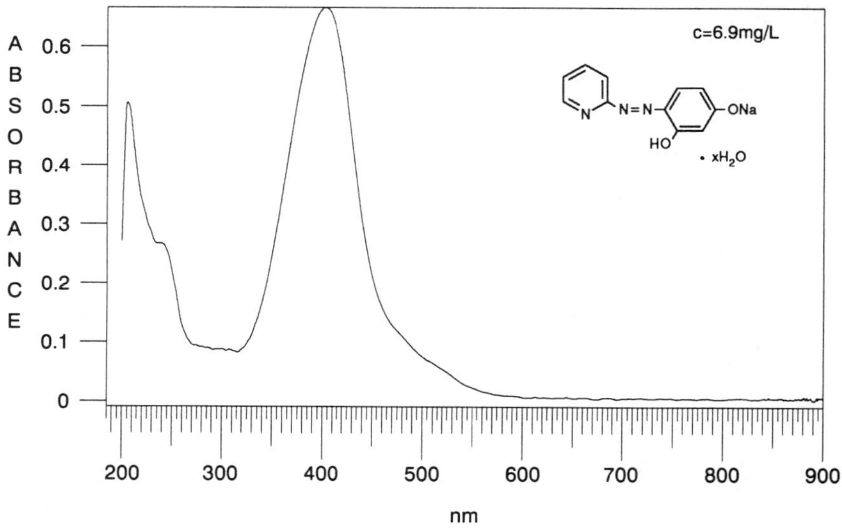

PAR is a monoazo chelometric dye that is made by coupling diazotized 2-aminopyridine to resorcinol and converting the coupling product to the monosodium salt. It is soluble in water, slightly soluble in ethanol and insoluble in ether and other low-polarity solvents. An alkaline solution of PAR exhibits a lemon-yellow color up to pH 10.5, changing color at this point to a bright red.

PAR forms complexes with many metal ions. Those with analytical significance are iron, copper, cadmium, cobalt, nickel, lead, zinc, silver, palladium (green in color), rare earths, thallium, gallium, indium, thorium, titanium, zirconium, vanadium and uranium.[1] It is a highly selective and sensitive reagent for cobalt and iron.[2] Kalyabaraman and Fukasawa[3] describe the spectrophotometric determination of zirconium and hafnium with PAR. Roston,[4] in a preliminary investigation, has shown that by precolumn chelation with PAR, metal ions can be determined simultaneously by liquid chromatography.

In fact, it appears feasible, using this technique, to determine copper(II), cobalt(II), nickel(II) and iron(II) as PAR chelates at the part-per-billion level. More recently,[5] trace levels of chromium(III) were determined by ion-pair, reverse-phase high-performance liquid chromatography.

References:

1) Wehber, P. *Z. Anal. Chem.* **1957,** *158*, 10.
2) Yotsuyanagi, T.; Yamashita, R.; Aomura, K. *Anal. Chem.* **1972,** *44*, 1091.
3) Kalyabaraman, S.; Fukasawa, T. *ibid.* **1983,** *55*, 2239-2241
4) Roston, D.A. *ibid.* **1984,** *56*, 241-244.
5) Hoshino, H.; Yotsuyanagi, T. *ibid.* **1985,** *57*, 625.

General references:

Aldrich FT-IR Library	**1**(2), 973B
Sigma FT-IR Library	**1**(2), 202B
IR Library	**3**, 1455D
Safety	**2**(2), 3002D

PYROCATECHOL VIOLET

Pyrocatecholsulfonephthalein

Aldrich Cat. No.:	10,173-7
Sigma Prod. No.:	P7884
Grade:	Indicator
CAS No.:	[*115-41-3*]
Mol. Form.:	$C_{19}H_{14}O_7S$
F.W.:	386.38
Appearance:	Dark-green powder
λ_{max}:	441nm in methanol
Solubility:	H_2O 60mg/ml
	EGME 60mg/ml
	EtOH 8mg/ml

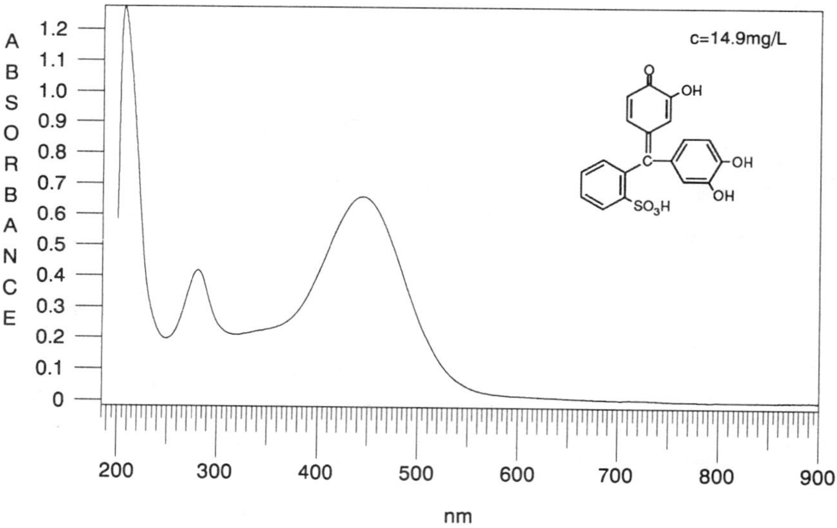

Pyrocatechol Violet is a sulfonephthalein dye that is made by condensing two moles of catechol with one mole of *o*-sulfobenzoic acid anhydride. It has no industrial utility because of its indicator properties and its fugitive nature. The indicator grade is the free-acid form which is slightly soluble in ethanol, soluble in water, more soluble in acetic acid and very soluble in aqueous sodium hydroxide solution. It is yellow in the presence of hydrochloric acid, and violet in the presence of sodium hydroxide. Corbin[1] describes its use as a reagent for the determination of many metals including bismuth, and a rapid and sensitive method for the determination of tin.

Reference:

1) Corbin, H.B. *Anal. Chem.* **1973,** *45*, 534.

General references:

Aldrich FT-IR Library	**1**(2), 1022D
Sigma FT-IR Library	**1**(2), 202C
IR Library	**3**, 1479G

PYROGALLOL RED

Pyrogallolsulfonephthalein

Aldrich Cat. No.:	22,323-9
Sigma Prod. No.:	P8759
CAS No.:	*[85531-30-2]*
Mol. Form.:	$C_{19}H_{12}O_8S$
F.W.:	400.37
Appearance:	Red powder
λ_{max}:	480(282)nm in ethanol
Solubility:	H_2O 1mg/ml
	EGME 3mg/ml
	EtOH 4mg/ml

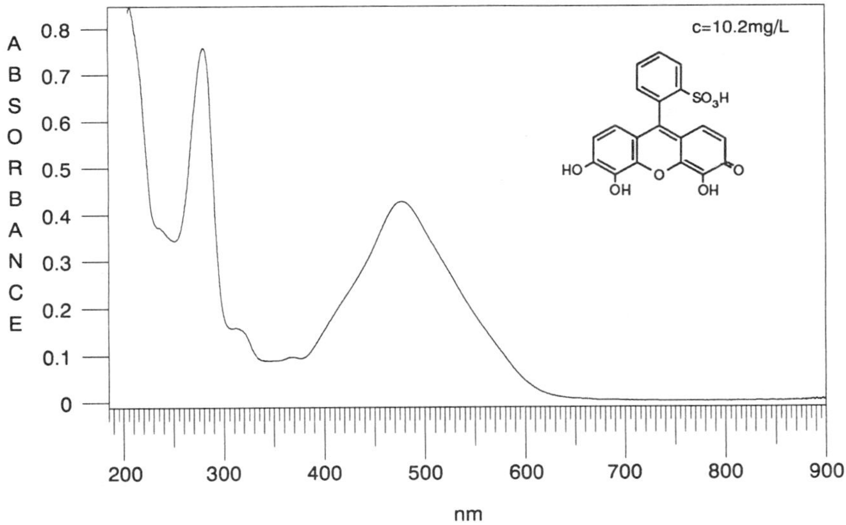

This sulfonephthalein indicator is made by condensing two moles of pyrogallol with one mole of *o*-sulfobenzoic acid anhydride. Hydration alters the physical appearance of the solid produced. Material containing four waters of crystallization displays a bright-red color, while the anhydrous product is a dull, brownish-red. Only very slightly soluble in water, Pyrogallol Red is somewhat more soluble in the lower aliphatic alcohols and in alkali solutions.

Pyrogallol Red is sensitive to pH change (orange-red to deep purple); however, it is not used as an acid-base indicator since the change is too broad and lacks sharpness. Instead, its ability to form complexes with various metals makes it a useful analytical reagent. The following metal ions are among those that complex: yttrium, titanium, gallium, lead, aluminum, germanium and vanadium(IV).

Pyrogallol Red was included in a study of DNA-intercalating compounds with planar triple-ring systems that were found to be potent phenylalanine ammonia lyase (PAL) inducers when applied to freshly incised pea pods.[1] The study, which included Acridine Orange, 6,9-diamino-2-ethoxyacridine lactate monohydrate, Quinacrine, Thionin, Methyl Green, Nile Blue, ethidium bromide, Spermine, 8-azaguanine, Mitomycin C and Azure A, indicated that newly synthesized RNA was necessary for induction.

Reference:

1) Hadwiger, L.A.; Schwochau, M.E. *Plant Physiol.* **1971,** *47*(3), 346-351; *Chem. Abstr.* **1971,** 74, 95597q.

General references:

Aldrich FT-IR Library **1**(2), 1027D

Sigma FT-IR Library **1**(2), 202D

PYRONIN B

Aldrich Cat. No.:	20,130-8
Sigma Prod. No.:	P6892
Grade:	Certified

CAS No.:	*[2150-48-3]*
C.I. No.:	45010
Mol. Form.:	$C_{42}H_{54}Cl_8Fe_2N_4O_2$
F.W.:	1042.28
m.p.:	175-176°C
Appearance:	Dark-green crystals
λ_{max}:	553nm in ethanol
Solubility:	H_2O 20mg/ml
	EGME 10mg/ml
	EtOH 7mg/ml

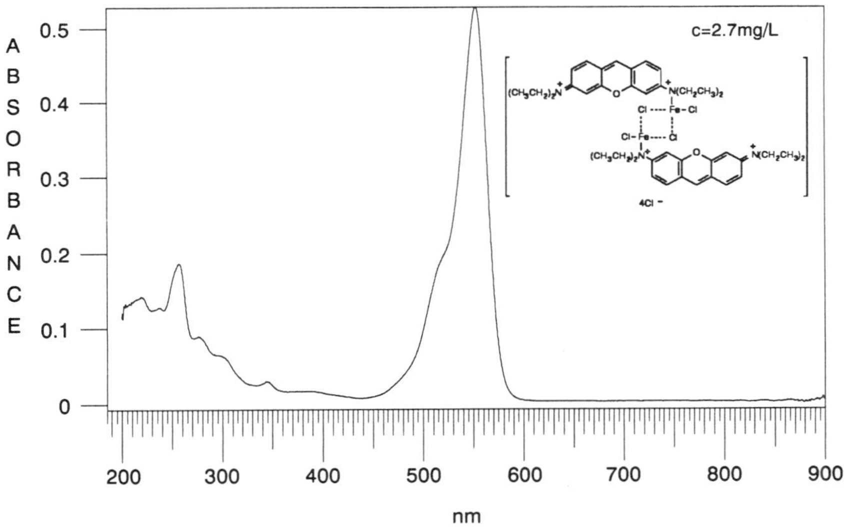

Pyronin B (bluish) is a bluish-red, cationic xanthene dye that is made by condensing *N,N*-diethyl-3-aminophenol with formaldehyde to form the leuco dye. The leuco form is then converted to the color by oxidation. If ferric chloride is used as the oxidant, and it usually is, Chamberlain *et al.*[1] have proposed a Pyronin B structure that embraces two dye molecules complexed with iron to form a super molecule. Pyronin B is soluble in water, slightly soluble in the lower alcohols and methyl Cellosolve and soluble in ethylene glycol, methyl acetate and ethyl acetate. It fluoresces in all of the solvents mentioned; however, the fluorescence is more intense in the organic solvents than it is in water.

Industrially, Pyronin B has been used in color photography. It is generally considered inferior but acceptable as a substitute for Pyronin Y in biological-stain applications including Pappenheim's stain for differentiating DNA from RNA.[2,3] Scudder and Lisa,[4] and Gurr[5] demurred from this position. The BSC certifies Pyronin B, as it does Pyronin Y, for use in Kurnick's Methyl Green-Pyronin stain for nucleic acids. Clark[3] provides a staining procedure utilizing Pyronin and Methyl Green designed to show nucleoli and cytoplasmic granules of liver cells (*i.e.*, to differentiate cytoplasmic and nucleolar basophilia from chromatin).

References:

1) Chamberlain, E.M.; Powell, B.F.; Williams, D.E.; Conn, J. *J. Org. Chem.* **1962,** *27*, 2263-2264.

2) *Conn's Biological Stains,* 9th ed.; Lillie, R.D., Ed.; Williams and Wilkins: Baltimore, 1977; pp 328-330.

3) Clark, G. *Staining Procedures,* 4th ed.; Williams and Wilkins: Baltimore, 1981; p 199.

4) Scudder, S.A.; Lisa, J.R. *Stain Technol.* **1931,** *6*, 99-105.

5) Gurr, E. *Encyclopedia of Microscopic Stains*; Leonard Hill: London, 1960; p 340.

General references:

Aldrich FT-IR Library	**1**(2), 1042C
Sigma FT-IR Library	**1**(2), 203A
IR Library	**3**, 1488F
Safety	**2**(2), 3010B

PYRONIN Y

Aldrich Cat. No.:	21,351-9
Sigma Prod. No.:	P6653
Grade:	Certified
CAS No.:	[*92-32-0*]
C.I. No.:	45005
Mol. Form.:	$C_{17}H_{19}ClN_2O$
F.W.:	302.81
m.p.:	250-260°C
Appearance:	Green powder
λ_{max}:	548(254)nm in 50% ethanol
Solubility:	H_2O 30mg/ml
	EGME 9mg/ml
	EtOH 3mg/ml

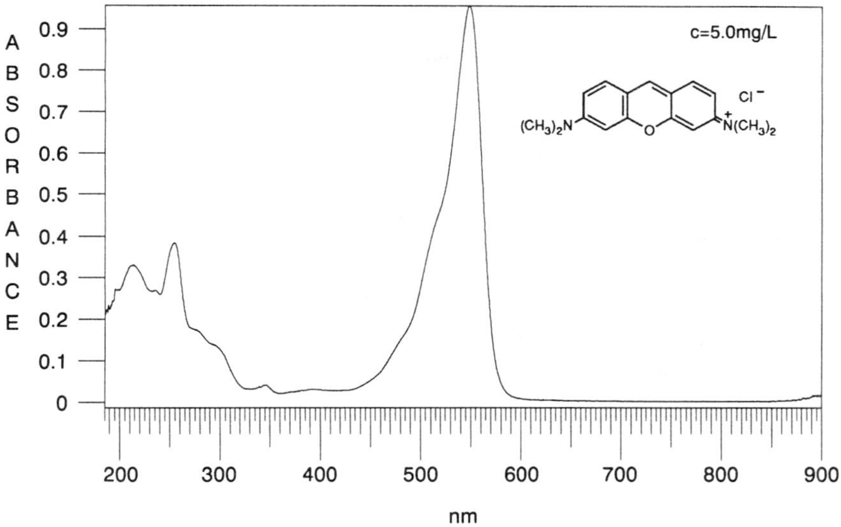

Pyronin Y (yellowish) is a yellowish-red, cationic xanthene dye that is made by condensing *N,N*-dimethyl-3-aminophenol with formaldehyde in the presence of sulfuric acid. The resulting leuco dye is then oxidized to the dye, which is soluble in water and less soluble in ethanol, methyl Cellosolve and ethylene glycol. All solutions are red in transmitted light, displaying a yellow fluorescence in reflected light.

There are no reported industrial uses for Pyronin Y. Undoubtedly, this is due to the expense involved in its manufacture.

Probably Pyronin Y is used most widely in combination with Methyl Green to selectively and differentially stain nucleic acids. The Pyronin Y, in the dichromic dye solution, stains ribonucleic acid red, while the Methyl Green stains deoxyribonucleic acid green.[1] Scudder and Lisa[2] used Pyronin Y in a combined Gram-Pappenheim stain. Kurnick[3] found the combined Methyl Green-Pyronin Y stain to be a useful histochemical reagent. The BSC uses the Kurnick method for evaluating samples of Pyronin Y submitted for certification, stating that nuclear chromatin should be blue-green, nucleoli red and cytoplasm pink to red. *Staining Procedures*[4] provides a method for staining animal tissue using a Methyl Green-Pyronin Y mixed stain. This stain is designed to show nucleoli and cytoplasmic granules of liver cells (*i.e.*, to differentiate cytoplasmic and nucleolar basophilia from chromatin).

Prior to 1930, the literature did not differentiate between Pyronin Y and Pyronin B, specifying only the Pyronins. Scudder and Lisa[2] specified Pyronin B in their Gram-Pappenheim stain and Gurr[5] stated that "while either of the two dyes will serve equally well for some purposes, only Pyronin G or Y is satisfactory for demonstrating RNA in sections and smears." The BSC will certify lots of Pyronin B with a minimum anhydrous dye content of 30% and lots of Pyronin Y with a minimum anhydrous dye content of 45%, if all other requirements, including the Kurnick's Methyl Green-Pyronin Stain on liver cells fixed in Carnoy's fluid, are met. These minimum dye content values may seem ridiculously low, especially for certified dyes, but in practice have been difficult to attain. In fact, only approximately a third of the lots of these materials examined in our laboratory would meet the BSC specifications for minimum dye content. Considering that the total impurities exceed the dye content in each product and that the impurities vary with each product, it seems reasonable to question whether the staining variations reported, except staining hues, were due entirely to the minor structural differences that exist between the dye homologs. Jakobsen *et al.*[6] reported a method for preparing chromatographically pure Pyronin Y using isopenyl nitrite as the oxidant to convert the leuco compound to the dye. Perhaps by making both Pyronin Y and Pyronin B by this method, substituting only the intermediate *N,N*-diethyl-3-aminophenol for *N,N*-dimethyl-3-aminophenol to make the latter, a more conclusive comparison can be made.

References:

1) Pappenheim, A. *Virchows Arch. Pathol. Anat.* **1899,** *157*, 19-76.
2) Scudder, S.A.; Lisa, J.R. *Stain Technol.* **1931,** *6*, 624.
3) Kurnick, N.B. *ibid.* **1952,** *27*, 233-242.
4) Clark, G. *Staining Procedures,* 4th ed.; Williams and Wilkins: Baltimore, 1981; p 199.
5) Gurr, E. *Encyclopedia of Microscopic Stains*; Leonard Hill: London, 1960; pp 340-341.
6) Jakobsen, P.; Andersen, A.; Lyon, H.; Treppendahl, S. *Microscop. Acta* **1983,** *87*(1), 41-47.

General references:

Aldrich FT-IR Library	**1**(2), 1015C
Sigma FT-IR Library	**1**(2), 203B
Safety	**2**(2), 3010C
Merck Index	**11,** 8017

QUINALDINE RED

Aldrich Cat. No.: 20,131-6
Sigma Prod. No.: Q2625

CAS No.: [*117-92-0*]
Mol. Form.: $C_{21}H_{23}IN_2$
F.W.: 430.33
Appearance: Greenish-black powder
λ_{max}: 528nm in ethanol
Solubility: H_2O 0.5mg/ml
EGME 2mg/ml
EtOH 0.6mg/ml

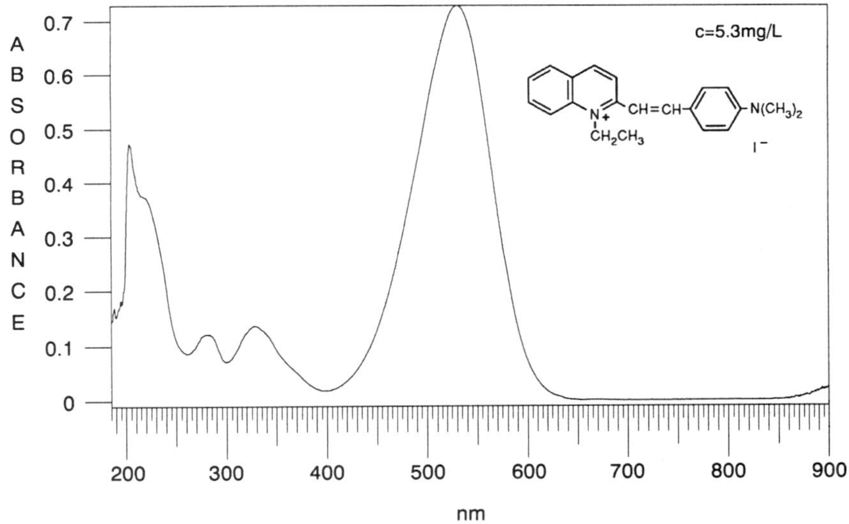

Quinaldine Red is a cationic, quinoline styryl dye that is very slightly soluble in water and ethanol. It is made by condensing quinaldine ethyl iodide with 4-dimethylaminobenzaldehyde in ethanol in the presence of piperidine. Despite its fugitive nature and salt error, it is a useful acid-base indicator with a visual-transition interval from pH 1.4 (colorless) to pH 3.2 (red). *Merck*[1] states that in the presence of chlorides the visual-transition interval is shifted to pH 1.2 to pH 3.0.

Reference:

1) *The Merck Index,* 11th ed.; Budavari, S. *et al.*, Eds.; Merck: Rahway, 1989; Monograph 8057

General references:

Aldrich FT-IR Library	**1**(2), 1045C
Sigma FT-IR Library	**1**(2), 204A
IR Library	**3**, 1490B
Safety	**2**(2), 3021D

QUINALIZARIN

Mordant Violet 26 • 1,2,5,8-Tetrahydroxyanthraquinone

Aldrich Cat. No.:	13,677-8
Sigma Prod. No.:	Q3125
CAS No.:	[*81-61-8*]
C.I. No.:	58500
Mol. Form.:	$C_{14}H_8O_6$
F.W.:	272.21
Appearance:	Orange powder
λ_{max}:	512(490)nm in methanol
Solubility:	H_2O 0.2mg/ml
	EGME 2mg/ml
	EtOH 0.4mg/ml

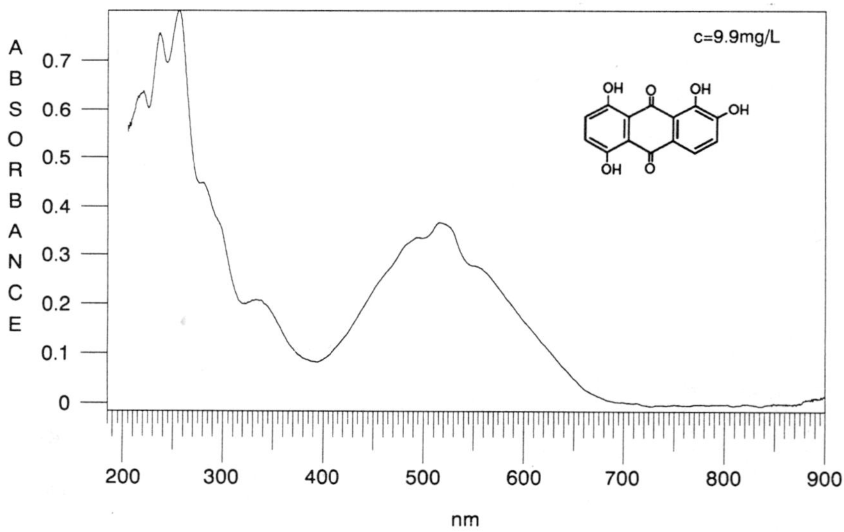

Quinalizarin is a red, mordant anthraquinone dye that is prepared by the controlled addition (35-40°C) of anthraquinone to an excess of 65% fuming sulfuric acid. The tetrasulfonated anthraquinone is isolated and finally subjected to an alkaline hydrolysis to form the desired product. After this product is washed with water and dried, it is further purified by recrystallization from acetic acid. Quinalizarin is soluble in alkali hydroxide solutions, very slightly soluble in ethanol and even less soluble in water. It forms chelates with various metals including aluminum (dark red), chromium (bluish violet) and magnesium and beryllium (blue).

Formerly, it was used to dye chromium-mordanted wool violet out of a sulfuric acid bath. Feigl[1] used it as a reagent for the detection of indium.

Broda,[2] in a comparative study with Thiazol Yellow and Azo Blue, used Quinalizarin as a reagent to demonstrate magnesium in plant tissue.

References:

1) Feigl, F. *Spot Tests in Inorganic Analysis*; Elsevier: New York, 1958.

2) Broda, B. *Mikrokosmos* **1939,** *32*, 184.

General references:

Aldrich FT-IR Library	**1**(2), 89B
Sigma FT-IR Library	**1**(2), 204B
IR Library	**3**, 900G
Merck Index	**11**, 8059
Colour Index	3rd ed., 1971

QUINIZARIN

1,4-Dihydroxyanthraquinone • Smoke Dye (orange)

Aldrich Cat. No.: Q90-6
Sigma Prod. No.: Q3500
Purity: 96%

CAS No.: [*81-64-1*]
C.I. No.: 58050
Mol. Form.: $C_{14}H_8O_4$
F.W.: 240.21
m.p.: 198-199°C
Appearance: Orange powder
λ_{max}: 250(485)nm in chloroform
Solubility: H_2O 0.2mg/ml
EGME 2mg/ml
EtOH 0.3mg/ml

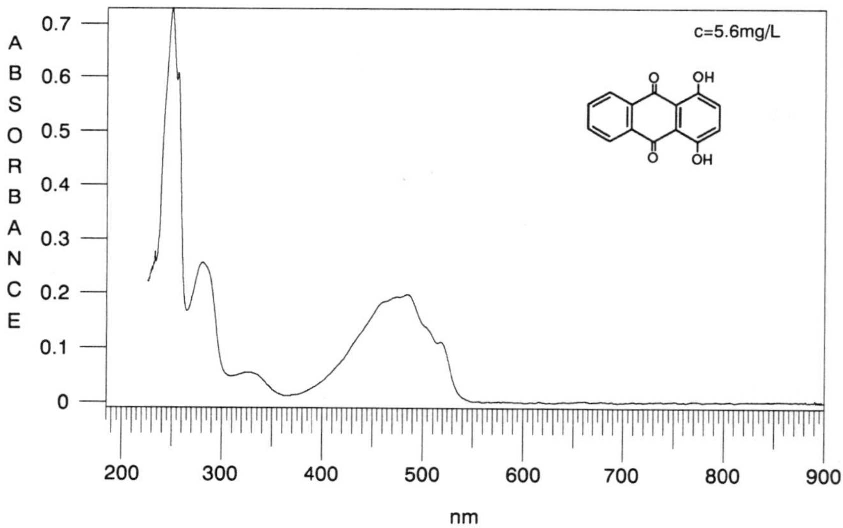

Quinizarin is an orange anthraquinone dye. However, with the possible exception of use as a smoke colorant, it is not used as a dye, since it has only one-tenth the tinctorial strength of its isomer Alizarin.

Bigelow and Reynolds[1] provided an excellent method for synthesizing Quinizarin from phthalic acid and 4-chlorophenol by gradually heating a mixture of the two to 200°C in the presence of sulfuric and boric acids.

Very slightly soluble in water and ethanol, and somewhat more soluble in ether, aqueous alkalis, ammonium hydroxide and acetic acid (1 part in 13 parts, boiling), it is used as an intermediate for making other dyes, such as

C.I. Acid Green 25 and C.I. Solvent Violet 13.

Quinizarin has rarely been used as a biological stain, if at all. Gurr[2] states that "it is probably employed in the study of sebaceous glands."

References:

1) Bigelow, L.A.; Reynolds, H.H. *Organic Synthesis*; Wiley: New York, 1941; Coll. Vol. 1, p 476.

2) Gurr, E. *Encyclopedia of Microscopic Stains*; Hill: London, 1960; pp 342-343.

General references:

Aldrich FT-IR Library	**1**(2), 87A
Sigma FT-IR Library	**1**(2), 205B
IR Library	**3**,900C
NMR Library	**2**(2), 91A
Merck Index	**11**, 8094

QUINOLINE YELLOW A, spirit soluble
Solvent Yellow 33

Aldrich Cat. No.: 23,413-3

CAS No.: [*8003-22-3*]
C.I. No.: 47000
Appearance: Yellow-gold powder
λ_{max}: 413(435)nm in methanol
Solubility: H_2O <0.1mg/ml
EGME 3mg/ml
EtOH 0.3mg/ml

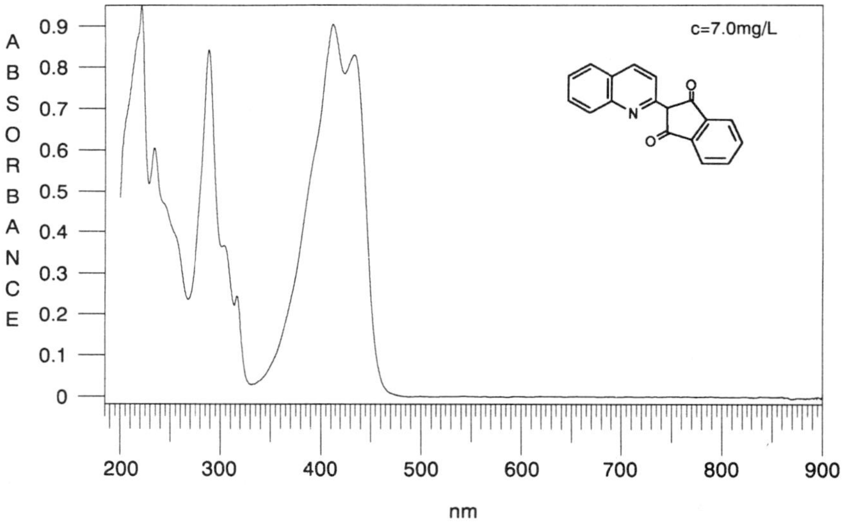

Quinoline Yellow A is a quinoline dye that is made by condensing quinaldine with phthalic anhydride at 200°C in the presence of zinc chloride. This synthetic route yields a dye mixture which is composed of approximately two parts of 2-(2-quinolinyl)-1,3-indandione and one part of 2-(6-methyl-2-quinolinyl)-1,3-indandione.

Soluble in acetone, benzene, toluene and xylene, slightly soluble in ethyl acetate, linseed oil, mineral oil, paraffin wax and stearic acid, very slightly soluble in ethanol and insoluble in water, it is used to color spirit lacquers, polystyrenes, polycarbonates, polyamides, cosmetics and smoke. In addition, it is the precursor of Quinoline Yellow S (C.I. 47005) which is made from the solvent dye by sulfonation.

Quinoline Yellow S is a water-soluble anionic dye that is used to dye wool; the aluminum salt is used as a pigment (Pigment Yellow 115), finding application in cosmetics.

Quinoline Yellow A is rarely used as a biological stain. However,

Conn's[1] indicates that an acid Quinoline Yellow is occasionally used as a plasma stain.

Reference:

1) *Conn's Biological Stains,* 9th ed.; Lillie, R.D., Ed.; Williams and Wilkins: Baltimore, 1977; p 364.

General references:

Sigma FT-IR Library	**1**(2), 205D
Merck Index	**11**, 8101
Colour Index	3rd ed., 1971

QUINOLINE YELLOW S, water soluble
Acid Yellow 3

Aldrich Cat. No.: 30,905-2

CAS No.: [*8004-92-0*]
C.I. No.: 47005
Appearance: Yellow powder
λ_{max}: 412(288)nm in water
Solubility: H_2O 100mg/ml
EGME 10mg/ml
EtOH 0.4mg/ml

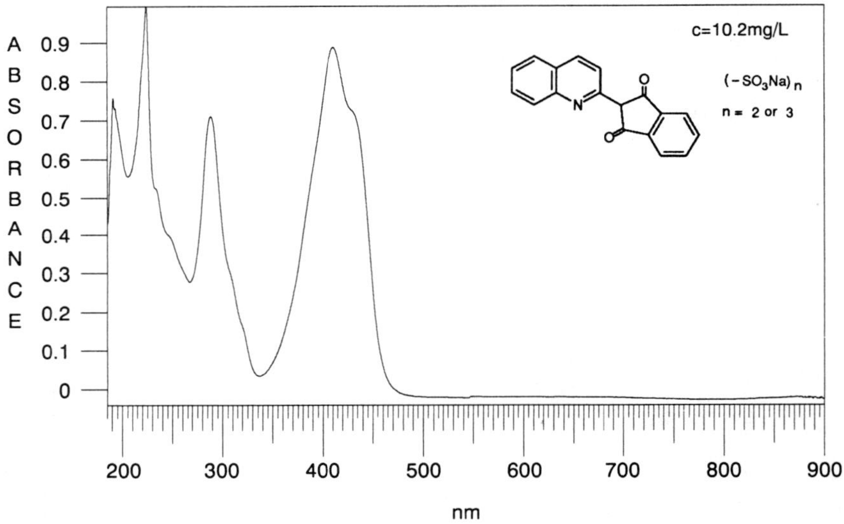

For description, see Quinoline Yellow A on pages 608 and 609.

General reference:
Merck Index **11**, 8100

REACTIVE BLACK 5

Remazol Black B

Aldrich Cat. No.: 30,645-2
Sigma Prod. No.: R6506

CAS No.: [*17095-24-8*]
C.I. No.: 20505
Mol. Form.: $C_{26}H_{21}N_5Na_4O_{19}S_6$
F.W.: 991.82
m.p.: >300°C
Appearance: Dark-maroon powder
λ_{max}: 597nm in water
Solubility: H_2O 60mg/ml
EGME 70mg/ml
EtOH 0.9mg/ml

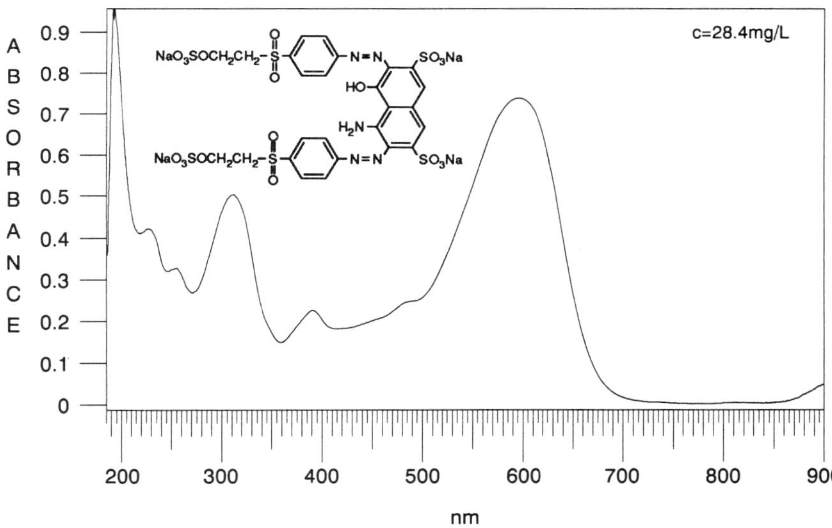

This black, disazo reactive (vinylsulfonyl) dye is soluble in water.

Commercially, it is used to dye urea- and sodium bicarbonate-padded cotton utilizing steam or dry heat for fixing. According to the *Colour Index*, cellulose is printed directly with sodium bicarbonate alginate thickener, utilizing neutral steam for fixing.

There is no evidence that it has been used as a biological stian.

General reference:

Colour Index 3rd ed., 1971

REACTIVE BLUE 2

Cibacron Blue 3G-A • F3GA

Aldrich Cat. No.: 24,222-5

CAS No.: [*12236-82-7*]
C.I. No.: 61211
Mol. Form.: $C_{29}H_{17}ClN_7Na_3O_{11}S_3$
F.W.: 840.12
Appearance: Dark-blue powder
λ_{max}: 607nm in water
Solubility: H_2O 5mg/ml
EGME 10mg/ml
EtOH 0.7mg/ml

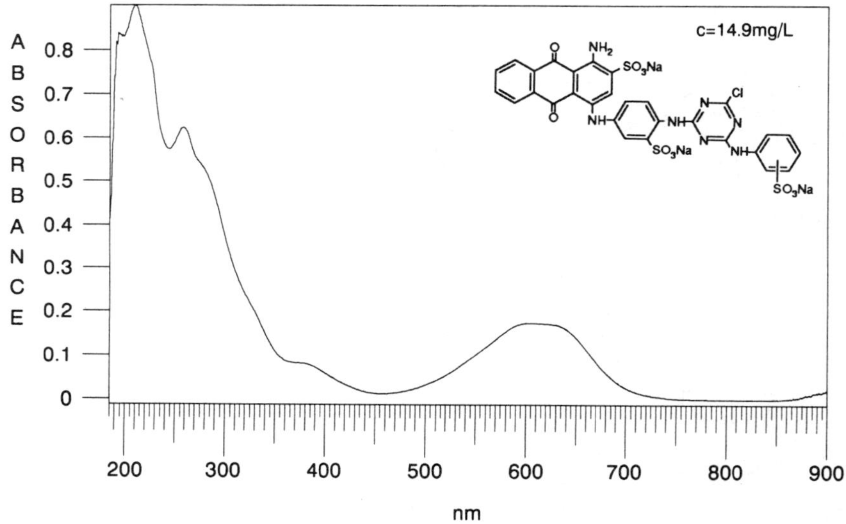

Reactive Blue 2 is a slightly water-soluble, substituted 1,4-diamino-anthraquinone monochlorotriazinyl dye. The monochlorotriazine moiety facilitates covalent bond formation between the dye and active hydrogen sites on both cellulosic and proteinaceous substrates, and this bonding mode is largely responsible for the dye's good (6-7) lightfastness.

Industrially, Cibacron Blue 3G-A has been used to dye and print cotton, nylon, silk and wool.

An important biochemical tool, Cibacron Blue 3G-A is used as a ligand for the affinity chromatographic separation and purification of enzymes and proteins.[1-4]

References:

1) Kopperschlager, G.; Freyer, R.; Diezel, W.; Hofmann, E. *FEBS Lett.* **1968,** *1*, 137.
2) Staal, G.; Visser, J.; Veeger, C. *Biochim. Biophys. Acta* **1969,** *185*, 39.
3) Roschlau, P.; Hess, B. *Hoppe-Seyler's Z. Physiol. Chem.* **1971,** *354*, 441.
4) Bohme, H.; Kopperschlager, G.; Schulz, J.; Hofmann, E. *J. Chromatogr.* **1972,** *69*, 209.

General references:

Safety **2**(2), 3028B
Colour Index 3rd ed., 1971

REACTIVE BLUE 4

Aldrich Cat. No.:	24,481-3
Sigma Prod. No.:	R9003

CAS No.:	[*13324-20-4*]
C.I. No.:	61205
Mol. Form.:	$C_{23}H_{14}Cl_2N_6O_8S_2$
F.W.:	637.44
Appearance:	Navy-blue powder
λ_{max}:	595nm in water
Solubility:	H_2O 100mg/ml
	EGME 50mg/ml
	EtOH 5mg/ml

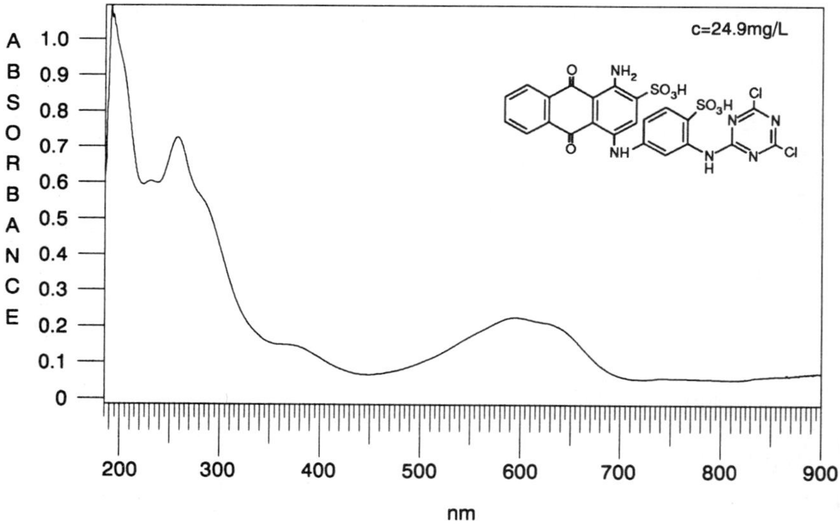

This bright-blue anthraquinone dye includes a dichloro-*s*-triazinyl moiety that is responsible for its reactivity, forming covalent bonds at room temperature with substrates that contain reactive hydrogen sites. In addition, it possesses two sulfonic acid functions which render it very water soluble and enhance its utility as a dye.

Demonstrating excellent lightfastness (7), it is used to dye and print cellulosics, nylon, silk and wool.

While there are no reported biological stain application, Reactive Blue 4 is typical of the reactive dyes used by Goland and Grand[1] as fixatives and markers in their studies demonstrating growth in teeth and bones.

Reference:

1) Goland, P.; Grand, N. *Am. J. Phys. Anthropol.* **1968,** *29,* 201-218.

General references:

Aldrich FT-IR Library	**1**(2), 851C
Sigma FT-IR Library	**1**(2), 206C
Safety	**2**(2), 3028C
Colour Index	3rd ed., 1971

REACTIVE ORANGE 16

Remazol Brilliant Orange 3R

Aldrich Cat. No.:	30,650-9
CAS No.:	[*12225-83-1*]
C.I. No.:	17757
Mol. Form.:	$C_{20}H_{17}N_3Na_2O_{11}S_3$
F.W.:	617.54
m.p.:	>300°C
Appearance:	Red-orange powder
λ_{max}:	494(388)nm in water
Solubility:	H_2O 100mg/ml
	EGME 20mg/ml
	EtOH 8mg/ml

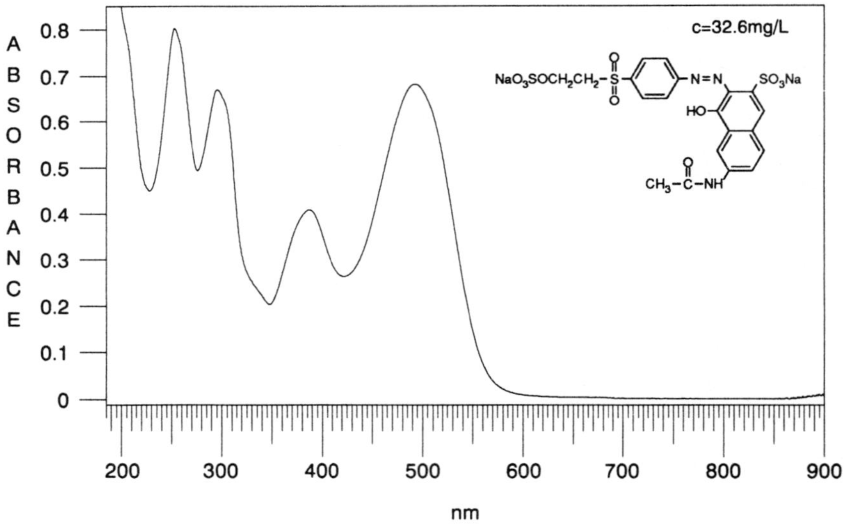

This bright, yellowish-red, monoazo reactive (vinylsulfonyl) dye is very soluble in water and only slightly soluble in methanol and ethanol.

Commercially, it is used to dye cotton by the pad-batch method. Resistance to light-fading is good (5-6).

There is no evidence that it has been used as a biological stain.

General reference:

Colour Index 3rd ed., 1971

REACTIVE RED 8

Aldrich Cat. No.: 24,482-1

CAS No.: *[25489-36-5]*
C.I. No.: 17908
Mol. Form.: $C_{20}H_{12}Cl_2N_6Na_2O_8S_2$
F.W.: 645.37
Appearance: Dark-burgundy powder
λ_{max}: 500nm in water
Solubility: H_2O 30mg/ml
EGME 40mg/ml
EtOH 1mg/ml

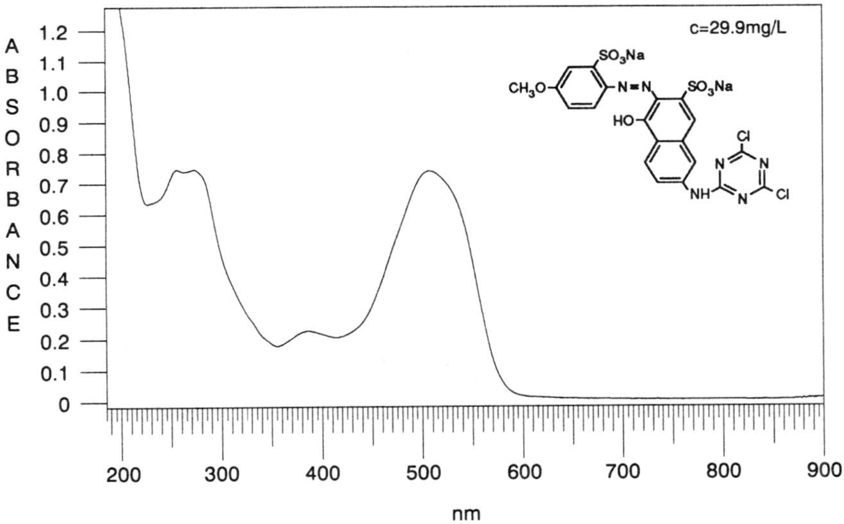

This bright, yellowish-red, monoazo dichloro-*s*-triazinyl dye is made by coupling diazotized 2-amino-5-methoxybenzenesulfonic acid with J acid (1-hydroxy-6-aminonaphthalene-3-sulfonic acid) and condensing the product with cyanuric acid and then converting the condensation product to the disodium salt. Since it has two reactive chlorines, it forms covalent bonds, at room temperature, with substrates that possess active hydrogen sites. This can be facilitated by the presence of sodium bicarbonate to neutralize the hydrogen chloride as it is formed.

Possessing good water solubility and moderate lightfastness (3-4), it is used to dye and print cellulose, nylon, silk and wool.

While there are no reported biological stain applications, Reactive Red 8 is typical of the reactive dyes used by Goland and Grand[1] as fixatives and markers in their study demonstrating growth in teeth and bones.

Reference:

1) Goland, P.; Grand, N. *Am. J. Phys. Anthropol.* **1968,** *29*, 201-218.

General references:

Aldrich FT-IR Library **1**(2), 1025B
Colour Index 3rd ed., 1971

REINECKE SALT

Ammonium tetrathiocyanatodiamminechromate(III)

Aldrich Cat. No.:	18,368-7
Sigma Prod. No.:	R7756
Grade:	A.C.S. reagent

CAS No.:	[*13573-16-5*]
Mol. Form.:	$C_4H_{10}CrN_7S_4 \cdot H_2O$
F.W.:	354.47
m.p.:	268-272°C (dec.)
Appearance:	Red crystalline powder and chunks
λ_{max}:	309nm in methanol
Solubility:	H_2O 8mg/ml
	EGME 8mg/ml
	EtOH 2mg/ml

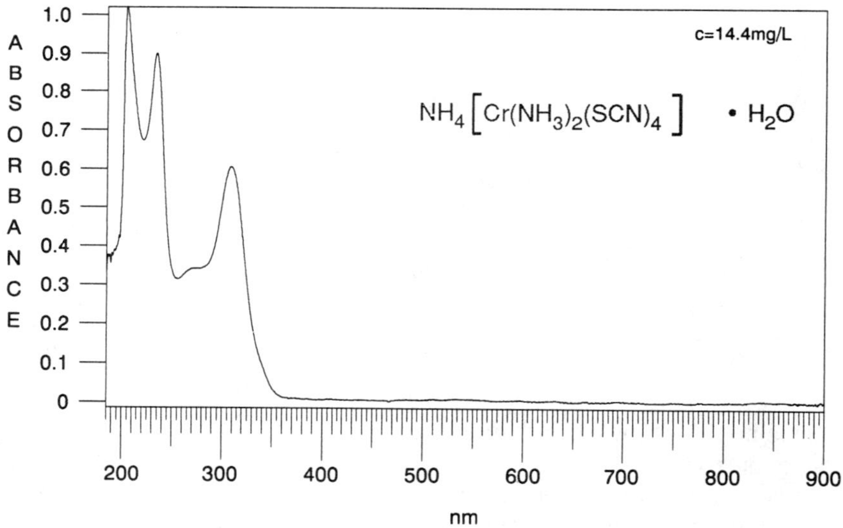

Reinecke salt is not a dye but an analytical reagent that is sparingly soluble in cold water. Solutions are not stable for more than two weeks at room temperature, and decomposition occurs more rapidly in solutions subjected to higher temperatures. Dakin[1] describes a method for preparing Reinecke salt by fusing ammonium thiocyanate with ammonium dichromate.

Merck[2] states that Reinecke salt is used to detect mercury and as a precipitant for primary and secondary amines, proline, hydroxyproline and certain amino acids. *Reagent Chemicals*[3] includes specifications and tests for Reinecke salt for use as a reagent for detecting choline chloride.

References:

1) Dakin, H.D. In *Organic Synthesis*; Blatt, A.H., Ed.; Wiley: New York, 1943; Col. Vol. II, p 555.

2) *The Merck Index,* 11th ed.; Budavari, S. *et al.*, Eds.; Merck: Rahway, 1989; Monograph 8135.

3) *Reagent Chemicals,* 7th ed.; American Chemical Society: Washington, DC, 1986; pp 552-553.

General references:

IR Library	**3**, 525C
Merck Index	**11**, 8135

REMAZOL BRILLIANT BLUE R

Reactive Blue 19

Aldrich Cat. No.: 22,324-7
Sigma Prod. No.: R8001

CAS No.: [*2580-78-1*]
C.I. No.: 61200
Mol. Form.: $C_{22}H_{16}N_2Na_2O_{11}S_3$
F.W.: 626.56
Appearance: Dark, blue-black powder
λ_{max}: 592(285)nm in water
Solubility: H_2O 20mg/ml
EGME 20mg/ml
EtOH 40mg/ml

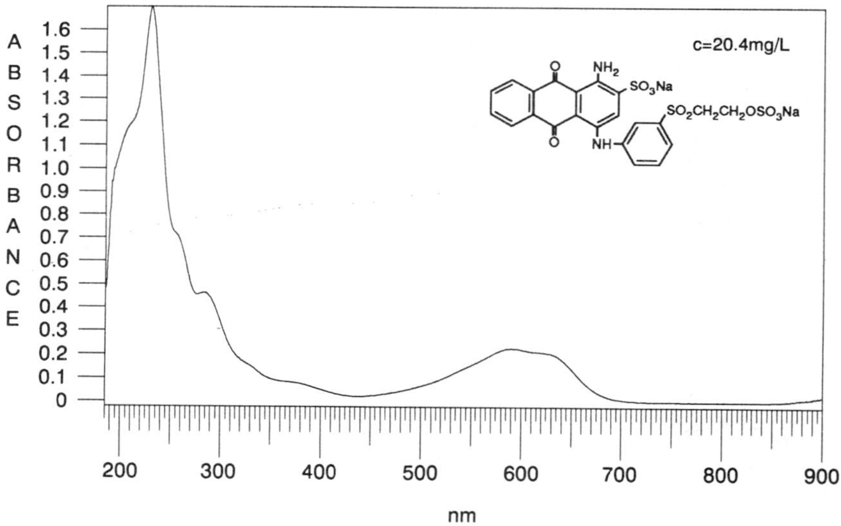

Remazol Brilliant Blue R is a vinylsulfonyl anthraquinone dye that is soluble in water, alcohol and methyl Cellosolve.

Industrially, it is used to dye urea- and sodium bicarbonate-padded cellulosic substrates by the application of steam or dry heat. The dye is extremely fast, since it becomes an integral part of the substrate through a covalent bond. These bonds occur when the reactive vinylsulfonyl function of the dye reacts with the conjugate base of the hydroxyl groups of the substrate in a conducive environment (steam or dry heat with a pH of 7.5-8.0).

While no specific applications as a biological stain are reported, Goland and Engel[1] and Goland and Grand[2] used similar dyes to stain tissue covalently.

The latter study demonstrated incremental bone growth in test rats that were administered different-colored reactive dyes, over a period of time, before they were sacrificed.

Uniblue A is a research grade of Reactive Blue 19 (RB 19), differing from the latter in two respects: The dye content is approximately 98% compared to 50% for RB 19 and, unlike RB 19, the vinylsulfone group is already activated and will react with reactive hydrogen sites in protein substrates without the addition of base. Datyner and Finnimore[3] describe a method for the use of Uniblue A as a protein stain in electrophoresis.

References:

1) Goland, P.; Engel, M.J.; *J. Histochem. Cytochem.* **1963**, *11*, 751-762.

2) Goland, P.; Grand, G. *Am. J. Phys. Anthropol.* **1968**, *29*, 201-208.

3) Datyner, A.; Finnimore, E. *Anal. Biochem.* **1973**, *52*, 45.

General references:

Aldrich FT-IR Library	**1**(2), 1019A
Sigma FT-IR Library	**1**(2), 208D
IR Library	**3**, 1477F
NMR Library	**2**(2), 831B
Colour Index	3rd ed., 1971

RESAZURIN

7-Hydroxy-3*H*-phenoxazin-3-one 10-oxide, sodium salt

Aldrich Cat. No.: 19,930-3
Sigma Prod. No.: R2127
Grade: Certified

CAS No.: [*550-82-3*]
Mol. Form.: $C_{12}H_6NNaO_4$
F.W.: 251.18
Appearance: Black powder
λ_{max}: 598(380)nm in water
Solubility: H_2O 20mg/ml
EGME 8mg/ml
EtOH 6mg/ml

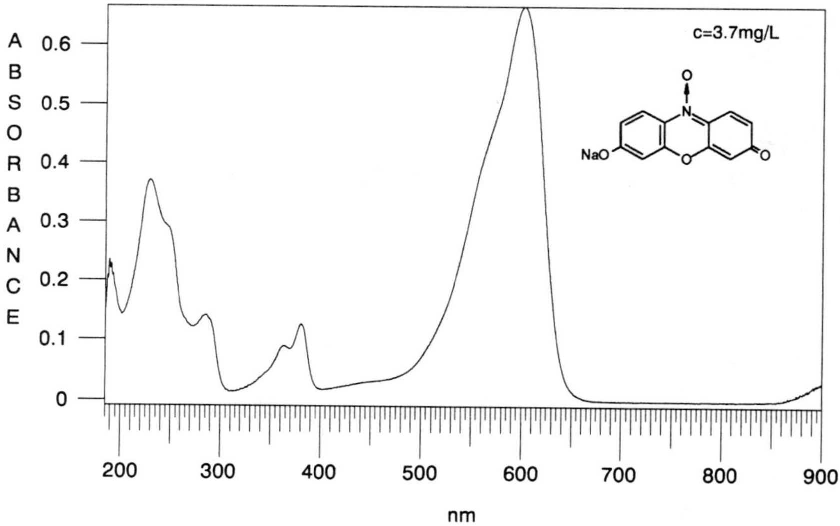

Resazurin and Resorufin are two closely related heterocyclic compounds that contain the oxazone chromophore. Resorufin is a reduced form of Resazurin. Neither has practical dye applications due to their great sensitivity to oxidation-reduction.

However, Ramsdell *et al.*[1] believed Resazurin to be superior to Methylene Blue as an indicator of reduction in milk. This observation was based on its greater speed, sensitivity to oxidation-reduction and the fact that the transition product of reduction, Resorufin, is red. This intermediate color change, a forewarning of the colorless endpoint (leuco form), is not displayed by Methylene Blue.

Most references list Resazurin as the free phenol with solubility peculiar to that form: insoluble in water and ether, slightly soluble in alcohol and glacial acetic acid, and soluble in dilute potassium or sodium hydroxide.

Resazurin, sodium salt, is the form the BSC has certified for more than 30 years. The sodium salt is soluble in water, insoluble in ether and slightly soluble in alcohol and acetic acid.

Resorufin is also available as the sodium salt (Aldrich Cat. No. 23,015-4) with solubility characteristics very close to those of Resazurin, sodium salt.

Kolthoff[2] described Resazurin as not very satisfactory as a pH indicator, listing a visual-transition interval from pH 3.8 (orange) to pH 6.5 (purple). Long[3] reported the same color change and visual-transition interval as Kolthoff.

References:

1) Ramsdell, G.A.; Johnson, F.R.; Evans. F.R. *J. Dairy Sci.* **1935**, *18*, 706-717.

2) Kolthoff, I.M. *Acid Base Indicators*; Macmillan: New York, 1937.

3) Long, E. *Biochemists Handbook*, Van Nostrand: Princeton, 1961.

General references:

Aldrich FT-IR Library	**1**(2), 919C
Sigma FT-IR Library	**1**(2), 209A
IR Library	**3**, 1424E
Merck Index	**11**, 8144

RESORCIN CRYSTAL VIOLET

Aldrich Cat. No.: 21,680-1
Sigma Prod. No.: R2628

CAS No.: [*81133-29-1*]
Appearance: Black powder
λ_{max}: 590(303)nm in methanol
Solubility: H_2O 7mg/ml
EGME 80mg/ml
EtOH 30mg/ml

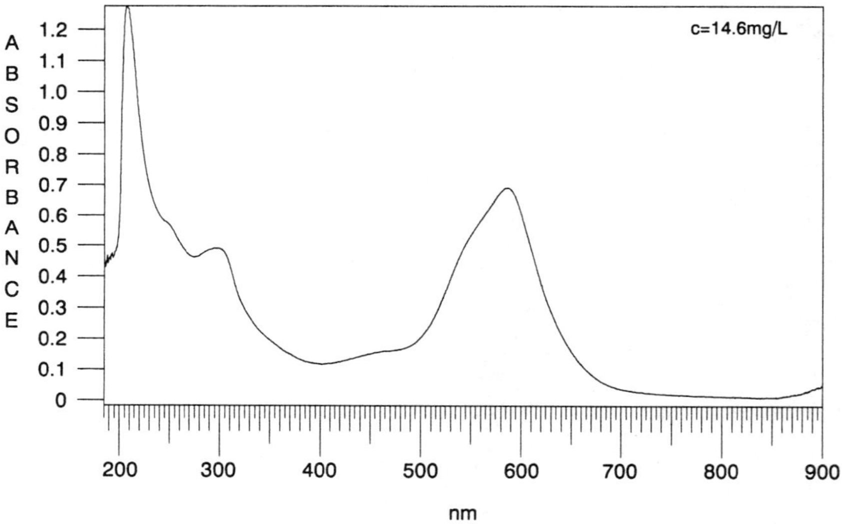

Resorcin Crystal Violet (RCV) is a collagen-fiber stain that is slightly soluble in water and soluble in the lower alcohols.

Lillie[1] believes that Volkmann and Strauss[2] may have used RCV and Orcein separately in a combined stain that utilized Acid Alizarin Blue GR (C.I. 63015) as one component. The combined stain was used in the Mallory-Heidenhain azan-collagen method.

There are no reported industrial uses for Resorcin Crystal Violet.

References:

1) Lillie, R.D. *J. Technol. Methods* **1945,** *25*, 1-47.

2) Volkmann, R.; Strauss, F. *Z. Wiss. Mikrosk*. **1934,** *51*, 244-249. (Abstracted in *Stain Technol.* **1934,** *10*, 70.)

General references:

Aldrich FT-IR Library	**1**(2), 1043B
Sigma FT-IR Library	**1**(2), 880A
IR Library	**3,** 1489A

RESORUFIN

7-Hydroxy-*3H*-phenoxazin-3-one, sodium salt

Aldrich Cat. No.: 23,015-4
Sigma Prod. No.: R3257

CAS No.: [*635-78-9*]
Mol. Form.: $C_{12}H_6NNaO_3$
F.W.: 235.18
Appearance: Purple powder
λ_{max}: 573nm in methanol
Solubility: H_2O 20mg/ml
EGME 20mg/ml
EtOH 2mg/ml

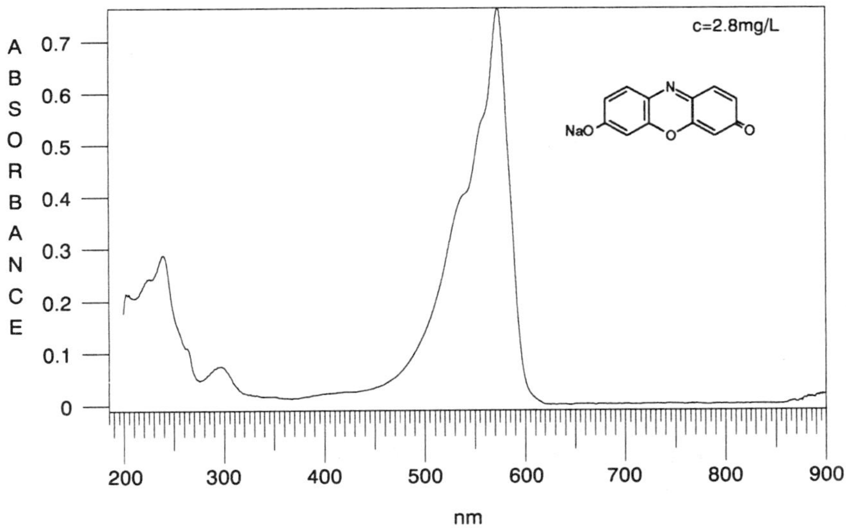

See description on pages 622 and 623 under Resazurin.

General reference:

Aldrich FT-IR Library **1**(2), 1035A

RHODAMINE 123 dihydrate

Aldrich Cat. No.:	28,394-0
Sigma Prod. No.:	R8004
CAS No.:	[*123333-76-6*]
Mol. Form.:	$C_{21}H_{16}N_2O_3 \cdot 2H_2O \cdot HCl$
F.W.:	416.87
Appearance:	Brownish-red crystalline powder
λ_{max}:	501nm in water

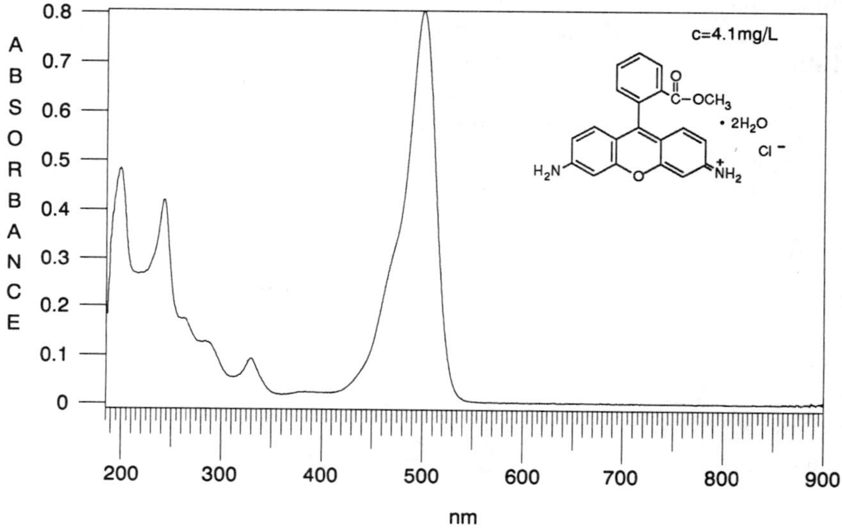

This fluorescent, cationic rhodamine dye is soluble in ethanol, ether, various fats and lipids.

Due to its comparatively low cytotoxicity, Rhodamine 123 is the dye of choice for visualizing mitochondria.[1] Either blue or green light will elicit excitation, with the former being observed as green and the latter as a red fluorescence. Generally however, blue excitation (470nm) provides a better visualization of structural details.[2]

Mitochondria[3-6] and cell[7] viability has been measured with Rhodamine 123 being used as a fluorescent indicator, and the dye has been used as a mitochondrial marker.[8,9]

More recently, cancer chemotherapy[10-12] and photochemotherapy[13,14] studies have been carried out using Rhodamine 123.

References:

1) Weiss, M.L.; Chen, L.B. *Kodak Lab. Chem. Bull.* **1984,** *55*(2), 1.

2) Johnson, L.V.; Walsh, M.L.; Chen, L.B. *Proc. Natl. Acad. Sci. U.S.A.* **1980,** *77*(2), 990.

3) Casey, C.A.; Anderson, P.M. *J. Biol. Chem.* **1982,** *257*(14), 8449.

4) Maro, B.; Marty, M.C.; Bornens, M. *EMBO J.* **1982,** *1*(11), 1347.

5) Sonka, J.; Stohr, M.; Voyt-Schaden, M.; Volm, M. *Cytometry (Baltimore)* **1983,** *4*, 141.

6) Staiano-Coico, L.; Darzynkiewicz, Z.; Melamed, M.R.; Weksler, M. *ibid.* **1982,** *3*(2), 79.

7) Evenson, D.P.; Darzynkiewicz, Z.; Melamed, M.R. *J. Histochem. Cytochem.* **1982,** *30*(3), 279.

8) Walker, C.; Shay, J.W. *Somatic Cell Genet.* **1983,** *9*(4), 469.

9) Tanabe, K. *J. Protozool.* **1983,** *30*(4), 707.

10) Bernal, S.D.; Lampidis, T.J.; Summerhayes, I.C.; Chen, L.B. *Science* **1982,** *218*, 1117.

11) Bernal, S.D.; Lampidis, T.J.; McIsaac, R.M.; Chen, L.B. *ibid.* **1983,** *222*, 169.

12) Lampidis, T.J. *et al.* *Ann. N.Y. Acad. Sci.* **1982,** *397*, 299.

13) Powers, S.K. *et al.* *J. Neurosurg.* **1986,** *64*, 918.

14) Baines, A.J. *et al.* *J. Chromatogr.* **1986,** *356,* 301.

RHODAMINE B
Basic Violet 10

Aldrich Cat. No.:	R95-3	25,242-5
Sigma Prod. No.:	R6626	——
Grade:	——	Laser dye
Dye content:	~80%	~90%

CAS No.:	[*81-88-9*]	
C.I. No.:	45170	
Mol. Form.:	$C_{28}H_{31}ClN_2O_3$	
F.W.:	479.02	
m.p.:	210-211°C (dec.)	
Appearance:	Green powder	
λ_{max}:	543(355)nm in methanol	
Solubility:	H_2O 30mg/ml	50mg/ml
	EGME 20mg/ml	100mg/ml
	EtOH 50mg/ml	70mg/ml

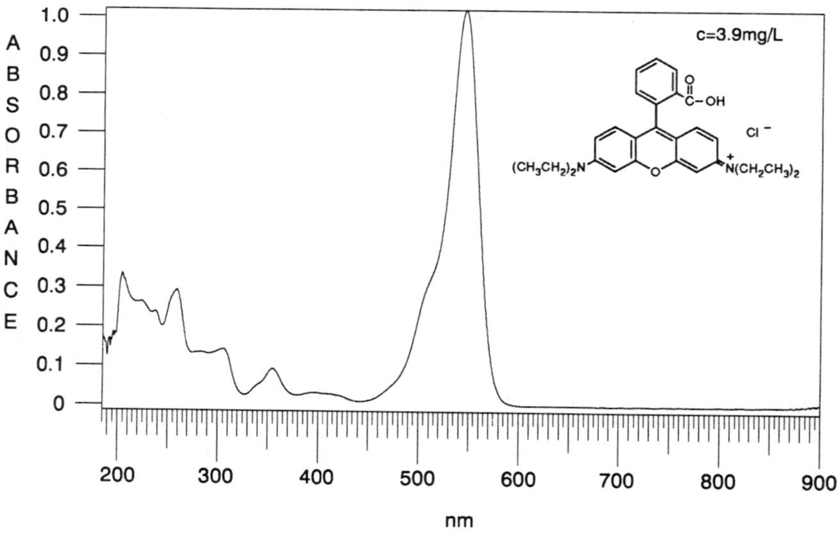

Rhodamine B (Aldrich Cat. No. R95-3 and 25,252-5) is a bluish-red, fluorescent (green) rhodamine dye that is prepared by condensing *N,N*-diethyl-*m*-aminophenol hydrochloride with phthalic anhydride.

Soluble in water, ethanol and methyl Cellosolve and slightly soluble in acetone, it is used to dye wool and cotton, the latter with a tannin mordant.

Conn's[1] traces the application of this dye as a biological stain back to Greisbach, who used it along with osmic acid to simultaneously fix and stain blood, and to Ehrlich's use as a component of "neutral-stain" complexes. Over the years since its discovery in 1887 by Ceresole, it has been used widely as a fluorochrome in fluorescence microscopy. *Staining Procedures*[2] describes methods for the cytochemical demonstration of peroxidase by fluorescence microscopy (from Papadimitriou *et al.*, 1978), for the fluorescence-microscopy, acid-fast procedure of Truant, Brett and Thomas developed in 1962 and for acid-fast organisms in tissue sections, modified from the previous procedure.

Rhodamine B (Aldrich Cat. No. 25,242-5) is 99% pure as demonstrated by HPLC and is suitable as an efficient laser dye for both pulsed and continuous-wave operation. The optimum lasing wavelength (in acidic ethanol) is 627nm and the fluorescence λ_{max} (also in acidic ethanol) is 580nm.[3]

Rhodamine B base (Aldrich Cat. No. 23,414-1) is soluble in methanol, ethanol, stearic and oleic acids and acetone, slightly soluble in ethyl acetate and very slightly soluble in water, benzene, xylene and mineral oil. It is used as a colorant in a number of solvent systems, ballpoint-pen inks, rotogravure inks and for waxes and fats.

References:

1) *Conn's Biological Stains,* 9th ed.; Lillie, R.D., Ed.; Williams and Wilkins: Baltimore, 1977; p 332.

2) Clark, G. *Staining Procedures,* 4th ed.; Williams and Wilkins: Baltimore, 1981; pp 95, 382, 450.

3) *Dye Lasers;* Schafer, F.P., Ed.; Springer: Berlin, 1973.

General references:

Sigma FT-IR Library	1(2), 209B
Safety	2(2), 3048A
Merck Index	11, 8182
Colour Index	3rd ed., 1971

RHODAMINE B base
Solvent Violet 49

Aldrich Cat. No.: 23,414-1
Sigma Prod. No.: R1879
Dye content: 97%

CAS No.: [*509-34-2*]
C.I. No.: 45170:1
Mol. Form.: $C_{28}H_{30}N_2O_3$
F.W.: 442.56
Appearance: Pink powder
λ_{max}: 544nm in methanol
Solubility: H_2O 1mg/ml
EGME 60mg/ml
EtOH 60mg/ml

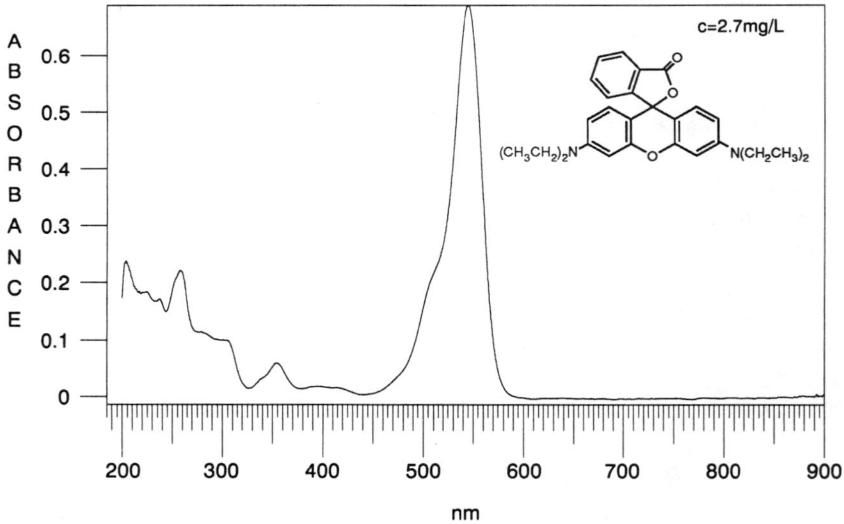

For description, see Rhodamine B on pages 628 and 629.

General references:

Aldrich FT-IR Library	**1**(2), 1014A
Sigma FT-IR Library	**1**(2), 209C
Safety	**2**(2), 3048B

RHODAMINE 6G
Basic Red 1

Aldrich Cat. No.:	20,132-4	25,243-3
Sigma Prod. No.:	R4127	—
Grade:	—	Suitable as laser dye
Dye content:	~95%	~99%

CAS No.:	[*989-38-8*]	
C.I. No.:	45160	
Mol. Form.:	$C_{28}H_{31}ClN_2O_3$	
F.W.:	479.02	
Appearance:	Maroon powder	
λ_{max}:	524nm in water	530nm in ethanol
Solubility:	H_2O 20mg/ml	20mg/ml
	EGME 20mg/ml	40mg/ml
	EtOH 100mg/ml	50mg/ml

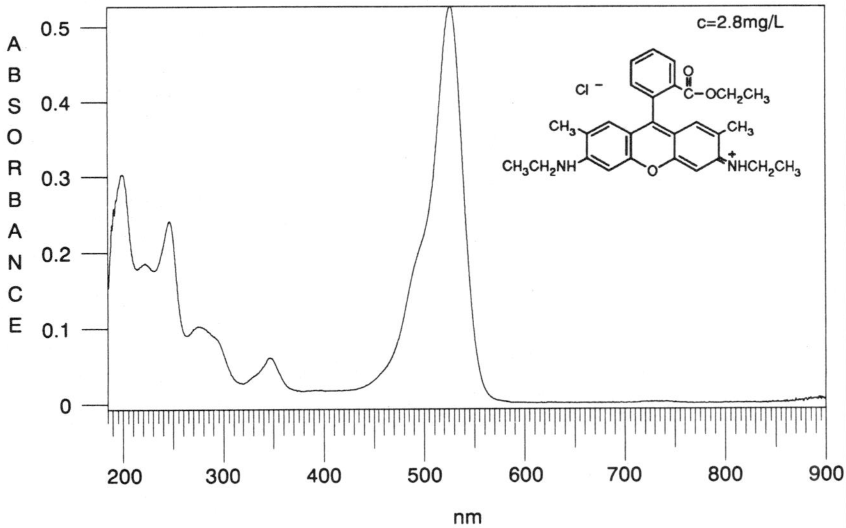

Rhodamine 6G (Aldrich Cat. No. 20,132-4) is a cationic xanthene dye, with a strong tinctorial capacity, that is made by condensing two moles of 3-ethylamino-*p*-cresol with one mole of phthalic anhydride and esterifying the product with ethanol in the presence of hydrogen chloride. It is soluble in water with a red color and a green fluorescence and more soluble in ethanol with a red color and a yellow fluorescence. Since it is fugitive (1-2), it finds little use in the textile industry. It is more widely used to dye paper and to prepare the phosphotungstomolybdic acid salt (PTMA) (Pigment Red 81, C.I. 45160:1) and the copper ferrocyanide complex (Pigment Red 169, C.I. 45160:2). The former shows good lightfastness (5), is soluble in both aliphatic and aromatic hydrocarbons and moderately soluble in ethanol and methyl Cellosolve. This pigment is used to prepare printing inks and to color crayons and make colored pencils. Pigment Red 169 is soluble in both aliphatic and aromatic hydrocarbons, ethanol, esters and ketones. Like PTMA, it is used to prepare printing inks. While it is not quite as lightfast (4) as PTMA, it is less expensive.

Rhodamine 6G has not been investigated in biological-stain applications nearly as much as Rhodamine B. *Conn's*[1] cites two studies that compare the staining performances of these dyes.[2,3] Rhodamine 6G (Aldrich Cat. No. 25,243-3; ~99%) is free of impurities as demonstrated by HPLC and is suitable as an efficient laser dye for both pulsed- and continuous-wave (CW) operation. The optimum lasing wavelength (in ethanol) is 590nm.

Rhodamine 6G perchlorate (Aldrich Cat. No. 25,244-1) is 97% pure as demonstrated by HPLC and is suitable as an efficient laser dye for both pulsed- and continuous-wave operation. The optimum lasing wavelength (in ethanol) is 590nm and the fluorescence λ_{max} is 552nm (in ethanol).[4]

Rhodamine 6G tetrafluoroborate (Aldrich Cat. No. 25,246-8) is 98% pure as demonstrated by HPLC and is suitable as an efficient laser dye for both pulsed- and continuous-wave operation. The optimum lasing wavelength (in ethanol) is 590nm.

References:

1) *Conn's Biological Stains,* 9th ed.; Lillie, R.D., Ed.; Williams and Wilkins: Baltimore, 1977; p 332.

2) Strugger, A. *Protoplasma* **1938,** *30,* 85-100.

3) Monné, L. *ibid.* **1939,** *32,* 184-192.

4) *Dye Lasers*; Schafer, F.P., Ed.; Springer: Berlin,1973.

General references:

Aldrich FT-IR Library	**1**(2), 1013D
Sigma FT-IR Library	**1**(2), 210A
IR Library	**3,** 1475B
Safety	**2**(2), 3048C
Colour Index	3rd ed., 1971

RHODAMINE 6G perchlorate

Aldrich Cat. No.: 25,244-1
Sigma Prod. No.: R9879
Grade: Suitable as a laser dye
Dye content: ~99%

CAS No.: [*13161-28-9*]
Mol. Form.: $C_{28}H_{31}ClN_2O_7$
F.W.: 543.02
m.p.: 263-265°C
Appearance: Shiny red crystals
λ_{max}: 528nm in methanol
Solubility: H_2O <0.1mg/ml
EGME 10mg/ml
EtOH 1mg/ml

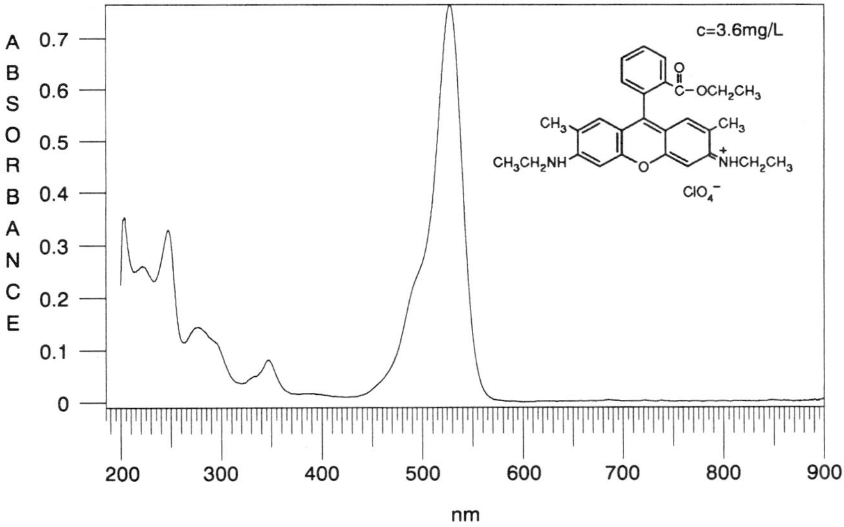

For description, see pages 631 and 632 under Rhodamine 6G.

General reference:

Safety **2**(2), 3049B

RHODAMINE 6G tetrafluoroborate

Aldrich Cat. No.:	25,246-8
Sigma Prod. No.:	R0130
Grade:	Suitable as a laser dye
Dye content:	~99%

CAS No.:	[*54854-14-7*]
Mol. Form.:	$C_{28}H_{31}BF_4N_2O_3$
F.W.:	530.38
m.p.:	264-266°C
Appearance:	Green and brown crystals
λ_{max}:	528nm in methanol
Solubility:	H_2O 0.2mg/ml
	EGME 30mg/ml
	EtOH 3mg/ml

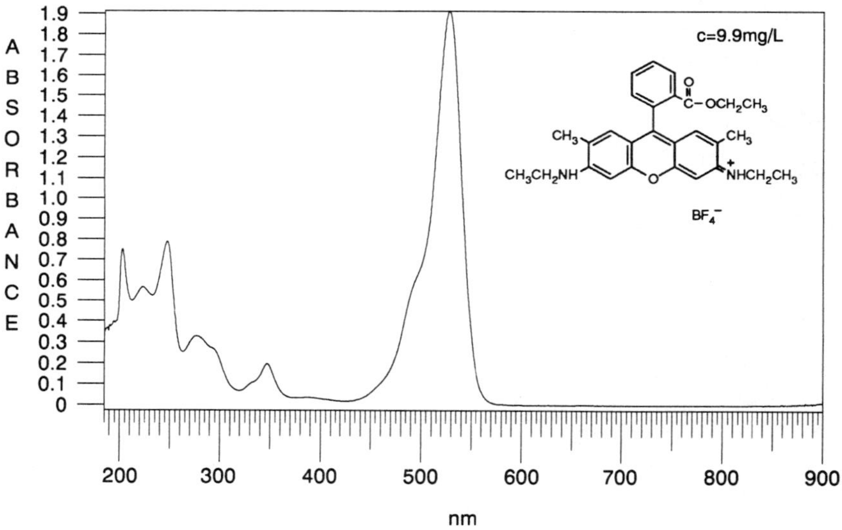

See description on pages 631 and 632 under Rhodamine 6G.

General reference:

Safety **2**(2), 3049C

RHODANILE BLUE

Aldrich Cat. No.: 12,149-5
Sigma Prod. No.: R9627

CAS No.: [*14969-56-3*]
Mol. Form.: $C_{48}H_{48}ClN_5O_3$
F.W.: 778.40
Appearance: Green crystalline powder
λ_{max}: 552(630)nm in water
Solubility: H_2O 20mg/ml
EGME 40mg/ml
EtOH 20mg/ml

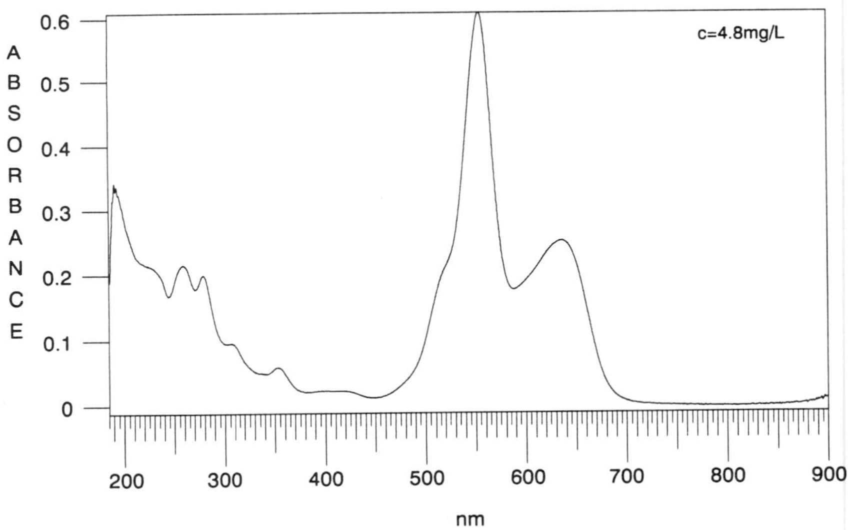

Rhodanile Blue (RB)[1] is a complexed dye formed by reacting equimolar quantities of the amphoteric dye Rhodamine B (C.I. Basic Violet 10) and the cationic dye Nile Blue. It is not a neutral dye despite the coulombic attachment between a cationic site on the Nile Blue molecule and an anionic function on the Rhodamine B molecule, since one of the three remaining unencumbered cationic sites is protonated. More correctly, RB is a cationic dye which is soluble in water, methanol, ethanol and methyl Cellosolve.

Gurr[1] described RB as a useful fibrin stain and a redox indicator. The redox property, according to MacConnail and Gurr,[2] is largely responsible for the differential-staining quality that RB demonstrates. Simpson, Carlisle and Mallard,[3] in a study that used RB as a supravital stain to demonstrate Heinz bodies in the fresh blood of a dog chronically poisoned with phenylhydrazine, may have reinforced this observation. These authors reported that the Heinz bodies were stained a dark purple while the erythrocytic-cytoplasm staining varied from orange-yellow to blue-green. *Staining Procedures*[4] (from Simpson, Carlisle and Mallard) provides a recipe and method for staining Heinz bodies with Rhodanile Blue.

Kass[5] used RB to identify human megakaryocytes. He stated that "the technique is simple, direct, and reproducible, possibly lending more information to the cytochemistry of megakaryocytes and complementing the more complex immunologic techniques currently available."

References:

1) Gurr, E. *Lab. Pract.* **1966,** *15*, 1128-1131.

2) MacConnail, M.A.; Gurr, E., Jr. *J. Med. Sci.* **1964** (June), 243-250.

3) Simpson, G.F.; Carlisle, J.W.; Mallard, L. *Stain Technol.* **1970,** *45*, 221-223.

4) Clark, G. *Staining Procedures,* 4th ed.; Williams and Wilkins: Baltimore, 1981; p 178.

5) Kass, L. *Arch. Pathol. Lab. Med.* **1985,** *109*, 320-324.

General references:

Aldrich FT-IR Library	**1**(2), 1035D
Sigma FT-IR Library	**1**(2), 210B
Safety	**2**(2), 3050A

ROSE BENGAL

Acid Red 94

Aldrich Cat. No.:	19,825-0
Sigma Prod. No.:	R4507
Grade:	Certified
CAS No.:	*[632-69-9]*
C.I. No.:	45440
Mol. Form.:	$C_{20}H_2Cl_4I_4Na_2O_5$
F.W.:	1017.65
Appearance:	Red-brown powder
λ_{max}:	548nm in water + 1ml 1% sodium carbonate
Solubility:	H_2O 100mg/ml
	EGME 60mg/ml
	EtOH 30mg/ml

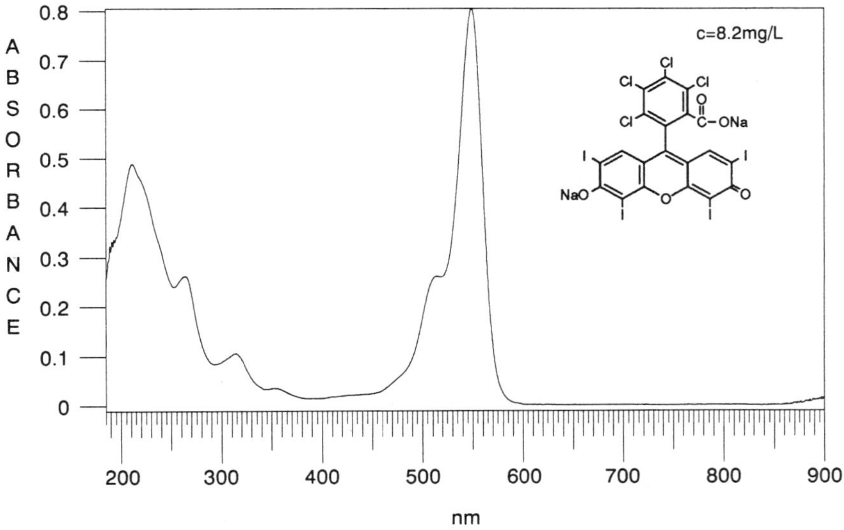

Rose Bengal, certified, is a deep-pink, anionic xanthene dye that is made by condensing two moles of resorcinol with one mole of tetrachlorophthalic anhydride in the presence of zinc chloride at approximately 190°C. The purified product of this condensation is tetraiodinated and converted to the disodium salt to complete the synthesis. It is very soluble in water and soluble in ethanol.

As a biological stain, Rose Bengal has been used as a contrast and cytoplasm stain. Metcalf and Patton[1] used it as a fluorochrome for studying fats under ultraviolet light. *Staining Procedures*[2] provides methods for using Rose Bengal in the following applications: with Bismarck Brown Y and Light

Green SF to demonstrate connective-tissue displacement as described by Clark and Barnes in 1979; with Bismarck Brown as a displacement stain for acidophilic structures as described by Clark in 1979; as a cytoplasm stain for foraminiferans as described by Walton in 1952; and for the staining of bacteria in soil as reported by Conn in 1928 and 1929.

Watson and others[3-5] used Rose Bengal as a photosensitizing dye model for induced thrombotic stroke. The same authors[6] found the method suitable for spinal-cord studies. By irradiating the intact spinal cord of a rat previously injected with Rose Bengal, they were able to reproducibly infarct the spinal-cord tissue. Utilizing the above techniques, the same investigators[7] then pursued controlled arteriovenous malformation therapy studies.

Lamberts *et al.*[8] prepared Rose Bengal, bis(triethylammonium) salt which is soluble in nonpolar solvents such as dichloromethane, chloroform and pentane, as well as methanol. They claim the altered solubility displayed by this salt form enhances the efficacy of the dye as a photosensitizer and marker in biological applications.

References:

1) Patton, R.L.; Metcalf, R.L. *Science* **1943,** *98*, 184.

2) Clark, G. *Staining Procedures,* 4th ed.; Williams and Wilkins: Baltimore, 1981; pp 121, 208, 285, 425.

3) Watson, B.D.; Dietrich, W.D.; Busto, R.; Watchtel, M.S.; Ginsberg, M.D. *Ann. Neurol.* **1985,** *17*(5), 497-504.

4) Dietrich,W.D.; Ginsberg, M.D.; Busto, R.; Watson, B.D. *J. Cereb. Blood Flow Metab.* **1986,** *6*, 184-194.

5) *Idem ibid.* **1986,** *6*, 195-202.

6) Watson, B.D.; Prado, R.; Dietrich, W.D.; Ginsberg, M.D.; Green, B. *Brain Res.* **1986,** 367.

7) *Idem J. Cereb. Blood Flow Metab.* **1988,** *8*, 186-192.

8) Lamberts, J.J.M.; Schumacher, D.R.; Neckers, D.C. *J. Am. Chem. Soc.* **1984,** *106*, 5879-5883.

General references:

Aldrich FT-IR Library	**1**(2), 1014B
Sigma FT-IR Library	**1**(2), 211D
IR Library	**3**, 1475D
Safety	**2**(2), 3054D
Merck Index	**11**, 8242

ROSE BENGAL, bis(triethylammonium) salt

Aldrich Cat. No.: 30,527-8
Sigma Prod. No.: R6631

CAS No.: [*91491-51-9*]
Mol. Form.: $C_{32}H_{36}Cl_4I_4N_2O_5$
F.W.: 1176.07
m.p.: 184°C (dec.)
Appearance: Brown crystalline powder
λ_{max}: 559(520)nm in dichloromethane
Solubility: H_2O 3mg/ml
EGME 20mg/ml
EtOH 10mg/ml

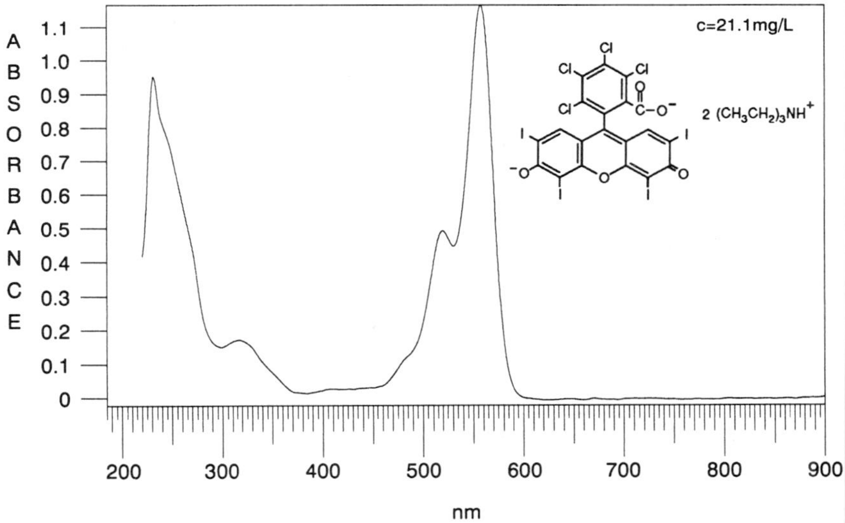

See description on pages 637 and 638 under Rose Bengal.

General reference:

Safety **2**(2), 3055A

ROSOLIC ACID

Aurin

Aldrich Cat. No.:	86,132-4
Sigma Prod. No.:	R9502
CAS No.:	*[603-45-2]*
C.I. No.:	43800
Mol. Form.:	$C_{19}H_{14}O_3$
F.W.:	290.32
Appearance:	Red powder
λ_{max}:	482nm in ethanol + 2ml 1*N* hydrochloric acid
Solubility:	H_2O 9mg/ml
	EGME 60mg/ml
	EtOH 60mg/ml

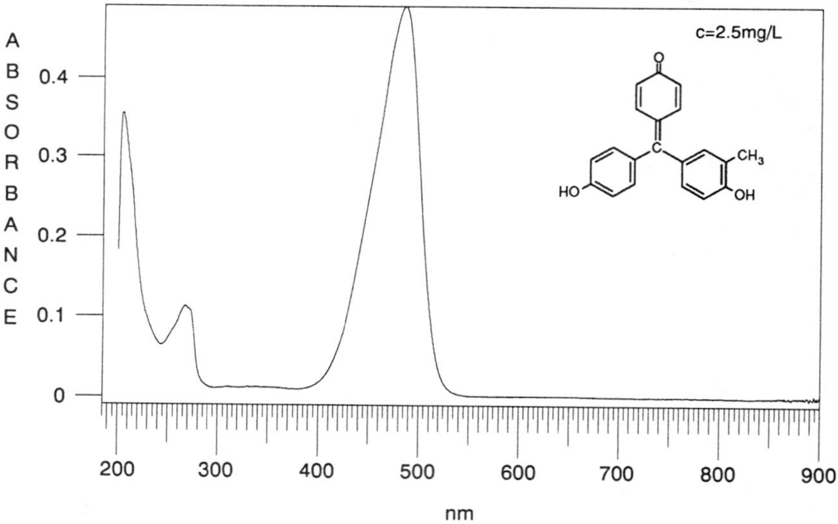

Rosolic Acid is a hydroxytriphenylmethane solvent dye that is made by heating phenol with anhydrous oxalic acid to 120°C in the presence of 100% sulfuric acid. It is soluble in ethanol (golden yellow), moderately soluble in glacial acetic acid and slightly soluble in water, ether and chloroform. The disodium salt is known as Yellow Coralline and is soluble in water (red) and ethanol (magenta-red).

Kolthoff[1] listed Rosolic Acid as a useful indicator for titrations performed in alcoholic solutions. The visual-transition interval is pH 6.6 (yellow) to pH 8.0 (red).

Conn's[2] cites Pappenheim (1898) who used Rosolic Acid in a decolorizing solution following Carbol Fuchsin in a method designed to distinguish between true tubercle bacillus and smegma and other acid-fast bacteria.

References:

1) Kolthoff, I.M. *Acid-Base Indicators;* Macmillan: New York, 1937; p 162.

2) *Conn's Biological Stains,* 9th ed.; Lillie, R.D., Ed.; Williams and Wilkins: Baltimore, 1977; p 303.

General references:

Aldrich FT-IR Library	**1**(2), 1028D
Sigma FT-IR Library	**1**(2), 212A
IR Library	**3**, 1483A
Safety	**2**(2), 3055B
Merck Index	**11**, 899

RUTHENIUM RED

Ammoniated ruthenium oxychloride

Aldrich Cat. No.:	20,173-1
Sigma Prod. No.:	R2751
CAS No.:	[*11103-72-3*]
Mol. Form.:	$H_{42}Cl_6N_{14}O_2Ru_3$
F.W.:	786.36
Appearance:	Red-brown powder
λ_{max}:	534(376)nm in water
Solubility:	H_2O 30mg/ml
	EGME 0.3mg/ml
	EtOH 0.1mg/ml

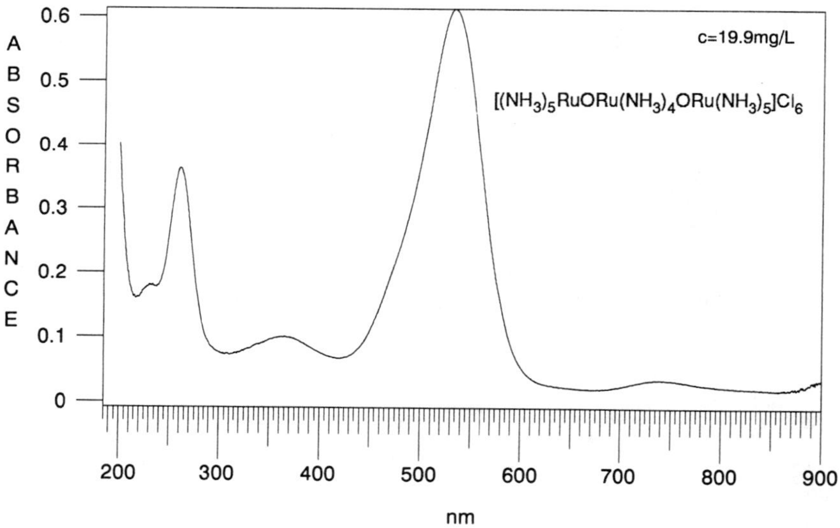

This water-soluble ruthenium salt is used as a reagent for staining pectin, animal tissue, bacteria and gum,[1-6] and as an electron-dense stain for electron-microscopy use, with or without osmic acid.[7]

Fletcher *et al.*[8] described a preparative method.

References:

1) Luft, J.H. *J. Cell Biol.* **1964,** *23*, 54A.
2) *Idem Anat. Rec.* **1965,** *151*, 380.
3) Benedeczky, I.; Smith, A.D. *Histochemie* **1972,** *32,* 213.
4) Burdett, I.D.J.; Rogers, H.J. *J. Ultrastruct. Res.* **1972,** *38*, 113.
5) Nakao, K.; Bashey, R.I. *Exp. Mol. Pathol.* **1972,** *17*, 6.
6) Hanke, D.E.; Northcote, D.H. *Biopolymers* **1975,** *14*, 1.
7) Hyatt, M.A. *Positive Staining for Electron Microscopy;* Van Nostrand: New York, 1975; pp 163-165.
8) Fletcher, J.M.; Greenfield, B.F.; Hardy, C.J.; Scargill, D; Woodhead, J.L. *J. Chem. Soc.* **1961,** 2000.

General references:

Aldrich FT-IR Library **1**(2), 1156B
Sigma FT-IR Library **1**(2), 212B
IR Library **3**, 1568C
Merck Index **11**, 8273

SAFFRON

Natural Yellow 6

Aldrich Cat. No.: 19,961-3
Sigma Prod. No.: S8381

CAS No.: [*42553-65-1*]
C.I. No.: 75100
Appearance: Red and yellow fibers
λ_{max}: 427(452)nm in ethanol
Solubility: H_2O 30mg/ml
EGME 4mg/ml
EtOH 1mg/ml

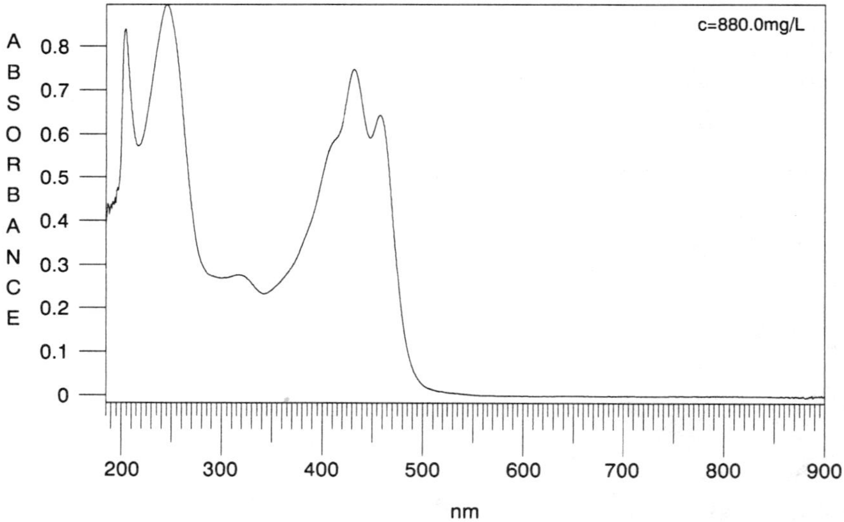

This natural, yellow pigment is present in the dried stigma of the flowers of saffron, *Crocus sativus*, a species of iris indigenous to Spain, Afghanistan and Kashmir. Crocetin (red needles, m.p. 285°C), a dicarboxylated polyene, is the active color present in saffron where it exists as the digentiobiose ester, crotin. Crotin is soluble in water and is readily hydrolyzed by dilute alkali to yield the acid dye crocetin.

Saffron dyes silk and wool a golden yellow directly and tin-mordanted cotton, orange. It has been used on a limited scale as a textile dye in India.

Although the use of Saffron as a biological stain dates back to 1714 and Leeuwenhoek, it has never been used widely by biologists. Masson[1] used it along with Erythrosin and Hematoxylin as a trichrome stain for connective tissue. van Hoeck and Sebruyns[2] used it for staining gastric glandular cells.

References:

1) Masson, P. *C.R. Soc. Biol.* (Paris) **1911,** *70*, 573-574.
2) van Hoeck, R.; Sebruyns, H. *Stain Technol.* **1952,** *27,* 263-265.

General reference:

Merck Index 11,8285

SAFRANIN O (Y, T)

Basic Red 2

Aldrich Cat. No.:	10,214-8
Sigma Prod. No.:	S8884
Grade:	Certified
CAS No.:	*[477-73-6]*
C.I. No.:	50240
Mol. Form.:	$C_{20}H_{19}ClN_4$
F.W.:	350.85
Appearance:	Dark-green powder
λ_{max}:	530nm in 50% ethanol
Solubility:	H_2O 50mg/ml
	EGME 80mg/ml
	EtOH 20mg/ml

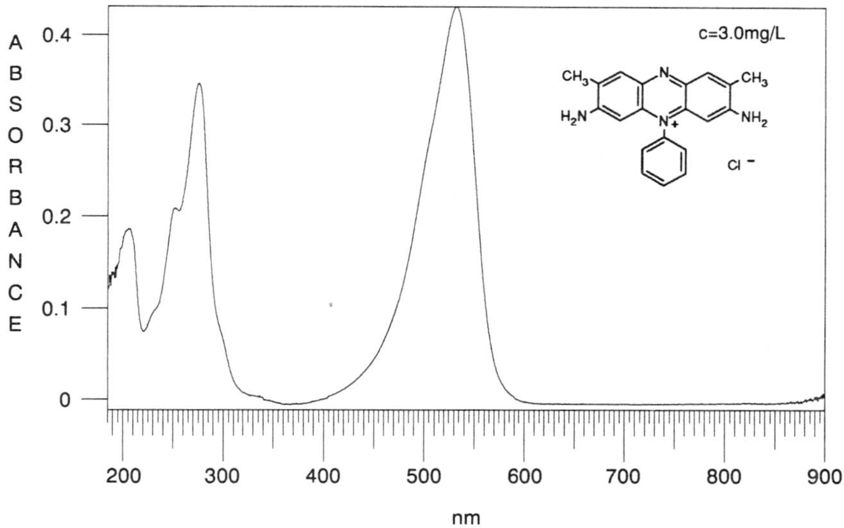

This red, cationic azine dye is soluble in water and methyl Cellosolve and somewhat less soluble in ethanol, ethylene glycol and pyridine. Industrial applications include dyeing cotton, polyacrylonitrile fibers, paper and leather and the preparation of alcoholic inks and pigments. The free base is a solvent dye yielding a dull bordeaux hue in alcohol.

Conn's[1] most recently described Safranin O as a BSC problem dye of long standing due to lot-to-lot color variations. These variations were reported to be particularly noticeable in the Flemming triple-stain technique where Safranin O is used along with Crystal Violet and Orange G in the study of plant mitosis. He stated that reds were sometimes too yellow and other times too blue. It was reported that the variations appeared not to be correlated with either dye concentration or spectral dissimilarity. In fact, the samples studied yielded identical spectra. Subsequently, the BSC[2] found that, by simply adjusting the pH of the working solution to 7.0, consistent results could be attained. Commercial lots of Safranin often display a pH as low as 5.6 when dissolved in deionized water. The GMP (Good Manufacturing Practice) for Safranin O prepared in our laboratory for use as a biological stain includes a pH specification of 7.0-7.5. Materials conforming to this specification obviate the need to adjust the pH of the working solution.

Safranin O long has been considered one of the most important nuclear stains available for histological studies. It is particularly useful to the botanist and bacteriologist. *Staining Procedures*[3] provides methods and a bibliography for the following applications involving Safranin O: use as a counterstain in the silver nitrate method for calcium deposits; as a counterstain in the methenamine method for argentaffin cells; as a counterstain to Iron Hematoxylin for general plant tissue and as a counterstain to Aluminum Hematoxylin Stain for cell walls and nuclei; combined use as Safranin/Fast Green for meristematic tissue and as Safranin/Orange G demonstrating cell walls and protoplasts. This reference also cites Safranin with various counterstains for use on many plant materials: Safranin with Picro Aniline Blue for woody tissue, fungi and bacteria in lignified tissue; in wet and dry procedures for the Gram staining of microorganisms in smears; and in various spore stain methods.

References:

1) *Conn's Biological Stains,* 9th ed.; Lillie, R.D., Ed.; Williams and Wilkins: Baltimore, 1977; p 386.

2) Willis, C.; Biological Stain Commission; personal communication, 1977.

3) Clark, G. *Staining Procedures,* 4th ed.; Williams and Wilkins: Baltimore, 1981; pp 187, 206, 238, 317-321, 324, 326, 329, 355, 372, 377, 390-393.

General references:

Aldrich FT-IR Library	**1**(2), 887B
Sigma FT-IR Library	**1**(2), 212D
IR Library	**3**, 1409B
Safety	**2**(2), 3063B
Colour Index	3rd ed., 1971

SOLVENT BLUE 35

1,4-Bis(butylamino)-9,10-anthracenedione

Aldrich Cat. No.:	30,643-6
Sigma Prod. No.:	S6140
CAS No.:	[*17354-14-2*]
C.I. No.:	61554
Mol. Form.:	$C_{22}H_{26}N_2O_2$
F.W.:	350.46
m.p.:	120-122°C
Appearance:	Purple powder
λ_{max}:	652(604)nm in chloroform
Solubility:	H_2O 6mg/ml
	EGME 20mg/ml
	EtOH 1mg/ml

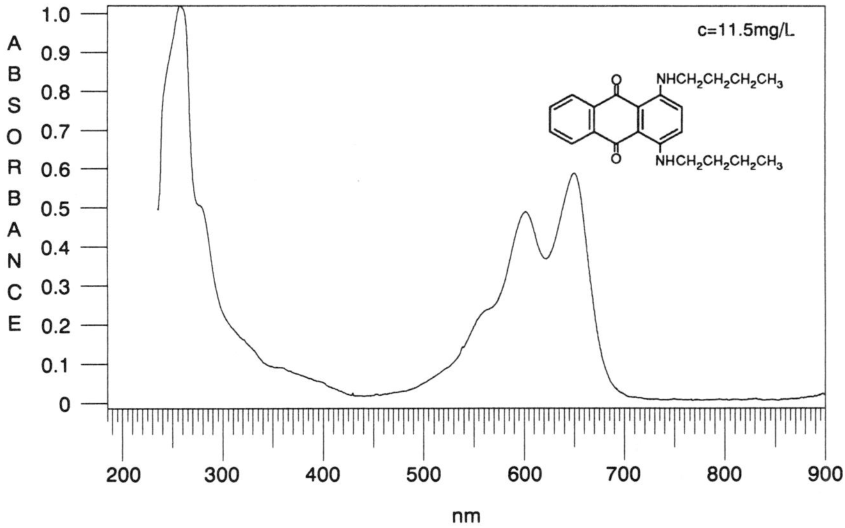

This bright greenish-blue, anthraquinone solvent dye is soluble in acetone (1%), benzene (1.8%), butyl acetate (0.7%), linseed oil (0.62%), mineral oil (0.16%), oleic acid (1.2%), paraffin wax (0.4%), stearic acid (2.5%), petroleum ether (3.0%) and xylene (1.2%). It is also very slightly soluble in ethanol and slightly soluble in water.

Industrially, it is used to color alcoholic, ester and hydrocarbon solvents, oils, fats and waxes and transparent spirit lacquers.

There is no evidence that it has been used as a biological stain.

General reference:

Colour Index 3rd ed., 1971

SOLVENT BLUE 59

1,4-Bis(ethylamino)-9,10-anthracenedione

Aldrich Cat. No.: 22,912-1
Purity: 98%

CAS No.: [*6994-46-3*]
C.I. No.: 61552
Mol. Form.: $C_{18}H_{18}N_2O_2$
F.W.: 294.36
Appearance: Blue powder
λ_{max}: 642(594) in ethanol

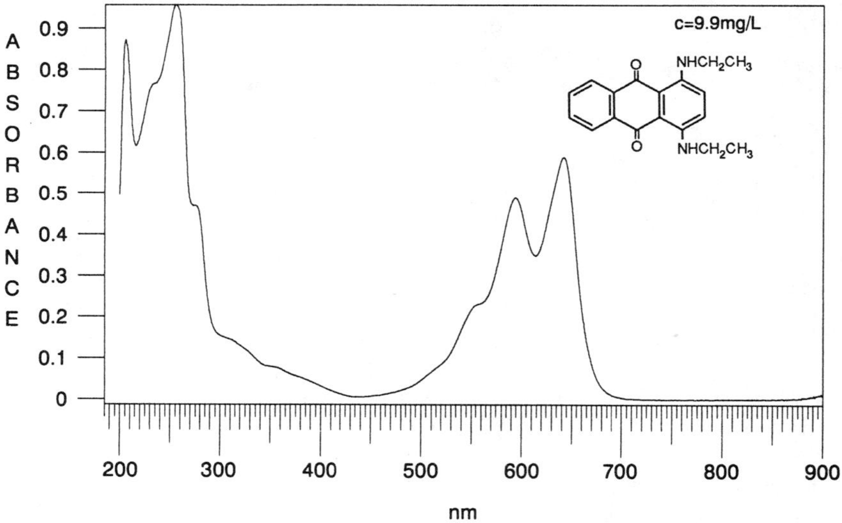

This reddish-blue, anthraquinone solvent dye is soluble in acetone, benzene, methyl Cellosolve, linseed oil, oleic acid, toluene, turpentine, stearic acid and xylene and slightly soluble in ethanol, methanol and paraffin oil.

Industrial uses include coloring hydrocarbon solvents, oils, fats, petroleum products, polystyrene and lacquers.

There is no evidence that it has been used as a biological stain.

General references:

Safety	**2**(2), 3160D
Colour Index	3rd ed., 1971

SOLVENT GREEN 3

Alizarin Cyanine Green, fat soluble

Aldrich Cat. No.: 21,198-2
Sigma Prod. No.: S8007
Grade: FIAT

CAS No.: [*128-80-3*]
C.I. No.: 61565
Mol. Form.: $C_{28}H_{22}N_2O_2$
F.W.: 418.50
m.p.: 220-221°C
Appearance: Dark burgundy
λ_{max}: 644(607)nm in acetone
Solubility: H_2O 0.9mg/ml
EGME 2mg/ml
EtOH 0.3mg/ml

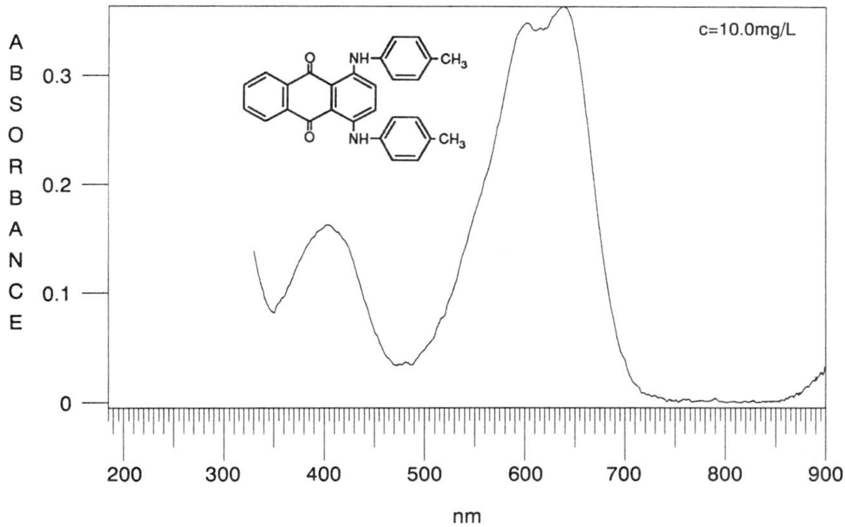

This neutral aminoanthraquinone dye is made by condensing either leucoquinizarin or a mixture of leucoquinizarin and quinizarin with two moles of *p*-toluidine in the presence of boric acid.

Solvent Green 3 is soluble in benzene and toluene, slightly soluble in acetone, oleic acid and stearic acid and very slightly soluble in water, ethanol and paraffin wax.

Industrial applications include coloring fats, oils, waxes, hydrocarbon solvents, hydrocarbon-base wood stains, cellulose acetate, polystyrene and soap, and in the preparation of oleic or stearic acid rotogravure inks.

There is no evidence that it has been used as a biological stain.

General references:

Aldrich FT-IR Library	**1**(2), 1019C
Sigma FT-IR Library	**1**(2), 213C
IR Library	**3**, 1477G
Safety	**2**(2), 3161A
Merck Index	**11**, 8094
Colour Index	3rd ed., 1971

STAINS-ALL

DBTC • 1-Ethyl-2-{3-(1-ethylnaphtho[1,2-*d*]thiazolin-2-ylidene)-2-methylpropenyl}naphtho[1,2-*d*]thiazolium bromide

Aldrich Cat. No.:	23,096-0
Sigma Prod. No.:	E9379
CAS No.:	*[7423-31-6]*
Mol. Form.:	$C_{30}H_{27}BrN_2S_2$
F.W.:	559.60
Appearance:	Dark-grey powder
λ_{max}:	575nm in ethanol
Solubility:	H_2O 1mg/ml
	EGME 5mg/ml
	EtOH 3mg/ml

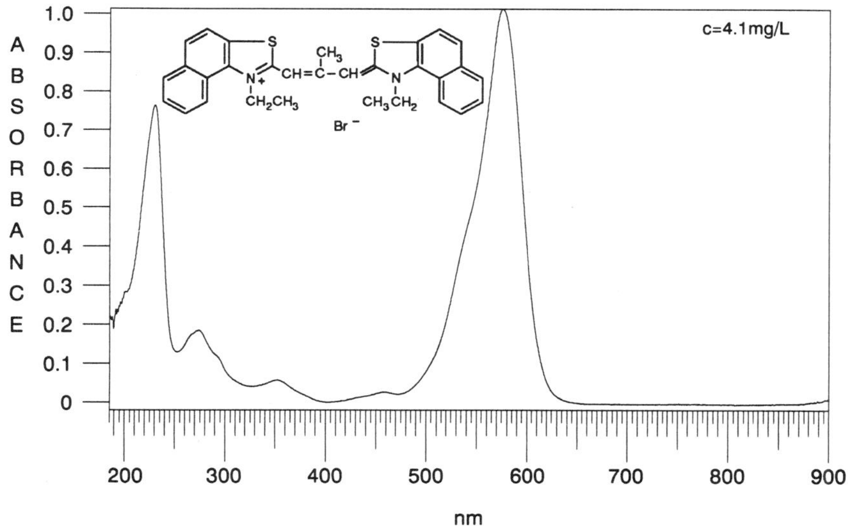

Stains-all, a trivial name proposed by Lillie,[1] is a cationic carbocyanine dye that is very slightly soluble in water. Stock solutions are prepared by making a 0.1% solution in formamide and adjusting the pH to 7.3-7.4 with HCl. This solution should be refrigerated and stored in the dark. It can then be used to prepare a working solution by diluting to 0.01% with the addition of 50% water and 50% buffered formamide.

Stains-all is a sensitive stain for the electrophoretic characterization of nucleic acids on composite agarose-acrylamide gels,[2] hyaluronic acid on polyacrylamide gel[3] and to differentiate phosphoproteins such as casein from nonphosphorylated proteins.[4] The dye forms a blue complex (λ_{max} 675nm) with DNA and a red complex (λ_{max} 510nm) with gelatin.

References:

1) Lillie, R.D. *J. Histochem. Cytochem.* **1974,** *22*, 1169.

2) Dahlberg, A.E.; Dingman, C.W.; Peacock, A.C. *J. Mol. Biol.* **1969,** *41*, 139.

3) Bader, J.P.; Ray, D.A.; Steck, T.L. *Biochim. Biophys. Acta* **1972,** *264,* 73.

4) Green, M.R.; Pastewka, J.V.; Peacock, A.C. *Anal. Biochem.* **1973,** *56*, 43.

General references:

Aldrich FT-IR Library	**1**(2), 1041A
Sigma FT-IR Library	**1**(2), 554A
Safety	**2**(2), 3168C

SUDAN I

1-Phenylazo-2-naphthol • Solvent Yellow 14

Aldrich Cat. No.:	10,362-4
Sigma Prod. No.:	S8506
CAS No.:	[*842-07-9*]
C.I. No.:	12055
Mol. Form.:	$C_{16}H_{12}N_2O$
F.W.:	248.29
m.p.:	131-133°C
Appearance:	Orange powder
λ_{max}:	476(418)nm in ethanol
Solubility:	H_2O 0.5mg/ml
	EGME 20mg/ml
	EtOH 4mg/ml

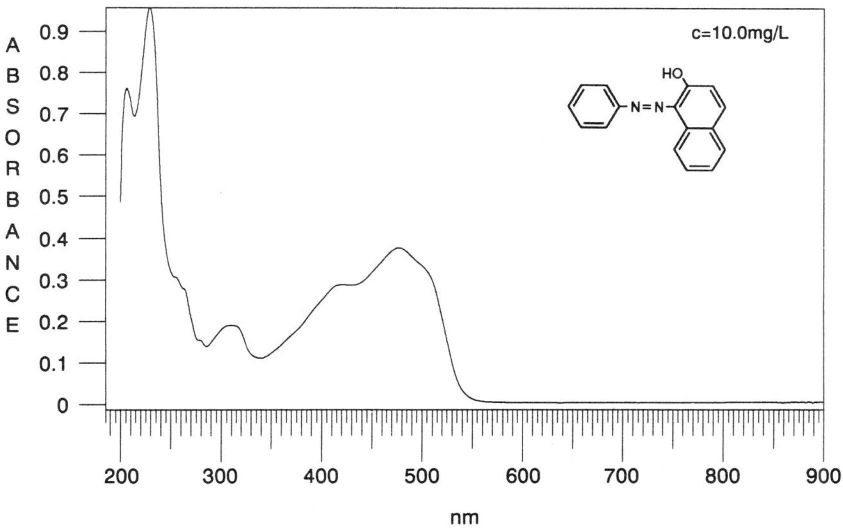

This reddish-yellow, monoazo solvent dye is made by coupling diazotized aniline to 2-naphthol. It is very soluble in stearic acid and toluene, soluble in benzene, xylene, turpentine and butyl acetate and slightly soluble in ethanol, methanol and acetone. Although not very fast to light, it has been used to color hydrocarbon solvents, oils, fats, waxes, shoe and floor waxes, cellulose ester varnishes, styrene resins and soap.

There is no evidence that Sudan I has been used as a biological stain.

General references:

Aldrich FT-IR Library	**1**(2), 969B
Sigma FT-IR Library	**1**(2), 214A
IR Library	**3**,1452B
Safety	**2**(2), 3193C
Colour Index	3rd ed., 1971

SUDAN II

Solvent Orange 7 • Sudan Orange RR

Aldrich Cat. No.:	19,965-6
Sigma Prod. No.:	S8631

CAS No.:	[*3118-97-6*]
C.I. No.:	12140
Mol. Form.:	$C_{18}H_{16}N_2O$
F.W.:	276.34
m.p.:	156-158°C
Appearance:	Brownish-orange powder
λ_{max}:	493(420)nm in methanol
Solubility:	H_2O 8mg/ml
	EGME 10mg/ml
	EtOH 1mg/ml

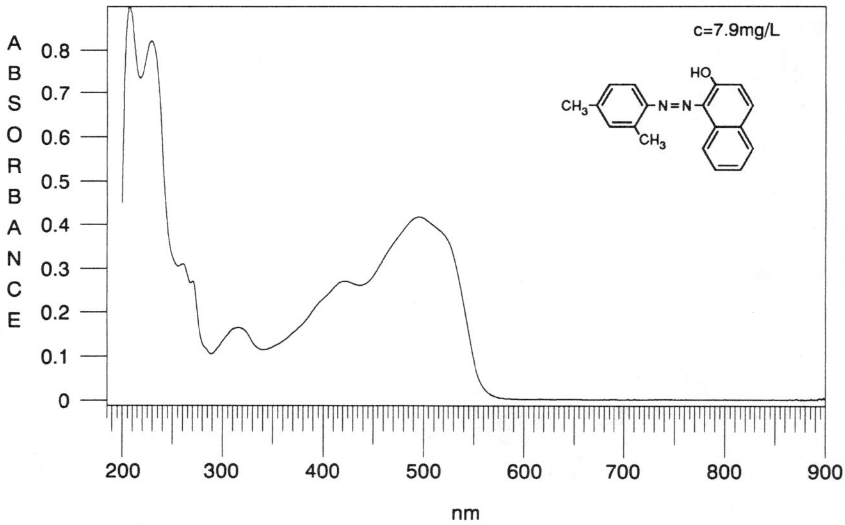

This reddish-orange, monoazo solvent dye is prepared by coupling diazotized 2,4-xylidine with 2-naphthol. It is soluble in turpentine, stearic acid, toluene and xylene, moderately soluble in mineral oil, butyl acetate and benzene, slightly soluble in methyl Cellosolve, acetone and mineral spirits and very slightly soluble in ethanol. Industrially, it has been used to color oil, fats, waxes, polystyrene resins and solvents used to prepare polishes and candles.

Conn's[1] reported that, according to Lillie and Ashburn,[2] Sudan II performed satisfactorily when substituted for Oil Red O in their supersaturated isopropanol technique for myelin and fats, although they found it less stable and more yellowish. (The names Sudan II and Oil Red O were used synonymously in the 1st edition of the *Colour Index* causing some confusion.) Lillie[3] reported it to be a good contrast stain when used with the blue-black Weigert's myelin-sheath stain for demonstrating degenerated myelinated nervous tissue. Lillie's method can be found in *Staining Procedures*.[4]

References:

1) *Conn's Biological Stains,* 9th ed.; Lillie, R.D., Ed.; Williams and Wilkins: Baltimore, 1977; p 93.
2) Lillie, R.D.; Ashburn, L.L. *Arch. Pathol.* **1943,** *36*, 432-435.
3) Lillie, R.D. *Histopathological Technic and Practical Histochemistry*, 3rd ed.; MacGraw-Hill: New York, 1965.
4) Clark, G. *Staining Procedures,* 4th ed.; Williams and Wilkins: Baltimore, 1981; p 107.

General references:

Aldrich FT-IR Library	1(2), 976A
Sigma FT-IR Library	1(2), 214B
IR Library	**3,** 1456F
Safety	**2**(2), 3193D
Colour Index	3rd ed., 1971

SUDAN III
Solvent Red 23

Aldrich Cat. No.:	19,811-0
Sigma Prod. No.:	S4136
Grade:	Certified
CAS No.:	[*85-86-9*]
C.I. No.:	26100
Mol. Form.:	$C_{22}H_{16}N_4O$
F.W.:	352.40
m.p.:	199°C (dec.)
Appearance:	Red-brown powder
λ_{max}:	507(354)nm in toluene
Solubility:	H_2O <0.1mg/ml
	EGME 6mg/ml
	EtOH 2mg/ml

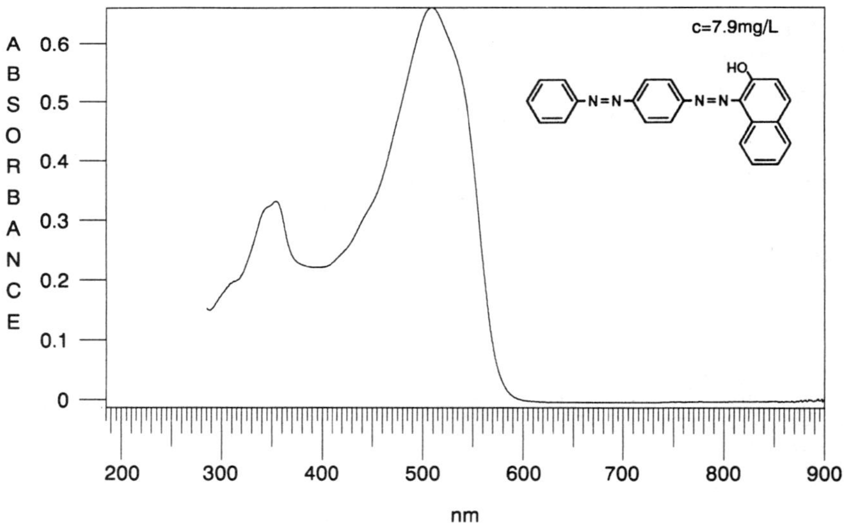

Sudan III is a red, disazo dye that is neither acidic nor basic. It is made by coupling diazotized *p*-phenylazoaniline to 2-naphthol. The *Colour Index* lists it as being very soluble in benzene, soluble in acetone and slightly soluble in ethanol. The same source lists it as a colorant for oils, waxes, hydrocarbons, polishes, cosmetics and occasionally polystyrenes.

Herxheimer,[1] following Daddi's[2] earlier investigation, used it in a comparative study with other fat-soluble dyes. *Staining Procedures*[3] lists it, along with Sudan IV and Oil Red O, in a method designed to show fat in animal tissue.

The use of Sudan III as a fat stain has diminished over the years, first in favor of Sudan IV; more recently, Oil Red O seems to have gained favor over both Sudan III and Sudan IV. Assuming this trend has taken place largely on the basis of comparative staining studies, it is interesting to note that, from a physical chemistry standpoint, Sudan IV should be a better oil colorant than Sudan III, and Oil Red O better than either one of the two. Oil Red O has more hydrocarbon character than Sudan IV which has more of the same than Sudan III. Stated differently, the hydrophobic character of Oil Red O (+8.99 Hansch *pi* value) is greater than Sudan IV (+7.67) which is greater than Sudan III (+6.35). Horobin's[4] study provides more insight into this aspect.

References:

1) Herxheimer, G. *Dtsch. Med. Wochenschr.* **1901,** *27*, 607-609.

2) Daddi, L. *Arch. Ital. Biol.* **1896,** *26*, 143-146.

3) Clark, G. *Staining Procedures,* 4th ed.; Williams and Wilkins: Baltimore, 1981; pp 194-195.

4) Horobin, R.W. *Acta Histochem.*, Suppl.-Band XXIV **1981,** *5*, 237-246.

General references:

Aldrich FT-IR Library	**1**(2), 975B
Sigma FT-IR Library	**1**(2), 214D
IR Library	**3**, 1456C
Safety	**2**(2), 3194A
Merck Index	**11**, 8858
Colour Index	3rd ed., 1971

SUDAN IV

Scarlet Red • Solvent Red 24

Aldrich Cat. No.:	19,810-2
Sigma Prod. No.:	S4261
Grade:	Certified
CAS No.:	*[85-83-6]*
C.I. No.:	26105
Mol. Form.:	$C_{24}H_{20}N_4O$
F.W.:	380.45
m.p.:	199°C (dec.)
Appearance:	Dark red-brown powder
λ_{max}:	520(357)nm in toluene
Solubility:	H_2O 0.7mg/ml
	EGME 10mg/ml
	EtOH 3mg/ml

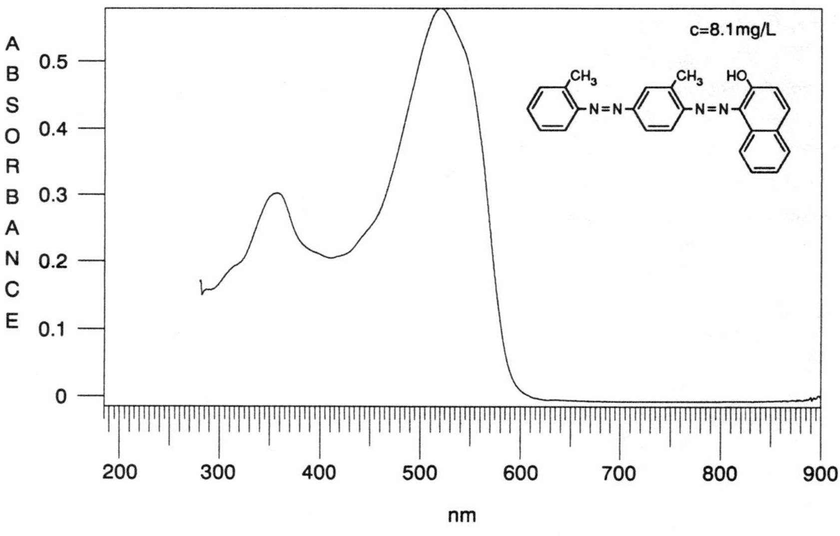

This red, disazo solvent dye has the same basic structure as Sudan III with two added methyl groups. These structure modifications, using Sudan III as a reference, cause the λ_{max} to shift bathochromically 12-15nm and increase the hydrophobicity by +1.37 Hansch *pi* units, with attendant increased hydrocarbon solubility. Sudan IV is prepared by coupling diazotized 4-(*o*-tolylazo)-*o*-toluidine to 2-naphthol. It is very soluble in benzene, soluble in methanol, isopropanol, acetone, hydrocarbon solvents, oils, fats and waxes and slightly soluble in ethanol. Industrially, it has been used to color oils, waxes, shoe polish, greases, oily-resin lacquers, varnishes, cellulose acetate and acrylic resins, and in wood stain.

As a biological stain, it is used primarily to stain lipids and fatty substances present in cells and tissues. The color imparted to these substrates is a function of the solubility of the dye in the substance affected and does not involve chemical bonding. *Staining Procedures*[1] provides procedures and a bibliography for the following methods using Sudan IV: method for staining fat (Herxheimer's technique); fat staining in supersaturated isopropanol performed by Lillie and Fullmer in 1976; and in fat-soluble dyes for plant material.

Reference:

1) Clark, G. *Staining Procedures,* 4th ed.; Williams and Wilkins: Baltimore, 1981; pp 189,194, 332.

General references:

Aldrich FT-IR Library	**1**(2), 976C
Sigma FT-IR Library	**1**(2), 215A
IR Library	**3**,1456H
Safety	**2**(2), 3194B
Merck Index	**11**, 8349
Colour Index	3rd ed., 1971

SUDAN BLACK B

Fat Black HB • Solvent Black 3

Aldrich Cat. No.:	19,966-4
Sigma Prod. No.:	S2380
Grade:	Certified
CAS No.:	[*4197-25-5*]
C.I. No.:	26150
Mol. Form.:	$C_{29}H_{24}N_6$
F.W.:	456.55
m.p.:	120-124°C
Appearance:	Dark-brown powder
λ_{max}:	598(415)nm in ethanol
Solubility:	H_2O 0.1mg/ml
	EGME 50mg/ml
	EtOH 30mg/ml

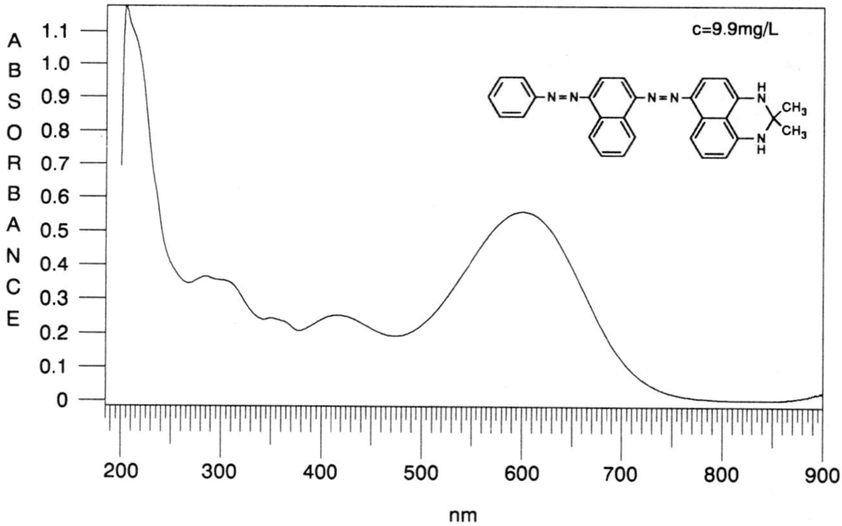

Sudan Black B is a neutral disazo dye that is soluble in ethanol, acetone, benzene, toluene, hydrocarbon solvents, fats, oils and paraffin. It is made by first coupling diazotized aniline to 1-naphthylamine. The product of this reaction is then diazotized and coupled to 2,3-dihydro-2,2-dimethylperimidine. Industrially, it is used to color alcoholic, ester and hydrocarbon solvents, in printing inks, in cellulose lacquers and in oils, fats and waxes.

Sudan Black B is an excellent fat stain for animal tissue, bacteria and other microorganisms, leucocyte granules and Golgi apparatus. *Staining*

Procedures[1] provides methods for the following: the Lillie and Fullmer (1976) method for demonstrating fat in animal tissue, the Lillie (1965) method for demonstrating fat droplets and lipid-rich structural components of tissue, the Burdon (1946) stain for demonstrating sudanophilic granules and the Hartman (1940) wet-mount demonstration of sudanophilic granules.

Reference:

1) Clark, G. *Staining Procedures,* 4th ed.; Williams and Wilkins: Baltimore, 1981; pp 193, 229, 393, 394.

General references:

Aldrich FT-IR Library	**1**(2), 991C
Sigma FT-IR Library	**1**(2), 215B
IR Library	**3**, 1464F
Safety	**2**(2), 3194C
Colour Index	3rd ed., 1971

SUDAN ORANGE G

4-(Phenylazo)resorcinol • Solvent Orange 1

Aldrich Cat. No.: 19,967-2
Sigma Prod. No.: S8881

CAS No.: [*2051-85-6*]
C.I. No.: 11920
Mol. Form.: $C_{12}H_{10}N_2O_2$
F.W.: 214.22
m.p.: 143-146°C
Appearance: Reddish-orange powder
λ_{max}: 400nm in methanol
Solubility: H_2O 0.7mg/ml
EGME 20mg/ml
EtOH 10mg/ml

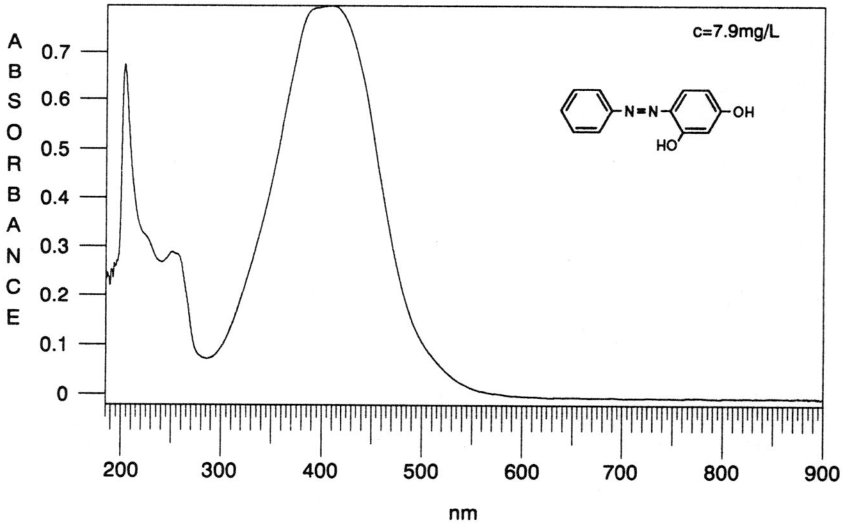

This orange, solvent monoazo dye is prepared by a carefully controlled coupling of diazotized aniline with resorcinol. The currently available product which, according to the *Colour Index,* is formed by the coupling of a second mole of aniline to the monoazo product, is essentially free of the disazo impurity.

Sudan Orange G is very slightly soluble in water, slightly soluble in ethanol and soluble in ether, oils, fats, waxes and alkaline solutions. Industrially, it is used to color fats, waxes, polishes, shoe creams and phenoplastic molding powder.

Sudan Orange G is not often used as a biological stain. Herxheimer[1] used it, in a comparative study with a number of other dyes, as a fat stain.

Reference:

1) Herxheimer, K. *Dtsch. Med. Wochenschr.* **1901,** *27*, 607-609.

General references:

Aldrich FT-IR Library	**1**(2), 965C
Sigma FT-IR Library	**1**(2), 215C
IR Library	**3**, 1450B
Colour Index	3rd ed., 1971

SUDAN RED 7B

Solvent Red 19

Aldrich Cat. No.:	20,161-8
Sigma Prod. No.:	F1000
CAS No.:	[*6368-72-5*]
C.I. No.:	26050
Mol. Form.:	$C_{24}H_{21}N_5$
F.W.:	379.47
m.p.:	130°C (dec.)
Appearance:	Dark-red powder
λ_{max}:	533(364)nm in toluene
Solubility:	H_2O 0.7mg/ml
	EGME 10mg/ml
	EtOH 3mg/ml

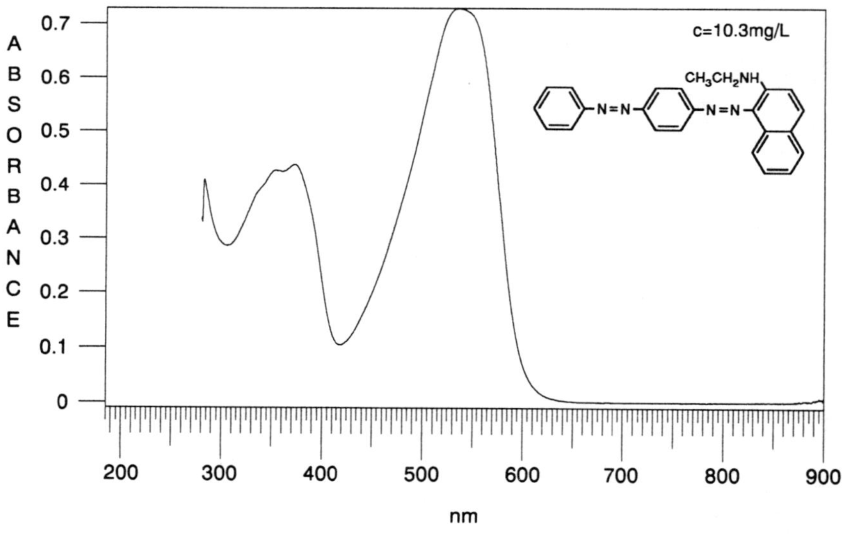

This cationic, disazo solvent dye is made by coupling diazotized 4-phenylazoaniline to *N*-ethyl-2-naphthylamine. It is virtually insoluble in water, slightly soluble in ethanol and soluble in acetone and benzene. Industrially, it is used as a colorant for alcohols, esters, hydrocarbon solvents, waxes, oils, fats, mineral oil and candles, for the preparation of double-tone printing inks and synthetic resin lacquers and in the mass coloring of cellulose.

Sudan Red 7B has been employed as a lipid stain. Pearse [1] considered Sudan Red 7B (Fettrot 7B, Ciba) superior to the Sudans and Oil Red O. He observed that Fettrot 7B has no tendency to crystallize when used from alcoholic solutions, and mounted specimens retain their clarity over a long period of time. Barrolier and Suchowsky[2] used Sudan Red 7B to stain fats in frozen tissue sections.

References:

1) Pearse, A.G.E. *Histochemistry;* Little, Brown: Boston, 1954; p 445.

2) Barrolier, J.; Suchowsky, G. *Acta Histochem.* **1958,** *5,* 294-296

General references:

Aldrich FT-IR Library	**1**(2), 975A
Sigma FT-IR Library	**1**(2), 146A
IR Library	**3**, 1456B
Safety	**2**(2), 3194D
Colour Index	3rd ed., 1971

SULFOBROMOPHTHALEIN sodium hydrate

Bromsulfalein • Bromthalein

Aldrich Cat. No.:	16,720-7
Sigma Prod. No.:	S0252
CAS No.:	[*123359-42-2*]
Mol. Form.:	$C_{20}H_8Br_4Na_2O_{10}S_2 \cdot xH_2O$
F.W.:	838.03
Appearance:	White powder
λ_{max}:	577(366)nm in 0.1*N* sodium hydroxide
Solubility:	H_2O 20mg/ml
	EGME 200mg/ml
	EtOH 3mg/ml

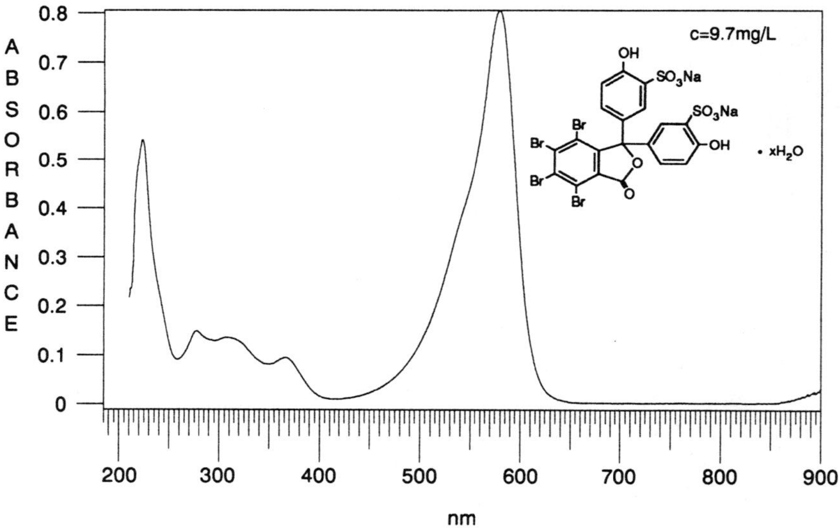

This hydroxytriarylmethane dye is soluble in water and very slightly soluble in ethanol and acetone. It is made by condensing two moles of phenol with one of tetrabromophthalic anhydride, followed by sulfonation and conversion to the disodium salt. There is no evidence that it has been used industrially or as a biological stain. However, a highly purified grade has been used as a diagnostic aid for hepatic-function testing. *The USP XX*[1] describes specifications and test methods for this grade.

Reference:

1) *The United States Pharmacopeia,* twentieth version; United States Pharmacopeial Convention; Mack: Easton, 1979; p 755.

General references:

Aldrich FT-IR Library	**1**(2), 1025A
Sigma FT-IR Library	**1**(2), 215D
IR Library	**3**, 1480H
Merck Index	**11**, 8933

SULFONAZO III, tetrasodium salt

3,6-Bis(2-sulfophenylazo)-4,5-dihydroxy-2,7-naphthalenedisulfonic acid, tetrasodium salt

Aldrich Cat. No.: 10,215-6
Sigma Prod. No.: S6381

CAS No.: [*68504-35-8*]
Mol. Form.: $C_{22}H_{12}N_4Na_4O_{14}S_4$
F.W.: 776.57
Appearance: Dark-green powder or flake
λ_{max}: 567(629)nm in water
Solubility: H_2O 40mg/ml
EGME 30mg/ml
EtOH 2mg/ml

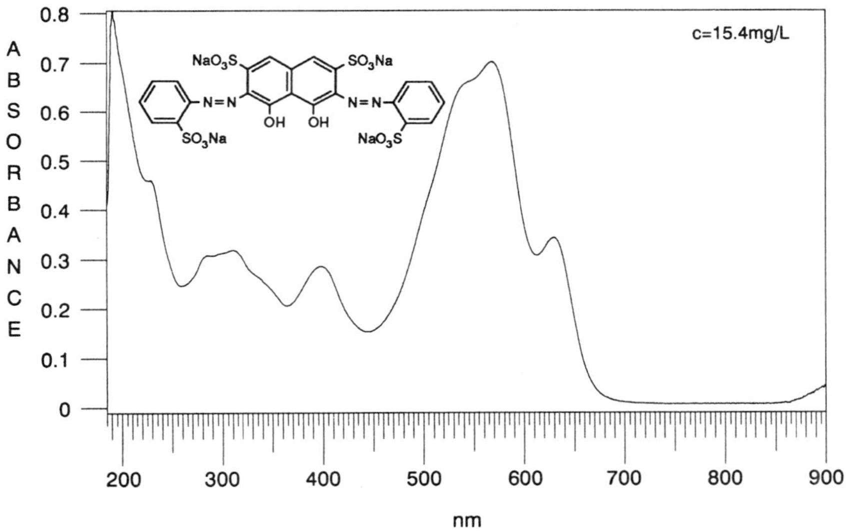

Sulfonazo III is an anionic disazo dye that is soluble in water, imparting a wine-red color (free-acid form). The solubility of the sodium salt listed above may vary from that of the free acid, but they can be used interchangeably and the information contained in this monograph applies to both forms. The barium complex formed with this dye is red with an absorption maximum at 640nm and a molar absorptivity of 11.2×10^4 L mole^{-1}cm^{-1}. The absorptivity of the calcium complex is negligible.

Sulfonazo III is not listed in the *Colour Index* and apparently has not been used industrially. However, it is a useful analytical reagent that Budesinsky[1] used as an indicator for the titrimetric determination of sulfates and sulfur (color change at the end-point, red to blue). The same author described it as an indicator for the volumetric determination of potassium with tetraphenylboron sodium (Aldrich Cat. No. T2,540-2) and benzalkonium chloride (Aldrich Cat. No. 23,442-7).

Reference:

1) Budesinsky, B. *Anal. Chem.* **1965,** *37*, 1159.

General references:

Aldrich FT-IR Library	**1**(2), 986C
Sigma FT-IR Library	**1**(2), 216B
IR Library	**3,** 1462C
Colour Index	3rd ed., 1971

2-(4-SULFOPHENYLAZO)-1,8-DIHYDROXY-3,6-NAPHTHALENE-DISULFONIC ACID, trisodium salt
SPADNS

Aldrich Cat. No.: 11,475-8
Sigma Prod. No.: S9256
Grade: Indicator

CAS No.: [*23647-14-5*]
Mol. Form.: $C_{16}H_9N_2Na_3O_{11}S_3$
F.W.: 570.41
Appearance: Red powder
λ_{max}: 507(370)nm in water
Solubility: H_2O 10mg/ml
EGME 3mg/ml
EtOH 1mg/ml

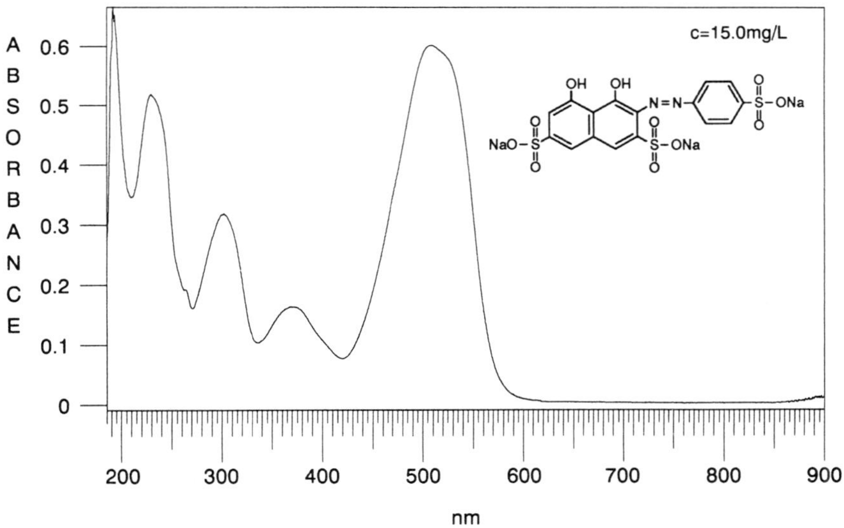

SPADNS is an anionic monoazo dye that is somewhat soluble in water (red solution) and only very slightly soluble in alcohol and other organic solvents, forming a blue-violet complex with thorium, a pink complex with zirconium, and a red-violet complex with rare-earth metals. Their absorption maxima all fall in the range of 500-550nm. There is no evidence that this dye has been used industrially, although it has varied applications as a laboratory reagent, being used as a colorimetric reagent for beryllium, thorium, zirconium, rare-earth metals and fluoride ion. It is especially valuable for the determination of fluoride ion in biological samples (*i.e.*, serum, urine, bone, teeth and the juice of fruits and vegetables).[1-9]

References:

1) Benergee, G. *Z. Anal. Chem.* **1955,** *146*, 147.
2) *Idem ibid.* **1955,** *147*, 105.
3) *Idem ibid.* **1955,** *148*, 349.
4) Benergee, G. *Anal. Chim. Acta* **1955,** *13*, 409.
5) *Idem ibid.* **1958,** *16*, 56.
6) Hollingsworth, R.P. *Anal. Chem.* **1957,** *29*, 1130.
7) Bellack, E.; Schouboe, P.J. *ibid.* **1958,** *30*, 203.
8) Peck, L.C.; Smith, V.C. *Talanta* **1964,** *11*, 1343.
9) Munshi, K.N.; Dey, A.K. *Microchem. J.* **1964,** *8*, 152.

General references:

Aldrich FT-IR Library **1**(2), 510B
Sigma FT-IR Library **1**(2), 216C
IR Library **3,** 1165C
NMR Library **2**(2), 830B

SULFORHODAMINE 101 hydrate

Aldrich Cat. No.: 28,491-2
Sigma Prod. No.: S7635
Grade: Suitable as a laser dye

CAS No.: [*123333-78-8*]
Mol. Form.: $C_{31}H_{30}N_2O_7S_2 \cdot xH_2O$
F.W.: 606.72
Appearance: Green powder
λ_{max}: 576nm in ethanol
Solubility: H_2O 20mg/ml
EGME 9mg/ml
EtOH 30mg/ml

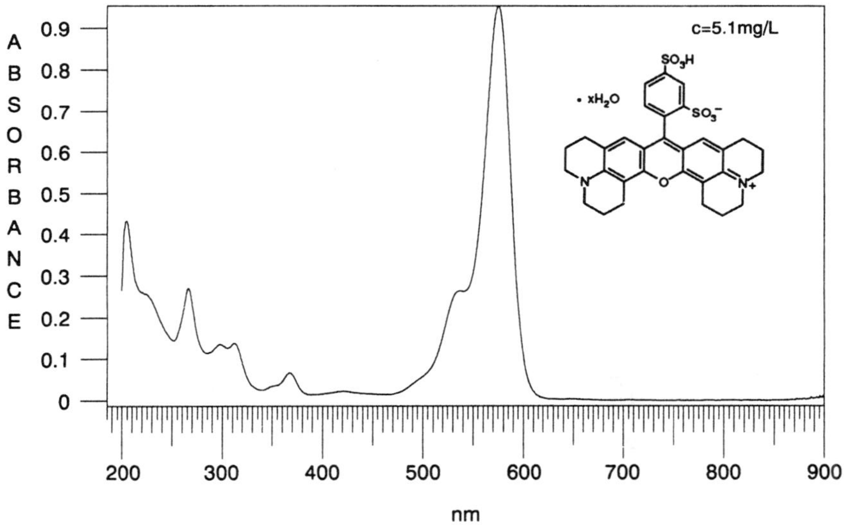

This red rhodamine dye is soluble in water and ethanol.

Essentially free of impurities as demonstrated by TLC and with a dye content of 99%, it is suitable as an efficient laser dye for both pulsed- and continuous-wave operation. The optimum lasing wavelength (in ethanol) is 648nm and the fluorescence λ_{max} (in ethanol) 602nm.

General references:

Aldrich FT-IR Library	**1**(2), 217A
Safety	**2**(2), 3202D

SULFORHODAMINE 101 acid chloride

Texas Red®

Aldrich Cat. No.: 28,406-8
Sigma Prod. No.: S5385

CAS No.: [*82354-19-6*]
Mol. Form.: $C_{31}H_{29}ClN_2O_6S_2$
F.W.: 625.16
Appearance: Black powder
λ_{max}: 578nm in ethanol

®Reg. trademark of Molecular Probes

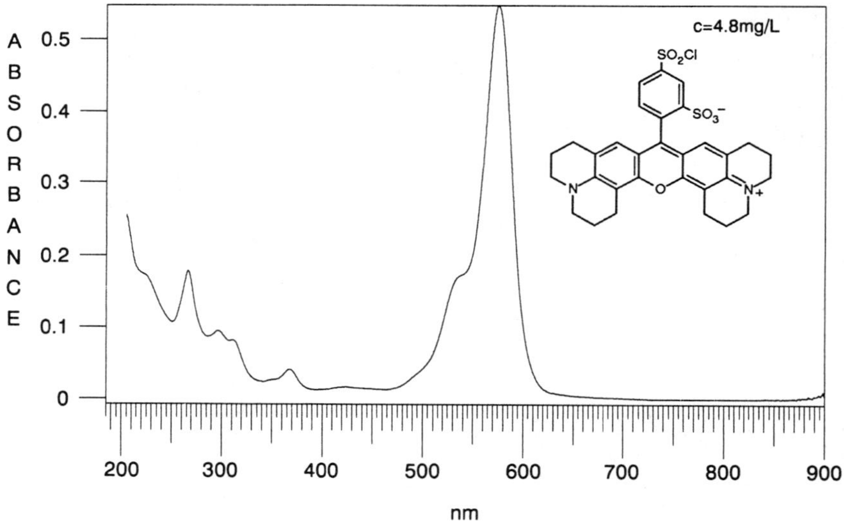

This red, fluorescent, reactive rhodamine dye is soluble in ethanol, acetone and water (the latter with hydrolysis).

Titus *et al.*,[1] after preparing this red-emitting fluorophore, utilized it to covalently bond to primary amino functions in proteins and other biological substrates. Since it is readily distinguished from fluorescein isothiocyanate (FITC), the authors reasoned it could be used as a stain in single- and double-label flow-microfluorometric and fluorescent microscopy investigations. This apparently stimulated interest by other workers and a flurry of papers appeared in 1984. Loken and Lanier[2,3] expanded to a three-color immunofluorescence system that utilized fluorescein and Phycoerythrin conjugates, in addition to those formed with the subject dye, to study lucine antigens. A dual-laser, fluorescence-activated cell sorter was employed with excitation sources from a dye laser, Rhodamine 6G (600nm), and an argon laser (488nm).

Hale and McCarthy[4] studied the effects of sublethal ionizing radiation on rat lymphocytes, employing much the same approach. Weichel[5] and Szejda *et al.*[6] reported studies utilizing Sulforhodamine 101 sulfonyl chloride conjugates and laser-flow cytometry.

References:

1) Titus, J.A.; Haugland, R.; Sharrow, S.O.; Segal, D.M. *J. Immunol. Methods* **1982,** *50*(2), 193.

2) Loken, M.R.; Lanier, L.L. *J. Cytometry* **1984,** *5*(2), 151.

3) Lanier, L.L.; Loken, M.R. *J. Immunol.* **1984,** *132*(1), 151.

4) Hale, M.L.; McCarthy, K.F. *Radiat. Res.* **1984,** *99*(1), 151.

5) Weichel, W. *Cytometry* **1984,** *6*(2), 116.

6) Szejda, P.; Parce, J.W.; Seeds, M.S.; Bass, D.A. *J. Immunol.* **1984,** *133*(6), 3303.

General references:

Sigma FT-IR Library	**1**(2), 217B
Safety	**2**(2), 3203A

SULFORHODAMINE B
Acid Red 52

Aldrich Cat. No.:	23,016-2
Sigma Prod. No.:	S9012
CAS No.:	[*3520-42-1*]
C.I. No.:	45100
Mol. Form.:	$C_{27}H_{29}N_2NaO_7S_2$
F.W.:	580.66
Appearance:	Red-brown powder
λ_{max}:	554nm in 0.1*N* sodium hydroxide in methanol
Solubility:	H_2O 20mg/ml
	EGME 20mg/ml
	EtOH 5mg/ml

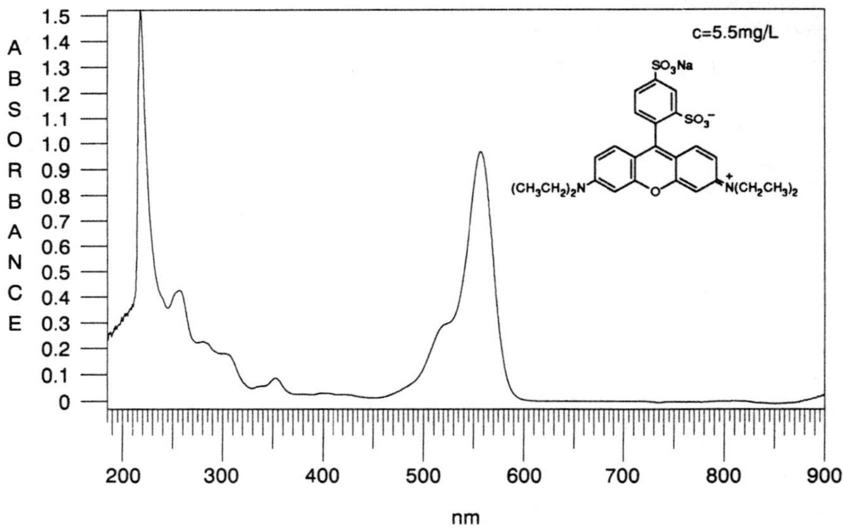

This bright, bluish-red, fluorescent, acid rosamine dye is soluble in water and slightly soluble in ethanol. It is prepared by condensing 4-formyl-1,3-benzenedisulfonic acid with *m*-diethylaminophenol, dehydrating the condensation product with sulfuric acid, oxidizing the leuco dye to the colored form with ferric chloride and converting this product to the sodium salt.

Fugitive to light, Sulforhodamine B is used rather sparingly as a dye for wool, silk and nylon and for printing wool and silk directly.

Conn's,[1] citing Chadwick *et al.*,[2] states that it is useful as a protein tag, since one of its sulfonic acid functions will readily condense with a proteinaceous substrate containing a primary amine function. Displaying a red fluorescence, it provides the necessary contrast to another fluorescent tag, the yellowish-green fluorescein isothiocyanate, so that they can be used for the simultaneous tagging of two proteins.

References:

1) *Conn's Biological Stains,* 9th ed.; Lillie, R.D., Ed.; Williams and Wilkins: Baltimore, 1977; p 330.

2) Chadwick, C.S.; McEntegart, M.G.; Nairn, R.C. *Lancet* **1958,** *1*, 412-414.

General references:

Aldrich FT-IR Library	1(2), 1037A
Sigma FT-IR Library	1(2), 216D
Colour Index	3rd ed., 1971

SULFORHODAMINE G
Acid Red 50

Aldrich Cat. No.:	23,056-1
CAS No.:	[*5873-16-5*]
C.I. No.:	45220
Mol. Form.:	$C_{25}H_{25}N_2NaO_7S_2$
F.W.:	552.61
Appearance:	Red powder
λ_{max}:	529nm in methanol
Solubility:	H_2O 20mg/ml
	EGME 4mg/ml
	EtOH 2mg/ml

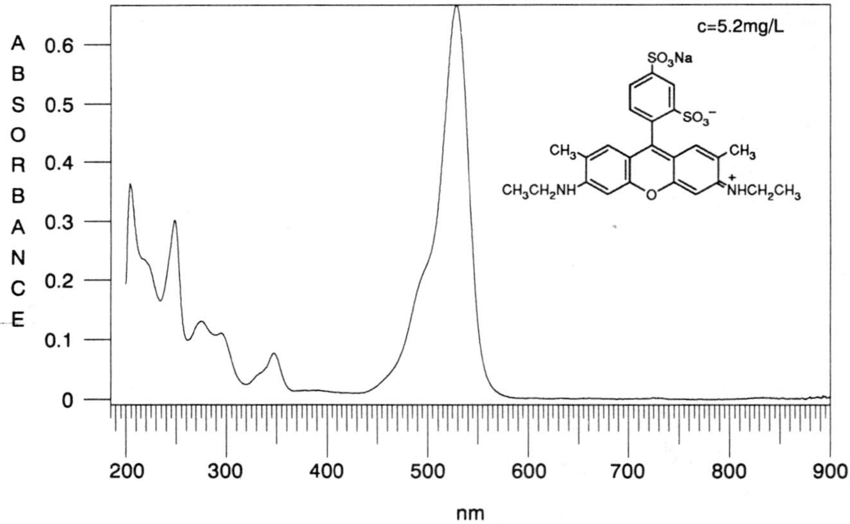

Sulforhodamine G (SG) is closely related to Sulforhodamine B (see pages 674 and 675), possessing essentially the same solubilities and reactivity. However, there is a hypsochromic shift of 25nm in both aqueous and alcoholic solutions of SG. These solutions are bright yellowish-red in transmitted light with a yellowish fluorescence. This is in contrast to comparable solutions of Sulforhodamine B which are bright bluish-red with a yellowish fluorescence.

Sulforhodamine G is prepared by condensing 4-formyl-1,3-benzenedisulfonic acid with 3-ethylamino-*p*-cresol, dehydrating the product with sulfuric acid, oxidizing with ferric chloride and converting this product to the sodium salt.

Industrial uses include dyeing wool, silk and paper and printing wool, silk and viscose rayon directly.

While no biological stain applications have been reported, it appears that this dye could be used as a protein tag in much the same manner as Chadwick *et al.*[1] used Sulforhodamine B.

Reference:

1) Chadwick, C.S.; McEntegart, M.G.; Nairn, R.C. *Lancet* **1958**, *1*, 412-414.

General references:

Aldrich FT-IR Library **1**(2), 1037B

Colour Index 3rd ed., 1971

TARTRAZINE

Acid Yellow 23

Aldrich Cat. No.:	20,195-2
Sigma Prod. No.:	T0388
CAS No.:	[*1934-21-0*]
C.I. No.:	19140
Mol. Form.:	$C_{16}H_9N_4Na_3O_9S_2$
F.W.:	534.37
Appearance:	Orange powder
λ_{max}:	425nm in water
Solubility:	H_2O 300mg/ml
	EGME 20mg/ml
	EtOH 0.8mg/ml

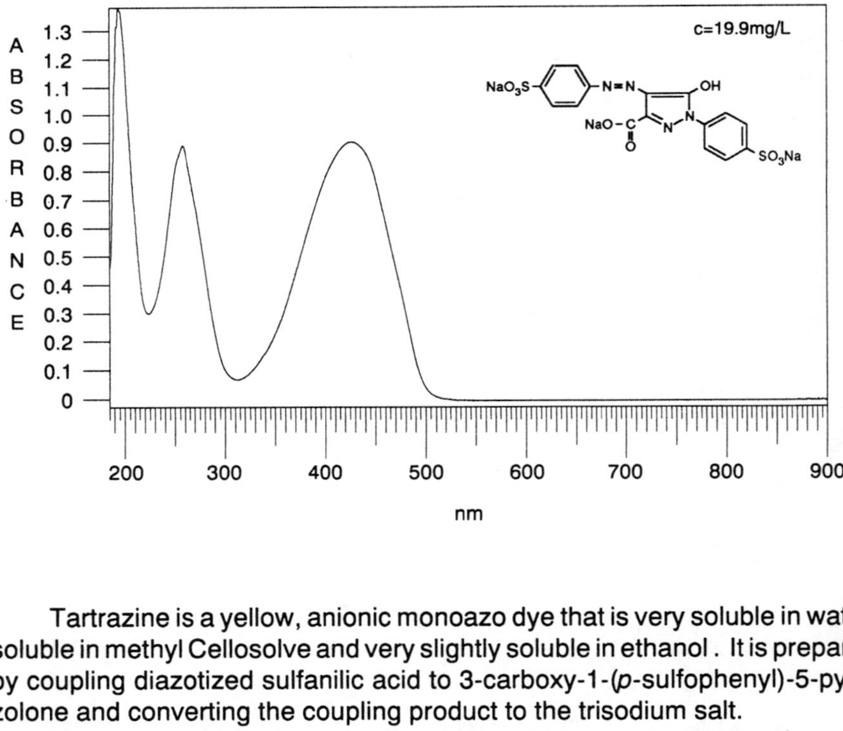

Tartrazine is a yellow, anionic monoazo dye that is very soluble in water, soluble in methyl Cellosolve and very slightly soluble in ethanol. It is prepared by coupling diazotized sulfanilic acid to 3-carboxy-1-(*p*-sulfophenyl)-5-pyrazolone and converting the coupling product to the trisodium salt.

Industrial uses include dyeing wool, silk, nylon, leather and paper, preparing writing inks and wood stains and coloring anodized aluminum, soap and casein plastics. A highly purified form (FD&C Yellow No. 5) is used as a food dye.

Tartrazine is used as a contrast stain that sometimes is substituted for Orange G in animal histology. Lendrum[1] used it, along with Phloxine and Hemalum, in a staining system designed to reveal cellular inclusion bodies.

Reference:

1) Lendrum, A.C. *J. Pathol. Bacteriol.* **1947,** *59*, 399-404.

General references:

Aldrich FT-IR Library	**1**(2), 1001A
Sigma FT-IR Library	**1**(2), 217C
IR Library	**3**, 1468G
Safety	**2**(2), 3214B
Merck Index	**11**, 9041
Colour Index	3rd ed., 1971

TETRABROMOPHENOL BLUE

Aldrich Cat. No.:	19,931-1
Sigma Prod. No.:	T1638
CAS No.:	[*4430-25-5*]
Mol. Form.:	$C_{19}H_6Br_8O_5S$
F.W.:	985.59
m.p.:	204°C (dec.)
Appearance:	Off-white to pale-lavender powder
λ_{max}:	610nm in 50% ethanol + 5ml pH 5.0 buffer
Solubility:	H_2O 3mg/ml
	EGME 50mg/ml
	EtOH 20mg/ml

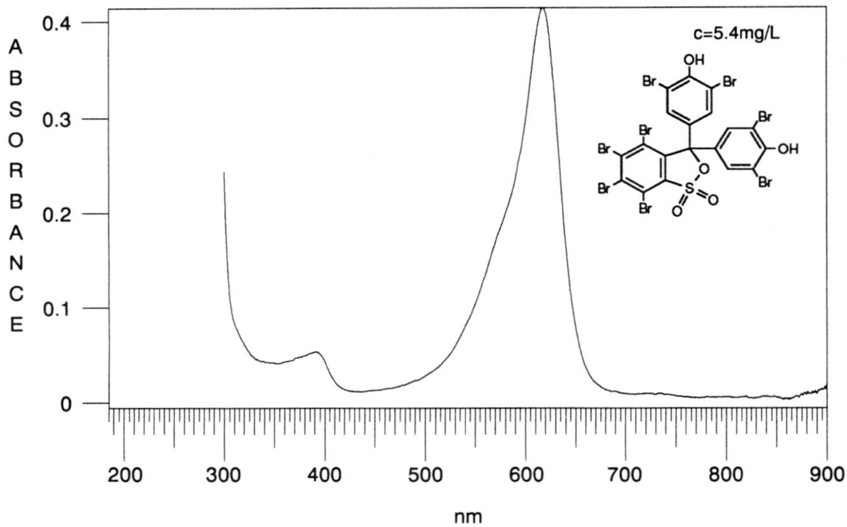

Tetrabromophenol Blue (TBPB) is a sulfonephthalein indicator that is made by condensing two moles of phenol with one mole of tetrabromo-2-sulfobenzoic acid cyclic anhydride (Aldrich Cat. No. 22,737-4) and tetrabrominating the condensation product. This sultone form (TBPB) of the indicator is slightly soluble in water, acetone and acetic acid and soluble in ethanol, methanol and alkali-hydroxide solutions, while the sodium salt (TBPB-WS) is soluble in water and slightly soluble in ethanol and methanol. Harden and Drake[1,2] and Long[3] reported a visual-transition interval of pH 3.0 (yellow) to pH 4.6 (blue), the same as displayed by Bromophenol Blue. Harden and Drake[1] observed that TBPB was free of dichromatism, giving a sharper color change with purer colors than Bromophenol Blue. Indicator test solutions can be be prepared as follows: (1) dissolveTBPB (0.01% w/v) in methanol; or (2) dissolveTBPB-WS (0.01% w/v) in deionized water.

TBPB is also used to estimate protein in biological systems.

References:

1) Harden, W.C.; Drake, N.L. *J. Am. Chem. Soc.* **1929,** *51,* 562-566.

2) *Idem ibid.* **1930,** *52,* 4954-4958.

3) Long, C. *Biochemist's Handbook*; Van Nostrand: Princeton, 1961.

General references:

Aldrich FT-IR Library **1**(2), 1025C

Sigma FT-IR Library **1**(2), 217D

IR Library **3**, 1481A

TETRABROMOPHENOL BLUE, water soluble

Aldrich Cat. No.:	23,604-7
Sigma Prod. No.:	T2518
Mol. Form.:	$C_{19}H_5Br_8NaO_5S$
F.W.:	1007.58
Appearance:	Dark-brown crystals
λ_{max}:	610nm in water
Solubility:	H_2O 50mg/ml
	EGME 10mg/ml
	EtOH 6mg/ml

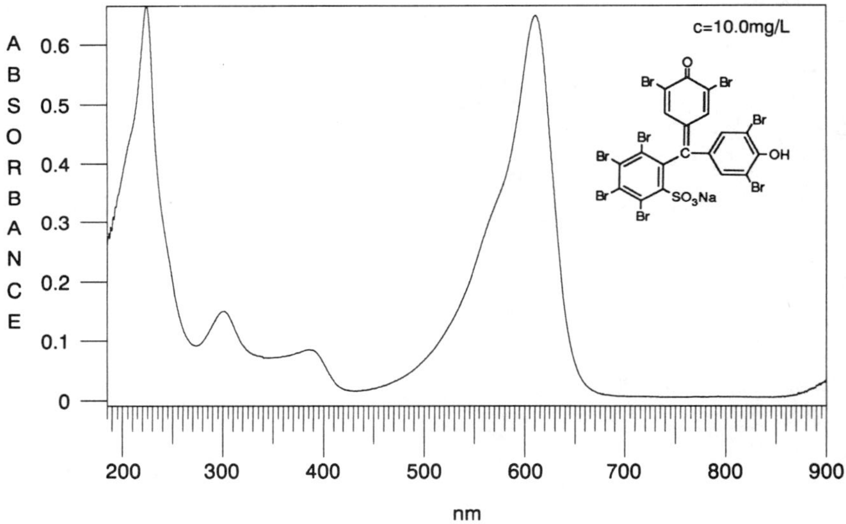

For description of Tetrabromophenol Blue, water soluble (TBPB-WS), see pages 680 and 681 under Tetrabromophenol Blue.

General references:

Aldrich FT-IR Library	1(2), 1024B
Sigma FT-IR Library	1(2), 218A

3',3",5',5"-TETRABROMOPHENOLPHTHALEIN

Aldrich Cat. No.: 23,416-8

CAS No.: [*1301-20-8*]
Mol. Form.: $C_{20}H_{10}Br_4O_4$
F.W.: 633.94
m.p.: 278-279°C (dec.)
Appearance: White powder
λ_{max}: 310nm in 0.01*N* sodium hydroxide
Solubility: H_2O <0.1mg/ml
EGME 20mg/ml
EtOH 20mg/ml

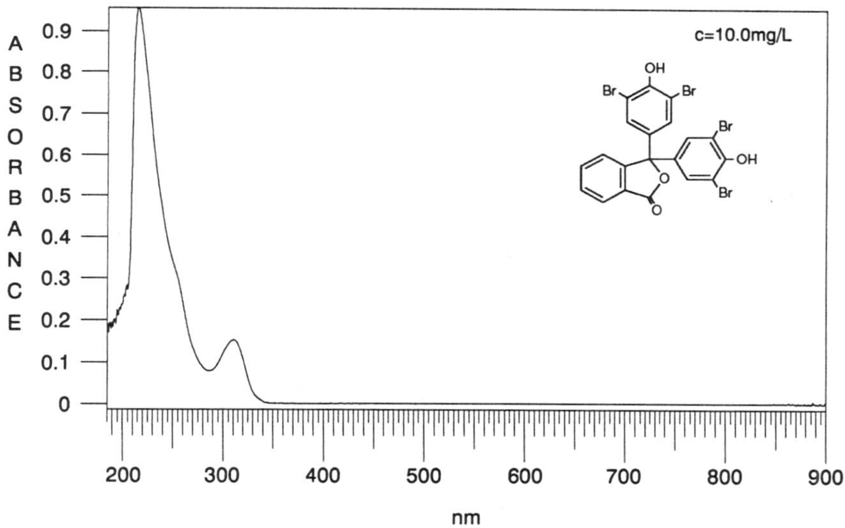

This phenolphthalein derivative is prepared by the addition of a slight excess of bromine to an ethanolic solution of phenolphthalein at room temperature.

Soluble in ether and alkali (violet), moderately soluble in alcohol and glacial acetic acid and insoluble in water, this indicator is listed by Kolthoff[1] as having a visual-transition interval of pH 7.6 (colorless) to pH 9.4 (violet). *The Merck Index*[2] describes the disodium salt (water soluble) as a diagnostic aid (radiopaque medium).

References:

1) Kolthoff, I.M. *Acid-Base Indicators;* Macmillan: New York, 1937; p 115.
2) *The Merck Index,* 11th ed.; Budavari, S. *et al.*, Eds.; Merck: Rahway, 1989; Monograph 9122.

General reference:
Aldrich FT-IR Library 1(2), 1008B

TETRABROMOPHENOLPHTHALEIN ETHYL ESTER, potassium salt

Aldrich Cat. No.:	11,477-4
Sigma Cat. No.:	T1889
Grade:	Indicator
CAS No.:	*[62637-91-6]*
Mol. Form.:	$C_{22}H_{13}Br_4KO_4$
F.W.:	700.08
m.p.:	270-276°C (dec.)
Appearance:	Purple powder
λ_{max}:	593(386)nm in methanol
Solubility:	H_2O 10mg/ml
	EGME 10mg/ml
	EtOH 8mg/ml

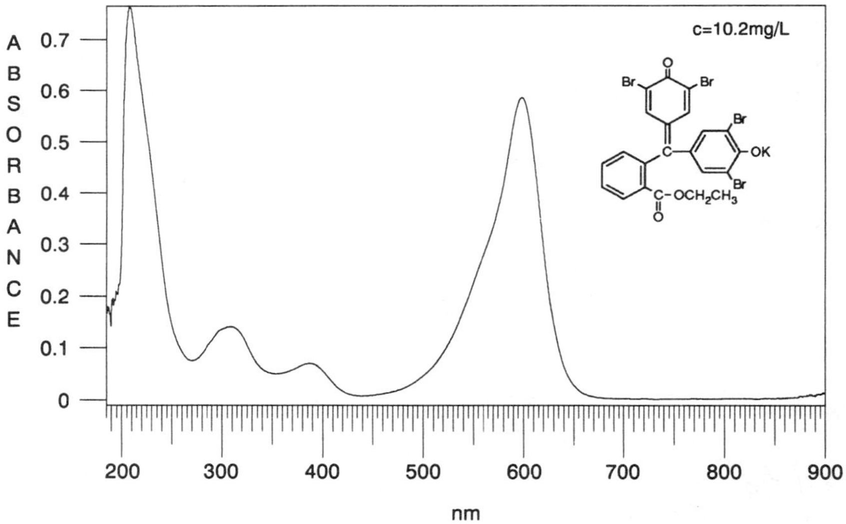

This phthalein indicator is prepared by condensing two moles of phenol with one mole of phthalic anhydride followed by tetrabromination, ethyl esterification and conversion to the potassium salt. It is slightly soluble in water and somewhat less soluble in ethanol and methanol. The visual-transition interval is pH 3.0 (yellow) to pH 4.2 (blue).

General references:

Aldrich FT-IR Library	**1**(2), 1008C
Sigma FT-IR Library	**1**(2), 218B
IR Library	**3**, 1472B

3,4,5,6-TETRABROMOPHENOLSULFONEPHTHALEIN

Aldrich Cat. No.: 22,735-8

Mol. Form.: $C_{19}H_{10}Br_4O_5S$
F.W.: 669.98
Appearance: Bright-red powder
λ_{max}: 426nm in methanol
Solubility: H_2O 30mg/ml
EGME 30mg/ml
EtOH 40mg/ml

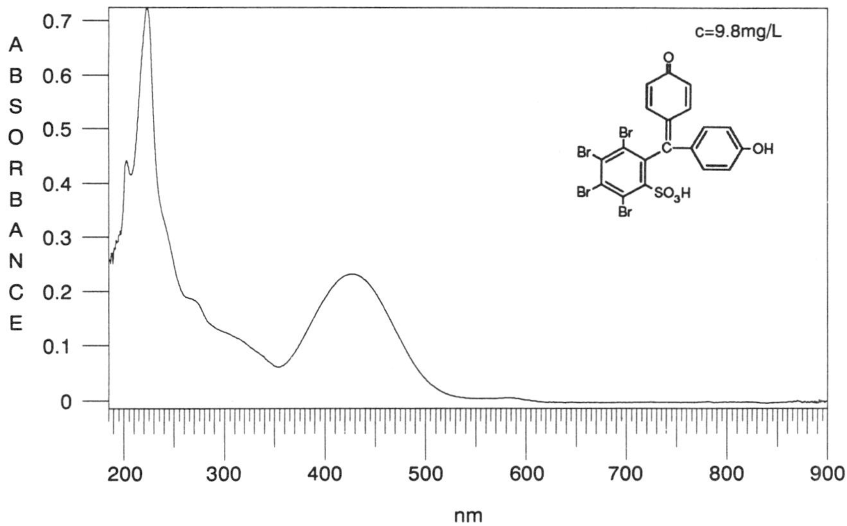

This sulfonephthalein indicator is made by condensing two moles of phenol with one mole of tetrabromo-2-sulfobenzoic acid cyclic anhydride (Aldrich Cat. No. 22,737-4).

It is slightly soluble in acetone and acetic acid and soluble in ethanol, methanol and alkali hydroxide solution. Kolthoff[1] reported its visual-transition interval to be from pH 7.2 (brown-yellow) to 8.0 (purple). Harden and Drake[2] listed it as pH 6.6 (yellow) to 8.2 (purple). Our laboratory agrees with the latter.

There is no evidence that it has been used as a biological stain.

References:

1) Kolthoff, I.M. *Acid-Base Indicators;* Macmillan: New York, 1937; p 133.

2) Harden, W.C.; Drake, N.L. *J. Am. Chem. Soc.* **1929,** *51*, 565.

General references:

Aldrich FT-IR Library **1**(2), 1008A
Safety **2**(2), 3233C

4,5,6,7-TETRACHLOROFLUORESCEIN

Aldrich Cat. No.: 25,005-8

CAS No.: [*6262-21-1*]
Mol. Form.: $C_{20}H_8Cl_4O_5$
F.W.: 470.10
Appearance: Orange powder
λ_{max}: 518nm in methanol
Solubility: H_2O 30mg/ml
EGME 30mg/ml
EtOH 10mg/ml

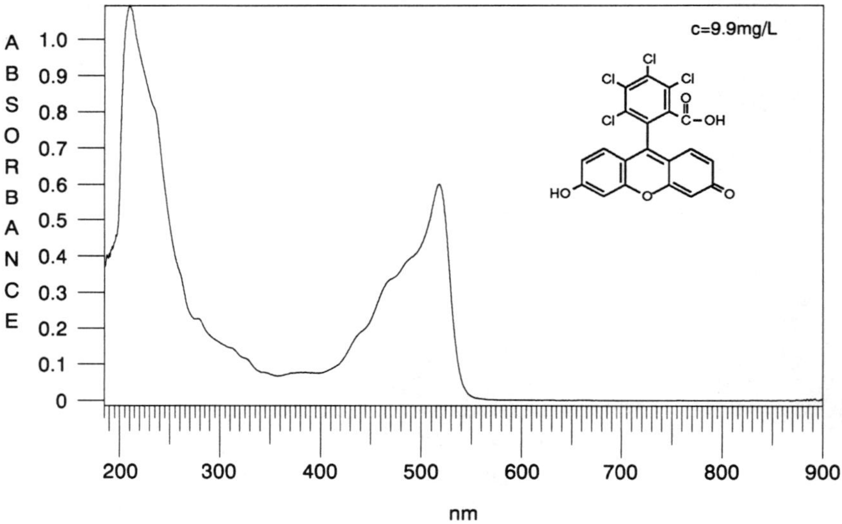

This yellowish-red, fluorescent (yellow), anionic xanthene dye is made by condensing one mole of tetrachlorophthalic anhydride with two moles of resorcinol in the presence of a condensing agent, such as zinc chloride.

Soluble in water and alkali and slightly soluble in ethanol, it is used commercially, primarily as an intermediate for the production of Phloxine B (C.I. 45410).

Tampion *et al.*[1] used the disodium salt as a fluorescent vital stain for demonstrating plant membranes.

Reference:

1) Tampion, J.; Wright, P.; Holt, J. *Stain Technol.* **1973,** *48*, 41-42.

General reference:

Aldrich FT-IR Library **1**(2), 1012D

TETRACHROME STAIN (MacNeal)

Aldrich Cat. No.: 20,019-0
Sigma Prod. No.: T2013

CAS No.: [*81142-52-1*]
Appearance: Green powder
λ_{max}: 640(524)nm in methanol
Solubility: H_2O 10mg/ml
EGME 30mg/ml
EtOH 4mg/ml

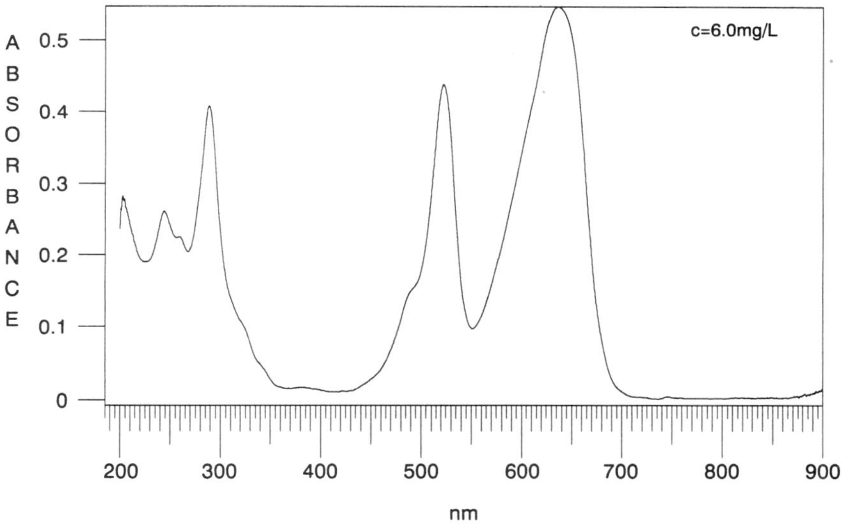

MacNeal,[1] in an effort to make a superior, less complex blood stain, blended certain thiazine dyes normally used to prepare Romanowsky-type blood stains with Eosin Y, all in their free state. Apparently, he intended to provide an option to the more costly and variable products, such as Wright and Leishman stains, made by blending various thiazine eosinate complexes.

MacNeal's recipe was considered to be a free-component blend of Methylene Blue chloride, Azure A, Methylene Violet (Bernthsen)[2] and Eosin. *Conn's*[3] provides the following formula as the one normally used for the blend: Methylene Blue chloride (about 90%), 1.0g; Azure A, 0.6g; Methylene Violet (Bernthsen)[2] free base, 0.2g; Eosin Y (80-85%), 1.0g. The yield of 2.8g of powder is the approximate amount needed to prepare 1 liter of stock solution.

Preparing a stable stock solution of Tetrachrome Stain is not accomplished as readily as one would predict, armed with the knowledge that all of the components of the blend are separately readily soluble in methanol. In practice, solution of the blend initially appears to be easily attained. However, with aging (24-48 hours), a crystalline precipitate usually forms, the absorbance drops 10-15%, the ratio of Eosin to thiazines changes and the staining quality declines. This occurs even when vacuum-dried powder and anhydrous methanol are used. *Conn's*[3] states that manufacturers recommended overcoming this problem by keeping the solution at 50° for 48 hours before filtering. It is true that most of the deterioration has occurred at this point and delayed filtration provides a solution free of precipitate and reasonably stabilized. Unfortunately, the damage has already been done. The suggestion to use a solvent system composed of 40% glycerol in methanol, similar to the Giemsa system, has merit and more recent studies have proposed other solvent variations. The International Committee for Standardization in Haematology (ICSH)[4] recommends a solvent system composed of 40% dimethyl sulfoxide in methanol as part of their recipe for a Romanowsky variation that uses pure, dye-free components (see the Azure B monograph on pages 112-114). We have had encouraging results using a solvent system containing 30% ethoxyethanol in methanol for a similar dye blend containing pure Azure B chloride and Eosin Y lactone. After 24 months, test solutions have shown no sign of deterioration.

Wittekind[5] summarizes his work with pure dyes in free-component blends used to stain blood films. He concludes that Azure B is superior to Azure A in Romanowsky-type stains and that the inclusion of Methylene Blue is not necessary to obtain a balanced stain. He further states that Methylene Violet (Bernthsen) is worthless as a component of Romanowsky-type blood stains.

References:

1) MacNeal, W.J. *J. Am. Med. Assoc.* **1922,** *78,* 1122.

2) Bernthsen, A. *Ber. Dtsch. Chem. Ges.* **1906,** *39*(II), 1804-1809.

3) *Conn's Biological Stains,* 9th ed.; Lillie, R.D., Ed.; Williams and Wilkins: Baltimore, 1977; pp 496-498.

4) International Committee for Standardization in Haematology (ICSH) *Brit. J. Haematol.* **1984,** *57*, 707-710.

5) Wittekind, D. *Clin. Lab. Haematol.* **1979,** *1*, 247-262.

General references:

Aldrich FT-IR Library **1**(2), 1043D

Sigma FT-IR Library **1**(2), 218C

IR Library **3**, 1489C

3',3",5',5"-TETRAIODOPHENOLPHTHALEIN

Iodophene • Nosophen

Aldrich Cat. No.:	22,498-7
Grade:	Indicator
CAS No.:	[*386-17-4*]
Mol. Form.:	$C_{20}H_{10}I_4O_4$
F.W.:	821.92
m.p.:	268°C (dec.)
Appearance:	Off-white to beige powder
λ_{max}:	315nm in 0.1*N* sodium hydroxide
Solubility:	H_2O 0.7mg/ml
	EGME 20mg/ml
	EtOH 5mg/ml

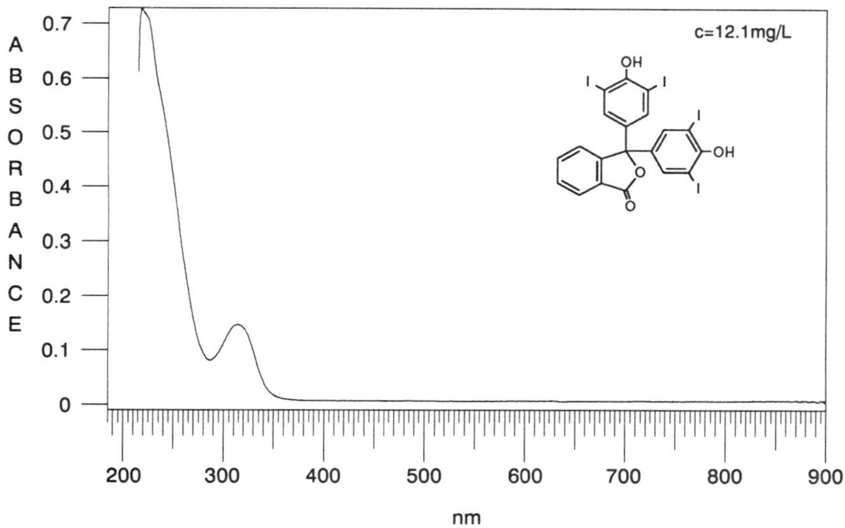

This tetraiodo derivative of phenolphthalein was prepared by Orndorff and Mahood[1] by the action of iodine (in a potassium iodide solution) on phenolphthalein in an aqueous alkali solution. The indicator grade (Aldrich Cat. No. 22,498-7) is practically insoluble in water, slightly soluble in ethanol and soluble in alkalis, ether and chloroform, while the water-soluble form (the disodium salt, Aldrich Cat. No. 22,497-9) is soluble in alcohol and completely soluble in water. The water-soluble form is hygroscopic and sensitive to decomposition by carbon dioxide uptake; therefore, it must be stored under an inert atmosphere in a tightly closed container.

While both forms of this indicator are sensitive to a change in pH (changing color), apparently the change lacks sharpness and is too fugitive for practical use.

Merck[2] reports that both forms of this indicator are used as a diagnostic aid in certain X-ray examination techniques.

References:

1) Orndorff, W.R.; Mahood, S.A. *J. Am. Chem. Soc.* **1918,** *40*, 937.

2) *The Merck Index,* 11th ed.; Budavari, S. *et al.*, Ed.; Merck: Rahway, 1989; Monographs 4931.

General reference:

IR Library **3**, 1472C

3',3",5',5"-TETRAIODOPHENOLPHTHALEIN, disodium salt

Aldrich Cat. No.:	22,497-9
Grade:	Water soluble
CAS No.:	[*632-73-5*]
Mol. Form.:	$C_{20}H_8I_4Na_2O_4$
F.W.:	865.88
Appearance:	Shiny, dark-blue powder
Solubility:	H_2O 60mg/ml
	EGME 10mg/ml
	EtOH 20mg/ml

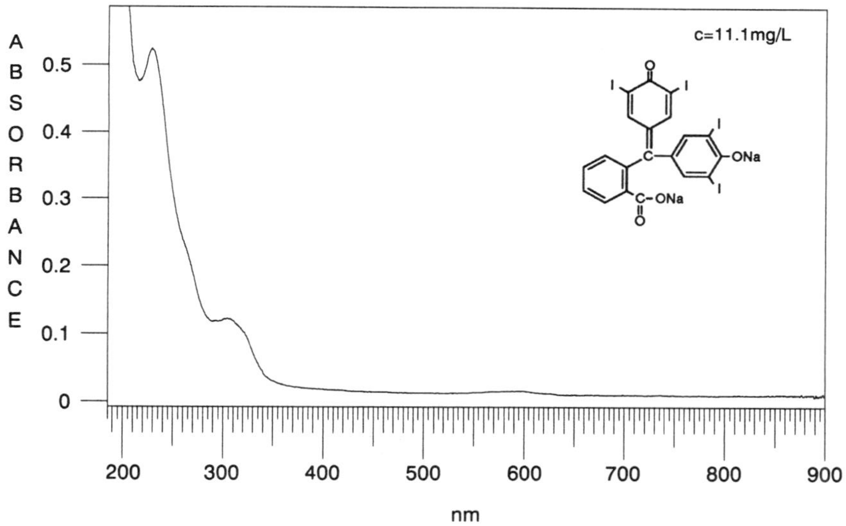

See description on pages 689 and 690 under 3',3",5',5"-Tetraiodophenolphthalein.

General reference:

Aldrich FT-IR Library 1(2), 1008D

3',3",5',5"-TETRAIODOPHENOLSULFONEPHTHALEIN

Aldrich Cat. No.:	23,097-9
Grade:	Indicator
CAS No.:	*[4430-24-4]*
Mol. Form.:	$C_{19}H_{10}I_4O_5S$
F.W.:	857.97
m.p.:	229°C (dec.)
Appearance:	Brown powder
λ_{max}:	433nm in methanol
Solubility:	H_2O 8mg/ml
	EGME 40mg/ml
	EtOH 40mg/ml

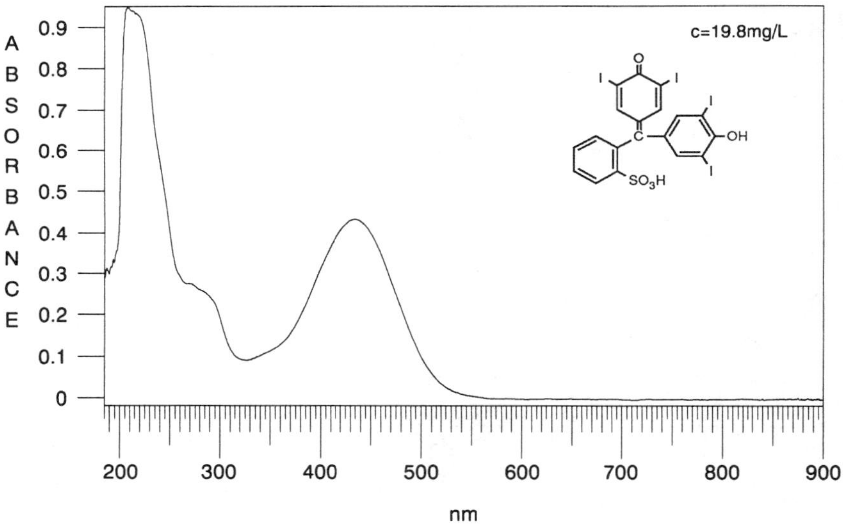

This sulfonephthalein indicator is slightly soluble in water and soluble in methanol, ethanol, acetone, ether and acetic acid. Sensitive to pH change, it has a visual-transition interval of pH 3.0 (yellow) to 4.8 (blue).[1]

There is no evidence in the literature that this indicator has been used in any industrial application.

Reference:

1) Long, C. *Biochemist's Handbook;* Van Nostrand: Princeton, 1961.

General reference:

Aldrich FT-IR Library **1**(2), 1026A

3,3',5,5'-TETRAMETHYLBENZIDINE
TMB

Aldrich Cat. No.:	86,033-6
Sigma Prod. No.:	T2885
Purity:	99+%

CAS No.:	*[54827-17-7]*
Mol. Form.:	$C_{16}H_{20}N_2$
F.W.:	240.35
m.p.:	168-169°C
Appearance:	White powder
λ_{max}:	288nm in ethanol
Solubility:	H_2O 2mg/ml
	EGME 7mg/ml
	EtOH 10mg/ml

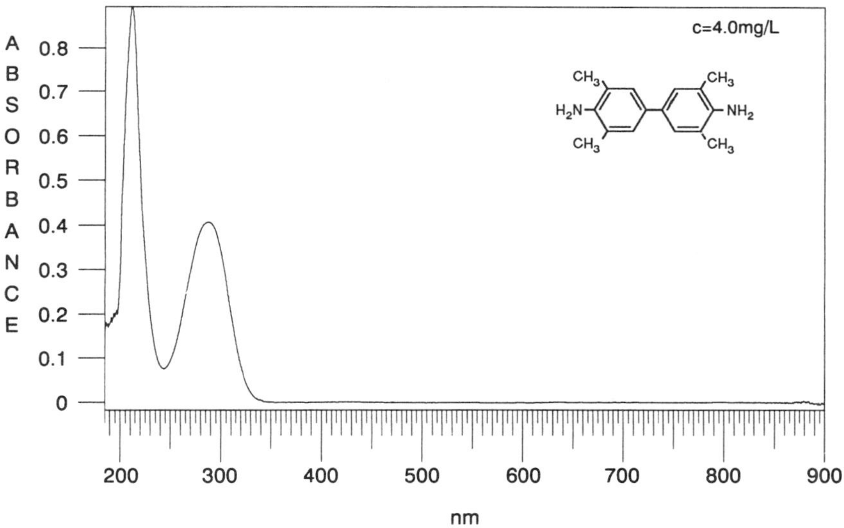

For many years, benzidine was used as a sensitive and specific chromogenic reagent for the detection of blood.[1] However, in 1974 the Occupational Safety and Health Administration restricted its manufacture and use in the United States because of its extreme carcinogenicity.[2]

One of the early hypotheses for the carcinogenicity of aromatic amines involved *ortho* hydroxylation.[3] Thus, it seemed that 3,3',5,5'-tetramethylbenzidine (TMB), in which *ortho* hydroxylation is impossible, might be an effective and safe substitute for benzidine.[4] Indeed, Holland *et al.*[5] reported that the subcutaneous injection of TMB into rats "produced no tumors specifically attributable to it, in doses greater than those which benzidine or *o*-

tolidine cause a high yield of neoplasms." The *Salmonella*/microsome test (Ames Test)[5] showed TMB to be nonmutagenic,[6-8] suggesting that it is noncarcinogenic.

Garner *et al.*[9] evaluated TMB as a presumptive test for blood in forensic work. In various concentrations of glacial acetic acid, TMB reacted with blood in the presence of H_2O_2 to form a colored product. Comparative studies with benzidine showed TMB to be equally sensitive in blood detection. Standefer and Vanderjagt[10] found that methods employing TMB for plasma-hemoglobin assay compared well in accuracy, precision and sensitivity with those using benzidine. Iron porphyrins, which can exhibit a peroxidase-like action, were detected at very low levels on paper chromatograms with TMB as a visualizing agent.[11]

TMB is only very slightly soluble in water, while its dihydrochloride salt is slowly soluble in water. The former is more readily discolored by light than the latter. However, both products should be protected from direct sunlight. After making appropriate adjustments for differences in solubility, both products can be used equally well as reagents in the tests described above.

References:

1) Culliford, B.J.; Nikolis, L.C. *J. Forensic Sci.* **1974,** *9*, 175.
2) Regulation H002, Occupational Safety and Health Administration, Washington, DC, Jan. 29, 1974.
3) Miller, E.C.; Miller, J.A.; Hartmann, H.A. *Cancer Res.* **1961,** *21*, 815.
4) Holland, V.R.; Saunders, B.C.; Rose, F.L. *Tetrahedron* **1974,** *30*, 3299.
5) Ames, B.N.; Durston, W.E.; Yamasaki, E.; Lee, F.D.; *Proc. Natl. Acad. Sci. U.S.A.* **1973,** *70*, 2281.
6) Garner, R.C.; Wapole, A.L.; Rose, F.L. *Cancer Lett.* **1975,** *1*, 39.
7) MacCann, J.; Choi, E.; Yamasaki, E.; Ames, B.N. *Proc. Natl. Acad. Sci. U.S.A.* **1975,** *72,* 5135.
8) MacCann, J.; Ames, B.N. *ibid.* **1976,** *73*, 950.
9) Garner, D.D.; Cano, K.M.; Peimer, R.S.; Yeshion, T.E. *J. Forensic Sci.* **1976,** *21*, 816.
10) Standefer, J.C.; Vanderjagt, D. *Clin. Chem.* **1977,** *23*, 749.
11) White, W.I. *J. Chromatogr.* **1977,** *138*, 220.

General references:

Aldrich FT-IR Library	**1**(1), 1236D
Sigma FT-IR Library	**1**(2), 218D
IR Library	**3**, 740D
Safety	**2**(2), 3292A

3,3',5,5'-TETRAMETHYLBENZIDINE dihydrochloride hydrate

Aldrich Cat. No.: 86,151-0
Sigma Prod. No.: T8768
Purity: 97%

Mol. Form.: $C_{16}H_{20}N_2 \cdot 2HCl \cdot xH_2O$
F.W.: 313.27
Appearance: White powder
λ_{max}: 288nm in methanol
Solubility: H_2O 10mg/ml
EGME 0.5mg/ml
EtOH 0.8mg/ml

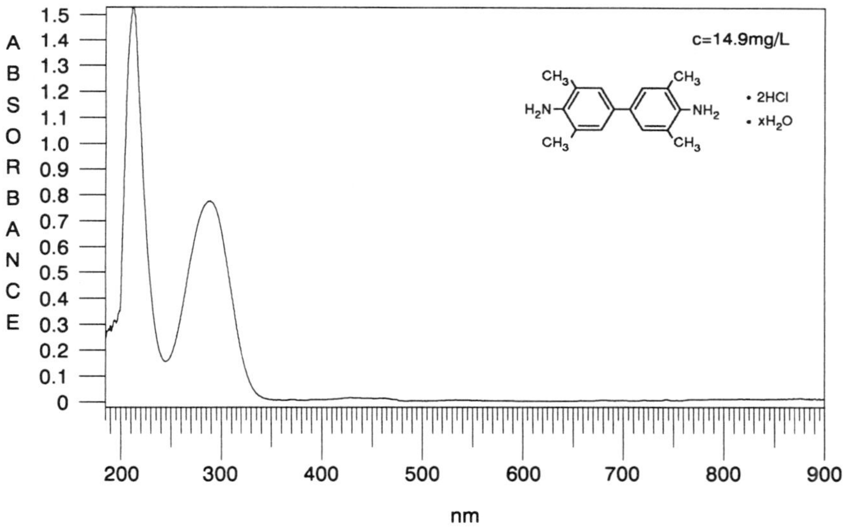

See pages 693 and 694, under 3,3',5,5'-Tetramethylbenzidine, for description.

General references:

Aldrich FT-IR Library	**1**(1), 1237A
Sigma FT-IR Library	**1**(2), 219A
IR Library	**3**, 740E
Safety	**2**(2), 3292B

TETRANITRO BLUE TETRAZOLIUM CHLORIDE hydrate

Aldrich Cat. No.: 13,316-7
Sigma Prod. No.: T4000

CAS No.: [*42798-98-1*]
Mol. Form.: $C_{40}H_{28}Cl_2N_{12}O_{10} \cdot xH_2O \cdot xC_2H_5OH$
F.W.: 907.64
m.p.: 165°C (dec.)
Appearance: Yellow powder
λ_{max}: 279nm in methanol
Solubility: H_2O 3mg/ml
EGME 9mg/ml
EtOH 3mg/ml

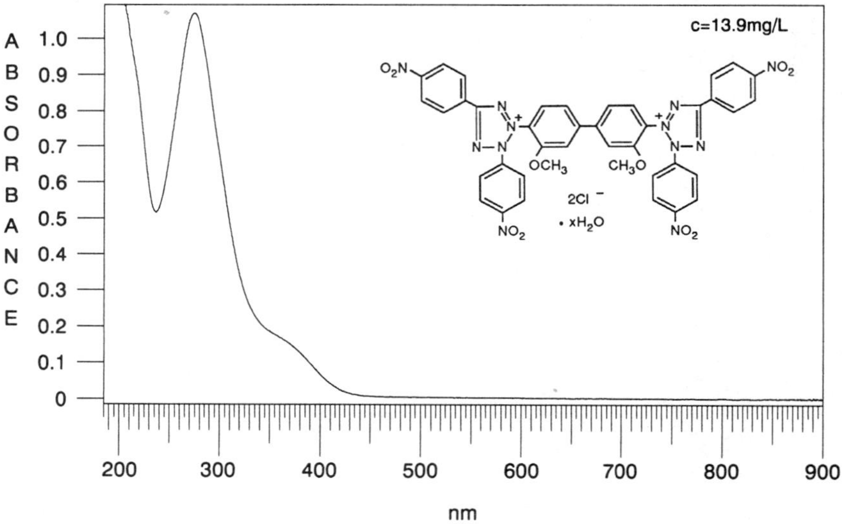

This tetranitrotetrazolium salt is the 5,5'-dinitro derivative of Nitro Blue Tetrazolium. It is readily reduced to the formazan yielding a fine, crystalline precipitate that is very slightly soluble in lipids. However, the formazan is brownish-red in color unlike the blue formazan produced by its precursor.[1]

Reference:

1) Nachlas, M.M.; Walker, D.G.; Seligman, A.M. *J. Biophys. Biochem. Cytol.* **1958,** *4,* 467-474.

General references:

Aldrich FT-IR Library	**1**(2), 634C
Sigma FT-IR Library	**1**(2), 219D
IR Library	**3**,1228D

TETRAZOLIUM VIOLET

2,5-Diphenyl-3-(1-naphthyl)-2*H*-tetrazolium chloride

Aldrich Cat. No.:	16,059-8
Sigma Prod. No.:	T0138

CAS No.:	[*1719-71-7*]
Mol. Form.:	$C_{23}H_{17}ClN_4$
F.W.:	384.87
m.p.:	247-249°C (dec.)
Appearance:	Tan powder
Solubility:	H_2O 30mg/ml
	EGME 20mg/ml
	EtOH 20mg/ml

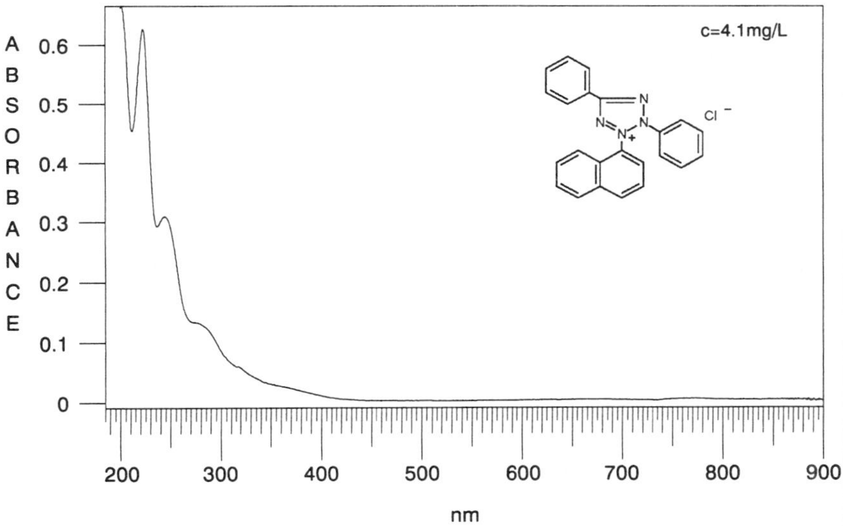

This colorless tetrazolium salt forms an oil-soluble, violet formazan dye on reduction. The reduction rate is slow and the dye crystals are large, making cellular localization impossible. These qualities limit the usefulness of this formazan-dye precursor as a histochemical reagent for establishing dehydrogenase activity.[1]

Reference:

1) Shelton, E.; Schneider, W.C. *Anat. Rec.* **1952,** *112*, 61-81.

General references:

Aldrich FT-IR Library	**1**(2), 632C
Sigma FT-IR Library	**1**(2), 220B
IR Library	**3**, 1227D
Safety	**2**(2), 500D

THIAZOL YELLOW G

Clayton Yellow • Direct Yellow 9 • Titan Yellow

Aldrich Cat. No.: 20,204-5
Sigma Prod. No.: T0513

CAS No.: [*1829-00-1*]
C.I. No.: 19540
Mol. Form.: $C_{28}H_{19}N_5Na_2O_6S_4$
F.W.: 695.73
Appearance: Rust powder
λ_{max}: 402nm in water
Solubility: H_2O 40mg/ml
EGME 40mg/ml
EtOH 3mg/ml

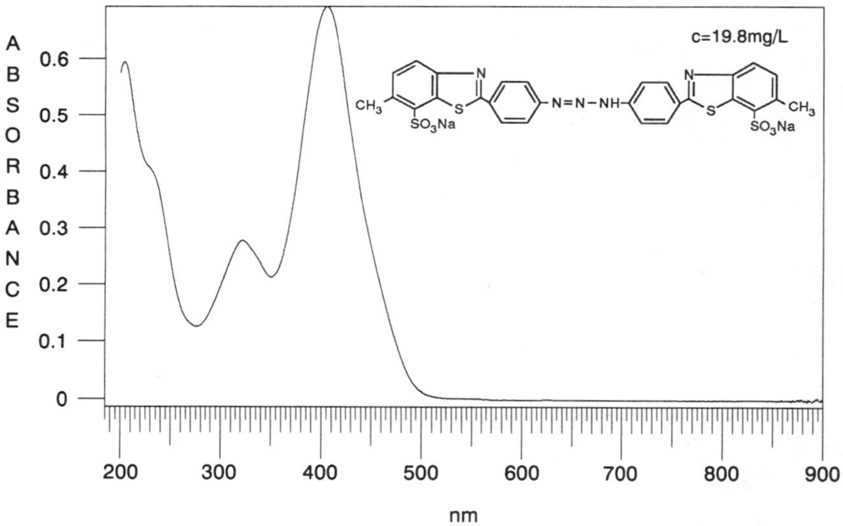

This yellow, anionic, monoazo thiazole dye is soluble in water and slightly soluble in ethanol. Prepared by coupling one mole each of diazotized and non-diazotized dehydrothio-*p*-toluidinesulfonic acid, Thiazol Yellow G displays acid-base indicator properties. The visual-transition interval occurs from pH 11 (yellow) to pH 13 (red).

Industrial uses include dyeing cotton, wool, silk, paper and chromed leather.

Welcher[1] listed it as a qualitative reagent for magnesium and *Conn's*[2] described its use as a reagent for the determination of magnesium in body fluids after calcium and organic matter have been removed. The latter source also listed references for its use as a fluorochrome in fluorescence microscopy.

References:

1) Welcher, F.J. *Organic Analytical Reagents;* Van Nostrand: New York, 1949; Vol. 4, p 391.

2) *Conn's Biological Stains,* 9th ed.; Lillie, R.D., Ed.; Williams and Wilkins: Baltimore, 1977; p 377.

General references:

Aldrich FT-IR Library	**1**(2), 1004D
Sigma FT-IR Library	**1**(2), 220C
IR Library	**3**, 1470F
Safety	**2**(2), 3321A
Merck Index	**11**, 9238
Colour Index	3rd ed., 1971

THIOFLAVIN S

Direct Yellow 7

Aldrich Cat. No.: 23,045-6
Sigma Prod. No.: T1892

CAS No.: [*1326-12-1*]
C.I. No.: 49010
Appearance: Light-brown powder
λ_{max}: 374nm in water
Solubility: H_2O 20mg/ml
EGME 4mg/ml
EtOH 2mg/ml

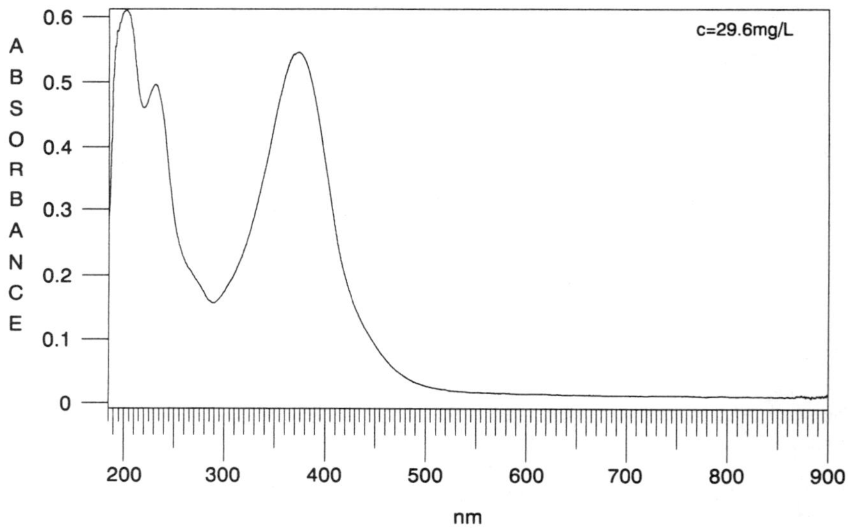

This fluorescent (greenish), yellow thiazole dye is made by sulfonating methylated Primuline base.

Soluble in water and slightly soluble in ethanol, it is used to dye cotton, wool and silk and to print silk directly.

Thioflavin S is useful in fluorescence microscopy as a fluorochrome. *Staining Procedures*[1] provides a recipe and method, after Schlegel[2] and Schlegel and Moses,[3] modified by Dohrmann and Wick,[4] for the intravital fluorescence microscopic visualization of blood vessels and patterns of blood flow, using Thioflavin S as the fluorochrome.

References:

1) Clark, G. *Staining Procedures,* 4th ed.; Williams and Wilkins: Baltimore, 1981; p 83.
2) Schlegel, J.U. *Anat. Rec.* **1949,** *105*, 433-443.
3) Schlegel, J.U.; Moses, J.B. *Proc. Soc. Exp. Biol. Med.* **1950,** *74*, 832-837.
4) Dohrmann, G.J.; Wick, K.M. *Stain Technol.* **1971,** *46*, 321-322.

General references:

Sigma FT-IR Library **1**(2), 221C
Colour Index 3rd ed., 1971

THIOFLAVIN T
Basic Yellow 1

Aldrich Cat. No.: 22,885-0
Sigma Prod. No.: T3516

CAS No.: [*2390-54-7*]
C.I. No.: 49005
Mol. Form.: $C_{17}H_{19}ClN_2S$
F.W.: 318.87
Appearance: Yellow powder
λ_{max}: 412nm in water
Solubility: H_2O 20mg/ml
EGME 60mg/ml
EtOH 100mg/ml

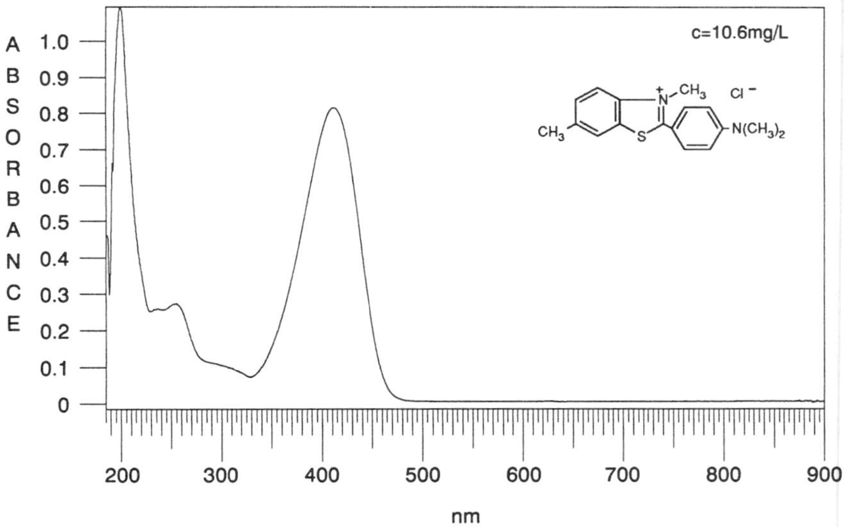

Thioflavin T is a cationic thiazole dye that is made by heating (160-170°C) dehydrothio-*p*-toluidine under pressure with methanol in the presence of hydrochloric or sulfuric acid.

Soluble in cold water and very soluble in ethanol and hot water, it is used to dye silk, tannin-mordanted cotton (lightfastness 1-2), wool, bast fibers, nylon and acetate (the latter two with a fluorescence), to color ballpoint-pen ink and to prepare the phosphotungstomolybdic acid salt (Pigment Yellow 18, C.I. 49005:1). The pigment is used to prepare ink for printing foil, tin plate and to color and coat paper.

Levaditi *et al.*[1] used Thioflavin T for the fluorescent staining of viruses. Medcalf and Patton,[2] utilizing fluorescence microscopy, studied entomological specimens stained with this dye.

References:

1) Levaditi, C.; Reinié, L.; Stamatin, L.; Bequignon, R. *Ann. Inst. Pasteur* (Paris) **1940,** *64*, 359-414.

2) Medcalf, R.L.; Patton, R.L. *Stain Technol.* **1944,** *19*, 11-27.

General references:

Aldrich FT-IR Library	**1**(2), 1004C
Sigma FT-IR Library	**1**(2), 221D
NMR Library	**2**(2), 579B
Colour Index	3rd ed., 1971

THIONIN
Lauth's Violet

Aldrich Cat. No.:	86,134-0
Sigma Prod. No.:	T3387
Grade:	Certified
CAS No.:	[*78338-22-4*]
C.I. No.:	52000
Mol. Form.:	$C_{14}H_{13}N_3O_2S$
F.W.:	287.34
Appearance:	Very dark-green powder
λ_{max}:	598nm in water
Solubility:	H_2O 30mg/ml
	EGME 4mg/ml
	EtOH 3mg/ml

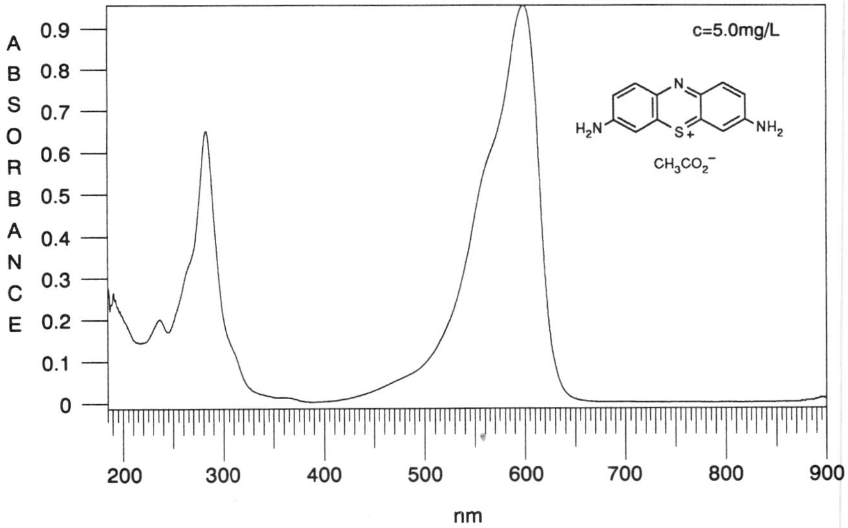

Thionin is a BSC-certified, metachromatic, cationic thiazine dye. Unlike some of its homologs, Thionin is available in an extremely pure form (99%) as determined by TLC or HPLC. This is accomplished by first 3,6-dinitrating phenothiazine. The nitro groups are then reduced to primary-amino functions and the resulting free-base form of the dye is converted to the acetate salt.

The acetate salt is the form that has been certified by the BSC for more than thirty years. It is much more soluble in water and the short-chain alcohols than the chloride. However, the literature is devoted exclusively to the chloride salt form and makes no mention of the acetate.

Currently, Thionin finds little use in commercial dyeing applications.

Merck[1] indicates that it has been used as an antioxidant for linseed oil, as a catalyst in the condensation of 1,3-butadienyl ethyl ether with maleic anhydride and for counting bacteria in milk.

Thionin is used widely as a biological stain and is particularly valuable in histology. Banny and Clark[2] concluded that the initial staining quality of Thionin was superior to Cresyl Fast Violet as a Nissl stain. However, the rapid fading they observed on aging made it their overall second choice to Cresyl Fast Violet.

Staining Procedures[3] described the following staining methods: stain for Nissl granules; Windle-Schmorl's Carbol Thionin for bone matrix; frozen section stain; Mallory-Lillie's stain for mast cells, mucin and cartilage; stain for temporary mounts of freehand plant-tissue sections and Diene's agar-fixation stain for *Mycoplasma*.

Donovick[4] reported excellent results when he used Thionin as a metachromatic stain for brain tissue.

References:

1) *The Merck Index,* 11th ed.; Budavari, S. *et al.*, Eds; Merck: Rahway, 1989; Monograph 9276.
2) Banny, T.M.; Clark, G. *Stain Technol.* **1950,** *25*, 195-196.
3) Clark, G. *Staining Procedures,* 4th ed.; Williams and Wilkins: Baltimore, 1981; pp 142, 182, 211, 327, 420.
4) Donovick, P.J. *Stain Technol.* **1974,** *49*, 49-51.

General references:

Aldrich FT-IR Library **1**(2), 1038B
Sigma FT-IR Library **1**(2), 222A
IR Library **3**, 1486H

THORIN I

Aldrich Cat. No.: 10,456-6
Sigma Prod. No.: T2138
Grade: Indicator

CAS No.: [*132-33-2*]
Mol. Form.: $C_{16}H_{11}AsN_2Na_2O_{10}S_2$
F.W.: 576.30
Appearance: Red powder
λ_{max}: 483nm in water
Solubility: H_2O 20mg/ml
EGME 10mg/ml
EtOH 0.1mg/ml

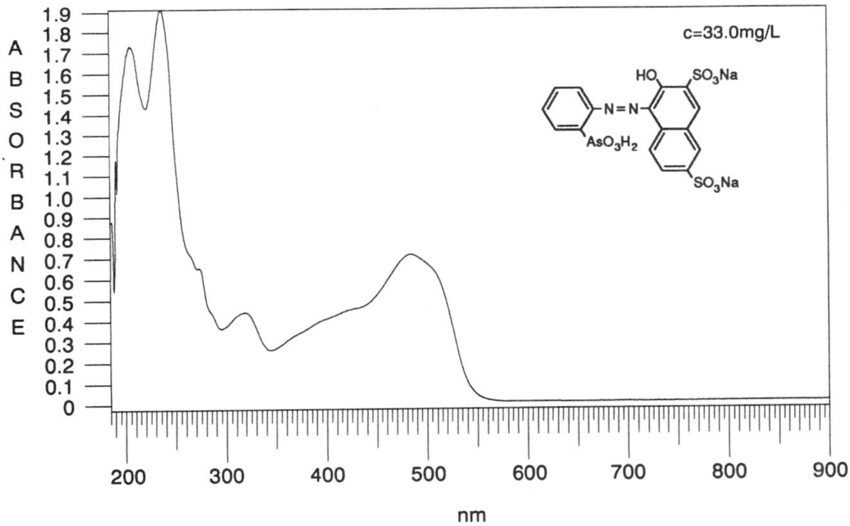

Thorin I is an anionic monoazo dye that is made by coupling diazotized 2-aminobenzenearsonic acid to R acid (2-naphthol-3,6-disulfonic acid) and converting the coupling product to the disodium salt. It is water soluble and forms complexes with various metals. Nietzel *et al.*[1] describe its use as an indicator for the determination of hafnium, thorium and zirconium. There is no evidence in the literature that it has been used industrially.

Reference:

1) Nietzel, O.A.; Wessling, B.W.; DeSesa, M.A. *Anal. Chem.* **1958,** *30*, 1182.

General references:

Aldrich FT-IR Library	**1**(2), 998D
Sigma FT-IR Library	**1**(2), 222B
IR Library	**3**, 1467H
Safety	**2**(2), 3340A

THYMOL BLUE

Aldrich Cat. No.:	11,454-5
Sigma Prod. No.:	T8527
Grade:	A.C.S. reagent
CAS No.:	*[76-61-9]*
Mol. Form.:	$C_{27}H_{30}O_5S$
F.W.:	466.60
m.p.:	221-224°C
Appearance:	Dark-green powder
λ_{max}:	594(376)nm in 0.1*N* sodium hydroxide
Solubility:	H_2O 0.6mg/ml
	EGME 100mg/ml
	EtOH 70mg/ml

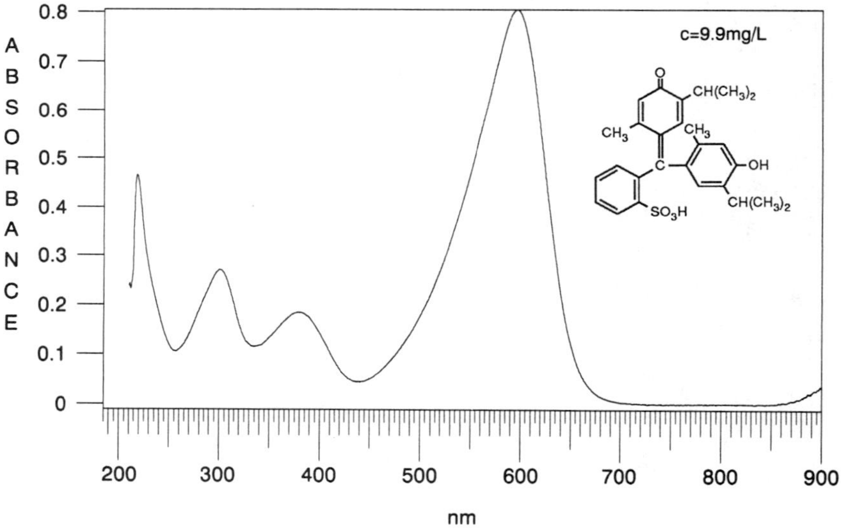

This indicator is a member of the sulfonephthalein series of acid-base indicators. It is made by condensing two moles of thymol with one mole of *o*-sulfobenzoic acid anhydride. The free-acid form is very slightly soluble in water, soluble in methanol, ethanol and acetic acid and readily soluble in alkaline solutions. The monosodium salt or water-soluble form is also soluble in the lower aliphatic alcohols. *Reagent Chemicals*[1] lists test methods and specifications for Thymol Blue including visual-transition intervals: acid range from pH 1.2 (red) to pH 2.8 (yellow); alkaline range from pH 8.0 (yellow) to pH 9.2 (blue). Please see the Indicator Chart in the back of this book for more detailed information on the preparation of test solutions for using Thymol Blue as an indicator.

Thymol Blue is the precursor for Bromothymol Blue, an indicator that is used widely in biological applications because it changes color in the physiological pH range (pH 6.0 to 7.6).

Reference:

1) *Reagent Chemicals,* 7th ed.; American Chemical Society: Washington, DC, 1986; pp 671-672.

General references:

Aldrich FT-IR Library	**1**(2), 1022C
Sigma FT-IR Library	**1**(2), 222C
IR Library	**3**, 1479F
Merck Index	**11**, 9334

THYMOL BLUE, water soluble

Aldrich Cat. No.:	86,136-7
Sigma Prod. No.:	T8652
Grade:	A.C.S. reagent

CAS No.:	[*81012-93-3*]
Mol. Form.:	$C_{27}H_{29}NaO_5S$
F.W.:	488.58
m.p.:	283-285°C (dec.)
Appearance:	Brown-black crystalline powder
λ_{max}:	590(378)nm in 0.1*N* sodium hydroxide
Solubility:	H_2O 60mg/ml
	EGME 60mg/ml
	EtOH 40mg/ml

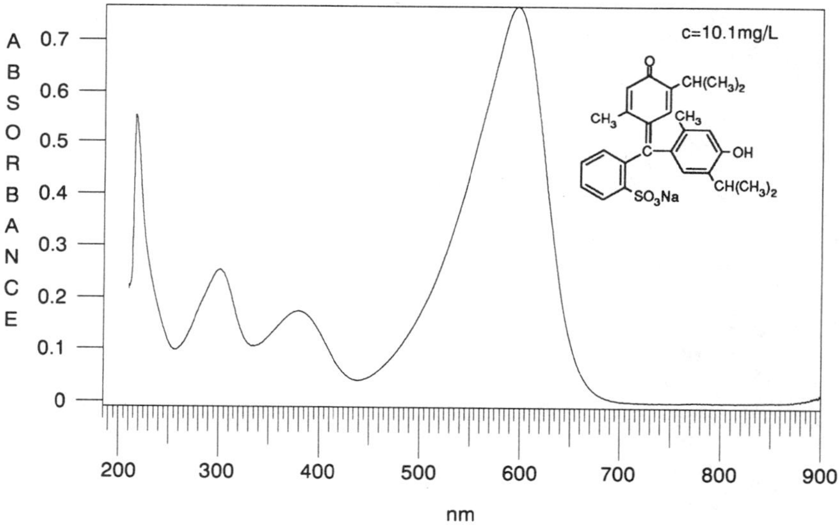

See description on pages 706 and 707 under Thymol Blue.

General references:

Aldrich FT-IR Library	**1**(2), 1026C
Sigma FT-IR Library	**1**(2), 222D
IR Library	**3**, 1481F

THYMOLPHTHALEIN

Aldrich Cat. No.:	11,455-3
Sigma Prod. No.:	T8402
Grade:	A.C.S. reagent

CAS No.:	[*125-20-2*]
Mol. Form.:	$C_{28}H_{30}O_4$
F.W.:	430.54
m.p.:	251-253°C
Appearance:	White powder
λ_{max}**:**	592(396)nm in 0.1*N* sodium hydroxide
Solubility:	H_2O 1mg/ml
	EGME 100mg/ml
	EtOH 70mg/ml

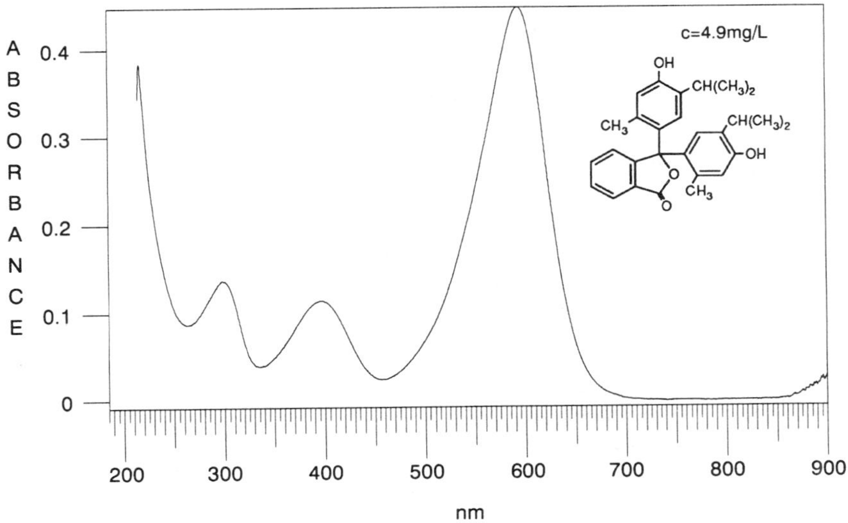

Thymolphthalein is a hydroxytriarylmethane indicator that is made by condensing two moles of thymol with one mole of phthalic anhydride in the presence of heat and a condensing agent, such as stannic chloride. It is soluble in the lower aliphatic alcohols, acetone and dilute alkalies. *Reagent Chemicals*[1] lists it with a visual-transition interval of pH 8.8 (colorless) to pH 10.5 (blue). *The Merck Index*[2] states that the leuco form of an alkali salt is used as a reagent to detect blood.

References:

1) *Reagent Chemicals,* 7th ed.; American Chemical Society: Washington, DC, 1986; p 672.

2) *The Merck Index,* 11th ed.; Budavari, S. *et al.*, Eds.; Merck: Rahway, 1989; Monograph 9336.

General references:

Aldrich FT-IR Library	**1**(2,), 1007A
Sigma FT-IR Library	**1**(2), 223A
IR Library	**3**, 1471F
Merck Index	**11**, 9336

THYMOLPHTHALEIN MONOPHOSPHORIC ACID, disodium salt hydrate

Aldrich Cat. No.: 20,136-7
Sigma Prod. No.: T7758

Mol. Form.: $C_{28}H_{29}Na_2O_7P \cdot xH_2O$
F.W.: 554.49
Appearance: Yellow powder
λ_{max}: 445nm in 0.1*N* sodium hydroxide
Solubility:

H_2O	30mg/ml
EGME	90mg/ml
EtOH	100mg/ml

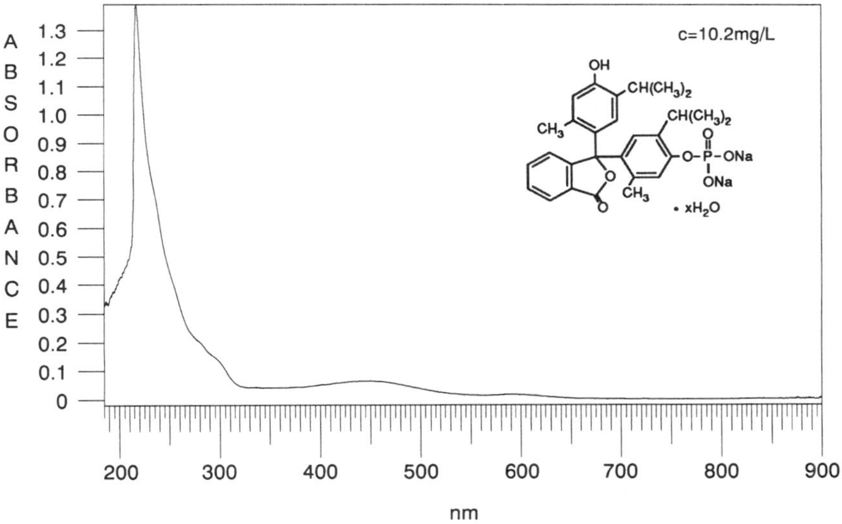

Thymolphthalein monophosphoric acid is a biochemical reagent made from the phthalein pH indicator, Thymolphthalein, by monophosphoric acid esterification. This compound, which is soluble in water, methanol and very soluble in ethanol, is used as a reagent for the determination of acid and alkaline phosphatases.[1,2]

References:

1) Van Rijn, H.J.M.; Boer, P.; Klosse, J.A. *J. Clin. Chem. Clin. Biochem.* **1980,** *18*(10), 627-630.

2) Roy, A.V.; Brower, M.E.; Hayden, J.E. *Clin. Chem.* **1971,** *17*(11), 1093.

General references:

Aldrich FT-IR Library	**1**(2), 1007C
Sigma FT-IR Library	**1**(1), 539A
IR Library	**3**, 1471H

TIRON

4,5-Dihydroxy-1,3-benzenedisulfonic acid,
disodium salt monohydrate

Aldrich Cat. No.: 17,255-3
Sigma Prod. No.: D7389

CAS No.: [*149-45-1*]
Mol. Form.: $C_6H_4Na_2O_8S_2 \cdot H_2O$
F.W.: 332.22
m.p.: >300°C
Appearance: Off-white powder
λ_{max}: 291nm in water
Solubility: H_2O 10mg/ml
EGME 2mg/ml
EtOH <0.1mg/ml

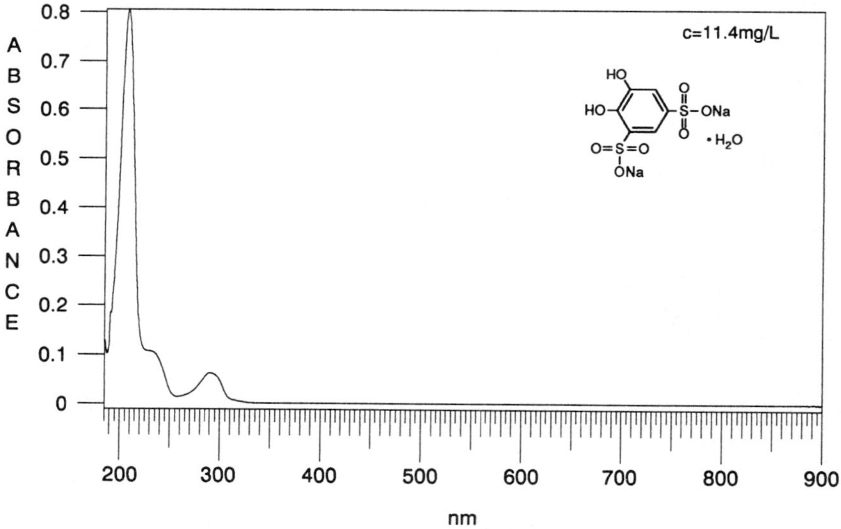

Tiron is a somewhat water-soluble, nonhygroscopic, colorless reagent that forms colored, water-soluble complexes with various metal ions. Iron(III) produces a deep-blue solution below pH 5, copper salts yield a green-yellow, molybdenum(VI) salts produce a light yellow and titanium salts display an orange hue when in solution.

Tiron is used as a colorimetric reagent for iron, manganese, titanium and molybdenum and as a complexing agent for zirconium.[1,2]

References:

1) von Schwarzenbach, G.; Willi, A. *Helv. Chim. Acta* **1951,** *34*, 528.

2) Intorre, B.I.; Martell, A.E. *J. Am. Chem. Soc.* **1960,** *82*, 358.

General references:

Aldrich FT-IR Library	**1**(2), 504D
Sigma FT-IR Library	**1**(2), 473B
Merck Index	**11,** 9392

TOLUIDINE BLUE O

Basic Blue 17 • Tolonium chloride

Aldrich Cat. No.:	19,816-1
Sigma Prod. No.:	T0394
Grade:	Certified
CAS No.:	[*92-31-9*]
C.I. No.:	52040
Mol. Form.:	$C_{15}H_{16}ClN_3S$
F.W.:	305.83
Appearance:	Dark-green powder
λ_{max}**:**	626nm in water
Solubility:	H_2O 10mg/ml
	EGME 7mg/ml
	EtOH 4mg/ml

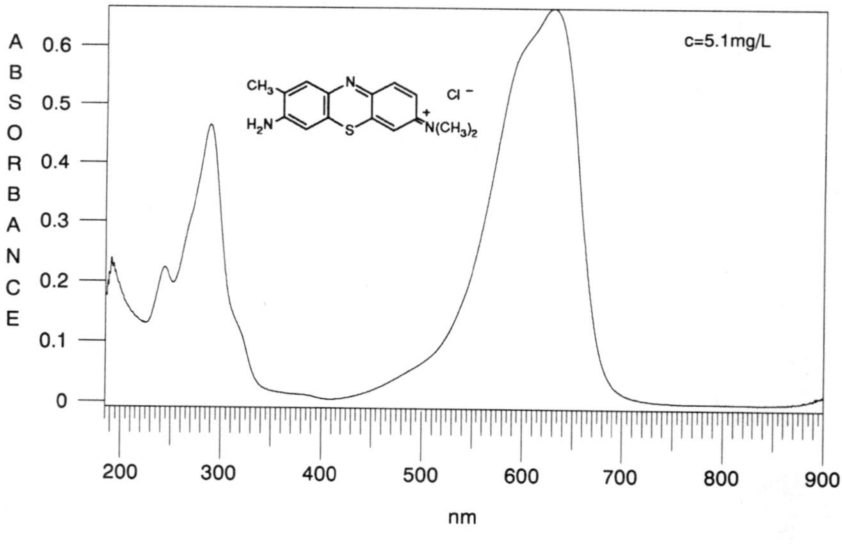

Toluidine Blue O is a metachromatic, cationic thiazine dye that is frequently used in biological-stain applications in place of Thionin or Methylene Blue. It is a nuclear stain similar in structure to Azure B, with the same empirical formula. However, its UV-visible absorbance spectrum and staining characteristics more closely resemble Azure A.

The BSC will certify material with a minimum dye content of 50%, if other requirements including the staining performance are met. The dye content of a typical commercial lot ranges from 50-60%. However, dye concentrations in this range (50-60%) usually indicate the lot is the zinc chloride double salt. The chloride salt form with a dye concentration of 90%, as described above, is the form of choice for precise applications and when a zinc-free product is indicated.

Toluidine Blue O has been used widely in *in vitro* biological applications. Snyder[1] used it in the bulk staining of Nissl granules. Williams[2] described a method for staining ganglion cells in mammalian viscera orthochromatically, while staining surrounding tissue metachromatically.

Toluidine Blue O finds wide application as a component of a tribasic stain along with Malachite Green and Basic Fuchsin. The tristain is used with thin sections of normal rat tissue fixed in osmium tetroxide embedded in plastic as for electron microscopy. Using this method, Grimley[3] claimed "exquisite details" were observed with light microscopy.

The Merck Index[4] lists tolonium chloride (Toluidine Blue O) as an anti-heparin compound. Niebel and Chomet[5] explored its use in screening methods for the recognition of squamous carcinoma. Mashberg[6] investigated the use of Toluidine Blue O in the form of a rinse to "detect unobserved asymptomatic oral-mucosal cancers." Koizumi *et al.*[7] found, in a photoreduction study of a number of cationic dyes, that Toluidine Blue O demonstrated a 100% yield on reoxidation. Smith and Squires[8] found that it was the most potent of the several dyes tested for 3*H*-flunitrazepam binding. Waller *et al.*[9] reported two improved Toluidine Blue O-mediated techniques for DNase detection.[10]

References:

1) Snyder, R.J. *Stain Technol.* **1943,** *18,* 35-39.

2) Williams, T. *ibid.* **1963,** *38*, 281-282.

3) Grimley, P.M. *ibid.* **1964**, *39*, 229-233.

4) *The Merck Index,* 11th ed.; Budavari, S. *et al.*, Eds.; Rahway, 1989; Monograph 9444.

5) Niebel, H.H.; Chomet, B. *J. Am. Dent. Assoc.* **1964**, *68*, 801-806.

6) Mashberg, A. *J. Am. Med. Assoc.* **1981**, *245*(23), 2408-2410.

7) Koizumi, M.; Obato, H.; Hayashi, S. *Bull. Chem. Soc. Jpn.* **1964**, *37*, 108-117.

8) Smith, T.M.; Squires, R.F. *Neurochem. Res.* **1983**, *8*(9), 1177-1183.

9) Waller, J.R.; Hodel, S.L.; Nuti, R.N. *J. Clin. Microbiol.* **1985**, *21*(2),195-199.

10) Shedd, D.P.; Hukill, P.B.; Bahn, S.; Ferraro, R.H. *Arch. Surg.* **1967**, *95,* 16-22.

General references:

Aldrich FT-IR Library	**1**(2), 1038C
Sigma FT-IR Library	**1**(2), 224D
IR Library	**3,** 1487A

TOLUIDINE RED

Pigment Red 3

Aldrich Cat. No.:	19,975-3
Sigma Prod. No.:	T0638
CAS No.:	[*2425-85-6*]
C.I. No.:	12120
Mol. Form.:	$C_{17}H_{13}N_3O_3$
F.W.:	307.31
m.p.:	270-272°C
Appearance:	Dark-red powder
λ_{max}:	507(398)nm in toluene
Solubility:	H_2O 0.8mg/ml
	EGME 0.9mg/ml
	EtOH 0.7mg/ml

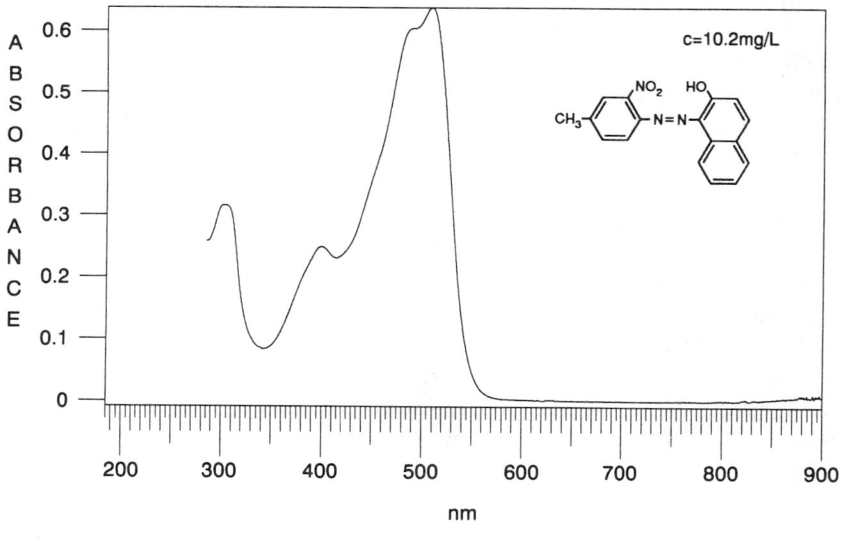

Toluidine Red is a monoazo pigment that is prepared by coupling diazotized 2-nitro-*p*-toluidine to 2-naphthol. It is very slightly soluble in ethanol and slightly soluble in acetone and benzene. Because of its bright-scarlet hue, tinctorial strength, fastness to acids, alkalis and light, and economy, Toluidine Red is used widely as a pigment in industrial applications. These include use in paints, printing inks, leather finishes, inks for foils and tinplate printing, paper coatings and dyeing, linoleum, wallpaper, concrete, carbon paper, typewriter ribbons and for textile printing.

There is no evidence in the literature that it has been used as a biological stain.

General references:

Aldrich FT-IR Library	**1**(2), 974D
Sigma FT-IR Library	**1**(2), 225A
IR Library	**3**, 1456A
Colour Index	3rd ed., 1971

2,3,5-TRIPHENYL-2*H*-TETRAZOLIUM CHLORIDE
TTC

Aldrich Cat. No.: T8,485-9
Sigma Prod. No.: T8877
Purity: 98%

CAS No.: [*298-96-4*]
Mol. Form.: $C_{19}H_{15}ClN_4$
F.W.: 316.80
m.p.: 250°C
Appearance: Off-white powder
Solubility: H_2O 200mg/ml
EGME 100mg/ml
EtOH 200mg/ml

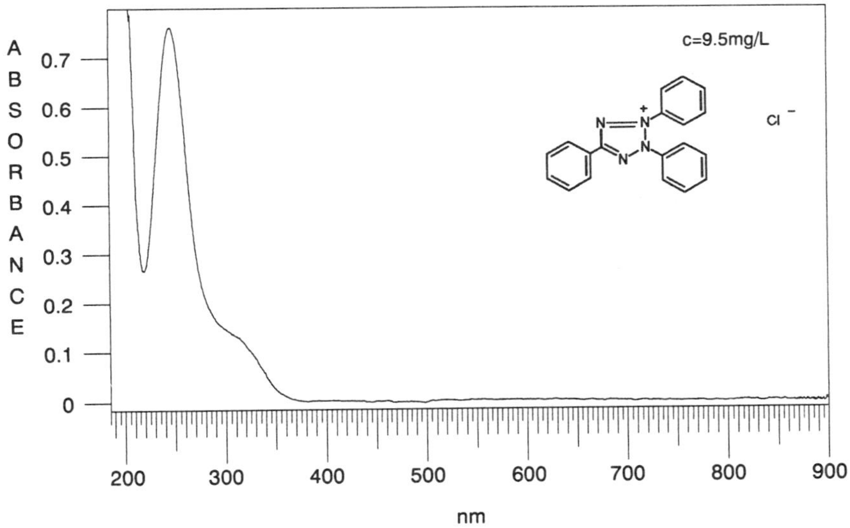

TTC is a colorless dye precursor that forms a red dye (triphenylformazan) on reduction.

Very soluble in water and alcohol, soluble in acetone and insoluble in ether, the dye (triphenylformazan) is extremely fat soluble and sensitive to light.[1] This light sensitivity eliminates TTC as a suitable reagent for definitive histological studies.

TTC is a sensitive reagent for reducing sugars and a germination indicator for seeds; the embryonic tissue present in live seeds hydrogenates the colorless reagent converting it to the red dye.[2]

References:

1) *Conn's Biological Stains,* 9th ed.; Lillie, R.D., Ed.; Williams and Wilkins: Baltimore, 1977; pp 226-227.

2) *The Merck Index,* 11th ed.; Budavari, S. *et al.*, Eds.; Merck: Rahway, 1989; Monograph 9658.

General references:

Aldrich FT-IR Library	**1**(2), 632A
Sigma FT-IR Library	**1**(2), 960D
IR Library	**3**, 1227C
NMR Library	**2**(2), 500B
Safety	**2**(2), 3522D

TROPAEOLIN O

Acid Orange 6

Aldrich Cat. No.: 19,968-0
Sigma Prod. No.: T2513

CAS No.: [*547-57-9*]
C.I. No.: 14270
Mol. Form.: $C_{12}H_9N_2NaO_5S$
F.W.: 316.27
Appearance: Rust powder
λ_{max}: 490nm in 0.1*N* sodium hydroxide
Solubility: H_2O 30mg/ml
EGME 40mg/ml
EtOH 2mg/ml

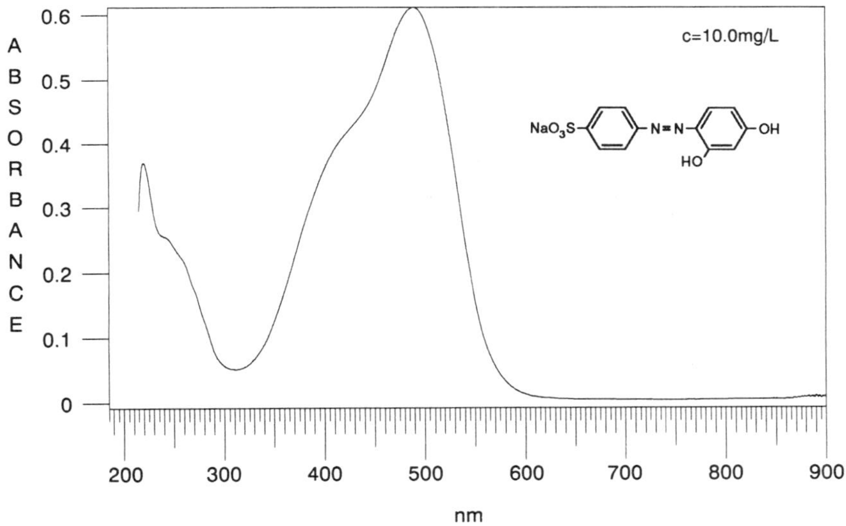

Tropaeolin O is an orange, anionic monoazo dye that is soluble in water, acetone, methyl Cellosolve and somewhat soluble in the lower alcohols. It is made by coupling diazotized sulfanilic acid to resorcinol in an acid bath. The product of this reaction is converted to the sodium or potassium salt to enhance its water solubility. Commercial uses include dyeing wool, silk and nylon, the latter from an acetic or formic acid bath. In the laboratory, it probably is used most widely as a pH indicator with a visual-transition interval of pH 11.0 (yellow) to pH 12.7 (red).

Gurr[1] reported that Emig used it to stain plant tissue, describing the vascular system as being stained a yellowish-brown from a dilute aqueous-acid bath. The same author reported that blue-green algae take on a clear, yellow hue when dyed with Tropaeolin O from a dilute, aqueous-alkali bath.

Reference:

1) Gurr, E. *Encyclopedia of Microscopic Stains*; Hill: London, 1960; p 394.

General references:

Aldrich FT-IR Library	**1**(2), 977B
Sigma FT-IR Library	**1**(2), 225C
IR Library	**3**, 1457C
Merck Index	**11**, 9687
Colour Index	3rd ed., 1971

TRYPAN BLUE

Direct Blue 14 • Niagara Blue 3B

Aldrich Cat. No.: 30,264-3
Sigma Prod. No.: T0776

CAS No.: [*72-57-1*]
C.I. No.: 23850
Mol. Form.: $C_{34}H_{24}N_6Na_4O_{14}S_4$
F.W.: 960.82
m.p.: >300°C
Appearance: Dark greenish-brown powder
λ_{max}: 607nm in methanol
Solubility: H_2O <0.1mg/ml
EGME 20mg/ml
EtOH 0.6mg/ml

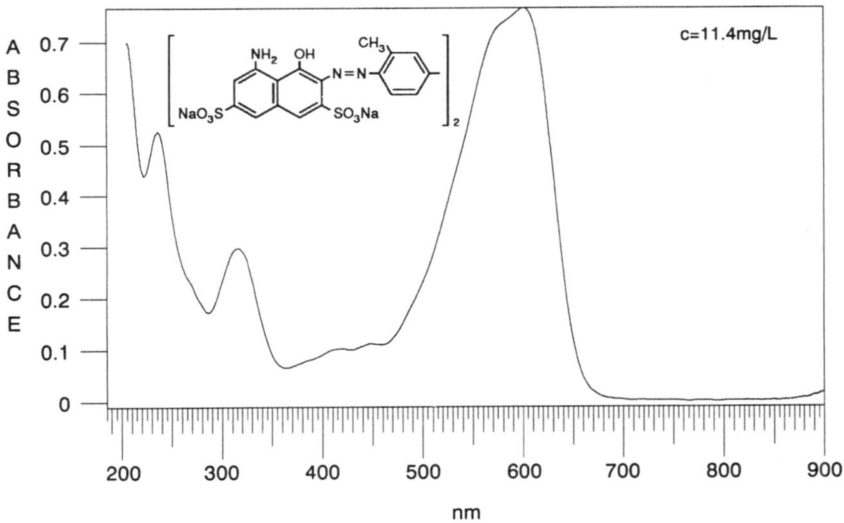

Trypan Blue (TB) is a blue acid dye with two azo chromophores. It is prepared by the alkaline coupling of one mole of tetraazotized *o*-tolidine to two moles of H acid (5-amino-4-hydroxy-2,7-naphthalenedisulfonic acid) and converting the coupling product to the tetrasodium salt. It is soluble in methyl Cellosolve and glycerol and has a strong affinity for cellulosic substrates such as cotton. However, its attraction to proteinaceous materials, including silk and wool, is considerably less. Commercially, it is used to dye and, to a lesser degree, discharge-print cotton. In addition, Trypan Blue dyes fully chromed leather an intense blue. HPLC analysis reveals that the product of commerce has a total dye content of approximately 25%. Of this, Trypan Blue constitutes

30-40%, or a net Trypan Blue content of approximately 7-10%. Fortunately, a laboratory grade of the dye, with a net TB content in excess of 80%, is available.

Like its isomer, Evans Blue, Trypan Blue is a teratogenic dye. Udenfriend[1] attributes this to the planar distortion that results when the H-acid moieties of TB are adsorbed onto protein. Both dyes find wide use as vital stains. A fair comparison of the effectiveness of the two dyes may not have been ascertained since Evans Blue has been available in reasonably pure form over the years while, until recently, the TB our laboratory has examined has been grossly inferior in quality. Clark[2] describes a method for demonstrating the teratogenic properties of TB through the fluorescence it elicits when it is adsorbed onto protein. The same source includes it as a component of a Van Gieson Stain solution variation.

References:

1) Udenfriend, S. *Fluorescence Assay in Biology and Medicine;* Academic: New York, 1962.

2) Clark, G. *Staining Procedures,* 4th ed.; Williams and Wilkins: Baltimore, 1981; pp 85, 115.

General references:

Safety	**2**(2), 3552D
Merck Index	**11**, 9701
Colour Index	3rd ed., 1971

TWORT STAIN

Aldrich Cat. No.: 21,679-8
Sigma Prod. No.: T7139

Appearance: Olive-green powder
λ_{max}: 634(440)nm in ethanol
Solubility: H_2O 0.7mg/ml
EGME 4mg/ml
EtOH 1mg/ml

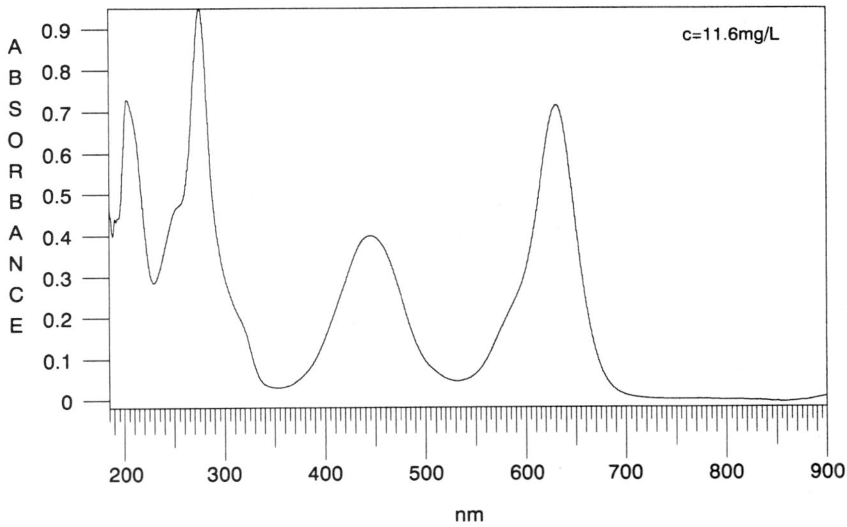

Twort Stain is a neutral dye that is made by complexing one mole of the anionic dye, Light Green SF, and three moles of the cationic dye, Neutral Red.

Twort[1,2] utilized it to stain parasites in tissue. Ollett[3,4] used it in a method for staining both Gram-positive and Gram-negative bacteria in sections.

There are no reported industrial applications.

References:

1) Twort, F.W. *J. State Med., London* **1924,** *32,* 351-355.
2) *Idem* *Br. J. Exp. Pathol.* **1924,** *5,* 350-351.
3) Ollett, W.S. *J. Pathol. Bacteriol.* **1947,** *59,* 257-258.
4) *Idem* *ibid.* **1951,** *63,* 166.

General references:

Aldrich FT-IR Library	**1**(2), 1043C
Sigma FT-IR Library	**1**(2), 226C
IR Library	**3**, 1489B
Safety	**2**(2), 3561B

UNIBLUE A, sodium salt

Aldrich Cat. No.:	29,840-9
Sigma Prod. No.:	U5753
CAS No.:	[*14541-90-3*]
Mol. Form.:	$C_{22}H_{15}N_2NaO_7S_2$
F.W.:	506.49
m.p.:	>300°C
Appearance:	Dark-blue powder
λ_{max}:	596nm in water
Solubility:	H_2O 40mg/ml
	EGME 20mg/ml
	EtOH 2mg/ml

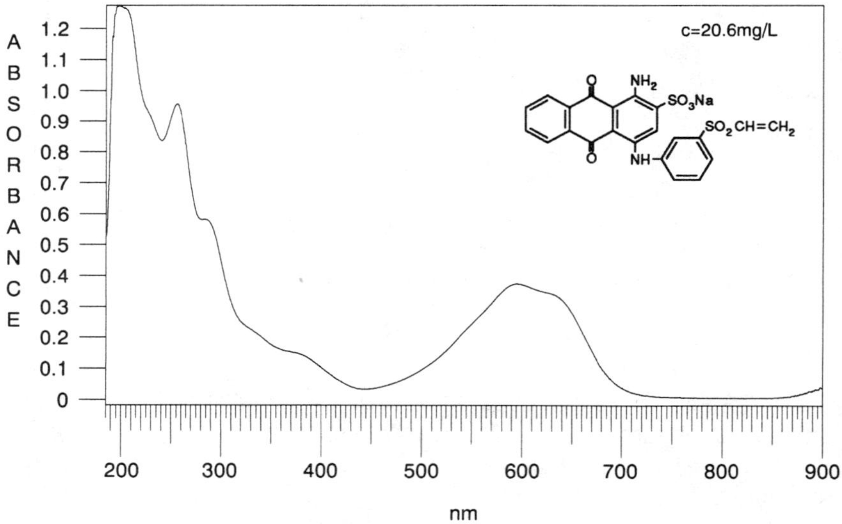

For description, see Remazol Brilliant Blue R on pages 620 and 621.

VARIAMINE BLUE RT salt

Azoic Diazo No. 22

Aldrich Cat. No.:	20,146-4
Sigma Prod. No.:	A5254

CAS No.:	[*4477-28-5*]
C.I. No.:	37240
Mol. Form.:	$C_{12}H_{11}N_3O_4S$
F.W.:	293.30
Appearance:	Gold powder
λ_{max}:	377nm in water
Solubility:	H_2O 30mg/ml
	EGME 70mg/ml
	EtOH 20mg/ml

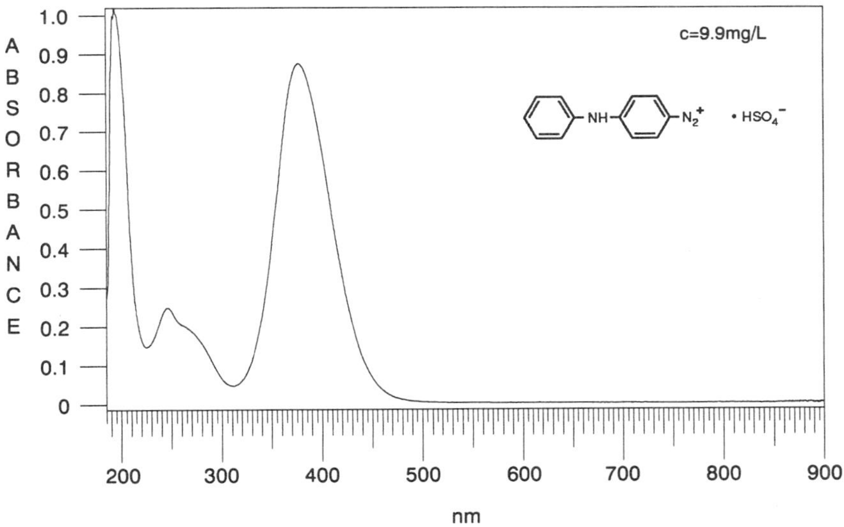

Variamine Blue RT (VBRT, diphenylamine-4-diazonium sulfate) is a stable diazonium salt which is made by first nitrosating diphenylamine. The resulting 4-nitrosodiphenylamine is reduced to *N*-phenyl-1,4-phenylenediamine which is diazotized and isolated as the sulfate.

This dye precursor is used commercially to dye both cotton and silk by the *in situ* formation of the dye on the substrate. This is accomplished by allowing Variamine Blue RT and an azoic coupling component, such as 2-naphthol or its derivatives, to penetrate the fabric before they are activated, usually by steaming, to facilitate coupling.

Although no specific references regarding the use of VBRT as a histochemical or cytochemical reagent were found, structurally it is very similar to other stable diazonium salts that have been investigated in these areas. *Conn's,*[1] under Fast Blue VB (Variamine Blue B), cites the work of Gomori in 1952 and 1954 and Gossner in 1958.

Reference:

1) *Conn's Biological Stains,* 9th ed.; Lillie, R.D., Ed.; Williams and Wilkins: Baltimore, 1977; p 215.

General references:

Aldrich FT-IR Library	**1**(2), 956B
Sigma FT-IR Library	**1**(2), 98B
IR Library	**3**, 1445D
Safety	**2**(2), 3594C

VICTORIA BLUE B
Basic Blue 26

Aldrich Cat. No.: 19,969-9
Sigma Prod. No.: V0753

CAS No.: [*2580-56-5*]
C.I. No.: 44045
Mol. Form.: $C_{33}H_{32}ClN_3$
F.W.: 506.10
m.p.: 206°C (dec.)
Appearance: Copper-colored chunks
λ_{max}: 599nm in ethanol
Solubility: H_2O 10mg/ml
EGME 20mg/ml
EtOH 20mg/ml

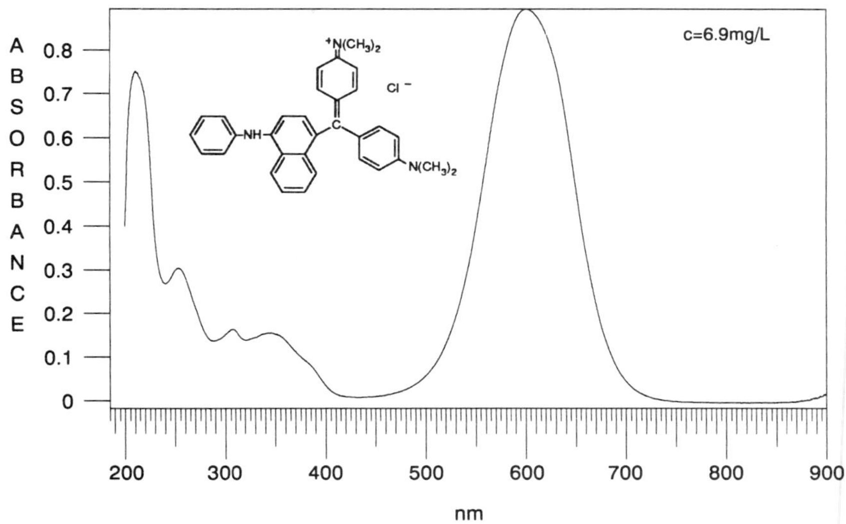

This blue, cationic diphenylnaphthylmethane dye is slightly soluble in water and somewhat more soluble in the lower aliphatic alcohols. Commercially, it is used to dye cotton blue using tannin as a mordant. Both silk and wool can be dyed or printed after proper pretreatment (discharged tannin mordant). Victoria Blue B (VBB) is used widely to dye paper, although it is fugitive to both light and alkalis. The phosphotungstic, phosphotungstomolybdic and phosphomolybdic acid salts of VBB are used as pigments in printing inks, copying paper, typewriter ribbons, plastics, paints and enamels.

Staining Procedures[1] describes a method that utilizes VBB and Phloxine B for the negative- or relief-staining of cilia and flagella.

Reference:

1) Clark, G. *Staining Procedures,* 4th ed.; Williams and Wilkins: Baltimore, 1981; p 287.

General references:

Aldrich FT-IR Library	**1**(2), 1032C
Sigma FT-IR Library	**1**(2), 226D
IR Library	**3**, 1485A
Colour Index	3rd ed., 1971

VICTORIA BLUE R

Basic Blue 11

Aldrich Cat. No.:	23,417-6
Sigma Prod. No.:	V3629
CAS No.:	[*2185-86-6*]
C.I. No.:	44040
Mol. Form.:	$C_{29}H_{32}ClN_3$
F.W.:	458.05
Appearance:	Dark-blue powder
λ_{max}:	615nm in methanol
Solubility:	H_2O 4mg/ml
	EGME 50mg/ml
	EtOH 20mg/ml

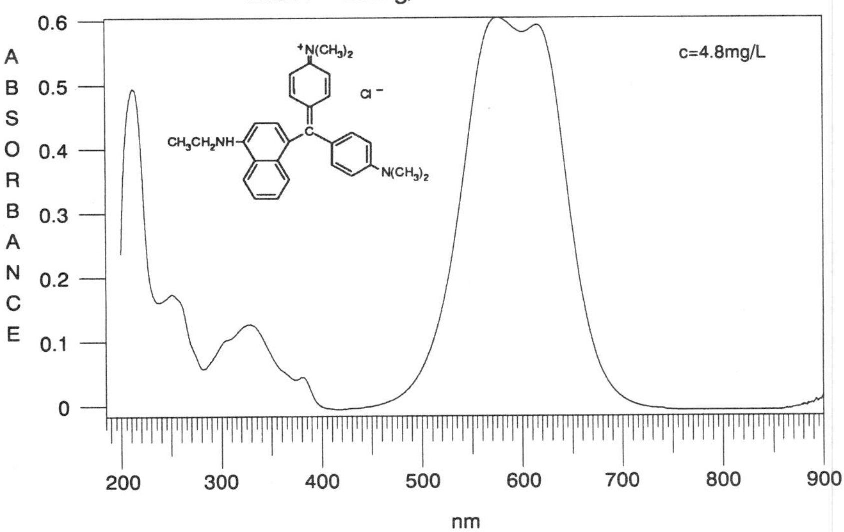

Victoria Blue R (VBR) is similar in structure to Victoria Blue B. If an ethyl group is replaced by a phenyl group, the former molecule is converted to the latter. While this structural variation causes a bathochromic shift in the visible spectrum of 16nm (VBB to VBR), little of practical value is altered. In fact, the *Colour Index* descriptions of properties and applications for these two are almost identical (see pages 727 and 728 for listing of Victoria Blue B). *Conn's*[1] suggests the foregoing as an explanation for the confusion that exists regarding the Victoria Blues in the life-science literature.

Reference:

1) *Conn's Biological Stains,* 9th ed.; Lillie, R.D., Ed.; Williams and Wilkins: Baltimore, 1977; p 298.

General references:

Aldrich FT-IR Library	**1**(2), 1034A
Sigma FT-IR Library	**1**(2), 227A
Colour Index	3rd ed., 1971

VICTORIA PURE BLUE BO
Basic Blue 7

Aldrich Cat. No.:	23,098-7
Sigma Prod. No.:	V3754

CAS No.:	[*2390-60-5*]
C.I. No.:	42595
Mol. Form.:	$C_{33}H_{40}ClN_3$
F.W.:	514.16
Appearance:	Gold, metallic chunks
λ_{max}:	619nm in water
Solubility:	H_2O 10mg/ml
	EGME 50mg/ml
	EtOH 10mg/ml

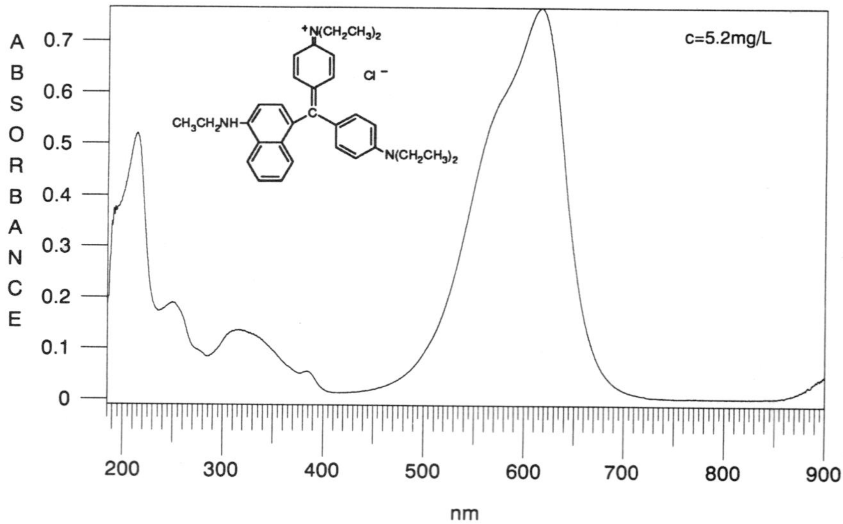

This blue, cationic diphenylnaphthylmethane dye is made by condensing 4,4'-bis(diethylamino)benzophenone with *N*-ethyl-1-naphthylamine in toluene in the presence of phosphorus oxychloride.

Slightly soluble in cold water and ethanol and moderately soluble in hot water, its use as a textile dye for silk, wool and cotton is limited due to poor lightfastness (1). However, it forms lakes with high-molecular-weight acids which are not light fugitive. The *Colour Index* states that Pigment Blue 1 (C.I. 42595:2), the phosphotungstomolybdic acid lake, is used primarily as a printing ink because of its exceptional brightness, clean tone and high tinctorial value. Other uses cited include coloring alkyd resin enamels, emulsion paints (if not too alkaline), plastics, paper coating, wallpaper, bookcloth, leather finishes, carbon paper, typewriter ribbons and artists' colors. The free base (Solvent Blue 5, C.I. 42595:1) is used in some of the same applications. Oleic and stearic acid solutions are used for carbon papers, typewriter ribbons, and stamping and ballpoint-pen inks.

Conn's[1] indicates that, while Victoria Pure Blue BO stains many substrates intensely from an aqueous solution, it is flawed because alcohol washes the stain out rapidly. The same source states that the stain may be stabilized by postchelation with phosphotungstic or phosphomolybdic acids.

Reference:

1) *Conn's Biological Stains,* 9th ed.; Lillie, R.D., Ed.; Williams and Wilkins: Baltimore, 1977; p 297.

General references:

Aldrich FT-IR Library 1(2), 1032D
Sigma FT-IR Library 1(2), 227B
Colour Index 3rd ed., 1971

VIOLAMINE R

Acid Violet 9 • Fast Acid Violet ARR

Aldrich Cat. No.:	22,938-5
Sigma Prod. No.:	V9753

CAS No.:	*[6252-76-2]*
C.I. No.:	45190
Mol. Form.:	$C_{34}H_{25}N_2NaO_6S$
F.W.:	612.64
Appearance:	Purple powder
λ_{max}:	529nm in water
Solubility:	H_2O 100mg/ml
	EGME 90mg/ml
	EtOH 20mg/ml

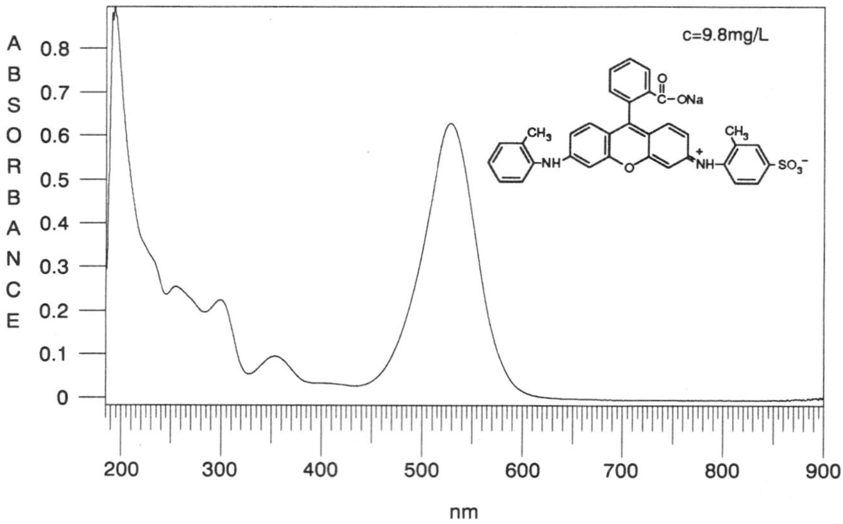

Violamine R is a violet, acid rhodamine dye that is made by condensing two moles of *o*-toluidine with one mole of 3',6'-dichlorofluoran to form Solvent Violet 10 (C.I. 45190:1), which is then sulfonated and converted to the sodium salt.

Very soluble in methyl Cellosolve and water, soluble in ethanol and slightly soluble in acetone, it is used to dye wool, nylon, silk and chromed leather, to print wool and silk directly, to stain wood and to color soap and anodized aluminum. Its heavy-metal salts serve as pigments that are used for coloring paper coatings and artists' watercolors.

Staining Procedures[1] describes a recipe and method, after Lillie (1965), for using Violamine R in place of Acid Fuchsin in a modified Van Gieson stain.

Reference:

1) Clark, G. *Staining Procedures,* 4th ed.; Williams and Wilkins: Baltimore, 1981; pp 114-115.

General references:

Aldrich FT-IR Library **1**(2), 1034C
Sigma FT-IR Library **1**(2), 227C
Colour Index 3rd ed., 1971

WRIGHT STAIN

Aldrich Cat. No.:	86,137-5
Sigma Prod. No.:	W3000
Grade:	Certified

CAS No.:	*[68988-92-1]*
Appearance:	Dark-green powder
λ_{max}:	640(522)nm in methanol
Solubility:	H_2O 3mg/ml
	EGME 20mg/ml
	EtOH 5mg/ml

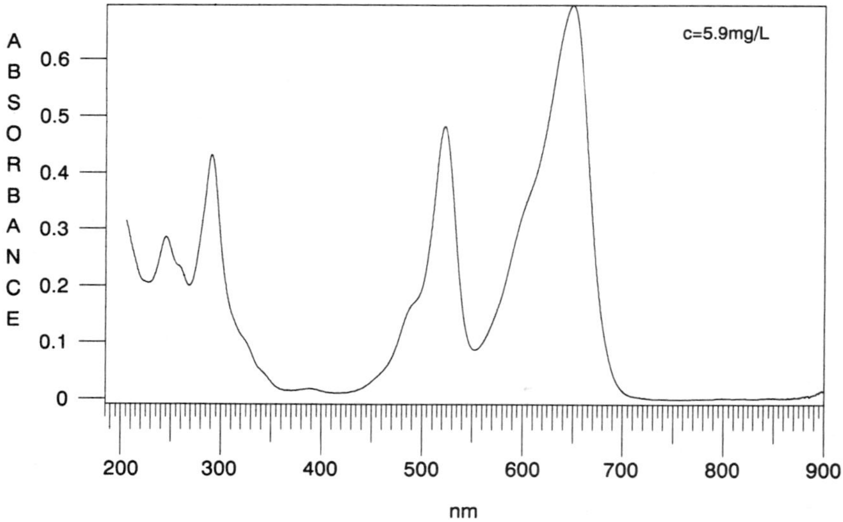

Wright Stain, a "neutral" stain, was first prepared by an Englishman, J.R. Wright,[1] by combining the cationic dyes Methylene Blue and its oxidation products, the azures, and the anionic dye Eosin Y. The concept of linking a basic dye to an acid dye through a coulombic or ionic bond to form a "neutral" stain was first described by Ehrlich about 1879. Wright Stain is a Romanowsky Stain[2] variation.

While Romanowsky is credited with being the father of all of the blood stains derived from combinations of thiazine and xanthene dyes, another Russian, Malachowski,[3] described a similar stain three weeks earlier. Since Malachowski's paper lacked details, it apparently was completely ignored for about a decade.

Wright Stain is used widely in the United States to stain blood cells and blood protozoa, while Leishman Stain, which was essentially the same and discovered independently by a fellow Englishman, W.B. Leishman,[4] finds favor in Europe in the same applications. With the possible exception of Hematoxylin, more Wright Stain is sold today, worldwide, than any other biological stain.

Although classic Wright Stain has been extremely useful for over 80 years, historically it has been plagued with an important flaw, lot-to-lot variability. This defect has not only caused laboratory-to-laboratory disagreement, but it has necessitated untold individual retesting as well.

Today, production lots of commercial, classic Wright Stain probably are monitored more thoroughly than in years past. Absorbance maxima and ranges are carefully controlled for both the blue components, Methylene Blue and the azures, and the red component, Eosin Y. The ratio of the absorbance of the red component to the absorbance of the blue components is determined by dividing the former into the latter. The optimum value obtained from such a calculation should fall between 1.65 and 1.75. Lots meeting this criterion routinely stain blood films optimally, yielding the desired balance between the blues and reds. More recently, reverse-phase HPLC has been utilized to monitor Wright Stain production. This technique readily separates dye components, yielding valuable information on the batch-to-batch variations. However, even with tighter controls, infrequent, subtle lot-to-lot variations occur. These small, unpredictable differences usually are related to solubility and solution stability rather than variability in the staining quality of freshly prepared solutions. They appear to be the result of minor, residual, inorganic impurities that promote oxidation, with its attendant dye deterioration and imbalance, when stock solutions of Wright Stain are stored for an extended period of time.

The universal dream of a reproducible, standardized Romanowsky-type blood stain, free of the aforementioned flaws of classic Wright Stain, appears to be closer to fruition since Lohr *et al.*[5] separated small quantities of the azures by column chromatography. This was followed by a paper by Marshall *et al.*[6] that described a balanced stain prepared from pure Azure B, Methylene Blue and Eosin Y. Subsequently, Wittekind *et al.*[7] found that a pure Azure B and Eosin Y blend, free of Methylene Blue gave equivalent results. (For monograph on Azure B, see pages 112-114.)

References:

1) Wright, J.H. *J. Med. Res.* **1902,** *7*, 138-144.

2) Romanowsky, D.L. *St. Petersburg Med. Wochenschr.* **1891,** *16*, 297-315.

3) Malachowski, E. *Cbl. Klin. Med.* **1891,** *12*, 601-603.

4) Leishman, W.B. *Br. Med. J.* **1901,** *2*, 757-758.

5) Lohr, W.; Grubhofer, N.; Sohmer, J.; Wittekind, D. *Stain Technol.* **1975,** *50*, 1949.

6) Marshall, P.N.; Bentley, S.A.; Lewis, S.M. *J. Clin. Pathol.* **1975,** *28*, 920-923.

7) Wittekind, D.; Kretschmer, V.; Lohr, W. *Blut* **1976,** *32,* 71-78.

General references:

Aldrich FT-IR Library	**1**(2), 1042D
Sigma FT-IR Library	**1**(2), 227D
IR Library	**3**, 1488G
Safety	**2**(2), 3616D

WRIGHT STAIN, solution in methanol

Aldrich Cat. No.: 29,937-5
Grade: Buffered

CAS No.: *[68988-92-1]*
Density: 0.778
Appearance: Blue liquid
λ_{max}: 647(522)nm in methanol

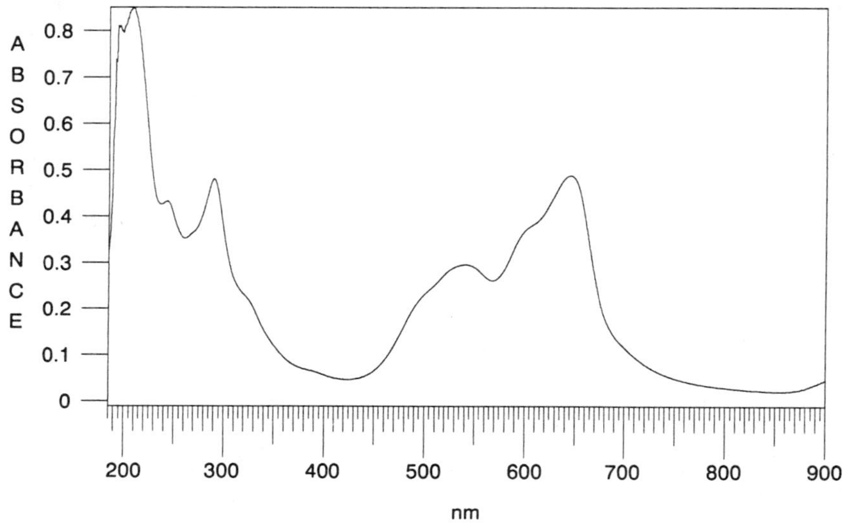

Wright Stain solution is an *in vitro* diagnostic aid that is used in hematology to differentially stain blood reticulocytes, leucocytes and certain blood parasites. It is an anhydrous methanol solution containing 0.25 to 0.3% dye. The dye used is a certified, classic Wright Stain blend composed of Methylene Blue and Methlyene Azures, present in both the free state and complexed with Eosin, and carefully blended to attain an optimum balance between the blue and red components. *Conn's*[1] provides test methods for Wright Stain. *Staining Procedures*[2] lists a method after Conn (1940) that uses Wright Stain solution to stain blood, and a method for staining the earthworm's ceolomic cells.

Wright Stain solution has no industrial application.

References:

1) *Conn's Biological Stains,* 9th ed.; Lillie, R.D., Ed.; Williams and Wilkins: Baltimore, 1977; pp 605-606.

2) Clark, G. *Staining Procedures,* 4th ed.; Williams and Wilkins: Baltimore, 1981; pp 171-172, 273.

General reference:

Safety **2**(2), 3617A

XYLENOL BLUE

p-Xylenolsulfonephthalein

Aldrich Cat. No.:	20,548-6
Sigma Prod. No.:	X6625
Grade:	Indicator

CAS No.:	[*125-31-5*]
Mol. Form.:	$C_{23}H_{22}O_5S$
F.W.:	410.49
m.p.:	212°C (dec.)
Appearance:	Dark-brown powder
λ_{max}**:**	424nm in methanol
Solubility:	H_2O 10mg/ml
	EGME 30mg/ml
	EtOH 30mg/ml

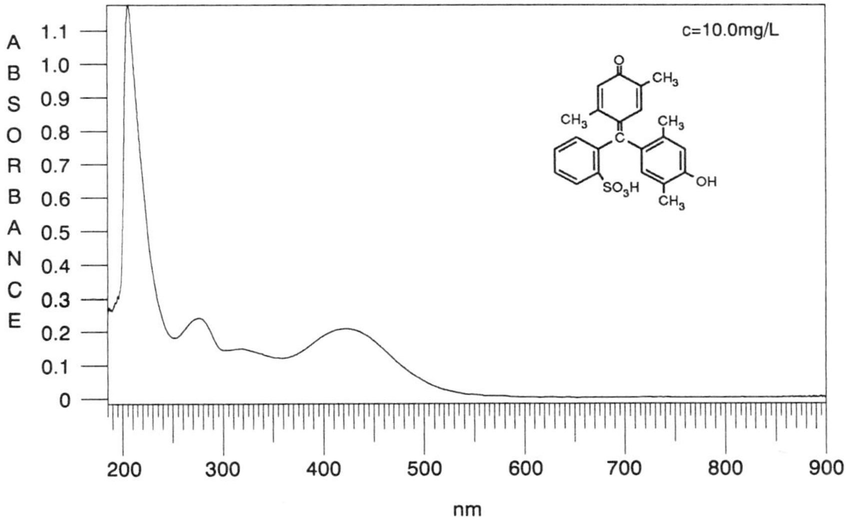

Xylenol Blue is a sulfonephthalein indicator that is made by the condensation of two moles of *p*-xylenol with one mole of *o*-sulfobenzoic acid anhydride. It is slightly soluble in water and soluble in alcohol and dilute alkali. The water-soluble (sodium salt) form is also slightly soluble in ethanol.

Xylenol Blue has the same visual-transition intervals as Thymol Blue: pH 1.2 (red) to pH 2.8 (yellow) and pH 8.0 (yellow) to pH 9.6 (blue). Cohen[1,2] stated that Xylenol Blue is twice as color intense as Thymol Blue. However, Kolthoff and Rosenblum[3] reported that they were not able to verify Cohen's observation in their laboratory, using commercial material.

This acid-base indicator has no other reported commercial application.

References:

1) Cohen, A. *Biochem. J.* **1922,** *16*, 31.

2) *Idem ibid.* **1923,** *17,* 535.

3) Kolthoff, I.M.; Rosenblum, C. *Acid-Base Indicators;* Macmillan: New York, 1937; p 131.

General references:

Aldrich FT-IR Library	**1**(2), 1022B
Sigma FT-IR Library	**1**(2), 228B
IR Library	**3**, 1479D
Merck Index	**11**, 9990

XYLENOL BLUE, water soluble

Aldrich Cat. No.: 11,456-1
Sigma Prod. No.: X3001
Grade: Indicator

Mol. Form.: $C_{23}H_{21}NaO_5S$
F.W.: 432.47
m.p.: >300°C
Appearance: Black crystalline powder
λ_{max}: 425nm in methanol
Solubility: H_2O 20mg/ml
EGME 30mg/ml
EtOH 70mg/ml

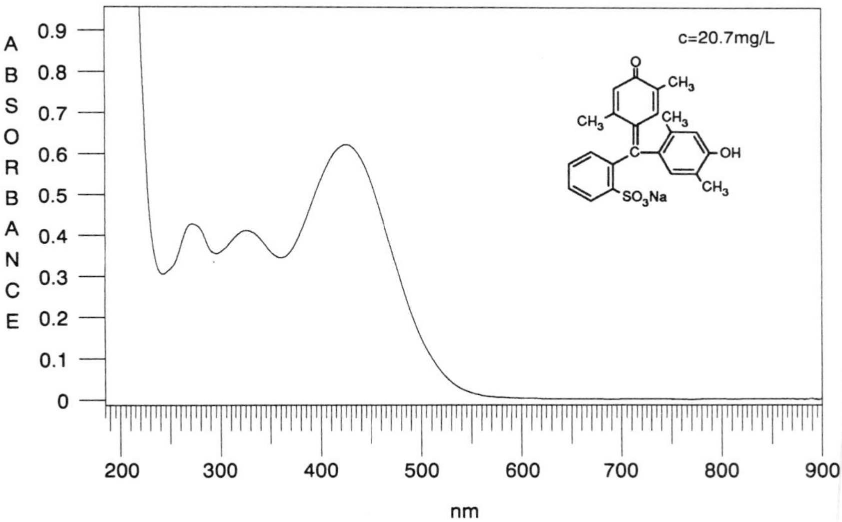

See description on pages 737 and 738 under Xylenol Blue.

General reference:

IR Library **3**, 1479E

XYLENOL ORANGE, water soluble

3,3'-Bis[*N,N*-di(carboxymethyl)aminomethyl]-*o*-cresolsulfonephthalein, sodium salt

Aldrich Cat. No.:	22,785-4
Sigma Prod. No.:	X3500

CAS No.:	*[63721-83-5]*	
Mol. Form.:	$C_{31}H_{32}N_2O_{13}S$	(free base)
m.p.:	210°C (dec.)	
Appearance:	Dark-brown powder	
λ_{max}:	580nm in 0.1*N* sodium hydroxide	
Solubility:	H_2O	200mg/ml
	EGME	7mg/ml
	EtOH	0.4mg/ml

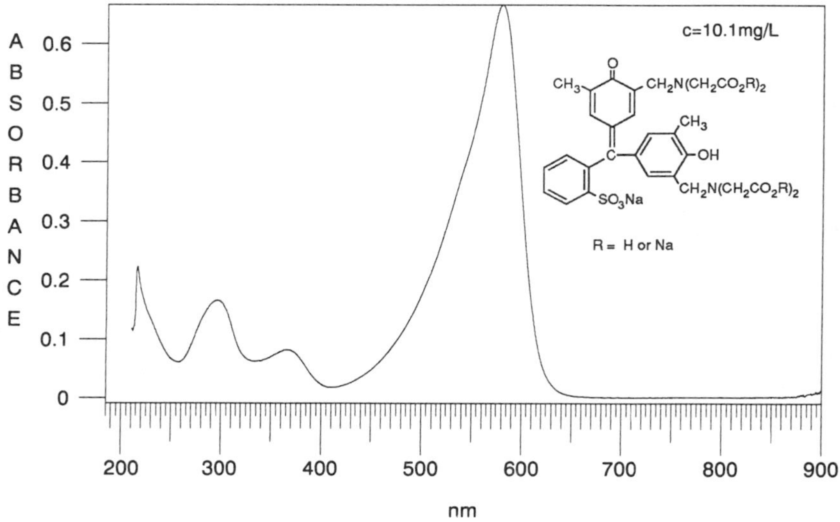

Xylenol Orange is a sulfonephthalein indicator that is very soluble in water and very slightly soluble in ethanol.

It finds little use as an acid-base indicator. It is used in the detection and determination of bismuth, thorium, scandium, lead, zinc, lanthanum, cadmium, mercury and zirconium because of its ability to form chelates with these metals.

In 1971, Rahn and Perren[1] used Xylenol Orange intravitally as a fluorescent label to demonstrate calcification in tissue. A good contrast is provided to most other fluorochromes that are used in the same application, namely: tetracycline, hematoporphyrin, Calcein, Calcein Blue and 2,4-bis[*N,N*-di(carboxymethyl)aminomethyl]fluorescein (DCAF). In conjunction with these, it can be used to demonstrate the incremental growth of teeth and bones (progressive calcification). *Staining Procedures*[2] states that Xylenol Orange is superior to other fluorescent labels because it demonstrates little, if any, toxicity at recommended doses, and thus does not interfere with progressive calcification.

References:

1) Rahn, B.A.; Perren, S.M. *Stain Technol.* **1971,** *46*, 125-129.

2) Clark, G. *Staining Procedures,* 4th ed.; Williams and Wilkins: Baltimore, 1981; p 84.

General references:

Aldrich FT-IR Library	**1**(2), 1028A
Sigma FT-IR Library	**1**(2), 228C
Safety	**2**(2), 3621D

XYLIDINE PONCEAU 3RS

Acid Red 26 • Ponceau 2R

Aldrich Cat. No.:	19,976-1
Sigma Prod. No.:	P2395
CAS No.:	[*3761-53-3*]
C.I. No.:	16150
Mol. Form.:	$C_{18}H_{14}N_2Na_2O_7S_2$
F.W.:	480.43
Appearance:	Red powder
λ_{max}:	503nm in water
Solubility:	H_2O 80mg/ml
	EGME 8mg/ml
	EtOH 0.7mg/ml

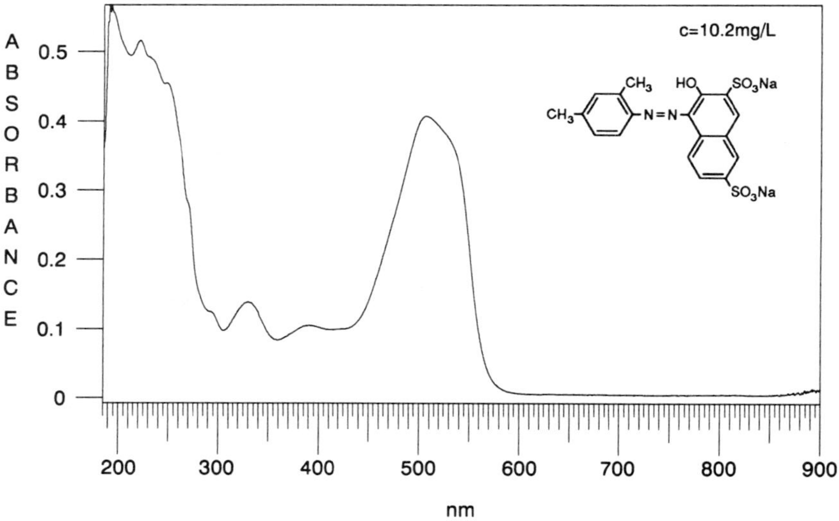

This bright-red, anionic monoazo dye is made by coupling diazotized 2,4-xylidine to R acid (2-naphthol-3,6-disulfonic acid). It is soluble in water and very slightly soluble in ethanol and acetone. Industrial uses include dyeing wool, nylon, silk, paper and chromed leather, printing silk and wool, in wood stain, as a colorant for inks and in the preparation of pigments (heavy-metal salts).

Xylidine Ponceau 3RS has limited use as a biological stain. Lillie[1] suggests other dyes that are superior to Xylidine Ponceau as a contrast stain in Masson's trichrome stain technique.

Reference:

1) Lillie, R.D. *Stain Technol.* **1941,** *15*, 17-22.

General references:

Aldrich FT-IR Library	**1**(2), 982A
Sigma FT-IR Library	**1**(2), 199D
IR Library	**3**, 1459G
Safety	**2**(2), 3622B
Colour Index	3rd ed., 1971

XYLIDYL BLUE I, water soluble

Aldrich Cat. No.: 24,826-6

CAS No.:	[*14936-97-1*]
Mol. Form.:	$C_{25}H_{20}N_3NaO_6S$
F.W.:	513.51
Appearance:	Dark-red powder
λ_{max}:	612nm in ethanol + 2ml 0.08*M* borax buffer
Solubility:	H_2O 6mg/ml
	EGME 10mg/ml
	EtOH 10mg/ml

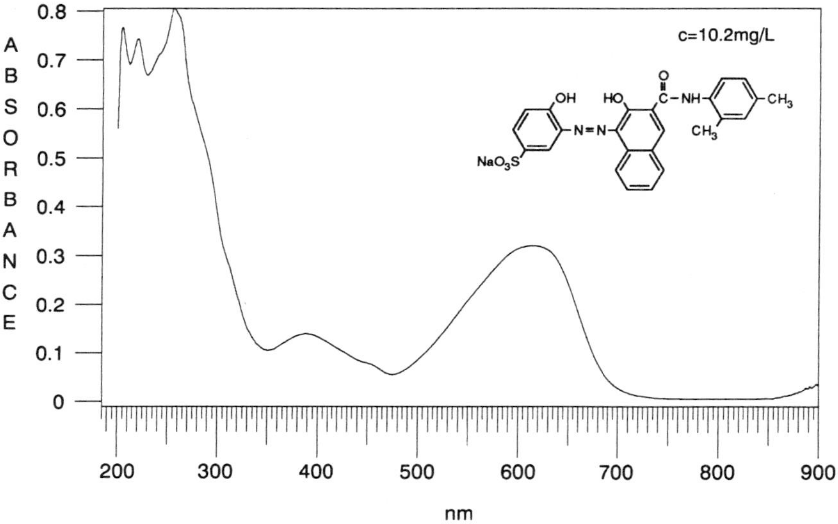

Xylidyl Blue I is an anionic monoazo dye that is somewhat soluble in water and ethanol.

Apparently it has not been used in industry; however, it has been used to advantage as a colorimetric reagent for the determination of magnesium in applications as diverse as "synthetic blood ash" and limestone.[1]

Reference:

1) Mann, C.K.; Yoe, J.H. *Anal. Chem.* **1956,** *28*, 202.

General reference:

Aldrich FT-IR Library **1**(2), 995A

ZINCON

2-[1-(2-Hydroxy-5-sulfophenyl)-3-phenyl-5-formazano]benzoic acid

Aldrich Cat. No.: 11,976-8
Sigma Prod. No.: Z3750

CAS No.: [*135-52-4*]
Mol. Form.: $C_{20}H_{16}N_4O_6S$
F.W.: 440.43
m.p.: 203°C (dec.)
Appearance: Purple powder
λ_{max}: 490nm in 0.1*N* sodium hydroxide
Solubility: H_2O 20mg/ml
EGME 40mg/ml
EtOH 10mg/ml

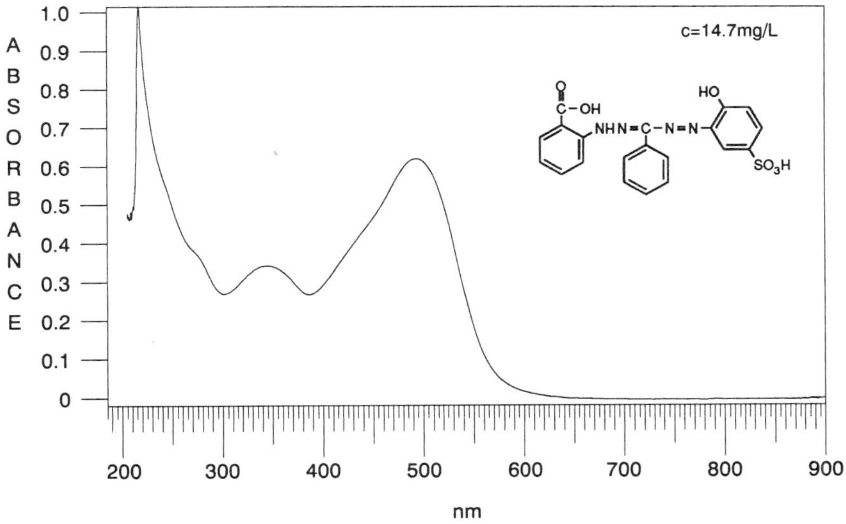

Zincon is a formazan dye whose free acid form (Aldrich Cat. No. 11,976-8) is sparingly soluble in water and the lower aliphatic alcohols and moderately soluble in acetone as well as in solutions of alkali hydroxides. The monosodium salt (Aldrich Cat. No. 20,133-2) is soluble in alkali hydroxide solutions, slightly soluble in water and ethanol and insoluble in most other organic solvents. The solubility variations described above are the only significant differences between these two products. Otherwise, they can be used interchangeably and the information that follows applies to both forms.

Zincon forms deep-blue-colored complexes with certain cations including zinc, copper, nickel, cobalt and mercury. It does not react with alkali earths or magnesium. The high sensitivity of Zincon to copper and zinc, with sufficient selectivity, makes it an especially useful reagent for the determination of these metals. Kinnunen and Merikanto[1] introduced Zincon into chelometric analysis as an indicator for the direct determination of zinc in an ammoniacally buffered medium (pH 9-10). Later Ringbom *et al.*[2] used it for the titration of calcium with a standard solution of ethylenebis(oxyethylenenitrilo)tetraacetic acid (EGTA). In this procedure, a small amount of zinc-EGTA complex must be added in addition to Zincon. Very precise results are obtained; magnesium can be determined in the same solution afterward using an Alizarin Chrome Black Pt endpoint color change, providing ascorbic acid and cyanide are added to the solution to mask other metals. More recently, Singhal and Tanden[3] described the direct chelometric determination of divalent mercury at pH 5.5-7.0. Reilley and Sheldom[4] used Zincon to titrate certain metal ions (*e.g.*, mercury, cadmium, zinc) with triethylenetetramine as a chelating agent. Colorimetric determinations for zinc, copper and mercury are described by Yoe and Rush.[5,6]

References:

1) Kinnunen, J.; Merikanto, B. *Chemist-Analyst* **1955**, *44*, 50.

2) Ringbom, A.; Pensar, G.; Vanninen, E. *Anal. Chim. Acta* **1958**, *19*, 525.

3) Singhal, G.K.; Tanden, K.N. *Talanta* **1967**, *2*, 1351.

4) Reilley, C.N.; Sheldom, M.V. *Chemist-Analyst* **1957**, *46*, 59-62.

5) Yoe, J.H.; Rush, R.M. *Anal. Chim. Acta* **1952**, *6*, 526-527.

6) Rush, R.M.; Yoe, J.H. *Anal. Chem.* **1954**, *26*, 1345-1357.

General references:

Aldrich FT-IR Library	**1**(2), 981A
Sigma FT-IR Library	**1**(2), 228D
IR Library	**3**, 1459A
Safety	**2**(2), 3631B

ZINCON, monosodium salt

Aldrich Cat. No.: 20,133-2
Sigma Prod. No.: Z2250

CAS No.: [*62625-22-3*]
Mol. Form.: $C_{20}H_{15}N_4NaO_6S$
F.W.: 462.41
Appearance: Maroon powder
λ_{max}: 490(523)nm in methanol
Solubility: H_2O 10mg/ml
EGME 20mg/ml
EtOH 2mg/ml

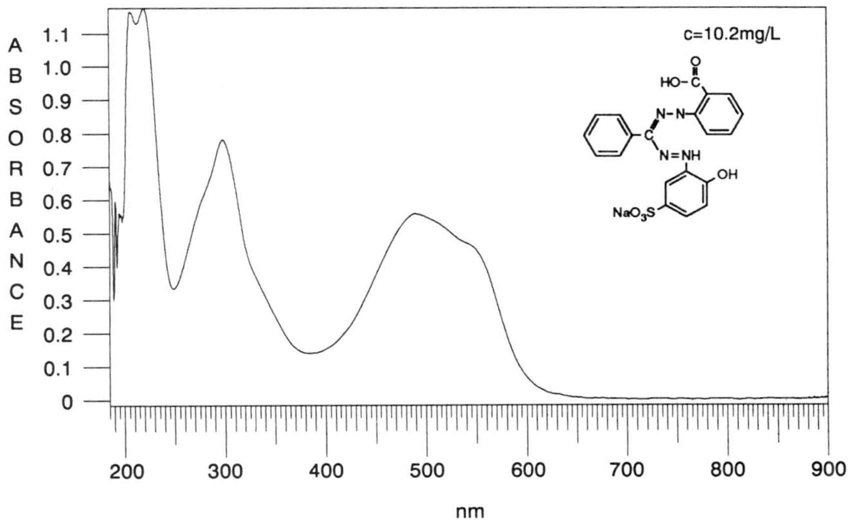

See Zincon on pages 745 and 746 for description.

General references:

Aldrich FT-IR Library **1**(2), 979A
Sigma FT-IR Library **1**(2), 1052B
IR Library **3**, 1458A

Alphabetical Index (including alternate names)

Chemical Abstract Service Registry Number

Colour Index Number

(Items marked with an * are similiar to the designated C.I. No.)

Absorption Maxima (λ_{max})

(in ascending order, including name and solubility)
(secondary peaks are in parentheses)

Product	λ_{max}	Solubility (mg/ml) H_2O	EtOH	EGME	Page
3,3'-(4,4'-Biphenylene)bis(2,5-diphenyl-*2H*-tetrazolium chloride)	250	20.0	40.0	30.0	139
4-(Dimethylamino)benzaldehyde	250	<0.2	<0.2	0.2	259
Quinizarin	250	0.2	0.3	2.0	606
Blue Tetrazolium	253	9.0	7.0	4.0	146
5,7-Dichloro-8-hydroxyquinoline	253	<0.1	3.0	10.0	253
Pyronin Y	(254)	30.0	3.0	9.0	601
Fast Red ITR base	256	0.5	30.0	100.0	351
Methyl Yellow	(256)	—	—	—	472
Malachite Green Carbinol base	(259)	0.3	10.0	20.0	436
Nitro Blue Tetrazolium Chloride monohydrate	260	10.0	5.0	20.0	523
Prussian Blue	(260)	6.0	20.0	10.0	590
Leuco Crystal Violet	261	0.3	0.6	5.0	418
Acriflavine	(262)	20.0	4.0	7.0	66
Acriflavine hydrochloride	(262)	20.0	7.0	6.0	68
Leucomalachite Green	262	<0.1	4.0	30.0	419
2-Nitroso-1-naphthol	262	0.5	6.0	20.0	528
4',6-Diamidino-2-phenylindole dihydrochloride hydrate	(263)	—	—	—	244
1,10-Phenanthroline	264	2.0	20.0	30.0	557
1,10-Phenanthroline monohydrate	264	2.0	20.0	30.0	559
1,10-Phenanthroline monohydrochloride monohydrate	264	600.0	40.0	80.0	560
Fast Red B, tetrafluoroborate salt	(266)	60.0	0.9	70.0	350
Fast Yellow GC salt	(268)	50.0	3.0	20.0	362
Phenanthrequinone	270	0.4	1.0	7.0	555
4-(Phenylazo)diphenylamine	(272)	<0.1	30.0	50.0	571
Coumarin	275	<0.1	<0.1	<0.1	224
Bathophenanthroline disulfonic acid, disodium salt hydrate	276	6.0	6.0	20.0	133
Lucifer Yellow VS	(277)	—	—	—	429
Tetranitro Blue Tetrazolium Chloride hydrate	279	3.0	3.0	9.0	696
Pyrogallol Red	(282)	1.0	4.0	3.0	597
Basic Red 29	(285)	—	—	—	131
Fast Red TR salt	285	50.0	10.0	20.0	354
Remazol Brilliant Blue R	(285)	20.0	40.0	20.0	620
4,7-Diphenyl-1,10-phenanthroline	288	0.2	8.0	10.0	265
Quinoline Yellow S, water soluble	288	100.0	0.4	10.0	610
3,3',5,5'-Tetramethylbenzidine	288	2.0	10.0	7.0	693
3,3',5,5'-Tetramethylbenzidine dihydrochloride hydrate	288	10.0	0.8	0.5	695
Tiron	291	10.0	<0.1	2.0	712
Hematoxylin	292	30.0	30.0	50.0	394
4-Methylesculetin	(292)	0.1	6.0	20.0	457
Fast Red Violet LB base	293	0.6	3.0	40.0	355